cod

THE ECOLOGICAL HISTORY OF THE NORTH ATLANTIC FISHERIES

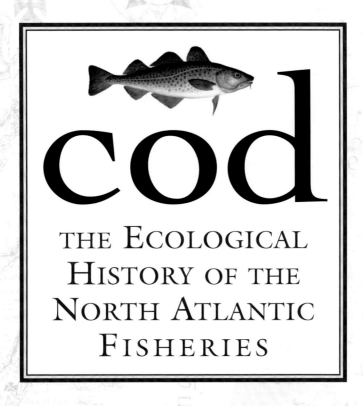

cod

THE ECOLOGICAL HISTORY OF THE NORTH ATLANTIC FISHERIES

GEORGE A. ROSE

BREAKWATER

BREAKWATER BOOKS LTD.
100 Water Street • P.O. Box 2188
St. John's • NL • A1C 6E6
www.breakwaterbooks.com

Library and Archives Canada Cataloguing in Publication

Rose, George A., 1948-
    Cod : An Ecological History of the North Atlantic Fisheries / George A. Rose.

Includes index.

ISBN 978-1-55081-225-1 (bound).   ISBN 978-1-55081-227-5 (pbk.)

1. Fisheries--Newfoundland and Labrador--History.
2. Fisheries--Environmental aspects--Newfoundland and Labrador--History.
I. Title.

SH224.N7R67 2007   338.3'72709718   C2006-905292-1

© Copyright 2007 George A. Rose

ALL RIGHTS RESERVED. No part of this publication may be reproduced, stored in a retrieval system or transmitted, in any form or by any means, without the prior written consent of the publisher or a licence from The Canadian Copyright Licensing Agency (Access Copyright). For an Access Copyright licence, visit www.accesscopyright.ca or call toll free to 1-800-893-5777.

We acknowledge the financial support of The Canada Council for the Arts for our publishing activities.

We acknowledge the support of the Government of Newfoundland and Labrador, Department of Tourism, Culture and Recreation for our publishing activities.

 We acknowledge the financial support of the Government of Canada through the Book Publishing Industry Development Program (BPIDP) for our publishing activities.

Printed in Canada.

FOR
Sarah

TABLE OF CONTENTS

ACKNOWLEDGEMENTS... 11 | PREFACE... 13

Ancient Beginnings . . .   17
The Contemporary Ecosystems . . .   53
6000 BC to AD 1450:
   The Arrival of Humans . . .   167
AD 1451 to 1576:
   The Beginnings of Export Fisheries . . .   177
AD 1577 to 1800:
   Migrant Fisheries and Liveyers . . .   205
AD 1801 to 1949:
   Settlement, Salt Fish, Seals and Science . . .   267
AD 1950 to 1992:
   The Loss of the Fishery . . .   377
AD 1993 to 2006:
   A Fishery Without Cod . . .   499

APPENDIX... 552 | REFERENCES... 555 | INDEX... 580

ACKNOWLEDGEMENTS

The following people each read parts of the manuscript and provided not only valuable insights into the history of the fishery and how to tell the story, but also helped me steer through many storms of controversy about past events. These are Drs. Arthur May, Peter Pope, Sarah Rose, Shannon Ryan, Mr. Les Dean, Mr. Aiden Maloney, and Ms. Andrea Carew. I thank them all — but point out that we could not always agree on all things or interpretations of history, and I take full responsibility for the work presented here. A special thanks to Susan Fudge, who assisted throughout the project and did much of the data and reference collection, and to Dr. Kristina Curren, who read and edited the entire book. The people at Breakwater Books were unfailingly supportive throughout this project. I especially wish to thank Rhonda Molloy and Annamarie Beckel for their invaluable work during the final production.

The original maps were drawn by Alan Grimwade and David Maltby, Cosmographics, U.K. Old maps are courtesy of the British Museum and the Memorial University Centre for Newfoundland Studies. The artwork was done by Brenda Guiled, Charting Nature, U.S.A. Photographs were supplied by Gordon King of St. John's, John Wheeler of Flatrock, and Tom Clenche and Tom Mills of Petley, Newfoundland and Labrador. All figures were drawn by the author.

Funding was provided by the Newfoundland and Labrador Department of Fisheries and Aquaculture and Fishery Products International (Victor Young, then Chief Executive Officer) through the NSERC Research Chair in Fisheries Conservation at the Fisheries and Marine Institute of Memorial University of Newfoundland.

Finally, thanks to Gord, who always managed to find a spot to sprawl contentedly near my feet within a never ending mass of documents and books. To all who contributed I express my deepest thanks, and hope that the book meets their expectations.

PREFACE

...being in latitude 54 degrees 30...we found great abundance of cod, so that the hooke was no sooner overboard, but presently a fish was taken. It was the largest and the best fed fish that ever I saw...

CAPTAIN JOHN DAVIS, from his journal off Labrador, 1586

There are places and species that have become inextricably linked in human history and in the human mind have taken on mythic proportions. This is not to say that these same species occur nowhere else, or that they are the most abundant or important, although they may be. But somehow the animal comes to symbolize and define the place. In this way, the wildebeest symbolizes the Serengeti Plains of Tanzania, the polar bear the Arctic, finches the Galapagos Islands, the panda China, and wild salmon the northern regions of the Pacific Ocean. But there is no stronger definition, no stronger attachment or symbolism, than Newfoundland and Labrador with the Atlantic cod.

Even now, with the cod almost gone, the knot holds. Defying economic rationalization, many fishermen remain more excited catching one cod worth two dollars than tonnes of crab worth thousands. They catch crab but dream of cod. This is why the decline of the once vast and seemingly boundless Newfoundland cod fishery in the late 1900s, after 500 years of fisheries that drove European economies and fed millions, has not only devastated a marine economy but perhaps more importantly, debased a unique culture. The collapse of the Newfoundland and Labrador cod stocks sent a shock wave through fisheries and resource management agencies around the world, and more than any other single event, has precipitated a worldwide concern about fisheries, fishing cultures, and the future of the world's oceans.

Indeed, the decline of the Newfoundland and Labrador fisheries has become an icon for human excess in a world being depleted of its natural bounty. In almost every recent scientific meeting and book about resource management, the Newfoundland cod is held up as *the* example of a pilloried resource.

Marine cultures dependent and partially based on cod date back to the Norse. Cod were never just a fish. Human cultures tend not to refer directly to a deity or object of great respect or authority. Such it was with cod. To Norse fishermen, cod were *thorskr*,[1] but most often referred to obliquely as the more generic *fiskr*. Conversely, fish, in many languages and cultures of the North Atlantic, is cod. In the Faroes, the answer to the question "what are we having for dinner?" could well be *fisk*, and "cod is the only possible meaning in this context."[2] The English word cod is of uncertain origin, but appears to have come into use in the 1200s. Nevertheless, in Newfoundland and Labrador, cod were referred to as fish, and fish means cod, and as with the ancient Norse, direct reference by naming the species was rare until modern times.

I started this book to tell the story of the Newfoundland cod fishery. This was the last of the great North Atlantic cod fisheries to be developed, after those of Norway-Russia and the Barents Sea, the North Sea, and Icelandic-Greenland waters. It was also the first to be depleted to the state of non or longterm recovery. I shortly realized that telling this story could not be done, at least not well, without telling the story of the marine ecosystems in which the cod and other fish lived, and in which the fisheries were prosecuted. Nor was this to be a straightforward fish story. Many social and political events have shaped the fate of the fisheries. Indeed, the full history of Newfoundland and Labrador has been written through the fisheries in one way or another, and it is impossible to separate the two into neat compartments. I have attempted to tell a comprehensive story, beginning with the formation of the ecosystems and ending in the early 2000s with cod reduced to a fragment of their former abundance – their fisheries near gone.

This book was written for anyone who might be interested, but especially for Newfoundlanders and Labradorians, of whom there are but few who do not have a tie to the fisheries. Much of this book was written at my cabin at Terra Nova Lake during a sabbatical year from teaching at Memorial University, broken only by frequent trips to the libraries and to keep an eye on the remaining large cod aggregation in nearby Smith Sound in Trinity Bay. The work spilled over into several more years.

The fisheries and biological information on which the book is based was researched from original documents and my own work over the past 25 years. I am indebted to the many fisheries and marine scientists who recorded their observations over the past century. For cultural and political historical information I have relied heavily on the research of others, but most original sources were checked where possible. The reference list is extensive, and I have tried to credit as many secondary sources as possible, in recognition of the rich historical literature on the history of Newfoundland and Labrador that I have drawn upon. The book is as scientifically accurate and up to date as possible, but to keep it readable and understandable, I have tried to avoid using unnecessary jargon in the main text. The endnotes contain additional detail and lead to the references on which the book is based.

It is my belief that the story of the Grand Banks and the marine ecosystems of Newfoundland and Labrador is important not only for fisheries management, but for all human societies. As with much of the natural world, cod were once thought to be so abundant and productive that no human intervention could alter the course of their populations or their ecosystems in the North Atlantic. We now know this to be false. The cod and other species have a story to tell – but the Newfoundland and Labrador fisheries should not simply be an icon for the narrow failures of fisheries management, but more deeply, for the broadly unsustainable relationship that exists between humans and Nature.

George Rose, Terra Nova, Newfoundland and Labrador, September 2006.

NOTES

[1] The Old Norse *Thorskr* became *Porskur* in Icelandic and *Trosc* in Irish (Gaelic).
[2] Lockwood (2006), page 13.

ANCIENT BEGINNINGS

*Ocean...is more ancient than the mountains,
and freighted with the memories and the dreams of Time.*

H.P. LOVECRAFT, *The White Ship*, 1919

Genesis of the Grand Banks . . .	19
Cooling of the North Atlantic . . .	24
Ocean Waters: Currents, Temperatures, and Salt . . .	26
North Atlantic . . .	26
Gulf Stream and North Atlantic Drift . . .	28
Origin of the Fishes . . .	29
Origin of Cod . . .	32
Ice Ages . . .	33
The Last 1000 Years . . .	38
North Atlantic Oscillation . . .	42
Labrador Current . . .	43

ECOSYSTEMS ARE EVER CHANGING: their present state is a product of their history and a transition to the future. To understand present and future ecosystems, we must first look to the past, for in the memory of an ecosystem is everything that makes it what it has been, what it is, and what it can come to be.

The Grand Banks are the centrepiece of a broad continental shelf that rings northeastern North America – the Banks extend 500 kilometres off the island of Newfoundland into the northwest Atlantic. From this commanding position, the Grand Banks intercept and steer two of the major currents of the North Atlantic: the Gulf Stream and the Labrador Current. Nutrients from the deep sea are brought to the surface by these currents. Over most of the Grand Banks, sea depths seldom exceed 150 metres, which allows light to penetrate to the seafloor over a vast expanse of ocean. The result of this unique combination of location, form, and depth is an elevated primary productivity that drove a fish abundance that startled early explorers. Historical catch rates of Atlantic cod in these waters were likely higher than in any fishery the world has ever known. The story of John Cabot's fishing with baskets in a sea of cod is renowned; less well known are the tales of the Russians who fished these waters centuries later in the 1960s. Like many before them, they had never experienced such rich fishing, not even in their home waters of the Barents Sea that holds a cod stock millions of tonnes strong.[1]

GENESIS OF THE GRAND BANKS

How did the Grand Banks and its adjacent continental shelf come to be? What forces created these huge undersea plateaus? The answers are to be found in the fundamental building and levelling processes of the earth's surface. To begin, we must understand tectonics.

Tectonics is the study of the shifting surface of the earth. The earth's crust began to cool some 4 billion years ago, forming 10-12 mobile pieces, or plates, like the stitched leather of a soccer ball. The plates moved around on the earth's surface at speeds of one to two centimetres per year, driven by convectional heat flows arising from the molten core of the earth. There were no spaces between the plates. Hence, they pivoted and pushed against one another, with new crust forming at spreading lines and old crust sinking into the earth's molten mantle where plates came together. Over the earth's history, the location of plates – and the spreading and sinking lines between them – changed. At spreading lines, the newly forming crust was heavy, dark molten rock, termed basalt after it cooled and solidified. Like a conveyor belt, the new crust moved towards the sinking edge where it slid back into the molten mantle. As a consequence of these movements, nowhere in the world is the ocean crust older than about 160 million years. In contrast, the continents are comprised of much lighter and older granite rocks with ages up to 3.8 billion years. The continental landmasses have floated like giant icebergs on the oceanic plates since the initial cooling of the earth.[2]

Continental shelves are just that: underwater shoulders of continental landmasses, with old and light rocks as distinct from the deep ocean basalt as chalk is from cheese. Over eons of time, these floating landmasses have moved around the earth, and endured many collisions and separations that formed new continents and oceans, mountain ranges and deep valleys, both on land and under water. Where plate movements tore the landmasses apart, the continental shelf at their margins tended to become stretched and much wider, as is the case along the east coast of North America. Conversely, where a continent was pushed against another plate, the leading edge became compressed and the continental shelf narrow, as on the west coast of North America.

As plates and continents separated or pushed against each other, cracks or faults developed, forming blocks that were pushed up or folded to become mountains, or that sank to form deep basins. Such movements caused the earth's surface to become uneven; over time, however, gravity and erosion worked against unevenness to flatten all landscapes, both above and below water, as old surfaces became levelled with broken and eroded sediments. All of these forces came into play in creating the Grand Banks.

The formation of the marine ecosystems off Newfoundland and Labrador began with the tectonics of long ago. Over 500 million years ago, the continental rock that now makes up the Grand Banks and the eastern part of the island of Newfoundland,

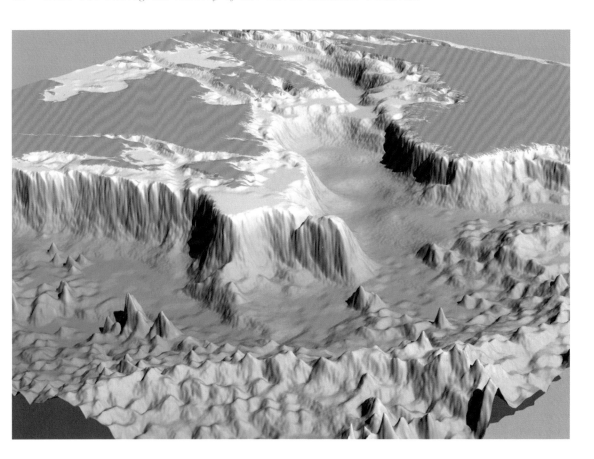

1.1 Northwest Atlantic and the Grand Banks. Elevation exaggerated.

as well as areas of coastal Nova Scotia and New England, was part of an ancient Euro-African landmass. The rocks in these now-separated locations are the same, as were the animals that lived there: the ancient fossils found on Newfoundland's Atlantic coast also occur in North Africa and are totally different from the proto-North American fossils of western Newfoundland. At that time, the continental landmass that would later become North America was located much farther to the west. An immense body of water called the Iapetus Sea separated these drifting protocontinents long before the Atlantic Ocean existed.[3]

Plate movements began to close the Iapetus Sea about 475 million years ago. Proto-North America, with the continental rocks of western Newfoundland at its leading edge, was driven towards the western edge of proto-Africa and Europe, whose front line was the rocks that would become eastern Newfoundland, the Avalon Peninsula, and the Grand Banks. As these great landmasses moved on a collision course, the earth's crust brokeand rose under great pressure. In what is now western Newfoundland, oceanic crust was forced westwards and overrode younger marine sediments. At Table Mountain in Gros Morne National Park, as well as on the Great Northern

Peninsula, huge broken slabs of 500-million-year-old basalt (termed ophiolites) sit on top of a jumble of younger rocks. This turnover of relatively young ocean-floor rock lies adjacent to some of the oldest exposed rocks on earth: the remnants of the 3-billion-year-old eastern coast of proto-North America. By 400 million years ago, the continental rocks along the collision line between proto-North America and Euro-Africa buckled and folded many times, giving rise to the Appalachian Mountains of eastern North America, and the twisted and folded masses of rock of central Newfoundland.

As proto-North America and Euro-Africa merged, the great Iapetus Sea closed and the rocks that now form the eastern and western parts of the island of Newfoundland were welded together. The weld is still highly visible, perhaps best observed near the town of Gambo. Geologists refer to the weld as the Dover Fault.[4]

For 200 million years, the rocks of present-day North America, Europe, and Africa remained united. Around 210 million years ago, during the late Triassic period, convection under the continents stirred again and the continental rock mass that would become North America began to move westward. Incredibly, parts of the old weld of proto-North America to Euro-Africa held. The inevitable rupture of the New World from the Old World occurred not where they had originally collided, but farther to the east. Former bits of Africa and the Iberian Peninsula (Spain and Portugal) became the eastern parts of the island of Newfoundland, Nova Scotia, and New England. The immense block of continental rock than now underlies the Grand Banks became the trailing edge of the westward moving New World. As spreading continued, the stretching of continental rock caused cracking, or faulting, on both the east and west sides, and a rift valley was formed, similar to the major rift that today divides eastern Africa. For more than 10 million years, in the early Jurassic period, the rift was landlocked with the Grand Banks on one side and the Iberian Peninsula on the other. Over the millennia, rivers from the new and old continents deposited layers of sediments on top of the continental rocks.

About 190 million years ago, the Atlantic rift was periodically invaded by seawater from the great southern ocean and the area that would become the Grand Banks got its first taste of saltwater. At first, the rift was the low point of the landscape and no water flowed out, so the seawater evaporated and left salt deposited in the valley. As the spreading of the rift continued, the rocks underlying the valley became stretched and thinner, and the rift sank deeper. More water poured in from the south. The basins, sediments, and hills of the last collapsing link between the Old World and New World became the foundation of the Grand Banks. At this time, an Arctic land bridge linking the northern Canadian islands to Greenland and to Norway was also above water. Twenty million years later, the spreading of the Old and New Worlds to the south became so severe that the stretched ocean crust in the middle of the newly forming Atlantic Ocean fissured to the molten mantle below. A ridge of volcanoes developed along the spreading line, forming the North Atlantic Ridge and the island of Iceland. At the trailing edges of the moving continental rock, intense pressures led to faulting and

1.2
The formation of the North Atlantic Ocean, with physical features in the early Jurassic period (150-180 million years ago), the early Tertiary period (60 million years ago), the mid-Tertiary period (25-30 million years ago), and the Quaternary period (last million years).

the formation of sunken regions, including the Jeanne D'Arc, Orphan, and Whale basins. Sediments began to accumulate within these basins – 50 million years later, in the late Jurassic period, rich oil reserves were formed.[5] As North America floated relentlessly westward, the failing land bridge between Newfoundland and Euro-Africa (Spain or Morocco) fractured and disappeared underwater.

The tropical seawaters that crept into the sinking rift between Europe and North America carried a myriad of planktonic animals and plants. Invertebrates and fishes followed, creating an ecosystem from what was essentially an ecological *tabula rasa*. The early North Atlantic and Arctic were tropical oceans, not cold ones as they are in the present day.[6] There were no ice caps. Temperate forests flourished in Greenland, at Spitsbergen north of Norway, and in Newfoundland and Labrador, and tropical forests grew in Great Britain and Europe.[7]

We can only speculate on what the earliest North Atlantic ecosystem

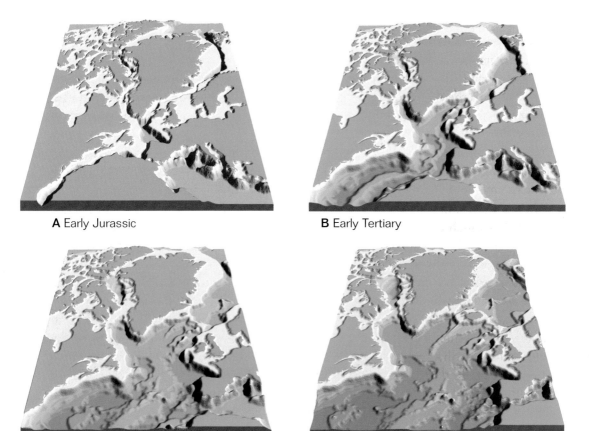

A Early Jurassic

B Early Tertiary

C Mid-Tertiary

D Quaternary

was like. The Caribbean or Mediterranean seas are perhaps the closest modern day analogues. The waters were likely warm and salty, and early plankton, invertebrates, and fishes would have been tropical species. Ancient deepwater species, such as the lanternfish, moved in. Some pelagic species – those that live in the surface waters – came north in summer to feed and retreated south with the seasons, as many still do today. Others stayed year round. There were probably many more species than exist today.[8] In terms of the species that inhabited the early northern seas, the Caribbean and Mediterranean are imperfect analogies – most early arrivals to the North Atlantic are now extinct, eliminated by time and changing environments.

The Grand Banks, underlain by continental rock, became the trailing edge of a westward-moving New World and the eastern shore of the new continent. A large knoll of uplifted rock, a sort of tectonic anchor, was dragged behind and would eventually become the Flemish Cap. The trailing edge contained some of the oldest rock on earth (up to 3.6 billion years old) floating on some of the youngest.[9]

Today, the Grand Banks have sharp edges, despite millions of years of erosion. Seamen know them from the swell and smell of their waters – you are either on the Banks or off them. At the edges of the Banks, the ocean floor drops over 1000 metres within a few nautical miles (1 nautical mile = 1852 metres). These steep slopes were moulded by two ancient transform faults in the continental and oceanic rock on the southern and northeastern edges. Transform faults occur as one tectonic plate slides against another, forming cracks in the crust that follow the direction of motion. Rock movements, whether uplifting or sinking, differ on opposing sides of a transform fault. The southern margin of the Grand Banks lies along a major transform fault: to the southeast, oceanic crust moved farther westward unimpeded by overlying continental rock, while, to the northwest, overlying continental rocks of the Grand Banks floated with the plate, resisting movement, and kept the Banks prominent above the underlying basalt.[10]

The northern limits of the Grand Banks formed more recently and in a different direction. For much of the early history of the North Atlantic, land extended seaward from the Avalon Peninsula over 400 kilometres from its present position. However, about 65-75 million years ago, the series of great calamities marking the end of the Cretaceous period occurred. Climates altered, resulting in major biological changes. Dinosaurs, the dominant large animals, became extinct. In the North Atlantic, the earth's crust shifted and an old transform fault, called Charlie, reactivated just to the north of the Grand Banks. The continental shelf from the Avalon Peninsula to the present Southeast Shoal of the Grand Banks, then dry land, uplifted farther, while a huge rock wedge to the north sank on a hinge to the Charlie transform. The sinking of this great block of continental crust, which came to underlie the oil-rich Orphan Basin, allowed seawater to flow in and separate the outer edges of North America from the receding and submerged lands that stretched from Ireland to Greenland. The Flemish

Cap had by then separated from Europe, and Atlantic waters were flowing into a newly opening Labrador Sea.

The centre of the Labrador Sea was a spreading ridge of new basalt crust, and the stretching of the land margins in Labrador and Greenland resulted in a widening of their continental shelves. These shelves, including the northeast Newfoundland and Labrador shelf, are smaller than the Grand Banks – they extend less than 200 kilometres from land in most places, and narrow to the north in proportion to the spreading between Labrador and Greenland. Eventually, seafloor spreading between Labrador and Greenland ceased, and a new centre of spreading developed between Greenland and the United Kingdom. During this epoch, the Orphan Basin settled into its modern and mostly deepwater position under the merged Atlantic Ocean and Labrador Sea. On the northeast Newfoundland shelf, its sunken area north of the uplifted Grand Banks would in time form the Bonavista Corridor, one of the great cod migration routes of the world.[11]

Spreading at the mid-Atlantic Ridge and continental sedimentation continued after the calamities of the end of the Cretaceous period, but the outline of what would become the Grand Banks and continental shelf of Newfoundland and Labrador was largely in place, although still mostly above water.[12] Ocean floor spreading continues to the present day at a rate of about one to three centimetres per year, and the underwater banks that trail behind westward-moving Newfoundland continue to pivot westward and subside.[13] The land sinks too: Signal Hill at St. John's is not quite as high as it was when Marconi received his famous message across the Atlantic from Poldhu, England, in 1901.

COOLING OF THE NORTH ATLANTIC

About 50 million years ago, the earth's climate began to cool. The initial and most extreme temperature declines occurred at the poles, and, within 10 million years, the first ice formed in the Antarctic seas. Arctic waters also cooled, but not as quickly as those in the Antarctic. Antarctic ice sheets began to form, perhaps intermittently at first, but by 25 million years ago a large ice cap became grounded.[14] Ice formation took large amounts of seawater out of circulation, lowering sea levels worldwide by perhaps as much as 300 metres.[15] For a time, the increasingly frigid Arctic and the warmer Atlantic waters were separated by the Arctic land bridge linking Greenland to Norway. About 38 million years ago, this bridge slipped beneath the rising seas, allowing colder waters to pour southward into the North Atlantic. There were inevitable and extreme biological consequences.

While the earliest North Atlantic was a species-rich tropical world, the cooling of the past 40-50 million years has produced a sparse and subarctic environment. Few of the original Atlantic tropical species survived, and those that did moved far to the south. Most species became extinct and are found today only as fossils.[16] To the north, the ecosystems became simpler, with fewer species and less complex food webs.[17] There are few ancient species in the North Atlantic – its environment has been too variable for species to survive over long periods of time. Although the long-term cooling has been interrupted from time to time with warmer climate episodes, it has resulted in the development of ice sheets over the Arctic and Greenland landmasses. More recently, ice ages have come and gone. Ancient fishes, such the 100-million-year-old coelacanth (*Latimeria chalumnae*), that have survived in the stable environments of subtropical southern oceans do not occur in the dynamic environments of the North Atlantic or Arctic oceans.

The ancient ecosystems of the North Atlantic probably did not include any of its modern species, although many of their ancestors were likely present. Early ecosystems may not have been as productive as the modern varieties, because the circulation upon which productivity depends was weaker. Nonetheless, the physical basement of the Grand Banks and continental shelf was in place long ago, and modern ecosystems have been profoundly influenced by their genesis. Of the species that currently inhabit the North Atlantic, most evolved only recently during the past 3-5 million years, and many came from other oceans, especially the North Pacific.[18]

The changes in the North Atlantic ecosystems over the millennia were not isolated from the rest of the world or its oceans. The genesis of the North Atlantic and the Grand Banks was part of a major tectonic event that separated the Old World from the New World and that brought western North America closer to Asia. About 80 million years ago, the circulation of warmer Pacific waters to the Arctic was impeded as North America (Alaska) shunted towards Asia, and eventually the two continents collided. Perhaps 40 million years ago, the newly formed basin of the North Pacific became a centre for marine evolution in the cooling northern hemisphere; this would eventually have great consequences for the North Atlantic.[19] The Bering land bridge blocked all flows from the Pacific to the Arctic and shielded the North Pacific somewhat from the extreme cooling of the Arctic region until about 6-12 million years ago.[20]

On the Grand Banks, huge masses of eroded sediment from land buried the underwater continental rocks and filled in most of the basins. Accumulated deposits in these areas are 10-15 kilometres deep. The sinking of the Grand Banks and the rise in sea levels submerged most, but not all, of what was once dry land: to the great surprise of early mariners, the Virgin Rocks of the Grand Banks still break water some 200 kilometres from shore. The edges of the Grand Banks blocked and redirected water flows. In particular, the steep edge of the southern Grand Banks opposed the northward-moving Gulf Stream, helping to turn it eastward, while the northern edge steered the southward-moving Labrador Current in a broad arc around its perimeter. At the southeastern corner of the Great Banks, these two moving titans collide.

OCEAN WATERS: CURRENTS, TEMPERATURES, AND SALT

The oceans are neither homogeneous nor randomly mixed waters. They are comprised of moving masses of water with differing properties and distinct boundaries between adjacent masses: some waters are warmer than others, with temperatures ranging from sub-zero to more than 30°C; some are saltier, with salinities from near fresh to very salty (36 psu[21]); and some contain far more nutrients than others, so that water masses range from nearly sterile to richly fertilized. Floating plankton accumulate at the borders, or fronts, between different water masses and attract fish.

Ocean water masses are in perpetual motion. Colder and saltier water masses are denser than warmer and fresher waters. Cold, salty waters sink, displacing lighter waters and driving much of the circulation in the oceans. Currents, formed from directed and narrowed flows driven by these differences in density as well as by orbital forces and wind, redistribute the oceans' nutrients and heat. Currents mould the biological parts of marine ecosystems. They bring heat or cold, and salt. They bring nutrients, or cause their upwelling from richer, usually deeper waters, and enhance the fluid environment that fosters the dispersal of plankton and fishes. In conjunction with depth and light, currents ultimately determine the patterns of ocean productivity and the limits of marine ecosystems.

North Atlantic

The ecosystems of Newfoundland and Labrador are strongly influenced both by the icy waters of the southerly flowing Labrador Current and the heated waters of the northeasterly flowing Gulf Stream, which meet near the southern edge of the Grand Banks. Exactly when these currents began to flow is not certain,[22] but their very existence resulted from several events in the earth's history: the joining of the Atlantic and Arctic oceans about 38 million years ago and the subsequent cooling of the North Atlantic; the partial closure of the Pacific-Arctic connection much later; and the emergence of the Panama land bridge that joined the North and South American continents.

For many millions of years after the Atlantic opened, prevailing easterly winds at the equator produced westerly ocean currents that flowed directly and without interruption to the Pacific through the gap between North and South America. There was little northward flow. Between 12 million and 3.5 million years ago, the equatorial flows changed radically when the Panama land bridge emerged to connect the Americas.[23] This event transformed the North Atlantic: after the closure, the buildup of water in the Gulf of Mexico turned north along the Florida coast in a huge jet of water called the Gulf Stream.

The currents of the present-day North Atlantic are complex. From the ocean surface to the seafloor, there may be two or three different water masses, each with its own temperature, salinity, flow direction, and speed. Although the deepest waters in the North Atlantic originate in the Antarctic Ocean, most of its cold, dense water comes from Arctic Ocean waters that spill over between Norway and Greenland, then flow

1.3
Circulation in the Atlantic Ocean.

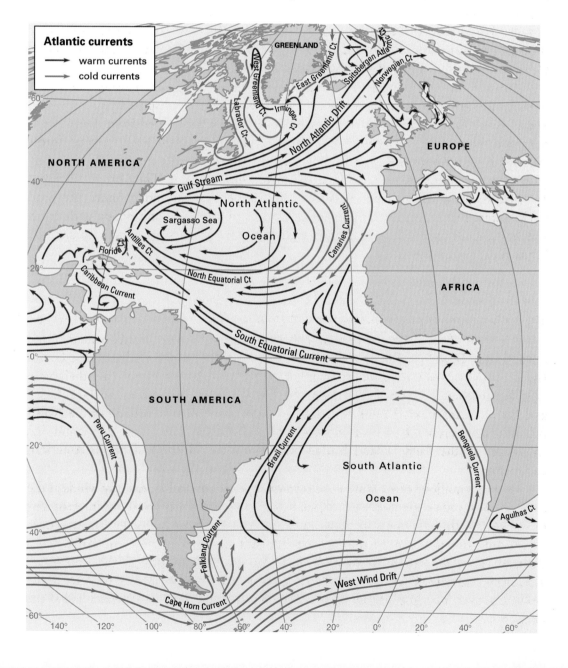

southwestward into the Labrador Sea. In its early history, currents in the North Atlantic were perhaps more modest than they are today. With the opening of the Labrador Sea and subsidence of the Grand Banks, however, a dominant southward flow of cold Arctic waters, ice melt, and recycled tropical waters developed in the northwest Atlantic.

In balance with these southward movements of cold Arctic waters, warm tropical surface waters flow northward and eastward in the Gulf Stream and North Atlantic Drift.[24] Together, these opposing currents underlie perhaps the largest heat exchanger on the planet, and result in remarkable differences in the ecosystems of the northwest and northeast Atlantic. In the northwest Atlantic, only the deep waters are warm; the surface waters are very cold. In stark contrast, both the deep and surface waters of the northeast Atlantic are relatively warm, and northern Europe owes its benign climate to this alone. Nowhere else on earth do such warm waters flow so far toward the poles. Beneath the surface of the ocean, the climate is no less remarkable. Coastal Iceland and Norway, despite their northern latitude (60-65° north latitude) have warm waters and ice-free oceans year round. Newfoundland and Labrador waters experience ice as far south as 40° north latitude, about the same latitude as the beaches of northern California.

Gulf Stream and North Atlantic Drift

By the time the Gulf Stream was set northward by the uplifted Panama Isthmus, the Arctic land bridge that had linked North America, Greenland, and Scandinavia had disappeared.[25] As the salty waters of the North Atlantic Drift cooled, they became heavier and sank, mixing with deeper waters from the Arctic shelves. The cold (-1°C) and dense water formed in the Norwegian-Greenland Sea spilled over into the Labrador Sea west of Greenland, becoming the source of most of the deep water of the Atlantic.

The exchange of cooled Arctic water for warmer Gulf Stream waters is the key to the ample productivity of the North Atlantic. The exchange makes the northern waters warmer overall, but even more importantly, results in nutrient enrichment through a natural fertilizer pump into the sterile surface Arctic waters.[26]

At present, the Gulf Stream off the Florida Peninsula carries 25 times as much water (19 million cubic metres per second) as all the rivers of the world combined.[27] Along the Florida coast, the Gulf Stream is 80 kilometres wide and nearly half a kilometre deep. This massive jet of tropical water flows north from the Florida Straits for thousands of kilometres, and then, partly because of the Coriolis force[28] and partly because of its collision with the massive Grand Banks and the opposing Labrador Current, strikes eastward across the North Atlantic. The Gulf Stream becomes more diffuse as it travels east: parts of this North Atlantic Drift strike north and then back east off Iceland, while other branches flow south towards the coast of Africa. However, the main Drift moves ceaselessly eastward towards Scandinavia and the Barents Sea. Eventually, its waters cool in the vast Norwegian-Greenland Sea, then sink and are recycled back to the south once again as Atlantic deep waters. These waters are dense from salt and cold, but less so than the seafloor-hugging Antarctic waters that flow as far north as 35° north latitude beneath the deep, southward-flowing waters from the Arctic. The Antarctic and Arctic flows pass each other, moving in opposite directions, in the deep mid-Atlantic near the equator. The Arctic waters finally mix with fresher surface waters from the Antarctic in the South Atlantic at 50-60° south latitude. The full cycle from polar to polar region takes about 1000 years.[29]

ORIGIN OF THE FISHES

The Age of Fishes occurred in the Devonian period, 345-405 million years ago, when North America was fused to Europe and Africa.[30] At that time there were two large landmasses – northern Laurasia (modern North America, parts of Europe, and Asia) and southern Gondwanaland (South America, Africa, India, and Australia) – separated by an ocean that stretched around the globe. Many types of fishes first appeared in the fossil record during the Devonian. Others were well established before that time: many specialized forms of the three main types of fishes – Cyclostomes (lampreys and hagfish), Selachians (sharks and rays), and pre-Teleosts (bony fishes) – were already present in the world's oceans. Only a few of these early fish species survived to modern times. Early nonparasitic Cyclostomes have modern descendants in the parasitic lampreys and hagfishes. Many sharks and rays exist now in much the same form as they did in the Devonian, and some other modern Selachian species, such as the monkfish (*Lophius americanus*), can be traced back 150 million years. A few ancient bony fishes also survived, such as the "living fossil" coelocanth of the Indian Ocean, freshwater forms of the sturgeons (*Acipenseriformes* spp.), the North American bowfin (*Amia calva*), and garpikes (*Lepisoteus* spp.). These early bony fishes also gave rise to the modern Teleost fishes in the late Cretaceous period, 75 million years ago, which in turn led to the evolution of most modern types of fishes.[31]

The cooling of the North Atlantic over the past 40 million years was part of a great change in ecosystems worldwide. The fate of older, warm-water-adapted species in the North Atlantic is unknown, but many presumably perished. The greatest biological changes in the North Atlantic, however, were yet to come.

A seminal event for the North Atlantic was the emergence of the Panama Isthmus and the reorganization of the Gulf Stream and North Atlantic Drift between 12 million and 3.5 million years ago. This was followed closely by the sinking of the Bering land bridge that joined Siberia to Alaska. The Bering Strait opened a western door to a northwest passage with waters sufficiently warm to allow migration of cold-water species.

The older and larger North Pacific has been an evolutionary engine for marine species[32] – a storehouse of cold-water fishes and the source of many fish, animal, and plant species of the present-day Arctic and North Atlantic communities.[33] The North Pacific houses more endemic families of fishes (at least twenty-one)[34] than the North Atlantic (which has none), and more endemic species of fishes (more than sixty, compared to the North Atlantic's twenty).[35] The differences in species abundance between the North Pacific and Atlantic are even greater with crustaceans: the North Pacific has at least one hundred endemic species of crustaceans, whereas the Atlantic has a mere eight or ten. With the opening of the Bering Strait, this great storehouse of cold-water-adapted species had access to an entirely new home in the North Atlantic.[36]

As the Bering land bridge sank beneath the sea, the North Pacific unleashed its influence on the North Atlantic. Many species migrated east, among them some that would dramatically change the young ecosystems: capelin and redfish, shrimp and snow crab, and lesser-known species of sculpins, alligatorfish, and seasnails all originated in the Pacific. Some fish species achieved much greater abundance and importance in their new home than they had in their native North Pacific. A good example is capelin: while it is not unimportant in the North Pacific, it has neither the abundance nor the seeming monopoly on the small pelagic fish niche that it does in much of the North Atlantic. Invertebrates also migrated from the Pacific to the North Atlantic. Of particular note are the shrimp, including the Pandalid (the so-called northern or ice shrimp) and the less well-known Crangid and Spirontocarid shrimp. There are 88 species of these types of shrimp in the North Pacific, but only six in the North Atlantic.[37] The few that made their way to the North Atlantic have, like capelin, achieved great abundance. The Chionocete crabs are also of Pacific origin – of four species that occur in the North Pacific, only the cold-loving snow crab made it to the North Atlantic, where it has thrived. Why this species migrated and survived to the present day, while others did not, is difficult to answer. Perhaps it was the cold environment, perhaps a bit of luck. Even some of the large species of kelp (*Laminaria* spp.) and eelgrass (*Zostera* spp.), known to shelter young fish in Newfoundland bays, came from the North Pacific.

A second source of marine species was the Arctic Ocean. A few species came south and stayed as the North Atlantic cooled. Others ventured south seasonally or during cold years, but then retreated north. Notable among these were the ice-pupping seals (harp and hooded), but several species of invertebrates, such as marine butterflies (known to Newfoundland and Labrador fishermen as slub), and a few fishes, such as the Arctic cod, also came south. Overall, however, few truly Arctic species made the North Atlantic their home and their contribution to the North Atlantic ecosystem was modest compared to that of the North Pacific.

1.4
Time lines, periods, epochs, and noted physical developments in the earth's history and the evolution of fishes.

	Period	Epoch (million years)		Physical	Fishes	Scale
Cenozoic	Quaternary	Recent Pleistocene	0-1	Ice ages Bering bridge sinks	Cod, capelin... Most living species evolve	
	Tertiary	Pliocene	1-13			
		Miocene	13-25			
		Oligocene	25-36			
		Eocene	36-58	Arctic bridge sinks	1st fossil gadoid	
		Paleocene	58-63			
Mesozoic	Cretaceous		63-135	Warm-no ice caps	Mass plankton extinction	
	Jurassic		136-181	Banks separate from Euro-Africa		
	Triassic		181-230	North America drifts west		
Paleozoic	Permian		230-280			
	Carboniferous		280-345			
	Devonian		345-405		First Gadiformes Age of fishes	
	Silurian		405-425			
	Ordovician		425-500	Closing of Iapetus Sea	First fishes	
	Cambrian		500-600		First marine invertebrates (hard parts)	
Precambrian	Proterozoic		600-2500		Algae and bacteria	
	Archean		2500-4000			

Compared with other oceans and large freshwater ecosystems, the North Atlantic is not rich in fish species – the present-day count is only about 188 in total. By comparison, there are hundreds of species of rockfish alone in the North Pacific and perhaps 500 species of a single fish family in Lake Malawi in Africa. Many fish families have very limited distributions in the North Atlantic. For example, sculpins (Cottidae), alligatorfish and poachers (Agonidae), seasnails (Liparidae), and eelpouts (Zoarcidae) are widespread in the deep waters of the North and South Pacific, but have limited ranges in the Atlantic. The simplest notion of why the Pacific has much wider distributions of species is that ecosystems become richer over time, and the North Atlantic is simply too young an ocean to be so richly endowed. Additionally, the North Atlantic may also be a more severe environment and more homogeneous than other oceans or lakes, both of which could limit diversity. Every cloud has a silver lining, however, and so it is in the North Atlantic: although the number of species in its simpler ecosystems may be fewer, they can achieve great abundance. This truth is reflected in what is known as Theinemann's Law (see Chapter 2).

Origin of Cod

Gadiform fish (codlike fish, referred to as gadoids) first appear in the fossil record about 40-50 million years ago.[38] Gadoids were not important fishes then, and most were small and deepwater fishes of the expansive southern oceans. Evolution often requires isolation, and the cooling temperatures and falling sea levels worldwide opened the door for cod evolution in the North Atlantic.[39] The eastern part of the North Atlantic in the North Sea became isolated from warmer waters, and gadoids first appeared as dominant fishes there.[40]

About 10 million years ago, the first large gadoids appeared in the North Sea; by 5 million years ago, the modern Atlantic cod, haddock, and pollock (or saithe) were all present in European waters. The modern cod are among the few fishes whose evolutionary home is the continental shelves of the North Atlantic.[41] From their beginnings in the European waters around the North Sea, cod likely spread to Iceland, Greenland, Labrador, and Newfoundland, and then south into the cool waters of coastal eastern North America. It was here, within the past 5-10 million years, that the modern gadoid fishes came into their own, and evolved to exploit and dominate the new boreal and subarctic ecosystems of the North Atlantic. More will be said about cod and their evolution in Chapter 2.

ICE AGES

> *Environmental changes associated with the last deglaciation had profound effects on the evolution of biotic communities. During glacial stages, species were physically forced southward by glacial ice.*
> — R.W. GRAHAM AND J.I. MEAD, *Environmental Fluctuations and Evolution of Mammalian Faunas During the Last Deglaciation in North America*, 1987

A general cooling trend in the earth's surface temperatures over the past 10 million years, with intermittent ice ages and sea level changes, helped to shape the modern ecosystems of Newfoundland and Labrador and the Grand Banks. The ice ages of the past few million years resulted in severe changes in the physical properties of the Newfoundland and Labrador ecosystems. During glacial periods, much of the Grand Banks and adjacent continental shelf were either covered in ice or reemerged above sea level. Water temperatures were colder, seas levels perhaps 100 metres lower than they had been, and the currents that pumped productivity into these ecosystems may have slowed or even stopped altogether.[42] Most coastal ecosystems within the bays were periodically smothered by ice and obliterated. Many species relied on these habitats for spawning or juvenile habitat, and these populations would have been eradicated or displaced.

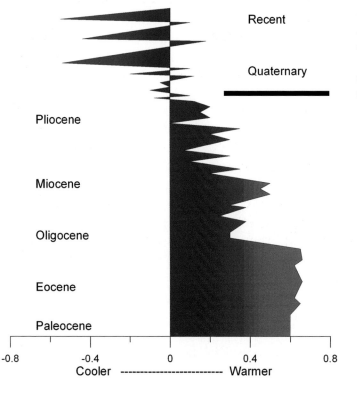

1.5
Relative temperature changes during the Tertiary and Quaternary periods (the past 63 million years), showing the recent cooling and the ice ages.

It is not possible to know if the ecosystems and species distributions prior to the Wisconsin glaciation, the last great ice age, were similar to those we know today. Perhaps they were – we can be reasonably sure that the same species have been here at least for the past few million years. But ecosystems perturbed or changed by climate may not always regenerate themselves in exactly the same way, and there is no certainty that identical configurations reoccurred during successive interglacial periods. All we really know is what developed since the last one.

Present day ecosystems of Newfoundland and Labrador, and of the North Atlantic, developed during the ice ages. No one knows exactly how many ice ages occurred before the last one – ironically, glaciers tend to erase all traces of previous ice ages.[43] Glaciation has likely occurred many times at regular intervals in the earth's recent history. Cold periods have been interspersed with warmer periods for at least the past 10 million years, but extensive glaciation began only about 2.5 million years ago.[44] Since then, the earth has alternated between glacial and interglacial periods at regular intervals of 41,000 and 23,000 years. During the most recent 700,000-900,000 years, a longer 100,000-year cycle has also been evident.[45] These cycles relate to changes in the earth's orbit, which cause temperature changes at the poles, especially in summer.

The most recent glacial period, the Wisconsin, began about 70,000 years ago.[46] After three ice advances, the Wisconsin finally retreated about 10,000-12,000 years ago. The extent and spreading patterns of the glaciers is a matter of debate among geologists.[47] A common misconception is that the glaciers spread from the north and slowly engulfed southern regions like the island of Newfoundland. It is more likely, however, that Wisconsin ice spread from several inland centres in Newfoundland, and it may have coalesced with the margins of the much larger Laurentide continental glacier in the Strait of Belle Isle area. Even considering the least likely extent of Wisconsin glaciation, the ice sheets were sufficiently large to overwhelm pre-existing terrestrial and coastal ecosystems. One to two kilometres of ice built up over much of the region,[48] its shear weight and abrasion displacing most sediments and soils. A few ice-free mountain regions and coasts may have persisted through the glaciation,[49] but thick ice scoured most bays and inlets. There is some opinion that glaciers created the bays and fjords of Newfoundland, but that is unlikely: as large and heavy as the glaciers were, they only rubbed the surface of an already strongly featured and much more ancient landscape.[50] The ice flows tended to follow existing major land and sea features, thus cleaning off, polishing, and reemphasizing previous land- and sea-scapes. This is not to minimize the effects of the glaciers – they left lasting imprints on both land and at sea, cleaning the slate of terrestrial and marine ecosystems. The weight of the ice forced even the oldest and hardest continental rocks downward into the crust. After the ice melted, the land rebounded – on the coast of Labrador, old beaches can still be seen, now stranded 150 metres above sea level. And although land-based, the effects of the glaciers were felt far out to sea.

The seaward extent of the ice remains controversial.⁵¹ There are two basic schools of thought. The first espouses that glaciation was minimal: under this scenario, the ice sheets barely spanned the bays and did not extend very far to sea. The second thinks that the ice sheets extended well out to sea and covered most of the Banks. There is evidence for both points of view. The minimalist view is supported by evidence of nonglaciated areas in Labrador and in western Newfoundland.⁵² On the other hand, on Hamilton Bank off Labrador, there is evidence of seaward extension of the ice sheets to the Bank's edge.⁵³ Moraines near the seaward edge of Hamilton Bank and in adjacent channels suggest the glaciers extended far offshore. On the Grand Banks things are less clear, but sediment evidence suggests that ice may have covered the shoreward third of the Banks.⁵⁴

Nearer to shore on the continental shelf, glacial ice pushed seaward, extending from the land into the sea. The Grand Banks were "glaciated by ice moving radially seaward from the central part of the Avalon Peninsula."⁵⁵ The moving ice gouged deep troughs through bays and along faults roughly parallel to shore as kilometre-thick ice flows dug into the soft sedimentary rocks laid down shoreward of the hardest rocks of the Banks. Southward-flowing currents added an additional push. Troughs as deep as 500 metres can be found from northern Labrador and the Hopedale Channel through the St. Anthony Basin and the Funk Island Deep to the Avalon Channel. In places,

1.6

Seabed on outer Hamilton Bank showing evidence of glaciation. Photo reprinted from Hamilton Bank, Marine Geology, Fillon, R.H., Labrador Shelf postglacial sediment dynamics and paleo-oceanography, 20, 7-25, 1976, with permission from Elsevier.

the ice flowed farther seaward through cross-shelf troughs. Just south of Hamilton Bank, the Hawke Channel was scoured over 500 metres deep to the shelf edge, and terminal glacial moraines marking the extent of the ice can be found there. Troughs in the Hawke Channel became deep enough to escape the cold surface waters of the Labrador Current. Much later, Atlantic waters in the troughs would provide warmed highways for

Above Sea Level But Now Submerged Ice Sheet

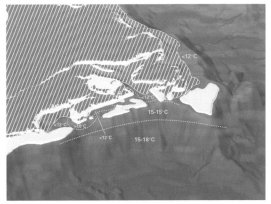

A 15,000 years ago: Ice sheet beginning to retreat. Sea level 150 m below present level.

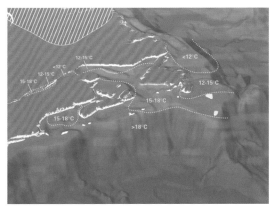

C 7000 years ago: Continental ice virtually gone. Sea level 30 m below present level.

B 12,500 years ago: Ice retreating. Sea level 100 m below present level.

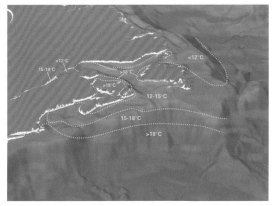

D 3000 years ago: Sea near present level.

1.7
Grand Banks and Newfoundland-Labrador sea level and land-sea changes since the Wisconsin ice age that began to retreat about 10,000 years ago. White areas are regions which are now submerged, but were previously above sea level. (Compilation of data from several sources.)

cross-shelf cod migration, away from the icy Current, and would become key areas for juvenile fish.[56] The Hawke Channel consequently became one of the most important fishing areas in the North Atlantic, a biological garden of zooplankton, shrimp, snow crab, capelin, cod, and the deeper water redfish and Greenland halibut.

The amount of ice in the North Atlantic and Arctic oceans was staggering. Even if land ice were only a few hundred metres to a kilometre thick, that is many times as deep as most of the Grand Banks. With so much seawater frozen into ice, it was inevitable that sea levels would fall. At the glacial maximum, sea levels were up to 200 metres lower than at present. As a result, the seaward half of the Grand Banks was above water: frozen flat islands off a glaciered coast. These islands formed an archipelago that extended from the Flemish Cap to Georges Bank off New England, with the largest island centred on the Southeast Shoal of the Grand Banks. As the glaciers receded and sea levels rose over the past 10,000 years, the archipelago slowly slipped beneath the sea. On the Grand Banks, the last vestiges of hilly peaks of these islands, the Virgin Rocks, still break water, and the Southeast Shoal is in places less than 40 metres deep, over 200 nautical miles (360 kilometres) from shore.

The extent of the sea ice in North Atlantic and Arctic waters between northern Europe and North America during the last glaciation is also controversial. Early investigations suggested that the winter ice margin stretched as far south as 45° north latitude (St. John's lies at 47.5° north latitude).[57] However, more recent studies suggest that, as a result of warm-water drift northward and eastward across the North Atlantic, ice-free areas may have reached much farther north in the central and eastern Atlantic and even to the Barents Sea.[58] Such areas would have provided refuges and feeding areas for fishes even during the glacial maximum. However, on the western side of the Atlantic, sea ice likely covered much of the area from Newfoundland to Iceland each winter.

However great the extent of the ice coverage in the North Atlantic, there is no doubt that sea temperatures, salinities, and currents were profoundly influenced by glaciation.[59] The Gulf Stream and Labrador Current were flowing long before the recent glaciations, but the presence of ice altered these currents and may have stopped them altogether at

1.8
Extent of winter ice cover and changes in sea surface temperatures (degrees Celsius) during the last glacial maximum (data from Ruddiman 1987). Note that the Newfoundland and Labrador region was covered with continental shelf and sea ice, and the North Atlantic suffered severe sea temperature declines.

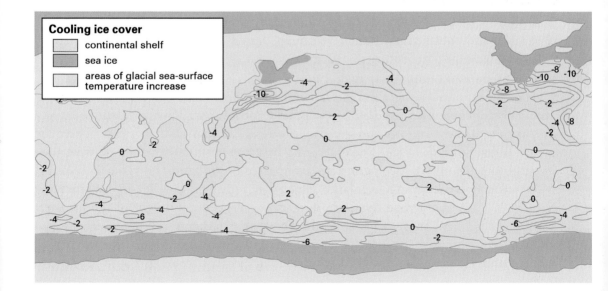

times. Over the period from 160,000 to 10,000 years ago, and since the end of the last ice age, temperatures and currents changed quickly and regularly, and differed greatly from their present configuration. The Labrador Current may have been diverted eastward from the coast of Newfoundland and Labrador during and after the last Wisconsin glaciation.[60] If so, warmer-water refuges for cod and other species may have persisted south of Newfoundland and the Gulf of St. Lawrence.

THE LAST 1000 YEARS

The variations in the ocean and climate of the past 1000 years have been relatively minor compared to those during the ice ages – Newfoundland and Labrador marine ecosystems have experienced far more extreme conditions in the past than they do today. Nevertheless, the variations of the last 1000 years are not irrelevant to ocean productivity or to fishes. At the very least, and of great interest, they have determined the state of the ecosystems during the relatively brief modern period of fisheries exploitation.

Periods of warm and cold temperatures have alternated since the last glacial retreat. Reconstructions of temperature indicate that conditions in the northern hemisphere were relatively mild 1000 years ago, but a slow cooling took place over the next 500 years. The "little ice age" occurred from the 1500s to the mid-1800s, with strong regional variations. In the Newfoundland and Labrador region, the later 1600s were cold, as was the beginning of the 1700s. An abrupt warming then appears to have taken place in the third decade of the 1700s; it lasted until the early 1800s, when cooling again occurred and continued until at least midcentury. The 1900s brought generally warmer conditions that peaked in most areas from 1920-1960. The only constant has been variability, with both short-term and, more importantly, long-term (periods of up to a century) episodes of warmer and colder years. Recent warming, caused mainly by the burning of fossil fuels, has made the past century the warmest on record for many thousands of years, and the rate of warming is now very high.[61]

The retreat of the last glaciers did not end the presence of ice on the Newfoundland and Labrador coast. Ocean ice off Labrador and northeastern Newfoundland remains a key feature of these regions and greatly influences the marine ecosystems. There are several different kinds of ice. The oldest and best known in North Atlantic waters are icebergs, large chunks of long-frozen and compacted precipitation from the high Arctic. Most icebergs come from Greenland, which has remained covered in ice since the Wisconsin glaciation period. New ice formation increases the weight of the ice sheet while gravity forces movement, with the result that icebergs calve in huge broken chunks from the Disko Bay glacier area of western Greenland. Like a line of moving white islands, icebergs are delivered by prevailing currents across to Baffin Island and

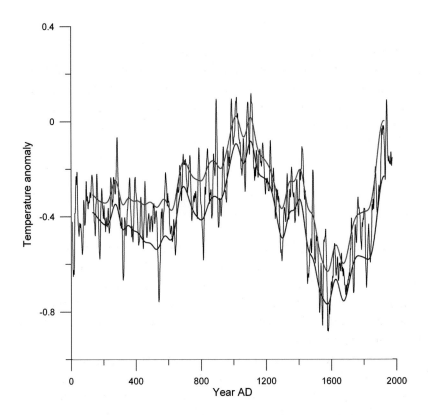

1.9 Reconstructed temperatures over the past 2000 years. The black line is the decadal average of the original data. The red line is the upper bound and the blue line is the lower bound of average temperatures with the high frequency variation removed. Data from Moberg et al. (2005).

southward along the Labrador coast. During their journey, their tracks are influenced by winds and pure fancy – their movements are not totally predictable. Some fishermen believe that icebergs attract fish, presumably because cold, fresh meltwater creates local water mass boundaries where plankton accumulate. In some years, there are many icebergs; in others, few. At times they travel as far south as 40° north latitude, about the same latitude as northern California or the Mediterranean Sea. In most years, even the largest icebergs melt in the warming waters of the southern Grand Banks, but they may occasionally venture farther. In the spring of 1912, a fleet of icebergs moved across the shipping lane well south of the Grand Banks and into history. Here the *Titanic* met its fate – but such is the variability of the North Atlantic.

Sea ice has none of the glamour of icebergs. It forms each year as a result of the freezing of the cold and relatively fresh waters of the Labrador Current. There are many types of sea ice: local first-year ice that melts each year; multiyear ice from farther north; landfast ice that freezes in the bays and differs from other types of ice because its forming waters are less saline; and larger pans of open-ocean ice formed from saltier waters. Sea ice is very important to local climates and ocean ecosystems: it limits heat transfer between the atmosphere and ocean, protects the waters from wind, may influence fish plankton and fish distributions, and provides the seasonal breeding habitat for hooded and harp seals, the top predators and most numerous marine mammals in these ecosystems.

1.10 Iceberg in Trinity Bay in 2005. Photograph by Tom Clenche.

1.11 Sea ice in the northern Gulf of St. Lawrence in winter 1984. Photograph by the author.

Average Ice Conditions, Spring

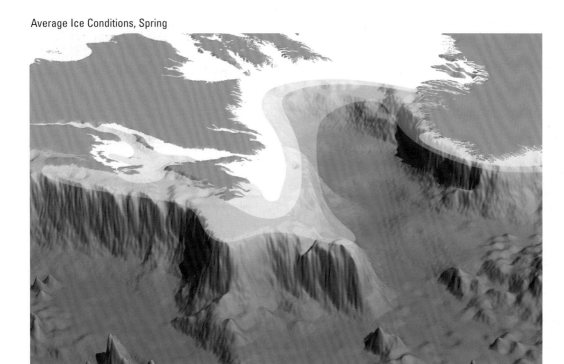

Close pack ice
Open pack ice
Max. extent, all ice

Average Ice Conditions, Summer

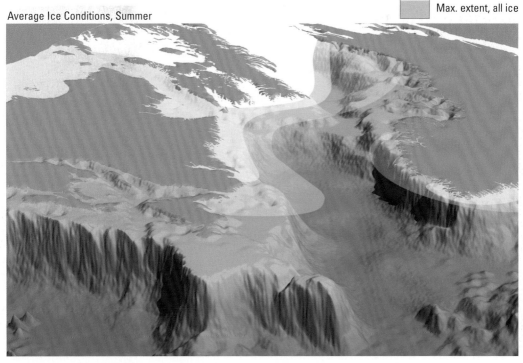

1.12 Sea ice extent in the northwest Atlantic in spring and summer.

North Atlantic Oscillation

A key feature of the North Atlantic climate is an atmospheric pressure feature called the North Atlantic Oscillation (NAO).[62] The NAO is the difference in atmospheric pressure between the Azores Islands high pressure and Icelandic low pressure systems, and links the climate and ocean conditions off Newfoundland and Labrador with the overall climate of the North Atlantic. When the NAO is positive, the flows of the Labrador Current tend to weaken, bringing warmer conditions to the continental shelf and Grand Banks. When the NAO is negative, the Labrador Current strengthens and colder conditions prevail. The variations in the NAO are related to many climatic, oceanographic, and ecological features in the marine ecosystems of Newfoundland and Labrador and the Grand Banks, including iceberg flows, ocean temperatures, the strength of the Labrador Current, and the distribution and biology of many species. It provides a useful index of ocean conditions related to warm and cold periods in North Atlantic and Newfoundland and Labrador waters.[63]

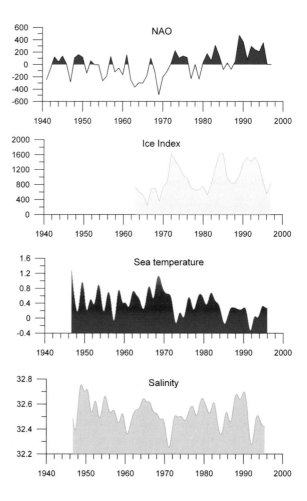

1.13
The North Atlantic Oscillation pressure difference between the Azores high and Icelandic low and other physical variables for the Grand Banks region (data from various sources). The ice index is a relative indicator of maximum ice extent, and sea temperature (degrees Celsius) and salinity (psu) are average measures from oceanographic station 27 near Cape Spear, Newfoundland.

Labrador Current

The currents that impact the marine ecosystems of Newfoundland and Labrador are part of a larger system of climate and ocean variation that spans the North Atlantic and comprises part of global circulation. The southward flowing Labrador Current dominates the oceanographic regime of the coastal areas of the Newfoundland and Labrador shelves, the Grand Banks, and the Gulf of St. Lawrence. Its influence can be felt as far south as New England. The heart of the current is stunningly cold, at -1.7°C (seawater freezes at about -2°C), and covers most of the continental shelf off Labrador.

There are two main parts to the Labrador Current, with different sources. The larger offshore branch results from the spin of the West Greenland Current, driven by cold Arctic flows in the Irminger Current along eastern Greenland. These flows are strengthened but also warmed and salted by offshoots of the North Atlantic Drift. The offshore branch carries ten times more water and is saltier and warmer than the inshore branch. The offshore branch skirts the continental shelf and the Grand Banks, and confronts the Gulf Stream southeast of the Banks. In a clash of the titans, the Gulf Stream turns east, while the lesser Labrador Current curls back to the north, reinforcing the counter-clockwise gyre in the Labrador Sea and eventually completing the circuit in the West Greenland Current.

The inshore branch of the Labrador Current is colder and fresher than the offshore branch. Its origins are in the Canada Current from the high Arctic, and along the way it is reinforced with freshwater inputs from Hudson Bay and many small rivers.[64] It meets little opposition on its journey south. It hugs the northeast coast of Newfoundland and Labrador and the Avalon Channel, chilling both sea and land, turns west along the south coast of the island, penetrating Placentia Bay to its head and out again, then ploughs westward into the Gulf of St. Lawrence. Some of the Labrador waters join cold waters flowing out of the Gulf and move farther south onto the Scotian Shelf, influencing the coastal plant and animal communities as far south as Cape Cod. The Labrador Current carries only one-tenth of the flow of the Gulf Stream, but still more than twice the water of all the rivers of the world combined.

Together, the inshore and offshore branches of the Labrador Current force a counter-clockwise gyre in the Labrador Sea, with the ice-cold and fresher inshore branch extending to the edge of the shelf break, and the saltier offshore branch holding position over deeper waters off the shelf. Steep gradients between these flows result in nutrient-rich waters being brought to the surface near the shelf break.[65]

If Labrador Current waters flowed to the bottom of the continental shelf, the marine ecosystems of Newfoundland and Labrador would be very different and far less productive. Labrador Current waters are relatively fresh; they do not sink deeper than 150-200 metres despite their cold temperatures.[66] Much of the Grand Banks and other banks are shallower than this, so Labrador Current waters dominate from surface to bottom. In the

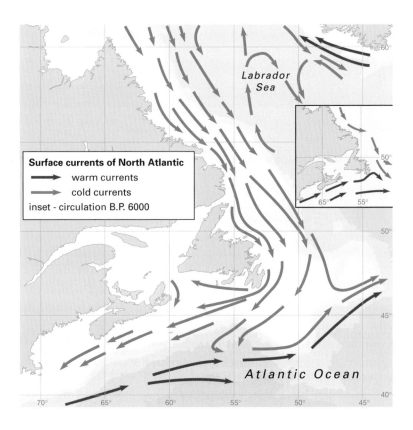

*1.14
Surface currents of the North Atlantic. The inset shows the probable currents 6000 years ago (Data from various sources; inset data from Fillon (1976)). B.P. indicates years before present.*

troughs and deeper basins, the warmer, saltier, and richer Atlantic Ocean waters cannot be displaced. Atlantic waters have year-round temperatures of 3-4°C, with higher salinities (34-35 psu) and nutrient contents than the Labrador Current.[67] In areas of mixing, Atlantic waters enrich the nutrient content of the more sterile surface waters. Enrichment occurs primarily along the shelf edge and in the cross-shelf channels such as the Hawke and Hopedale channels. The coverage of the Grand Banks by Labrador water varies from year to year. This influences not only productivity, but also the extent of the Grand Banks with conditions favourable to cod and other species. A few organisms have evolved to live in the coldest waters, but only a very small proportion of ocean species can live at -1.7°C.

Ocean waters move not only in currents, but also in tides that slosh back and forth in response to the gravitational pulls of the earth and moon on the water. Tides influence the flows and mixing of water masses, especially in shallow coastal areas, which in turn affect the ecology of many species. Tides in the North Atlantic run counter-clockwise around a central point about halfway between Newfoundland and the Azores. This rotation results in more or less synchronous tides from northern Labrador to southern Newfoundland. The Gulf of St. Lawrence, on the other hand, has its own tidal rotation: tides on the west coast of Newfoundland are four to five hours out of synch with those of the rest of the Newfoundland and Labrador coast. In general, tides in Newfoundland

and Labrador are small. The tidal range in most areas is less than 1.7 metres, with ranges over 2 metres occurring only in a few places such as Placentia Bay and northern Labrador. Exceptions to this can occur where tides force water into confined areas – such as Swift Current in Placentia Bay, where tidal ranges may approach 3 metres. Nowhere in Newfoundland and Labrador can the tides match those of Ungava Bay to the north or the Bay of Fundy to the south, where tidal heights can reach 10 metres.

The Grand Banks and adjacent continental shelves around Newfoundland and Labrador formed as a result of over a hundred million years of continental drift. The past forty million years of cooling has culminated in ice ages that radically changed the geography and the ocean climate of the North Atlantic. The closing of the Panama Isthmus and the opening of the Bering Strait resulted in major changes in ocean circulation and the distribution of species. Of primary importance to the developing ecosystems of the North Atlantic were the migration of species from the species-rich North Pacific and the evolution of the gadoids in the North Atlantic. Newfoundland and Labrador marine ecosystems house only a relatively small number of species, and during the ice ages, have had their slates repeatedly wiped clean. Despite these handicaps, and perhaps in part because of them, these ecosystems produced an abundance of marine life the likes of which the world has seldom seen.

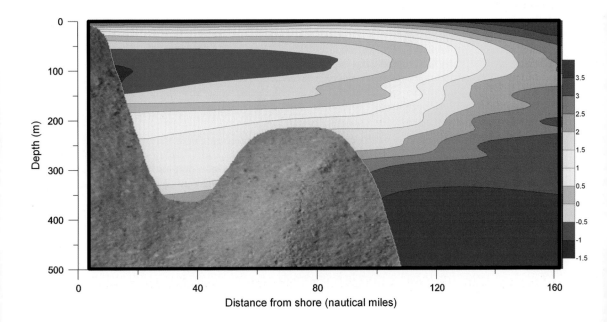

1.15
Cross-section of the Labrador Current across an inner channel and outer bank. The scale on the right shows water temperatures in degrees Celsius. (Compilation of data from several sources.)

NOTES

1. Travin and Pechenik (1963, 48-52) described the earliest developments of the Soviet Union fisheries in Newfoundland and Labrador waters in the 1950s: "Soviet commercial activities in the Northwest Atlantic began in August, 1956, after the large trawler *Sverdlovsk* discovered great stocks of *Sebastes mentella* [redfish] near the southwestern slope of the Fleming Cap Bank… By 1960 the Flemish Cap region was no longer most important; the most intensive fishing sites became the North Newfoundland Bank (3K), the Hamilton Bank (2J) and the southeastern slope of the Grand Newfoundland Bank (3N)…the average catches of Soviet trawlers per unit of effort in the Northwest Atlantic over the years were very high…they considerably exceeded the catches in the Barents Sea." One reason for this was that the Russians fished the overwintering and spawning concentrations of cod off Newfoundland and Labrador, but not in the Barents Sea. More will be said of this in later chapters.

2. Anderson (1986) provided a detailed description of the development of the theory of continental drift and plate tectonics.

3. Wilson (1966) asked an important question: "Did the Atlantic Ocean close and then reopen?" It helped to resolve what earlier geologists had observed in Newfoundland and Labrador and in North America, Europe, and North Africa: that the joining of rocks in central Newfoundland and the ophiolites of western Newfoundland made little sense without this theory. Much of this has now been accepted as a part of tectonic science, in large part as a result of the work of Newfoundland geologist Hank Williams, working at the Memorial University of Newfoundland.

4. For more detail on the geological history of Newfoundland, see Hodych et al. (1989). I drive over the fault on my trip to my cabin at Terra Nova Lake.

5. Keen et al. (1990) provided details of the separation, subsidence, and sedimentation of the Grand Banks. Although much has been learned, there is still some uncertainty: "Portions of this margin are among the most studied and best understood passive margins in the world – yet the more we study it, the more questions are posed."

6. Well before the acceptance of plate tectonics as the rational geological explanation for the positioning of the continental land masses and the changing oceans, there was indisputable evidence that a cold Arctic had not always been a feature of the northern hemisphere. In a 1963 publication, despite arguing against continental drift, Schwarzbach (1963) stated that: "All these occurrences prove the same thing: that for a very long time there was a climate completely different from today. There was no Arctic; not just a moderate, but a warm or very warm climate existed… the hot desert belt of the earth was situated there. In the Mesozoic, the temperatures were already somewhat lower, e.g. in Greenland, and the northerly boundary of the reef corals shifted farther to the south. But the Mesozoic floras prove that very favorable climatic conditions were still present."

7. Ekman (1953, 164) stated: "Forests containing a considerably greater amount of species than in the present Central Europe were found on Spitsbergen during the early Tertiary period and the same is roughly true for Greenland. It is said, however, that there is no completely convincing proof for a tropical or sub-tropical climate at any period in the Arctic. But this cannot be valid for regions which are at present boreal. During the upper Eocene the flora of England and Ireland had a character which is now to be found in the Malay archipelago and in tropical America."

8. Rose (2005) presented data that show that species diversity decreases exponentially from the equator to the polar latitudes.

9. Some of the oldest rocks on earth are located in western Greenland, dated at 3.8 billion years (see Erickson 1996), and at Saglek Fjord in Labrador, dated at 3.5 billion years (see Ryan 1989). Samples can be viewed at the Johnson GEO Centre next to Signal Hill National Historic Park in St. John's, Newfoundland and Labrador.

10   See Erickson (1996) for general marine geology.
11   Rose (1993) described the migration highway used by cod to cross the northeast Newfoundland shelf towards the coast and called it the Bonavista Corridor.
12   Detailed descriptions of the geology of the North Atlantic have been given by Jansa and Wade (1974), Sclater et al. (1977), Sclater and Tapscott (1979), and Emery and Uchupi (1984).
13   King (1974, 59) presented data resulting from analyses of magnetic anomalies showing that the Atlantic Ocean has spread at rates of 1.0-2.25 centimetres per year in geologically recent times (the past 3.6 million years). The Indian Ocean has spread at similar rates, but the Pacific has spread faster, at rates of 5.1-6.0 centimetres per year on the East Pacific Rise. Spreading does not occur at the same rate over time or from place to place. The Juan de Fuca Ridge off Vancouver Island shows a very symmetrical spreading at 3 centimetres per year, as does the Reykjanes Ridge southwest of Iceland in the North Atlantic. As the North American continent moves west, it pushes against a spreading Pacific plate moving north along the west coast, causing earthquakes from Mexico to Alaska.
14   See Briggs (2003).
15   Erickson (1996) gave the maximum sea level change as 300 metres, but it was likely somewhat less than this. In any case, this represents a large change that would elevate most of the Grand Banks above sea level. Only the channels and troughs are deeper than this.
16   See Nolf (2004).
17   Food webs, also called food chains, are a description of who eats whom in an ecosystem. In ecosystems with few species, the food web will be simple (e.g., despite their diverse tastes, Newfoundland cod depend to a large extent on only a few prey species, such as capelin, shrimp, and sand lance). In complex food webs, each species may feed on many other species.
18   See Ekman (1953) and Briggs (1974).
19   See Briggs (2006).
20   Briggs (2003) summarized biological evidence for the timing of the opening of Bering land bridge (6-12 million years ago); previous studies indicated a later opening (3.5-4 million years ago).
21   The practical salinity unit (psu) is the international unit of salinity, equivalent to parts per thousand.
22   Gaskell (1972, 1) stated that "the Gulf Stream has always affected life on earth." This is almost certainly not true. The Gulf Stream and North Atlantic Drift did not develop until the Panama land bridge joining the North and South American landmasses prevented the westward equatorial flows from passing into the Pacific Ocean. This occurred between 12 and 3.5 million years ago, but a deflection of the Gulf Stream to the north along the eastern coast of North America may have begun somewhat earlier. Volcanic islands were likely already in place between present-day Central and South America prior to the joining of the continents. Emery and Uchupi (1984, 794-803) discussed this possibility and showed a series of maps of the changing land mass.
23   Most estimates of the timing of the emergence of the Panama Bridge are within this range, but see also "Emergence of the Isthmus of Panama: Paleoceanographic implications in the Atlantic and Pacific Oceans" by Lizette Leon-Rodriguez at http://www.fiu.edu/~sukopm/seminar0405/leon.pdf.
24   The forward to Gaskell (1972) stated: "The more one thinks about the Gulf Stream the more it becomes apparent that this is a feature of our earth with a two-fold meaning. In the first instance there is the observable effect of a fast-moving body of water traversing the North Atlantic, an effect which has been of importance to navigators…in the past, while secondly there is the more nebulous fact that the Gulf Stream is only part of a whole, a general circulation of water in the ocean."
25   Erickson (1996) stated that Greenland separated from Europe 57-60 million years ago. However, Barry (1989) indicated that the land bridge between Greenland and Svalbard did not disappear until about 38 million years ago, at approximately the same time as the Bering Strait narrowed between

Alaska and Siberia. Baffin Bay had been open for some time by then, as it started spreading about 53 million years ago, although the shallow Davis Strait prevented flows between it and the Labrador Sea until 36 million years ago. Moreover, Barry cited earlier sources to state that: "There was probably no significant opening for Atlantic-Arctic water exchange in the Greenland Sea until after 27 million years ago and even during the Pliocene the Greenland-Svalbard passage remained narrow. The Iceland-Faeroes Ridge also separated a shallow Norwegian Sea from the Atlantic. Similarly, despite an opening of Baffin Bay 53 million years ago, Baffin Bay remained separated from the deep Labrador Sea 36 million years ago by volcanic rocks in Davis Strait. This may account for observed differences between marine faunas in the Arctic and Labrador seas." Barry's last observation was based on a study of whales (Hopkins and Marincovich 1984) and is debatable.

[26] See Farmer (1981) and Rasmussen et al. (2003).

[27] Gaskell (1972) stated that "the dimensions of the jet stream are so great that… it maintains its entity for some thousands of miles – like the jet from a hose squirted into a pond, it mixes only slightly with the water through which it flows because its volume is so large."

[28] The Coriolis force is an apparent force that results from the rotation of the earth. It tends to shift moving objects to the right in the northern hemisphere and to the left in the southern hemisphere.

[29] Carbon-14 dating shows that the deep water of the mid-Atlantic is about 650 years old, while Antarctic bottom water is 900 years old. Greater ages are found in the Pacific Ocean, ranging from 1300 to 2000 years. Hence, the deep waters of the Atlantic are thought to originate in the Atlantic. Also, there is an increase in age in bottom waters from south to north of 600 years in 6000 km, which indicates an average movement of deep waters of about 10 kilometres per year (Gaskell 1972).

[30] The Devonian is generally considered to be the Age of Fishes, when most species developed. However, Norman (1963) stated that: "The geological record has so far provided no evidence as to the origin of the fishes, and shortly after the time when fishlike fossils first made their appearance in the rocks the Cyclostomes, Selachians, and Bony Fishes are not only well established, but are represented by a number of diverse and often specialized types, a fact suggesting that each of the classes had already enjoyed a respectable antiquity."

[31] Colbert and Morales (1991) provided a detailed account of the evolution of the vertebrates through geological time.

[32] Briggs (2003) summarized recent knowledge of marine evolutionary centres in the Antarctic, North Pacific, East Indies, and the tropical West Atlantic.

[33] Briggs (1974, 248) stated that: "When one considers the diversity and general distribution of the marine faunas that occupy the cold waters of the world, it becomes clear that the North Pacific has functioned as a center of evolutionary radiation. It has supplied species to, and to a large extent has controlled, the faunal complexion of the Arctic and the North Atlantic. Furthermore, North Pacific species have, by means of isothermic submergence, bypassed the equatorial region to invade the cold waters of the Southern Hemisphere."

[34] Classification of species begins with the broadest level groups, such as plants and animals, then proceeds towards more closely related species. A family is a higher group of related genera, such as species of cod or salmon. A genus is a group of closely related species, such as the redfish or the Atlantic and rock cod.

[35] Each species is sufficiently different from any other that there should be no interbreeding.

[36] Ekman (1953, 158) advanced the notion that the North Atlantic may have been too warm for the evolution of cold-water forms. However, this does not explain the Atlantic cod, which is a cool-water, but not Arctic, species, or the Arctic cod, which has the most northerly distribution of any fish and is a gadoid. It may be that evolution was sufficiently rapid in the North Atlantic to produce some species with cold tolerance.

37 Ekman (1953, 164) summarized earlier evidence that the North Pacific never developed the warm conditions of the Atlantic during the Cenozoic period. The progression of these climate shifts is not entirely clear, but it led to the assertion that "during the whole of the Tertiary period, the North Pacific offered much more favourable conditions for the development of a fauna adapted to a cold-temperate climate than the North Atlantic."

38 See Table 22 in Ekman (1953, 158).

39 See Howes (1991). Eastman and Grande (1991) reported a hake- or codlike skull from the late Eocene (40 million years ago) from the Antarctic, but this is now thought to have been a notothenioid (Briggs 2003).

40 See Cohen et al. (1990) and Howes (1991). The 12 families and 482 known species of the order Gadiformes are found mostly in the northern hemisphere and North Atlantic.

41 Nolf (1995, 529) gave an account of the otolith record of gadoids in European waters and in the North Sea in particular. He concluded that "it is probable that the earlier evolution of these genera [referring to those of Atlantic cod, haddock, and pollock that showed up 5-10 million years ago in the North Sea] took place in cooler waters of the Arctic Seas."

42 Ekman (1953) presented one of the early arguments that the cod may be the exception to the rule that most North Atlantic species originally came from the North Pacific. This argument highlighted the uniqueness of the cod in the North Atlantic.

43 See Pflaumann et al. (2003).

44 Rogerson (1989, 117) stated: "Glaciers tend to erode and destroy unconsolidated sediments and landforms over which they move. Thus, evidence of earlier glaciations seldom survives except in the lee of resistant obstructions and at the very limits of the ice advance." Ruddiman and Wright (1987, 1) agreed: "Despite the geological youthfulness of the Quaternary period, most of the surficial geologic record in high-latitude North America has been repeatedly erased by the erosive power of ice sheets."

45 Ruddiman and Wright (1987) provided a detailed introduction to the ice ages, the origins and timing of comings and goings of the ice ages, and their effects on North America and adjacent oceans, including the North Atlantic.

46 Ruddiman and Wright (1987, 2) stated: "In summary, the first 1.6 to 1.7 m.y. [million years] of the northern hemisphere ice age was a sequence of some 40 full climatic cycles at the 41,000-year rhythm of orbital obliquity, whereas the last 0.9 to 0.7 m.y. was a time comparably dominated by the 100,000 year period of orbital eccentricity. In addition, numerous separate advances at periods of 41,000 and 23,000 years have been superimposed on the basic 100,000-year cycle."

47 Ruddiman and Wright (1987) provided geochemical and biological data from ocean sediments showing clear beginnings of the ice ages about 2.5 million years ago, and the beginnings of the Wisconsin glaciation about 70,000 years ago and its recession from 10,000 years ago. Therefore, the Wisconsin lasted for 60,000 years. The rise of *Homo sapiens* and development of human culture, fishing, and agriculture has all taken place within the most recent interglacial period. No one can say for sure if or when the next glaciation will occur; if the cycle holds, it will be another 50,000 years or so before the next one. This assumes no major change in the processes that influence the global heat budget, which is not certain.

48 Andrews (1987) gave a broad description and many maps of data and models of glacial spreading and extent across North America. There is still no universal agreement about all the details, but the basic patterns are generally accepted by geologists.

49 Andrews (1987) presented profiles of the Laurentide ice sheet with maximum depths of 3000 metres in the central continental regions, including central Labrador. It seems unlikely that ice achieved such depths in Newfoundland.

50 Rogerson (1989) presented evidence for the pattern of dispersal of local ice from eight centres on the island of Newfoundland (Long Range mountains, Northern Peninsula; Lapoile mountains; Indian Lake; Maelpeg Lake; Mount Sylvester; central Avalon Peninsula north of St. Mary's Bay; and on the Bay de Verde and northeast Avalon peninsulas), and one or two branches of the main continental Laurentide flows in Labrador.

51 Rogerson (1989) presented both sides of this debate, but argued that many of the large fjords and bays are bounded by faults and some deep fjords have rock formations incompatible with recent ice scouring.

52 Additional details of the arguments about the extent of glaciation in the Newfoundland and Labrador region can be found in Rogerson (1986).

53 There are various points of view on ice extension across the Grand Banks. There are glacial terminal moraines approximately half way across the Banks that could indicate extensions of the ice sheets from land. In addition, the islands on the Banks during glaciation could have had their own small glaciers, and been the focal points for additional glaciation.

54 Evidence of glacier extension from land to the edge of the Hawke Channel is strong. Fillon (1975, 429) stated that: "Identification of probable end moraines and the ubiquitous distribution of coarse gravel and cobbles, as well as the excellent state of preservation of glacial landforms, strongly imply that the entire continental shelf was glaciated by the Laurentide ice sheet during the Wisconsin. Coarse sediments, which appear in places on the continental slope to at least a depth of 400 metres, may have been deposited beneath a floating ice shelf. Sills covered by a coarse gravel and boulder lag at the seaward end of Cartwright and Hawke channels are interpreted to be mounds of detritus deposited beneath the ice shelf at a point just seaward of the grounding line. Although Hamilton Bank was completely ice covered, the greatest ice discharge seems to have been through these transverse channels. There is on average about 100-150 metres of glacial drift in the channels compared to 50 metres on the bank."

55 Flint (1943) thought that ice covered all of the Grand Banks. This is not supported by more recent studies.

56 See Rose (1952, 8).

57 Rose (1993) demonstrated how cod migrate across the shelf. Cod migration will be highlighted in a later chapter. These areas are also important to juvenile groundfish such as turbot, plaice, and, in the deeper waters, redfish.

58 Ruddiman (1987, 140) stated: "Both geologic data and modeling results indicate that North American ice sheets chilled and seasonally froze the North Atlantic Ocean north of 45-50°N [latitude] at the 41,000-year and 100,000-year rhythms and cooled the ocean southward into the middle latitudes at the 100,000-year rhythm. These two signals were transferred from the ice sheets to the ocean via the atmosphere with little or no lag."

59 Norgaard-Pedersen et al. (2003, 8) concluded on the basis of an extensive analysis of 52 sediment cores that open water may have existed far north in the northeast Atlantic. During the last glacial maximum, "a production region [existed] in the eastern to central Fram Strait and along the northern Barents Sea continental margin characterized by Atlantic Water advection, frequent open water conditions, and occasional local meltwater supply and iceberg calving from the Barents Sea Ice Sheet." Also, "although the total flow of Atlantic Water into the Arctic Ocean may have been reduced during the LGM [last glacial maximum], its impact on ice coverage and halocline structure in the Fram Strait and southwestern Eurasian Basin was strong."

60 Norgaard-Pedersen et al. (2003, 8) stated: "During the LGM [last glacial maximum], environmental boundary conditions in the Arctic were profoundly different from the present interglacial. Sea level was lowered and the huge Arctic shelf areas were either exposed or covered by ice sheets. Consequently, important gateways for the exchange of water masses with the world ocean were closed,

except the > 2500 metres deep Fram Strait connection to the Nordic Seas. A huge ice sheet was centered over the Barents Sea blocking Atlantic water supply to the Arctic Ocean through this presently important branch. Most of the Canadian archipelago was covered by the Laurentide Ice Sheet. The Innuitian Ice Sheet over Ellesmere Island coalesced with the Greenland Ice Sheet and blocked the Nares Strait connection to the Baffin Bay and Labrador Sea. The shallow Bering Strait connection to the Pacific and the huge Siberian shelves were subaerially exposed and covered by a periglacial tundra steppe."

[61] Fillon (1976) presented sediment evidence from Hamilton Bank that suggests the Labrador Current may have been pushed off the shelf during the last glacial maximum, and that the present landward position and strength of the Current is a result of deepening of waters over the inner shelf (due to subsidence of this region) and a cooling trend in the Arctic surface waters.

[62] The American Geophysical Union has stated that it is highly unlikely that human activities are not having an influence on the global warming and climate change so evident in the twenty-first century (American Geophysical Union Science and Policy, "Human Impacts on Climate," http://www.agu.org/sci_soc/policy/positions/climate_change.shtml).

[63] See Hurrell and Dickson (2004).

[64] See Ottersen et al. (2004).

[65] Farmer (1981) summarized the formation of the Labrador Current, citing earlier work by Lazier (1979). Salinities over Nain Bank showed little gradient at 32.5-32.75 psu, but increased considerably seaward of the Bank to 33.0-34.5 psu, reflecting an increasing influence of the West Greenland Current.

[66] Citing an unpublished work by Alan Longhurst, Farmer (1981, 67) stated: "A specific phenomenon related to the interaction of shelf water and slope water is the formation of a shelf-break front. Investigation reveals this to be a zone of ascending water which contributes to the nutrient enrichment of surface waters, and particularly evident in phytoplankton concentrations. It is interesting to note that, as a direct consequence of this enrichment, biota at various levels in the food chain, for example whales, while distributed widely over the shelf region, show a linear concentration corresponding to the shelf edge." Farmer provides no data to support this claim.

[67] The salinity of the Labrador Current ranges mostly from 30 psu to 33 psu.

(Bedford Institute of Oceanography, Deep Ocean Group data:
www.mar.dfo mpo.gc.ca/science/ocean/woce/labsea/LS_SAL).

THE CONTEMPORARY ECOSYSTEMS

It would be difficult to find in the whole world an equivalent area of the ocean bed which equals in historic interest and economic importance the Grand Bank of Newfoundland.

RAYMOND MCFARLANE, *A History of the New England Fisheries*, 1911

The Assembly of Ecosystems . . .	55
Post-Wisconsin Ecosystems . . .	59
Thienemann's Law . . .	61
Modern Species of the North Atlantic . . .	63
Atlantic Aboriginal Species . . .	64
Atlantic Cod . . .	67
Greenland Cod . . .	78
Haddock . . .	81
Pollock . . .	82
Arctic Cod . . .	83
Hake . . .	84
Atlantic Herring . . .	86
Harp and Hooded Seals . . .	89
Widely Distributed Families with Species in the Noth Atlantic . . .	91
Lumpfish . . .	91
Grenadier . . .	92

Sand Lance . . .	93
Myctophids . . .	96
Flatfishes . . .	98
Other Species . . .	105
Pacific Invaders . . .	106
Capelin . . .	106
Wolffish . . .	110
Redfish . . .	111
Eelpout . . .	115
Seasnails . . .	115
Atlantic Walrus . . .	116
Southern Migrants . . .	117
Atlantic Mackerel . . .	117
Atlantic Saury . . .	118
Bluefin Tuna . . .	119
Dogfish . . .	120
Anadromous Fish . . .	122
Atlantic Salmon . . .	122
Arctic Char . . .	123
Catadromous Fish . . .	125
Crustaceans . . .	126
Copepods . . .	127
Euphasiids . . .	128
Decapods . . .	129
Cetaceans . . .	135
Seabirds . . .	138
Murres . . .	140
Atlantic Puffins . . .	141
Northern Gannets . . .	142

THE ASSEMBLY OF ECOSYSTEMS

The physical geography of the Grand Banks and the adjacent coastal ecosystems of Newfoundland and Labrador developed over hundreds of millions of years, leaving a huge continental shelf with shallow, productive waters stretching far out to sea. The currents of the North Atlantic drove nutrient-rich waters upward from the depths and onto the Grand Banks, where sunlight penetrated to the ocean floor. As a consequence, productivity in the phytoplankton – the microscopic plants of the ocean – soared, which fed the zooplankton – microscopic animals – and, in turn, the plankton-eating fishes, the fish-eating fishes, seabirds, and marine mammals.

The exact composition of the ancient ecosystems or how they came to be is difficult to know with certainty, but there had to have been an order of sorts. Ecosystems are not assembled at random: there are principles that can be used to bind, within reasonable limits, the inevitable speculations about historical ecosystem structure. A prominent feature of North Atlantic ecosystems in general – and Newfoundland and Labrador ecosystems in particular – was that there are relatively few species and, as a result, fairly direct food chains or webs.[1] Species higher up the food web depended on those below them. If an important zooplankton species, such as *Calanus finmarchicus*, was not abundant, the consumers of this plankton – notably capelin, herring, and sand lance – had no other equally abundant and nutritious food source to turn to. Farther up the web, larger fishes and marine mammals fed on these small pelagic fishes and there were consequences if they were scarce. For example, in Norwegian and Icelandic waters, declines in capelin have led to difficulty for cod.[2] The same was true in Newfoundland and Labrador waters.[3] The first principle that can be applied to a developing ecosystem is that food must be available before expansions can occur in species that eat that food. Therefore, phytoplankton must have arrived first to feed the zooplankton, zooplankton must have been abundant before capelin could increase in numbers, and capelin must have been abundant before cod achieved large-scale dominance. This is a rule of order – some ecologists refer to rules of order as assembly rules,[4] although it is probably better to consider them as guidelines rather than as rules, which imply inflexibility.

A second principle supposes that large marine ecosystems are expressions of physical oceanographic features and processes.[5] The Newfoundland and Labrador marine ecosystems were greatly influenced by the cool waters of the Labrador Current and by the ice that formed seasonally and drifted past these shores. These were cold-water ecosystems: the species that inhabited them had to develop physiological or behavioural adaptations for life in a cold-ocean climate.

Resident species developed many specialized abilities to deal with the subarctic conditions of Newfoundland and Labrador marine ecosystems. The evolution of blood antifreeze in some species enabled year-round residency, while others migrated locally to

2.1

Annual productivity of carbon in the Atlantic Ocean (data from Berger (1989) and others). Carbon productivity lies at the base of the food web and measures the activity of the plankton in converting inorganic nutrients to organic (living) material.

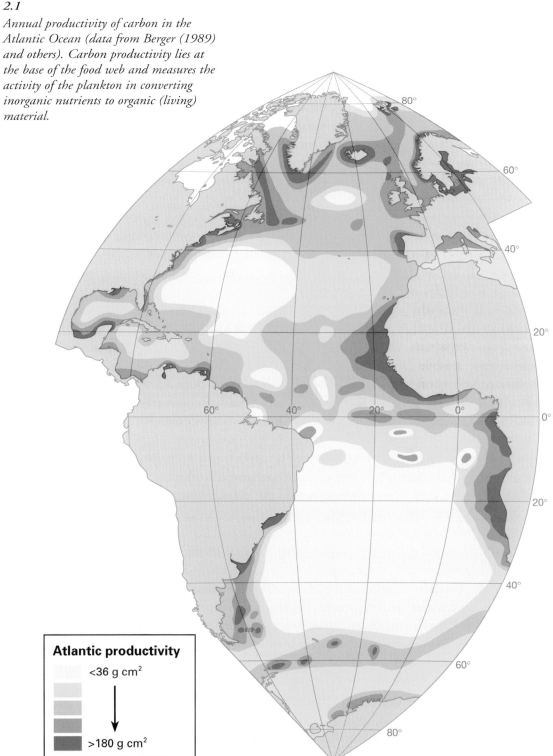

Atlantic productivity
<36 g cm²
↓
>180 g cm²

the warmest available waters. Still others did both. Resident winter flounder, a small coastal flatfish with limited range, could not escape the cold, but produced high levels of blood antifreeze that reduced the possibility of freezing.[6] These fish could withstand temperatures as low as -1.7°C with few problems; these temperatures would kill most animals. The northern races of cod could also produce antifreeze, but they did so less efficiently than winter flounder, and lost this ability as they grew older and more able to move to warmer waters in winter. At times this strategy could be risky: if temperatures declined suddenly, as sometimes they do in spring, cod could be trapped in the freezing inshore waters and suffer mass mortality. Capelin also evolved to live in and near cold water without antifreeze. They rely on speed and migration. Capelin have been known to swim rapidly into the coldest waters to escape predators, especially cod that prefer not to enter the heart of the Labrador Current.[7] But the capelin do not stay long in such waters, using them mainly as a "safe-house" that offers temporary protection.

Other species simply moved south in the coldest seasons. They took advantage of the luxuriant seasonal productivity in these northern waters, but developed long-distance migrations which brought them north in summer, then back to more temperate climes in winter. Among these are species of cetaceans (whales), large pelagic fishes (tunas and sharks), invertebrates (squid), and smaller pelagic fishes (Atlantic mackerel, saury).

The most mobile species, such as large pelagic fishes, marine mammals, and seabirds, fed mainly on small fishes and large invertebrates. Their feeding became a form of wild card in these ecosystems. According to the food chain assembly guideline, their abundance could be established only after the resident system was up and running, with their influence on the resident community slight. However, these species did not depend entirely on the local resident system, so their impact on it may periodically have become greater if their numbers became large based on productivity elsewhere.[8]

A third principle is a very simple one: the abundance of one species may affect the abundance of others. In simple systems, there may be species that influence or even control the levels of others. These are often called "keystone" species and can be either predators or prey. If a species changes in the ecosystem, especially a keystone species, the effects may be hard to predict; however, we can make some straightforward suppositions.

Before the invasion of the North Atlantic by North Pacific species, the ecosystems of the North Atlantic were very different than they are today. There were no capelin, no redfish, no Pandalid shrimps, and no snow crab. Cod and herring were present, with herring likely to have been the first species to achieve superabundance in the North Atlantic. In contrast, the big cod stocks – those of the Barents Sea, Iceland, Greenland, and the northern and Grand Banks stocks of Newfoundland and Labrador – could exist as we know them only after the coming of capelin, their main food.[9] During the earlier evolutionary period, cod probably developed in smaller populations along the coasts of first the northeast and then the northwest Atlantic. To this day, these more sedentary populations likely contain most of the genetic diversity of the species and are the most numerous of the cod groups.[10] Before capelin achieved a keystone status in these northern ecosystems, the Atlantic cod was likely more southerly distributed and preyed on other foods. At present, in the species-rich and warmer waters of Georges Bank and the Irish and North seas, sand lance, herring, and invertebrates take prominence in the cod diet. These southern stocks are small in comparison to the more northern "capelin stocks" and occupy smaller ranges, but individual fish growth is very high; perhaps food abundance is so great that quality is less important. In the northern and species-poor ecosystems, food quality may take prominence, and only capelin and herring are available as fat-rich food in sufficient quantity to support millions of tonnes of cod.

In the marine ecosystems of Newfoundland and Labrador, capelin fed the cod and supported their large numbers. Prior to the coming of the capelin, the small cod groups that occupied coastal and southern waters may not have been much different, but the large stocks – the northern cod, the Grand Banks cod, and the northern Gulf of St. Lawrence cod – could not have existed as we know them. Herring may have been the traditional food of cod long before capelin or northern shrimp arrived in the northwest Atlantic, and herring has remained a key prey of coastal stocks of cod and older, more sedentary fish. The largest of the cod stocks – the northern and Grand Banks stocks of Newfoundland and Labrador, and the Icelandic and Arcto-Norwegian stocks – set their clocks and yearly migrations in time with the movements of the capelin. While shrimp and other fishes and invertebrates also became an important food of the large cod stocks, they could not support large numbers of cod in these stocks on their own.[11] Shrimp populations were kept in check by large numbers of cod, and never reached numbers as high as they would have if cod were not eating so many of them. Although cod became by far the largest consumer in these ecosystems, other species that fed on capelin and shrimp may have reacted similarly, if less spectacularly, to the coming of capelin. Near the apex of the food web, harp seals feed on all of these species – particularly capelin and shrimp – on their annual migrations southward to pup on the ice floes off Newfoundland and Labrador and in the Gulf of St. Lawrence. But even the rapacious and migratory seals consumed far fewer capelin and shrimp than did the resident cod.

Abundance should not be confused with productivity. With the exception of the southern Grand Banks, the seas around Newfoundland and Labrador were cold-water ecosystems with high, but seasonal, productivity. Hence, cod did not grow as quickly or to as large a size as those in warmer, more productive waters off New England or in the North Sea. Ironically, the colder ecosystems typically had higher, not lower, levels of overall abundance.[12] In general, for cod stocks, the carrying capacity of the ecosystem varied considerably, "reaching its highest levels in the large migratory stocks (e.g., north-east Arctic, northern, Icelandic)…with…stocks inhabiting generally cooler environments [of lower productivity] having higher carrying capacity [maximum stock size]."[13] This has profound implications for how these ecosystems can be harvested sustainably, as sustainability depends on production, not on abundance or biomass.[14] The marine ecosystems of Newfoundland and Labrador cannot be harvested at as high a rate as more productive waters; if they are, stock crashes are more likely.

A fourth assembly rule of ecosystems is that few species are evenly distributed throughout an ecosystem. Most develop spotty, or what ecologists call "patchy," distributions, with individuals concentrated within a fraction of the total sea area. That area itself may change with the seasons as a function of reproductive, feeding, and overwintering requirements. As seasonal movements and migrations develop, the patches of one species may come to depend on the patches of another. For example, Atlantic cod migration patterns became partially determined by the seasonal patterns of capelin, so much so that many observers have considered that cod "follow" capelin.[15] Conversely, cod may have consumed so many shrimp as to create a patchiness in their distribution, with few survivors in areas where cod migrated to feed and many where they did not.[16] Patches of each species move through the ecosystem in part to fulfill their own biological requirements and in part in response to the moving patches of others.

By trial and error, the species of the North Atlantic worked out a balance of sorts, always tilted by the vagaries of the northern environment.[17] Over the past few million years, species from several sources, both locally evolved and immigrants, came to dominate these ecosystems, with many species expanding their ranges and pushing north during warm periods, only to retreat southward again during cold periods. In doing so, they became exquisitely adapted to the highly variable ocean environments of the North Atlantic.

Post-Wisconsin Ecosystems

During the ice ages, species were affected and redistributed in different ways. The least impacted were the cold-water and deepwater fishes that inhabited the shelf edge and off-shelf deep waters, where ice would not have penetrated to the ocean floor even during the glacial maximums. These species included the grenadier, the eelpout, and the Greenland halibut. However, even these species may have adapted to the presence of ice. The habitat of juvenile Greenland halibut is on the continental shelf, albeit in the deepest basins, and it may be that these areas provided refuges from ice. They are the warmest areas on the shelf in the present day. Northern lanternfish may have been pushed south – even though their habitat would not have been entirely iced over – because there are no cold-water species. On the other hand, the cold-water krill, which are small pelagic shrimp-like animals that occur in vast swarms in the Antarctic and Arctic oceans, were likely abundant adjacent to the ice fronts and even under the sea ice. Krill may have provided most of the food for the remaining fishes. The ice ages must have been good times for the cold-loving Arctic cod. For other, less cold-adapted species, the strategy was simple: go south or go deep.

On the Grand Banks and continental shelves, the effects of glaciation were far more severe than on the northeast coast or Labrador – much of the Grand Banks became islands. Shallow-water and continental shelf species did not fare well: their habitats were obliterated, or at best severely altered, and would have to be redeveloped from scratch after each recession of ice. In keeping with the organizing principles of physical oceanography and communities, zooplankton communities formed first and became highly abundant because of the phytoplankton richness in the zones of retreating ice.[18] Among the most important zooplankton were *Calanus finmarchicus*, an ancient but highly successful copepod, and the shrimplike euphasiids – *Thysanoessa inermis* and *Thysanoessa raschii* – a primary food for many whales and fishes. As the ice receded, sea levels and water temperatures rose, and the Gulf Stream and Labrador Current took their present configurations. As warming intensified and the ice retreated north, a deflection of the Labrador Current seaward off Hamilton Bank may have allowed warmer southern waters to reach southern Newfoundland and the Gulf of St. Lawrence. Enhanced warming would quicken the pace of recolonization by species pushed south during the glaciation.

The present marine ecosystems of Newfoundland and Labrador developed in the wake of retreating Wisconsin ice beginning about 10,000 years ago. The post-Wisconsin ecosystems were repopulated by the same species that had been present before the Wisconsin ice age. These species waited out the glaciation either farther south or in deeper, warmer waters – as ice retreated, they moved north and onto the newly submerged continental shelf. Expanding their range in the wake of the ice was nothing new for these species: many were colonizing species that had evolved to exploit cold northern systems and that had survived previous ice ages in the same manner. Some of the populations became massive, as the richness of nutrients and newfound sunlight encouraged blooms of plankton that in turn enabled the small, pelagic, plankton-feeding fishes – capelin, herring, and sand lance – to explode in abundance.

Capelin assumed their keystone role in the ecosystems of Newfoundland and Labrador as the ice receded, invading the region in response to the burgeoning plankton community. Their home base during the Wisconsin ice age is thought to have been the stranded beaches on the Southeast Shoal of the Grand Banks, south of the farthest reaches of the ice.[19] After the Wisconsin period, they recolonized the continental shelf as far north as mid-Labrador. Capelin found many suitable spawning beaches on the northeast coast of the island of Newfoundland and the coast of Labrador, as well as in the northern Gulf of St. Lawrence. Entirely new life histories and cycles developed as drifting capelin larvae spread out over the Grand Banks and continental shelf in accordance with current flows. As these capelin matured, they migrated back to the coast to spawn. Their abundance may have reached more than 10 million tonnes in short order – in modern times, capelin have been observed to respond quickly to oceanographic changes and are capable of increasing their numbers spectacularly within a decade.[20] Increasing abundances of oil-rich capelin would not go unnoticed by predators who shifted their migrations to feed on them.[21] Some of these species were temporary residents like the southern whales; however, the major consumer became the resident cod that followed the capelin north. In modern times, similar but smaller-scale phenomena of cod moving north and south with capelin distribution changes have been observed in the variable ocean climate in Iceland and Greenland.[22] The expansion of capelin enabled large cod groups to develop on the northern Grand Banks, the northeast Newfoundland and Labrador shelf, and south and west of the island in the northern Gulf of St. Lawrence.

As the ecosystems developed, several species – in particular capelin, Atlantic cod, and redfish – achieved incredible abundance levels numbering many billions of fish. The historical total biomass in the ecosystems, including the plankton and bottom organisms, likely approached several hundred million tonnes.

Thienemann's Law

> Thienemann's Law: With specialization of environmental conditions the number of species decreases while the populations of surviving species increase. – AUGUST FRIEDRICH THIENEMANN

Ecosystems with specialized environments have fewer types of habitats and fewer potential food types, hence there are fewer opportunities for different species, but species that find an ecological niche[23] find fewer competitors and predators and may become very numerous. This is Thienemann's Law, and in keeping with it, the "cold-specialized" North Atlantic has few species, but many have achieved great abundance. Thienemann's Law applies to both plankton and fish. One of the dominant zooplanktons, *Calanus finmarchicus*, can become so numerous that the colour of the sea changes during its blooms. The greatest zooplankton abundance occurs on the continental shelves – here, nutrients from the deep ocean rise to the light-filled surface waters and nourish an uncountable number of marine plants that in turn feed the zooplankton. Zooplankton-feeding fishes are very abundant in the North Atlantic. The Arcto-Norwegian herring stock migrates from spawning grounds on coastal Norway to feeding grounds at the Icelandic-Arctic front, and may have reached a size of 20 million tonnes (perhaps 400 billion fish) in its heyday.[24] Few fish stocks anywhere in the world's oceans have attained such abundance.

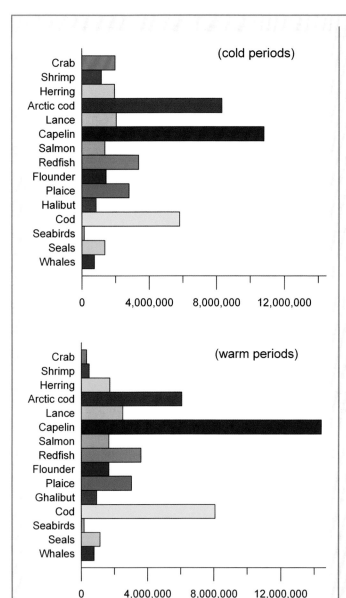

2.2

Newfoundland and Labrador species abundance during cold and warm periods, 5000 BC–AD 1450. Abundances are measured in biomass (tonnes) as output from the ecosystem model under cold and warm environmental conditions.

The biomass of 15 species or groups of species is estimated with a model that uses both "bottom-up" and "top-down" forcing, which means that the abundance of what a species eats, as well as what eats it, affects its biomass. There are also fishery inputs, which were zero in the ancient ecosystem. It should be stressed that the model is an abstraction and should be interpreted as a demonstration of how an ecosystem might work, not how it did work. Every attempt has been made to verify the outputs, but this is impossible for some species. This model will be used throughout the book to demonstrate structural changes in the ecosystem and the changing relationships among the linked species. According to this model, no species can change, either as a result of natural or fishery-related causes, without impacting others. The strength of each impact is uncertain and should be regarded as an "informed guesstimate." The model has two starting points. The first is capelin. The average ancient biomass of capelin was 12 million tonnes, based on a scaling up of the results of Bundy et al. (2000) to an average cod biomass of 6 million tonnes (Robichaud and

> Rose, 2004). Capelin biomass increases with temperature and is negatively impacted by seal and cod predation. All capelin feeders are impacted by capelin biomass, and all prey of seals are impacted by seal abundance. The second starting point is Arctic cod, which increases as temperatures decline and declines as temperature increases. Seals are dependent on Arctic cod.
>
> The ancient ecosystem contained about 50 million tonnes of the 15 groups of species. Biomass was higher during warm periods. The dominant species during the warm periods were capelin and cod. Capelin biomass ranged from 5-14 million tonnes, with cod biomass at 6-8 million tonnes. There were 2-3 million tonnes of sand lance, about 4 million tonnes of redfish, a similar amount of plaice, and lesser amounts of flounder. Shrimp and crab comprised only one or two million tonnes during warm periods due to severe predation by cod and other species. During cold periods, capelin declined and shrimp and crab increased, largely as a result of the declines in cod and other predators that occurred after capelin numbers fell. As Arctic cod increased, the numbers of seals increased, causing further declines in capelin, Atlantic cod, and several other species.

The results of the ecosystem model are as close as we can get to a description of the Grand Banks and adjacent Newfoundland and Labrador shelf marine ecosystems since the end of the Wisconsin ice age some 10,000 years ago. Although the precise biomass levels must remain uncertain, there is little doubt that the ecosystems were luxuriant and that the dominant species were very abundant. It was a fat system, not a lean one, and its productivity not only attracted tropical migrants to feed, but altered their migrations and life histories. From the floors of the deepest basins, where bottom invertebrates flourished and the flatfishes and shrimp numbered in the millions of tonnes, to the shores and beaches where herring and capelin spawned by the billion, this system was fundamentally rich. Although few in species number, it was exuberant in abundance, with many species demonstrating Thienemann's law. Among them, the aboriginal cod became highly numerous, their combined masses likely weighing in at over 7 million tonnes.[25]

MODERN SPECIES OF THE NORTH ATLANTIC

The species presently inhabiting the North Atlantic and marine ecosystems of Newfoundland and Labrador have four different origins. Some, including the Atlantic cod[26] and the herring,[27] are descendants of aboriginal Atlantic fishes. A few, such as the harp seal, are of Arctic origin, their ancestors having ventured south into the cooling North Atlantic during the past 38 million years. The majority of species, including capelin and Pandalid shrimp, came from the North Pacific, most only very recently, during the intermittent warmer times of the past 4 million years. The remaining species are southern and seasonal migrants. Once established in their new home, fishes and invertebrates evolved into the familiar modern species.[28]

The North Atlantic has fewer species than most other oceanic or major freshwater ecosystems. Canadian waters of the North Pacific have 482 known species of fishes, 313 of which occur in neither the Atlantic nor Arctic oceans. The North Atlantic has only 318 species, 137 of which are found only in the Atlantic. The Arctic Ocean has fewer unique species still, with a total of 89 species and only 2 that occur nowhere else. Newfoundland and Labrador waters are home to only 188 species of marine fishes. This is a generous estimate, as a number of these species are seasonal migrants that do not reside year round in Newfoundland waters. Seventy-two of these species are shared with the North Pacific and about the same number with the Arctic (Appendix 1). By comparison, the North Pacific shares only 57 of its more numerous species with the Arctic. These numbers are consistent with the geological and oceanographic history of the North Atlantic region: the Pacific-Arctic connection was limited both by the Bering land bridge and the dissimilarity of environmental conditions in the North Pacific and Arctic oceans, where waters do not mix extensively as they do at the Atlantic-Arctic interface.

ATLANTIC ABORIGINAL SPECIES

Gadoids – codlike fish – are among the few species that originated in the Atlantic Ocean. The gadoids had humble beginnings: their ancient ancestors were likely small meso-pelagic fish[29] of the southern Atlantic. Millions of years later, in the cold North Atlantic, their descendents would increase in size and function to achieve a dominant station among predatory fishes. Nowadays, in no other oceans but the North Atlantic and adjacent Arctic are gadoids very important; the only exception is the North Pacific, where, in a twist of the more common movement from west to east, the Pacific walleye pollock evolved from immigrant cod from the Atlantic.[30] In keeping with its North Atlantic heritage, the walleye pollock has achieved tremendous abundance and now supports the world's largest whitefish fishery.

There is also a Pacific cod, which is very similar to the Atlantic Ocean's Greenland cod and may, in fact, be the same species.[31] Both of these species look much like Atlantic cod, but are in reality very different. The Pacific and Greenland cod spawn eggs that stick to the bottom of the ocean and hatch there, whereas eggs of the Atlantic cod and Pacific walleye pollock are broadcast to the currents where they float and drift, hatching perhaps tens or even hundreds of kilometres from the spawning site. Pacific and Greenland cod have never gained the numerical dominance of their more prolific cousins. Pacific cod almost certainly developed from immigrant cod from the Atlantic, but here the story gets more complex. Did the Pacific cod evolve from immigrant Atlantic cod and then very recently migrate back to the North Atlantic to become the Greenland cod? Or did the Greenland cod evolve first and then move westward to the Pacific? The jury is still out on these questions regarding cod evolution in the northern oceans.[32] One thing is certain, however: no true species of cod inhabit the southern or tropical oceans.[33]

The most likely evolutionary centre of modern-day cod is the eastern Atlantic – i.e., the North Sea between the United Kingdom and continental Europe. At present, there are about 20 species of gadoids in the North Sea area, 15 species in the Norwegian/Iceland region, and 10 species near North America. All other areas of the world have only a few species of gadoids each, and the southern hemisphere only a few species overall, mainly distantly related hakes.[34] There is only one truly Arctic species of cod.

Cod evolution in the North Atlantic took several directions and filled many ecological niches. The Atlantic cod became the generalist, living in many different habitats, at times testing even the coldest waters, and developing swimming speeds fast enough to catch pelagic prey fishes while still retaining a generally "reserved" disposition and sedentary habits that conserved energy. Atlantic cod was the most widespread, abundant, and adaptable of the cod, with populations that ranged from thousands to billions of fish and that occupied almost every possible habitat of the continental shelves of the North Atlantic.[35] The other two dominant gadoids of the North Atlantic were modified versions of the basic cod design. The haddock, a cod with a bottom-oriented and almost sucker-like mouth, became a bottom feeder that preferred more productive and warmer waters. The Atlantic pollock became the speedster, combining the lines of a salmon with the soul of a cod and developing swimming speeds great enough to feed on even the fastest pelagic prey. Neither the haddock nor the pollock adapted to the coldest waters as well as the Atlantic cod, but all three species retained the high reproductive rates and predatory and social behaviour that characterize true gadoid fishes. Within the past 5-10 million years, these cod came to be the dominant large predatory fish of the continental shelves on both sides of the North Atlantic.

2.3

The geographical origins of the species of the contemporary marine ecosystems of Newfoundland and Labrador.

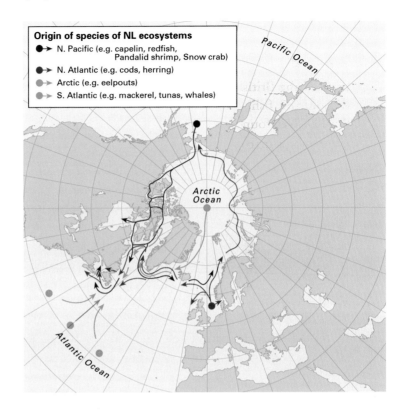

2.4

The modern distributions of Atlantic cod (Gadus morhua), Pacific cod (Gadus macrocephalus), and Greenland cod (Gadus ogac).

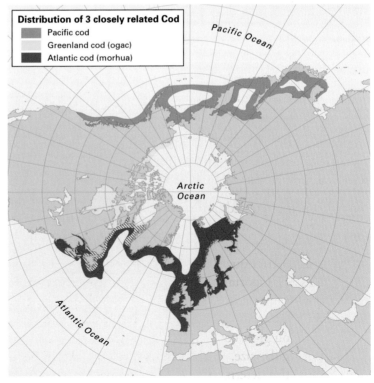

Atlantic Cod

> For tho' the British Channel and the German ocean [North Sea] are not without this fish, their numbers are so inconsiderable comparatively to those of Newfoundland, that they may rather be looked upon as stragglers.
> — ANTONIO DE ULLOA, *A Voyage to South America*

Atlantic Cod

Atlantic cod (*Gadus morhua*) are widespread and numerous – they are the dominant large predatory fish in most regions of the North Atlantic. Cod may live for 25 years; they reach maturity at 3-7 years of age and spawn most years after that.[36] In Newfoundland and Labrador waters, Atlantic cod are most highly aggregated in the winter months (January-March), with densities at times exceeding one fish per cubic metre of seawater. At these densities, one thousand tonnes of average-sized cod would fit over an area of the seafloor measuring approximately 125 metres by 125 metres in an aggregation some 30 metres deep.[37] As the spawning season approaches, the aggregations reorganize and the fish disperse somewhat from their overwintering state.[38] Individual cod shake off their winter torpor and engage in extensive and active courtship behaviour: males and females perform elaborate dances, circling and rising and falling in the water, with males flashing white bellies while females select potential mates.[39] Male cod establish the spawning grounds and are ready to spawn more or less on demand. Females arrive somewhat later, coming to the grounds when interested in spawning.[40] Their eggs are not all released at one time, but in batches over several weeks. A large female may hold 5-10 million eggs;[41] smaller females have many fewer eggs, perhaps less than a million. Older females not only carry more eggs, but larger and better quality eggs, and are very important to overall reproductive output.[42] Large males are important too: they have much greater sperm capacity and large females may favour, even require, large males for breeding. Cod tend to sort themselves out by size within a breeding aggregation, so that females and males of similar size are in close proximity.[43] Cod grow continuously throughout their life cycle and generally form schools with fish of similar size. This tendency was noted by the earliest fishermen, as the size of an individual fish determined how it could best be cured and marketed.[44]

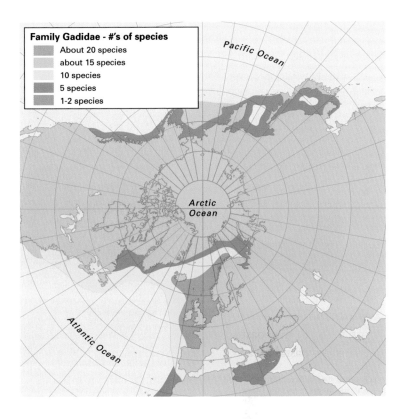

2.5
Modern distribution of cod species (gadoids) in the world's oceans.

After spawning, cod in the more northerly regions are typically emaciated and exhausted.[45] Although there is some dispersal of fish after spawning, most fish tend to stay with the aggregation, now intent on feeding. Offshore aggregations move shoreward and inshore aggregations seaward in search of capelin, with additional spawning continuing as the migration proceeds.[46] Cod spawn on the continental shelf and in coastal waters at many locations: spawning depths range from 20-450 metres and temperatures from below zero to 4°C.[47] Many spawning locations are used repeatedly, while others are more ephemeral.[48] At other times of the year, cod may be found in waters with temperatures ranging from below zero to 10°C.[49]

Cod spawning results in the release of billions of fertilized eggs into the sea. The eggs rise to the near-surface waters, being just buoyant enough so as not to sink to the seafloor and a sure death.[50] The tiny translucent eggs (about one millimetre in diameter) float in the currents for periods of weeks to months, depending on sea temperatures.[51] Cod produce many more eggs than they need to replenish their numbers, but few will survive longer than a few months. The reproductive strategy of the cod is equivalent to a bombing blitz: the more eggs, the greater chance that some will hit the target. At its peak, the large northern cod stock included 300 million adult females that released about 500 trillion eggs each year into the waters of Newfoundland and Labrador. If conditions in the ocean were favourable, perhaps one out of a million of these eggs would survive and grow to be an adult. If conditions were unfavourable, all but a very few would be recycled back to the sea, providing food for zooplankton, pelagic and bottom invertebrates, and even their own surviving larvae.

After a period of days to weeks, depending on sea temperature, the little cod bursts out of its egg and becomes a larva nourished by an attached sac of yolk. The larva can thrash about, but can't really swim. The yolk sac will be used up within a few days to a week and then the larva must feed independently for the first time. This is a crucial event in the life cycle of the Atlantic cod – many fail and die. Those that are successful in finding small zooplankton or fish eggs on which to feed grow quickly. They are still pelagic, floating and starting to swim in the upper waters. For most cod spawned in spring, it is now late summer and they may be far from the spawning grounds. By early fall, those that have survived begin to show signs of deserting their pelagic world and going to the seafloor. They are too large now to escape detection by their many predators, including other larger cod. They descend to the ocean floor, many in the coastal waters, and seek out cover in the form of rocky seabeds or coastal areas with eelgrass (*Zostera* spp.).[52]

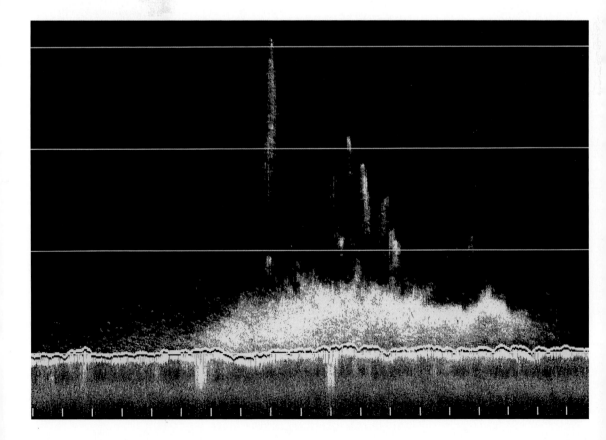

2.6
Echogram of the last large spawning aggregation of northern cod. Recorded in the Bonavista Corridor in June 1992 at a depth of 325 metres. The horizontal lines are 50 metres apart, while the vertical dashes are about 500 metres apart. Spawning columns are visible above the main body of fish.

As the young cod grow and enter their second year, they begin to move out of their cover more regularly to feed. Many will be taken by predators from both above (birds) and below (older cod and other fishes).[53] They begin to form shoals – size-structured aggregations – a habit they will keep for the rest of their lives.[54] At two years of age, cod leave their nursery areas.[55] At three to four years of age, the young cod begin to follow the spawning and feeding migrations of adults – in so doing, the routes that sustain the group are learned and maintained.[56] They may "home" to spawn in the region in which they were born.

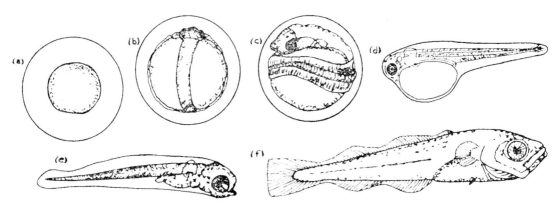

2.7 Early life stages of cod: (a) to (c) show developing embryo within an egg; (d) hatched larvae with yolk sac; (e) first feeding larvae; (f) early juvenile ready to "settle." (Drawings by Nancy Frost.)

The numbers of juvenile cod that join the adults vary from year to year.[57] In good years, it can be hundreds of millions; in others, perhaps only ten percent of that. During recolonization after the glacier retreats, there were likely too few adults to achieve high levels of survival. However, over the centuries, as populations grew and the carrying capacity of the ecosystem was approached, there were far more adults available than were necessary to produce the eggs required to replace the inevitable deaths that occurred each year.[58] In general, the maximum population size of a stock correlates with the area that it occupies, with larger stocks occupying larger ranges.[59] The numbers of surviving juveniles depend mostly on the number of adult spawners and whether or not ocean conditions are favourable – in general, warm years are better. Historically, most surviving cod would begin spawning by five to seven years of age.[60]

Off Labrador, warm years enable the most northerly spawning sites to be utilized, which leads to a more favourable drift pattern for eggs and larvae and a higher likelihood of ending up near the best juvenile grounds near the coast. In cold years, the cod have little choice but to spawn farther south, and survival rates of eggs and larvae are lower.[61]

Like many fishes, cod grow throughout their life and may reach lengths of 1.5 metres and masses of 75 kilograms or more. Growth is mostly dependent on the sea temperatures and food supply; hence, cod from different regions grow at different rates and achieve different sizes. In general, more southerly cod grow faster and reach larger sizes than more northerly cod.[62] The largest cod in the northwest Atlantic occur in the most southern stocks off New England, where a 6 foot (1.8 metre) monster that weighed over 211 pounds (96 kilograms) was caught in 1895.[63] The smallest and slowest-growing cod are the Labrador fish. Males and females generally have similar growth patterns and are difficult to distinguish, although females may grow a bit larger.

Cod are opportunistic feeders, especially as adults. Juvenile cod initially feed on zooplankton and eventually take larger prey as they grow in size. There are many records of cod consuming strange objects dropped into the water[64] and fishermen know that, at times, no bait at all is necessary to catch them. Despite this, the large northern and Grand Banks cod stocks have very specific food budgets and consume huge quantities of capelin.[65] The historical growth of their populations to hundreds of millions of fish weighing millions of tonnes depended on an ample supply of these small prey.

Herring are another important prey species for cod. Cod catch their prey one by one: unlike some whale species that also feed on capelin and herring, cod are not bulk feeders. Catching a larger herring makes more sense for a large cod than chasing down a small capelin. Herring may have been the rich fish food of cod prior to the arrival of

*2.8
Echogram of juvenile cod in Bay Bulls on the Avalon Peninsula in August, 1995. Water depths were about 20 metres. Each fish is seen as a smudge of colour with various shapes.*

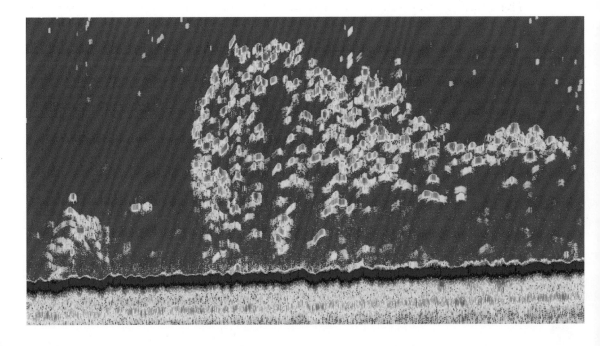

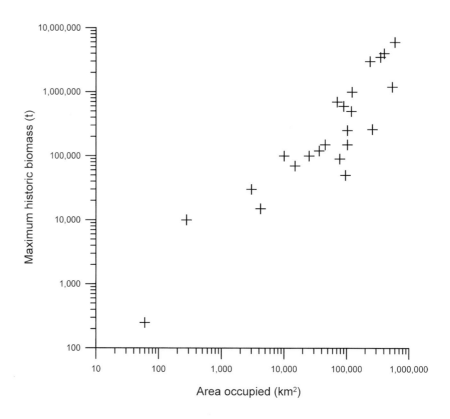

2.9 Maximum historic biomass and stock area of Atlantic cod (data from Robichaud and Rose (2004)).

capelin in the North Atlantic. Herring are still of prime importance to cod south of the capelin zone on the southern Nova Scotia and Georges banks, as they are in the warmer waters of the northeast Atlantic and the North Sea. Even within the more northerly capelin zone, herring may sustain the condition and growth of the largest cod.[66]

Atlantic Cod Populations. The range of the Atlantic cod shifted south during the ice ages to the southern Grand Banks, the banks of Nova Scotia, and Georges Bank off New England. With each ice retreat and attendant lowering of sea levels, potential new grounds re-emerged to the north. Over the past 10,000 years, as the Wisconsin ice retreated, cod advanced into the major bays of the island of Newfoundland and all the way to northern Labrador. A genetic cline[67] developed from north to south, with northern fish better able to resist the cold temperatures through superior antifreeze production and differing migration habits.[68] A complex stock structure of populations developed over this huge area: some populations spent much of their lives within one bay or along a coast, while other populations developed large-scale migration patterns that took them from the Grand and Labrador banks to the coastal areas and back again each year.[69]

The cod that remained on the southern Grand Banks during the Wisconsin ice age developed into a large stock after the retreat of the ice, and may have provided the seed stock for the recolonization of other areas. The Grand Banks stock did not undertake a long-distance migratory life cycle: it didn't need to. Grand Banks cod lived within the huge counter-clockwise oceanographic gyre formed by the Labrador Current as it hugged the shelf break of the Grand Banks. Within this gyre, cod fed on capelin and sand lance on top of the Banks and simply swam a short distance to deeper waters at the Banks' edge to avoid the cold in winter. The nutrient-rich waters of the southern Grand Banks and the abundance of capelin, sand lance, and redfish enabled these fish to reach huge size, much greater than in any other Newfoundland and Labrador stock. Hundred-pound (50 kilogram) cod grew quickly on the southern Grand Banks, but only slowly or not at all off Labrador.

Farther west, another stock developed on the St. Pierre Bank and its companion Halibut and Hermitage channels that lead to shore off Placentia and Hermitage bays. This region was likely among the first to be recolonized after the retreat of Wisconsin ice, but the small size of the area limited stock growth. Food was plentiful here: capelin, sand lance, and redfish abounded. However, lacking the direct influence of the Gulf Stream, the waters were colder and the fish did not grow quite as large as on the southern Grand Banks. Still, growth was much better here than in the northern stocks or in the Gulf of St. Lawrence.

Still farther west, another cod stock developed in the Gulf of St. Lawrence. The Gulf of St. Lawrence froze over in winter and its waters

2.10
Large cod from the southern Grand Banks in the 1940s (Gordon King Collection).

became very cold. In summer, however, it was rich in capelin and herring, and cod developed a migration pattern to exploit the patterns of these pelagic fishes. Most fish overwintered in deep waters around the Burgeo Bank, and in spring migrated north to the warmest waters available in the deep parts of the Laurentian and Esquiman channels. By May or June of each year, the fish had moved to Bay St. George, where most spawned, although spawning also took place as far north as Port-au-Choix, at the northern end

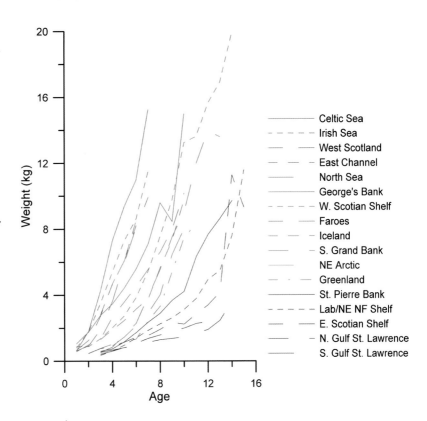

2.11
Cod growth expressed as weight versus age for most major world cod stocks. General groupings are made into fast (red), moderate (green), and slow (blue) growers. All of the Newfoundland and Labrador stocks except the southern Grand Banks fall in the slow category (data from Brander (1995)).

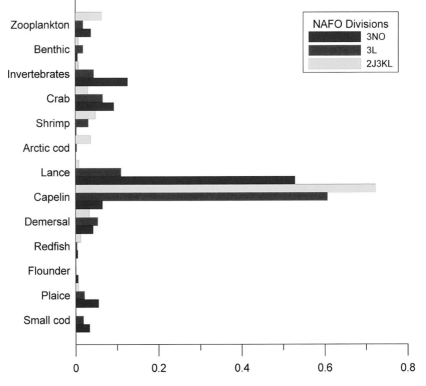

2.12
Prey species of Atlantic cod diet in the 1980s, expressed as fractions of total cod diet (data from Bundy et al. (2000)).

of the Esquiman channel. Eggs and larvae became broadly distributed in the counter-clockwise circulation of the Gulf of St. Lawrence, and juvenile cod were found in many coastal areas. After spawning, adult cod continued their pursuit of migrating capelin. In most years, cod followed capelin across the Gulf of St. Lawrence to the Lower North Shore region of Quebec and the Labrador Straits, at times venturing even farther north through the Strait of Belle Isle and around the tip of the Great Northern Peninsula of Newfoundland[70] to some of the best capelin grounds in the North Atlantic. In late summer and fall, after the capelin had gone, the cod dispersed and migrated back southward to their winter home in the deep waters around Burgeo Bank.

In the northeast, between the Funk Island Bank and northern Grand Banks, another great population of migratory cod developed. The fish overwintered in the warm (3-4°C) Atlantic waters on the edge of the Banks. Their home ground was the Bonavista Corridor – the depression south of the Charlie transform fault that bounds part of the Orphan Basin – and the northern edge of the Grand Banks. In spring, cod spawned on the shelf in large aggregations 50-100 metres deep and spanning tens of kilometres.[71] During and after spawning, these fish migrated shoreward to the northeast coast of the island – the "highway" through the Bonavista Corridor took them to the outer margins of Trinity Bay. From there, the search for capelin brought them into Trinity, Bonavista, and Notre Dame bays, and as far south as Petty Harbour on the Avalon shore. Their migratory pattern was similar from year to year, but never exactly the same because it depended on water conditions and on the movements of the capelin. Capelin migrated in spring to spawn, and the cod migrations became timed and spaced to intercept them. As with everything in this ecosystem, however, the abundances and the routes of the capelin varied. As went the capelin, so went the cod.

Labrador cod are the most northerly of the Atlantic cod stocks. They came to inhabit the two main banks off Labrador – the Nain and Hamilton banks – where they overwintered and spawned at depths greater than about 300 metres in waters that remained at 3-4°C year round.[72] These fish migrated in summer in search of capelin through the warm waters of the Hawke and Cartwright channels to the shores of Labrador, with some continuing south through the St. Anthony basin to northeast Newfoundland.[73] They migrated back to the Labrador banks and deeper water in late fall, thus taking advantage of the northern migrations of capelin that occurred at that time in most years and avoiding the coldest coastal waters in winter. After overwintering in dense aggregations at the edge of the banks, these cod spawned in late winter and spring on the slopes of the Hamilton Bank and Hawke Channel shelf.[74] The buoyant cod eggs and larvae floated in near-surface waters, drifting with the Labrador Current to the south and shoreward. Many young cod that were spawned off Labrador found their first home in the bays of Labrador and the northeast coast of Newfoundland.[75]

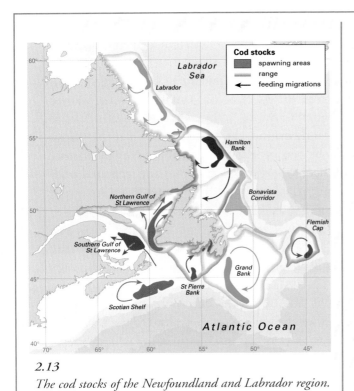

2.13
The cod stocks of the Newfoundland and Labrador region.

The historical biomass levels of the various cod stocks in Newfoundland and Labrador waters ranged from thousands to millions of tonnes. The largest of these was the northern cod stock that occurs off Labrador and south to the northern Grand Banks; it reached a historical maximum of 3.5 million tonnes. Coastal stocks occurred in most of the large bays around the island of Newfoundland. Those centred in Trinity and Placentia bays have recently been estimated at 30 thousand tonnes and 70 thousand tonnes, respectively; these stocks were always much smaller than the large migratory northern and Grand Banks stocks.[76]

A distinguishing feature of the large cod stocks of Newfoundland and Labrador is that most spawning occurs in the relatively warm waters (2-4°C) of the offshore Banks. There has been much confusion about this in the past. The earliest accounts reported that all spawning occurred along the coast, but later research suggested that all spawning took place near the edge of the continental shelf.[77] Here there were two extremes – and neither description was accurate, although spawning does occur in both places. Recent studies using historical survey data and observations at sea indicated that spawning is indeed widespread, but that most spawning takes place in large aggregations well offshore but still on the shelf.[78] The latitude of spawning concentrations may vary from year to year, depending on water temperatures, and there may be overlap or even mixing of stocks in some years. In 1963, "great concentrations" of spawning cod were observed on the southeast corner of the Hamilton Bank, but in 1964, a very warm year, the main spawning grounds appeared to be off northern Labrador.[79] In the cold years of the early 1990s, the largest concentrations of spawning cod occurred far to the south in the Bonavista Corridor.[80] Spawning in the bays occurs in much smaller aggregations.

The preponderance of offshore spawning in Newfoundland and Labrador waters in not typical for cod, but is likely a product of the cold coastal waters of the area. In contrast, the warm waters of the North Atlantic Drift bathe the coastlines and major cod grounds of southern Iceland, the Faroe Islands, the North Sea, and the Barents Sea. In these regions, cod stocks spawn mostly near the coast, where waters are warm and where currents carry eggs and larvae seaward to juvenile grounds on adjacent or sometimes distant banks.

Smaller coastal populations of cod developed in most of the larger bays and inlets around the island of Newfoundland and in some inlets in coastal Labrador. These fish spawned inshore and did not migrate long distances. These may be similar to the original cod populations prior to the coming of capelin. The largest of these coastal populations developed in Placentia Bay; its many islands were sheltered and offered calm overwintering waters and spring spawning grounds. Fish here were exposed to very cold waters in winter and spring, but there was little ice except where fresh waters flowed into the ocean and, therefore, little danger of cod freezing.[81] Local currents dispersed eggs and larvae within the bay and to the nearby St. Pierre Bank. Adults migrated out of the bay, although usually not too far, in search of capelin and sand lance. In some years, these fish followed capelin around the Avalon Peninsula as far north as Trinity Bay. Inevitably some intermixing occurred with fish there, although perhaps not for spawning.

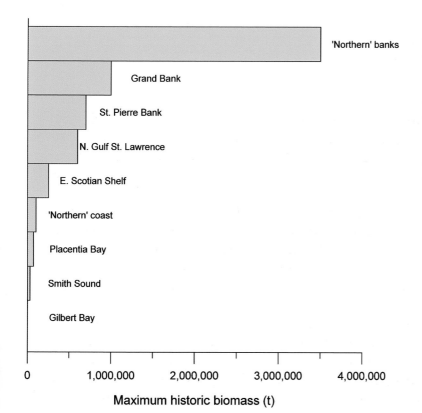

2.14
Maximum historic biomasses (tonnes) of the cod stocks in the Newfoundland and Labrador region (data from Robichaud and Rose (2004)). The 'Northern' designation refers to the northern cod stock of Newfoundland and Labrador.

Local populations also developed in Trinity and Bonavista bays and possibly in Notre Dame Bay. A small population survived in a remote bay in coastal Labrador, enduring very restricted food sources and extremely cold waters nearly year round, and exhibiting very slow growth rates as a result.[82] There may also have been a coastal population in Groswater Bay – many early fisheries reported a high concentration of fish there. In short, there was an Atlantic cod population for each environment.

Greenland Cod

The Greenland cod (*Gadus ogac*) is often called rock cod in Newfoundland and Labrador. It occurs in the cold waters of the northwest Atlantic from Greenland to Newfoundland, in coastal waters south to New England, and as far west as Alaska. It does not occur in the northeast Atlantic. The Greenland cod is a close relative of the Atlantic cod, but has adapted a very different life strategy. It inhabits only shallow coastal waters, where it lives a sedentary life among the rocks and marine vegetation, and spawns sticky demersal eggs[83] that attach to rocks or kelp. Juvenile Greenland cod can be found intermixed with young Atlantic cod in most bays and look so much like them that they can easily be confused.[84] However, Greenland cod populations never grow very large, even though young fish are abundant.

The Greenland cod is so similar to the Pacific cod (*Gadus macrocephalus*) that their separation as species likely occurred recently, perhaps during the last ice ages. Which species came first is not clear, but they likely had a common ancestor in the North Atlantic.[85] Unlike the Atlantic cod, which is a fully pelagic spawner with floating eggs, both the Greenland and Pacific cod spawn on the seafloor, and although both can achieve sizes of 75 centimetres and masses of at least 10 kilograms, neither attains the size of the Atlantic cod. They have also never achieved the Atlantic cod's dominance of their ecosystems.

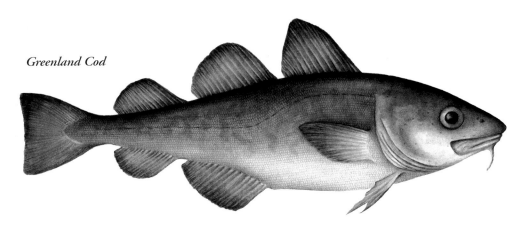

Greenland Cod

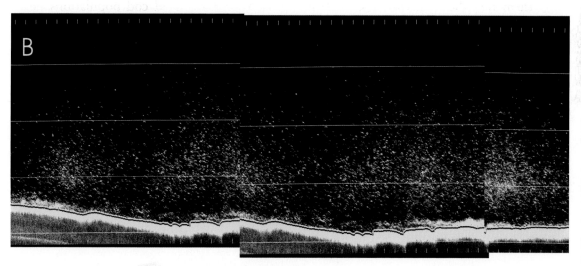

2.15 Echogram of cod in spring of 1962 taken from the research vessel A.T. Cameron *showing vertical school structures reaching well off the bottom. Cod catches from these huge aggregations (scale is uncertain but likely covers many km) were some of the largest ever recorded in any of the North Atlantic cod stocks. B) Last of the great migrating cod schools in the early 1990s. This composite echogram taken from the research vessel* Gadus Atlantica *spans more than 20 km and shows the bottom 150 m (total depth of the ocean was about 325 m).*

2.16
Greenland cod distribution in Newfoundland and Labrador waters.

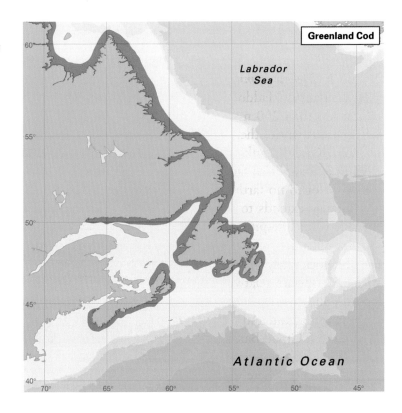

2.17
Haddock distribution in Newfoundland and Labrador waters.

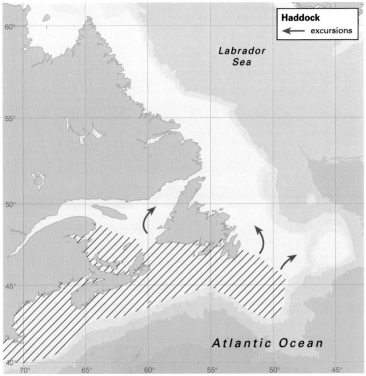

Haddock

Haddock (*Melanogrammus aeglefinus*) are bottom-feeding gadoids that occur only in the North Atlantic. Haddock are most abundant in relatively shallow bank waters with depths less than 250 metres. Unlike Atlantic cod, haddock avoid cold waters.[86] In the northwest Atlantic, their range is restricted to the warmer waters extending from the southern Grand Banks to New England. Large haddock sometimes make summer feeding excursions as far north as the Strait of Belle Isle, but permanent populations have developed no farther north than the southern Grand Banks and St. Pierre Bank. Their range extends to the Barents Sea and Spitsbergen in the central and northeast Atlantic where, despite the northern latitude, sea temperatures are warmer than around Newfoundland.

Haddock likely shared a common ancestor with Atlantic cod and evolved to become a separate species relatively recently.[87] During the ice ages, haddock would have been forced south, perhaps out of Newfoundland waters entirely; their stronghold was likely near New England and the southern shores of the Georges Bank. Haddock recolonized the southern Grand Banks, including the St. Pierre Bank region, during the past 10,000 years. They have not ventured north of these areas to any great extent, nor into the Gulf of St. Lawrence where waters are too cold. There are no haddock in the Pacific Ocean or any species like them – although haddock are capable of extensive migrations, the northern passage to the Pacific was too far north and too cold for haddock to traverse.

Haddock are similar to Atlantic cod in some ways, but differ markedly in others. They share the gadoid reproductive behaviour of forming large spawning aggregations, although the haddock is more oriented towards the ocean floor. In the past, large overwintering and spawning aggregations of haddock dominated the southwestern edges of the Grand and St. Pierre banks. The haddock schools dispersed to the north in

Haddock

summer and, depending on the distribution of warm waters, may have aggregated near the Southeast Shoal. In occasional years, they would migrate inshore like Atlantic cod, but that was unusual. Like Atlantic cod, haddock are long-lived and produce large numbers of drifting eggs; however, they specialize in bottom foraging and hence are much more tied to specific ocean floor habitats than Atlantic cod. Haddock prefer gravel, pebbles, sand, or even clay, avoiding kelp or mud. Their diet consists mostly of small, bottom-dwelling animals, though they will at times eat the same prey as Atlantic cod. Haddock never grow as large as the more piscivorous Atlantic cod, seldom exceeding 90 centimetres or a mass of 10 kilograms.[88]

Pollock

Pollock (*Pollachius virens*) are the "speedster" cod: they are specialized for cruising and live a more pelagic life than either the Atlantic cod or haddock. They feed almost entirely on small pelagic fish and zooplankton. Like the haddock, they do not tolerate very cold waters. Although they range over large areas, the heartland of the pollock is south of Newfoundland and Labrador in the much warmer waters off New England and southern Nova Scotia. The only place where pollock are common in Newfoundland waters is on the southern Grand and St. Pierre banks; even there, however, their occurrence is sporadic and dependent upon incursions of warm water from the Gulf Stream. Pollock are large fish and can reach lengths of up to one metre and masses over 15 kilograms. Farther south, pollock may achieve even greater size.[89]

During the ice ages, pollock were forced south to the margins of Georges Bank. With the melting of the Wisconsin ice, glacial retreat, and warming of Newfoundland waters, pollock were able to take advantage of the ample pelagic fish abundance on the southern Grand Banks. However, their intolerance of colder waters stopped more northerly movements and made even Newfoundland's warmest and most southern waters too cold most of the time; few if any permanent populations of this species developed. Nevertheless, schools of juvenile pollock are often found in the bays of the south coast of

Pollock

the island of Newfoundland, although adults seldom frequent these areas. Pollock also occur in the warmer waters around Iceland and in the northeast Atlantic; they are known there as saithe. The species does not occur in the North Pacific or in any other ocean.

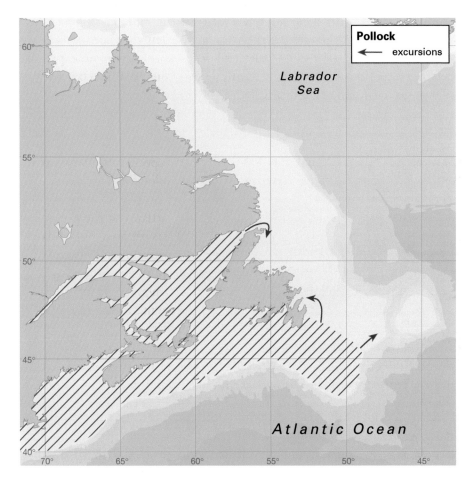

2.18
Pollock distribution in Newfoundland and Labrador waters.

Arctic Cod

The Arctic cod (*Boreogadus saida*) is a small pelagic fish that inhabits Arctic and sub-Arctic waters. Despite its small size and different habits, it is a close relative of the Atlantic cod. It is unique in being the only cod that occurs in both the North Atlantic and North Pacific oceans. Arctic cod are restricted to the coldest seas – their distributional limits are determined by the presence of warm water rather than by cold. Arctic cod likely evolved in the Arctic and spread outwards during and after the ice ages. The species probably split from the Atlantic cod and colonized the North Pacific from the Arctic during the past

3.5 million years. Since then, Arctic cod have become established in the Bering Sea and been found as far south as Japan.[90] Off eastern North America, Arctic cod are typically uncommon south of Labrador; however, during cold periods, such as the early 1990s, the species may be found as far south as the Grand Banks.

Abundance levels of Arctic cod have not been determined, but, in keeping with Theinemann's Law, may be considerable. Schools may be large or small, but are generally made up of similarly sized fish.[91] Arctic cod feed on zooplankton, especially calanoid copepods and larger crustaceans such as amphipods, and spawn in late winter. They are relatively short-lived and grow relatively quickly given the climate, but attain only a small body size.[92] Arctic cod are an important part of the diet of the harp seal, some whales, and Arctic seabirds, but not of the Atlantic cod or other gadoids as there is little overlap in their ranges.

Arctic Cod

Hake

Hake and other gadoid species likely have a common ancestor, but most hake species are sluggish bottom feeders that inhabit much warmer and more southerly waters than Atlantic cod. After the Wisconsin glaciation, a few hake species invaded the Grand Banks region from the south, but cold water temperatures limited their spread much farther north. The white hake (*Urophycis tenuis*) is common on the southern Grand Banks and is the largest of the hake species – large fish can reach over a metre in length and masses of 15 kilograms. They are most often found over soft, muddy bottoms in the deep basins of the Grand Banks. Like Atlantic cod, they are broadcast spawners. Their young are found in the same coastal eelgrass habitats of bays on the northeast coast, but the location of the spawning grounds of these fish remains unknown. White hake feed primarily on shrimps, small crustaceans, and fish.

Other species of hake are found only rarely in Newfoundland and Labrador waters. These include the spotted hake (*Urophycis regius*), the semipelagic silver hake (*Merluccius binlinearis*), and the cusk (*Brosme brosme*). At times these fish invade the southern Grand Banks, but they are typically much more abundant farther south in warmer waters.

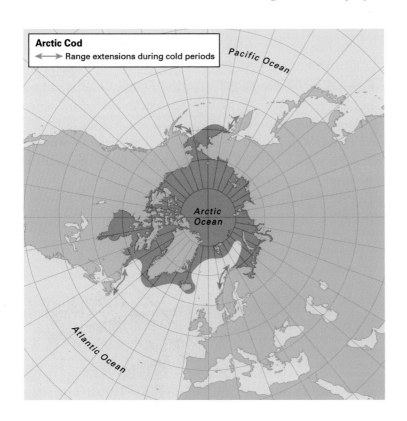

2.19
Arctic cod distribution worldwide.

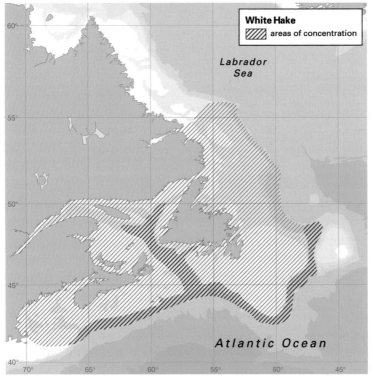

2.20
White hake distribution in Newfoundland and Labrador waters.

Atlantic Herring

The herrings are the original species of the North Atlantic, the pelagic counterpart to the cod. Unlike cod, however, the great majority of herring species are warm-water ishes: of approximately 190 species of herring worldwide, 150 are restricted to the tropics. Only a few species inhabit northern waters, and although the Atlantic herring migrates well to the north to feed in summer, especially in the northeast Atlantic, there are no truly Arctic species of herring.[93] Most herring species live exclusively in saltwater, but there are a number of freshwater species, most of them in the tropics, and an equal number of anadromous species (fishes that spawn in freshwater but spend much of their life in the ocean).[94]

Atlantic Herring

There are 14 species of herring in the North Atlantic and an additional 10 cool-water species that occur only in the Mediterranean region. There are only two resident herring species in the North Pacific: the Pacific herring (*Clupea harengus pallasi*), which is a subspecies of the Atlantic herring,[95] and a subtropical species of sardine. A few other subtropical species of herring occasionly venture into the North Pacific. All cold-water herring species are concentrated in the North Atlantic; for this reason, it can be concluded that herrings are the original endemic pelagic fishes of the North Atlantic, with the Atlantic herring the most abundant and widely distributed.

The Atlantic herring (*Clupea harengus harengus*), like the Atlantic cod, countered the trend of North Pacific fishes colonizing the North Atlantic after the collapse of the Bering land bridge. None of the anadromous herring species of the Atlantic occur in the North Pacific. Only the Atlantic herring, the species with the greatest tolerance for cold, made the journey through the frigid northern passage.[96] The present distribution of the Atlantic herring reflects its migration route into the North Pacific. The most northerly distribution of the Atlantic herring occurs in the northeast Atlantic, from where it invaded the freshwater estuaries of northern Russia. The Atlantic herring likely colonized the North Pacific from the west, moving through Russian waters to Siberia during the past 3.5 million years and perhaps even more recently during interglacial periods.[97]

Atlantic herring rank near the top of the class in forming structured fish schools. Hundreds of thousands of similarly sized Atlantic herring can swim in synchrony, retaining an almost perfect formation and thus constant density across a school.[98] Atlantic herring are plankton feeders. In winter, their school structure appears to break down: at this time, they are found in close contact with the bottom, with large fish (40 centimetres) in close proximity to much smaller fish (15 centimetres).[99]

Some Atlantic herring stocks migrate many hundreds of kilometres through open ocean, while others are more sedentary. All Atlantic herring stocks, however, spawn at specific sites in shallow coastal waters. As is often the case with small pelagic fishes, they are bottom spawners, with eggs that stick to vegetation and to the seafloor. They may live for up to 20 years and spawn each year after maturing, many returning to the same spawning sites year after year. Over their range, there are spring-, fall-, and winter-spawning Atlantic herring. They spawn in protected coastal backwaters and fjords in shallow waters usually less than 10 metres deep. The timing of spawning may be influenced by the ocean temperatures of the previous year.[100] Where Atlantic herring are plentiful, massive spawning aggregations leave conspicuous masses of eggs on kelp and other leafy marine plants, and during spawning the waters become whitish

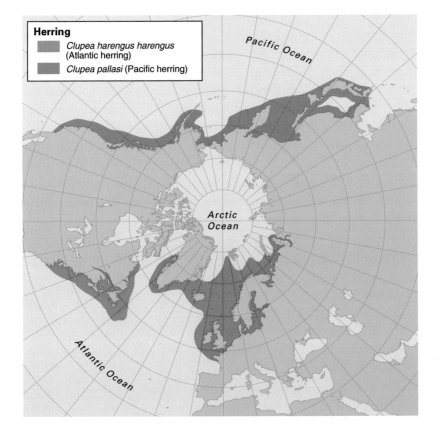

2.21
Herring distribution worldwide.

2.22 Echogram of juvenile Atlantic herring over-wintering in Bloody Reach, Bonavista Bay, in January 2000. Water depths are about 200 metres. The herring aggregation spans about two kilometres.

with the male reproductive product, termed milt. Migration patterns tend to be repeated from year to year and young fish may learn these routes from older fish.[101]

Atlantic herring populations evolved to work best at high numbers. Their reproduction is a hit and miss affair, largely because of the uncertainties of their coastal environment from year to year. Successful reproduction may only occur once every 5-10 years; as a result, Atlantic herring populations often have many fish of the same age and few others. If an Atlantic herring population becomes too small, their reproductive efforts may fail entirely, and it may take a long time for them to regain a high level of abundance.[102] The Atlantic herring found in Newfoundland waters are primarily spring spawners. There are spawning groups in White Bay, Notre Dame Bay, Trinity Bay, the Southern Shore, St. Mary's Bay, and, on the west coast, St. George's Bay. After spawning, herring schools move north to feed, and concentrations originating from several groups may be found in the vicinity of the Strait of Belle Isle during the summer.[103]

In the northwest Atlantic, the largest Atlantic herring stocks occur south of Newfoundland in the warmer waters of the southern Gulf of St. Lawrence, the Bay of Fundy, and near New England. Newfoundland and southern Labrador waters are as far north as Atlantic herring venture in the northwest Atlantic. Nevertheless, Atlantic herring colonized most coastal Newfoundland waters after the ice ages, and most of the bays of Newfoundland developed herring populations. Some of these fish appear to have evolved at least a partial tolerance for cold waters, as herring can be found on the northeast coast in very cold waters in winter and springtime.

Harp and Hooded Seals

All seals are relatively recent additions to marine ecosystems. Hair seals, a group which includes the harp (*Phoca groenlandica*) and hooded seals (*Cystophora cristata*) of the North Atlantic, are thought to have evolved from otterlike ancestors, while eared seals and walruses are believed to have evolved from bearlike ancestors.[104] The origin of hair and eared seals has not been firmly established, but both groups may have initially taken to the sea in the Pacific. However, there is an alternative opinion that the ancestral harp seal evolved during the Pliocene epoch in the North Atlantic,[105] with a common ancestor migrating to the Pacific with the opening of the Bering Sea some 3.5 million years ago. The harp seal of the North Atlantic and the ribbon seal of the North Pacific are very similar, which suggests that their separation was recent. In the North Atlantic region, there is fossil evidence that ancestral harp seals shifted their distributions as far south as the Champlain Sea of the St. Lawrence Valley and Maine in North America during the ice ages, and stayed there after the melting of the last glaciers. A few thousand years ago, these habitats became too warm and harp seals retreated northward. The same pattern occurred in Europe, with Norwegian harp seals moving into the Baltic Sea and perhaps as far south as France, where they were hunted by Neolithic humans.[106] Over the ages, the harp seal has played an important role in the lives and diets of several human communities. No less than half a dozen languages have names for the life stages of the harp seal. In Newfoundland and Labrador, harp seals are referred to as swiles, whitecoats, raggedy-jackets, beaters, bedlamers, old dogs, and bitches.[107] The much larger hooded seal, which may attain masses up to 435 kilograms, is far less numerous (perhaps 10% of the numbers of the harp seal), but still has been of importance to the ecosystems and to humans. Its evolutionary history is uncertain, but may be similar to that of the harp seal.

Hooded Seals

Harp Seals

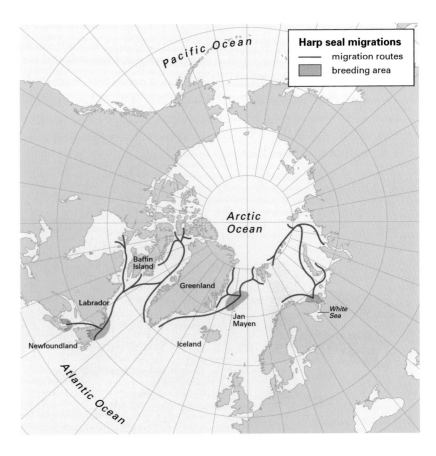

2.23
Harp seal distribution and breeding areas in the North Atlantic.

WIDELY DISTRIBUTED FAMILIES WITH SPECIES IN THE NORTH ATLANTIC

Lumpfish

The lumpfish (*Cyclopterus lumpus*) is a unique North Atlantic species related to the more widespread snailfishes and lumpsuckers.[108] In the northwest Atlantic, lumpfish are widely distributed, being found from New Jersey to western Greenland, as well as in the Gulf of St. Lawrence and around the island of Newfoundland. Concentrations of lumpfish may exist off the southwest coast of Newfoundland, south of the Grand Banks, and off southern Labrador. Lumpfish also occur in Icelandic waters, in the northeastern Atlantic around the United Kingdom, in the North and Baltic seas, and as far north as Spitsbergen.[109] Lumpfish migrate from deep waters on the continental shelf to the coast to spawn in spring. A return migration seaward occurs in the early fall. They are often caught far from bottom and well up in the pelagic zone.[110] Females spawn in waters of about 4°C and are thought to return to the same area to spawn in subsequent years.[111] They deposit relatively small numbers of large eggs in masses measuring 20-24 centimetres in diameter on the ocean floor. The eggs are prized as caviar and form the basis of directed lumpfish fisheries.[112] After spawning, the females retire seaward, leaving the males to guard the eggs. The males assume stunning protective colourations that fade once the eggs have hatched. Young lumpfish look similar to adults, except without the distinctive lumps that develop later in life. Lumpfish can grow as large as 60 centimetres in length, with females generally much larger than males. Lumpfish likely live for 10-15 years, but no fully reliable aging method exists.[113] Lumpfish feed on euphasiids, zooplankton, jellyfish, worms, other invertebrates, and at times on larval fish. Lumpfish are preyed upon by seals, and some are undoubtedly taken by cod and other large predatory fishes.[114]

Lumpfish

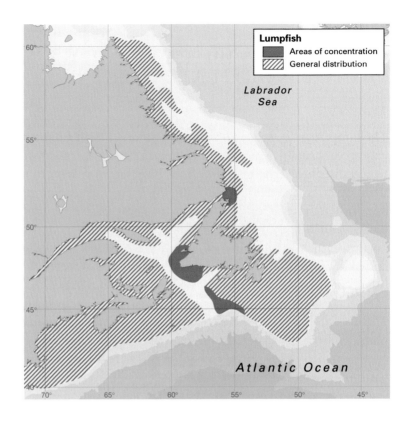

2.24
Lumpfish distribution in Newfoundland and Labrador waters.

Grenadier

There are approximately 300 species of rattail fish, also called macrourids, in the world's oceans. They are relatives of the cod and hake, and include the North Atlantic grenadiers. Rattails evolved to inhabit deep waters: most species live on the continental slopes at depths of 200-2000 metres. Some may even descend to over 5000 metres.[115] Rattails are adapted to life in the dark where speed is less important than persistence in finding prey. As a result, they have large eyes and a whip for a tail, a fish design that evidently works well in those environments. There are 180 species of rattails in the Pacific Ocean, but only 65 species in the Atlantic. Most rattails are tropical fishes: half of the Atlantic species live in the Gulf of Mexico and the Caribbean.[116] There are no Arctic species. A curious aspect of the distribution of rattails in the Atlantic is the close relationship between the continental-slope-dwelling rattails of the eastern and western continental slopes. Many species occur in both areas, separated by the wide and deep Atlantic Ocean basin.

Several species of rattails are Atlantic originals. Only two of these have adapted to subarctic waters: the roundnose grenadier (*Coryphaenoides rupestris*) and onion-eye or roughhead grenadier (*Macrourus berglax*), both of which are widely distributed from the northwest to northeast Atlantic. Unlike many other fish species, there is no evidence of continuing speciation in either grenadier in these widely separated areas, although some population structuring may have occurred.[117] The tectonic geology of the North

Atlantic may be responsible for this, as grenadier habitat occurs in a continuous swath of deep waters (450-1000 metres) from Labrador to Greenland to Iceland and then via the Faroe Ridge to the northeast Atlantic. It is more difficult to explain the lack of speciation in the more southerly distributed, continental-slope rattails, for which there are huge breaks in distribution. It is possible that they have a widely dispersing life stage, but there is little evidence of this. Another explanation is that rattails are ancient species, with populations separated by the spreading Atlantic 50 million years ago that have evolved only modestly to form still closely related subspecies.[118]

Little is known about the life histories of Atlantic species of grenadier.[119] They are thought to spawn near the seafloor and to produce large floating eggs[120] that hatch into larvae during a slow ascent from deep waters. Older larvae and juveniles are believed to settle to the seafloor and take up deepwater habits. Both the roundnose and roughhead grenadiers can attain lengths of over one metre in Newfoundland and Labrador waters. Grenadier feed on crustaceans, including Pandalid shrimp and euphasiids, and on bottom-dwelling invertebrates, capelin, jellyfish, and myctophids.[121] The roundnose and roughhead grenadier were two of the most successful fishes of the deepwater shelf edges around the North Atlantic. Although they attained reasonably high abundance levels in the past, they do not have high reproductive or growth rates (Thiennemann's Law again). Predictably, early grenadier fisheries initially encountered high densities of fish, especially of the roundnose grenadier; equally predictably, they did not last long.

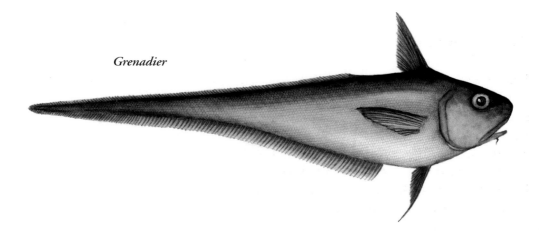

Grenadier

Sand Lance

Sand lance are small, spear-shaped fishes that inhabit the northern waters of the Atlantic and Pacific oceans. Although they are pelagic schooling fishes, they also burrow into the seafloor, mostly in sand. There is confusion about how many species of sand lance exist – estimates range from 6 to 23. Only six species are well described – two each in the North Pacific, northwest Atlantic, and northeast Atlantic.[122] Their evolutionary

2.25
Grenadier distribution in Newfoundland and Labrador waters.

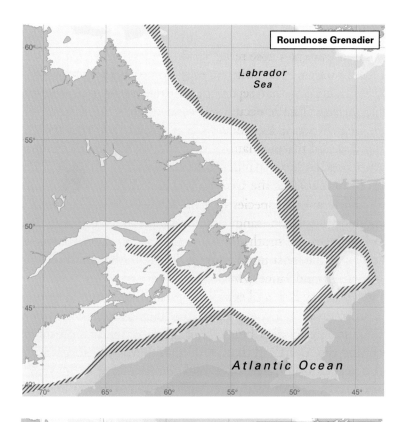

2.26
Sand lance distribution in Newfoundland and Labrador waters.

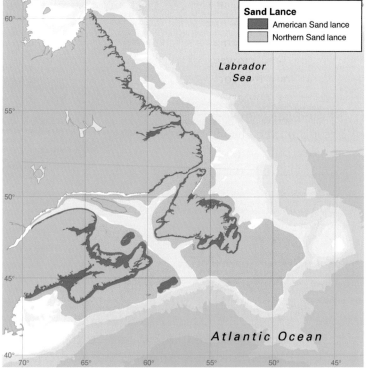

origin is unclear. Close relatives of the sand lance occur in tropical waters and may have a common ancestry with the more northern species. The difficulty in designating species suggests that separations between the northeast and northwest Atlantic and the northern Pacific occurred recently in geological time. Sand lance are important prey for many predators, including Atlantic cod, haddock, pollock, American plaice, seabirds, and marine mammals. In the northwest Atlantic and on the Grand Banks, they are second in importance only to capelin,[123] and they are of primary importance in ecosystems such as the Gulf of Alaska and North Sea.

There are two species of sand lance in Newfoundland and Labrador waters.[124] The common northern sand lance (*Ammodytes hexapterus*) inhabits the Grand Banks, especially to the south, as well as the Green and St. Pierre banks, where it supplants capelin as the most important food of Atlantic cod, haddock, and American plaice. The American sand lance (*Ammodytes americanus*) occurs in coastal waters from Placentia Bay to Trinity Bay, and sporadically northward, and is important to the many predators that feed near the coast, including seabirds, whales, and Atlantic cod. Although sand lance are widely distributed, the Grand Banks became an especially important habitat. On the eastern half of the Grand Banks, millions of years of eroded sediments from the adjacent hard continental rocks created a sandy seafloor ideal for sand lance. This is where the sand lance is most numerous and where it feeds the abundances of flatfish and cod. The sand lance inhabits the tops of the Grand Banks in spring and summer, where waters are cool, moving to deeper and warmer waters in winter, but seldom to depths greater than 250 metres.[125]

Sand lance are generally thought to be a short-lived (3-4 years) and relatively fast-growing fish. However, in the cold waters of the Grand Banks, they grow more slowly and may live for up to 10 years.[126] Sand lance spawn in the winter months (November to January) on the Grand Banks and also on sandy coastal beaches. Young sand lance grow quickly and feed almost exclusively on zooplankton. Together with capelin and herring, they complete the pelagic triumvirate that converts zooplankton to energy-rich food destined to be consumed by predatory fishes and mammals. Of these three species, sand lance are likely the easiest to catch and swallow, simply because of their cylindrical shape. They have been called the quintessential forage fish of the northern hemisphere,[127] although in Newfoundland and Labrador waters that role is more properly given to the capelin.

Sand Lance

Myctophids

Myctophids are commonly called lanternfish. Their bodies have small luminous organs, referred to as lanterns, which glow in the water.[128] Myctophids are mostly tropical and subtropical fishes that inhabit the deeper waters of the open-ocean regions of the world in vast but unknown numbers. Their origin was the tropical waters of ancient seas[129] and there are no exclusively Arctic water or cold-water species. Although the precise evolutionary progression of these fishes is not well known, it is likely that they were present in all the world oceans tens of millions of years ago, well before the ice ages. Their cautious advance into the North Atlantic almost certainly came after the retreat of the Wisconsin ice, and only the most cold-hardy species have survived in any numbers.

Lanternfish

There are 82 species of myctophids in the North Atlantic, but only a few species dominate in Newfoundland and Labrador waters.[130] Theinemann's Law holds well with myctophids: the number of species in North Atlantic waters is small, but their abundance is great. Sampling performed by the Woods Hole Oceanographic Institute between 1961 and 1974 indicated that the sub-Arctic region of the North Atlantic, which includes Newfoundland and Labrador, contained the highest density of myctophids recorded in all regions of the Atlantic.[131] Off the shelf break of the northeast coast of Newfoundland, 500-metre-thick layers of mixed myctophids and euphasiids can be found over deep water.[132]

The role of myctophids in the ecosystems of the Grand Banks and adjacent shelves is unclear. They may be dismissed as being important only to open- and deep-ocean ecosystems. However, their distribution and apparent abundance on the edge of the continental shelf suggests some importance. They must consume a considerable amount of zooplankton, and likely provide a potential food for deepwater fishes like the grenadier and redfish.[133]

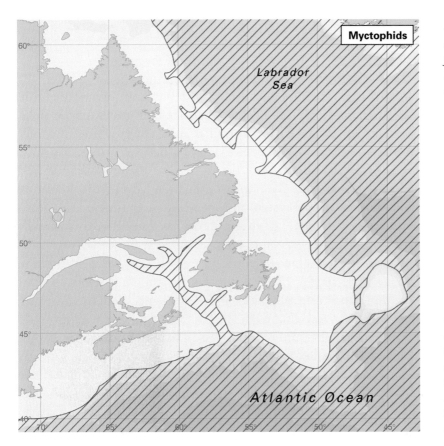

2.27
Myctophid (lanternfish) distribution in Newfoundland and Labrador waters.

2.28
Echogram of lanternfish on the outer edge of the continental shelf just north of the Grand Banks. Water depths were about 600 metres, and the panel covers about 5 kilometres.

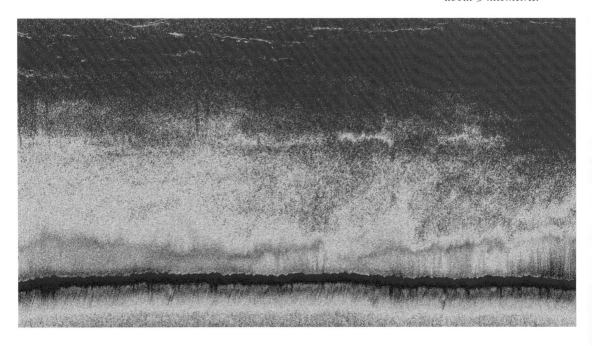

Flatfishes

The flatfishes are an easily recognized but surprisingly diverse group that has lost the upright symmetry of most fishes. Flatfishes start off life with the shape of a normal fish. As they grow, their eyes migrate to one side of their head – some species to the right, others to the left. The side that becomes the bottom is most often white in colour, while the top is darker or camouflaged. Most species adopt a flattened body shape and are adapted to a life on or near the seafloor. Most flatfishes have small mouths and feed on small organisms. However, there are exceptions to all these generalities: there are flatfishes that have recolonized the upper waters and spend no more time on the ocean floor than regular-shaped fishes; there are flatfishes with coloured undersides; and some flatfishes have become very predatory, with large mouths and very big teeth.

Flatfishes are thought to have originated in the tropical oceans some 50 million years ago.[134] There are over 500 species of flatfishes, falling into four tribes: halibut (which include the Newfoundland turbot), plaice, turbot, and sole. Halibut and plaice are typically cold-water, northern species, while turbot and sole are warmer-water species and more southerly distributed into the tropics. There are no sole or turbot species in Newfoundland and Labrador waters. The names of these fishes vary greatly from place to place: what is called a sole in one locale may really be a plaice; the Newfoundland turbot is really a halibut; and a dab may be either a sole or halibut. The classification system used here appears in Appendix 1 and an attempt will be made to identify these fish with the local names used in Newfoundland and Labrador.[135]

There are 27 genera (groups of closely related species) of flatfishes in the northern and Arctic oceans. All of these species occur in the North Pacific, while only 10 are represented in the North Atlantic. In Newfoundland and Labrador waters, only a few species are present. It is likely that the Atlantic species came from the Pacific.[136] The timing of their arrival is uncertain, but the close relationship between the species of the North Pacific and North Atlantic suggests that it was relatively recently, likely within the past 3.5 million years.

Atlantic halibut. Halibut are the largest of the flatfishes; the Atlantic halibut (*Hippoglossus hippoglossus*) can grow to lengths of two metres and masses of over 100 kilograms.[137] Atlantic halibut left the seafloor some time ago and have adapted a semipelagic distribution, especially when feeding. Small pelagic fishes and large zooplankton form the bulk of their diet, although organisms living on the seafloor are also eaten. Despite their flattened shape, Atlantic halibut are strong, fast swimmers capable of hunting even the most mobile prey. Atlantic halibut are widely distributed in the waters of Newfoundland and Labrador. Although they are nowhere really numerous, their abundance is greatest in the regions of the southern Grand Banks and the Gulf of St. Lawrence.

Greenland halibut (*Reinhardtius hippoglossoides*), known as turbot in Newfoundland and Labrador, are derivatives of close relatives in the North Pacific.[138] Since the arrival of their ancestors in the North Atlantic, Greenland halibut have become widely distributed in the deep waters off the continental shelves of both the northeast and northwest Atlantic.[139] This species, like the Atlantic halibut, is semipelagic, a strong swimmer, and a fierce predator. Greenland halibut prey heavily on small pelagic fishes, especially capelin and Arctic cod, and on crustaceans and small fishes. The Greenland halibut is a northern species that also occurs around Iceland and into Arctic waters off west Greenland and Spitsbergen. In the waters of Newfoundland and Labrador, Greenland halibut are most numerous off Labrador and the northeast coast of Newfoundland.

The life history of the Greenland halibut is closely related to the circulation of the Labrador Sea and the flow of the Labrador Current. Their major spawning grounds are located in the deep northern waters of the Labrador Sea, but there may also be spawning grounds farther south, especially in the Gulf of St. Lawrence. Small spawning groups may also have become established in the deeper bays of Newfoundland (e.g., Trinity Bay). Greenland halibut spawn tens to hundreds of thousands of floating eggs, with the larger females producing the most. From the northern spawning grounds, the eggs and larvae drift long distances; most end up on the continental shelf from northern Labrador to the northern Grand Banks. More southerly spawning may result in dispersal as far south as Georges Bank. Juvenile Greenland halibut are small, leaflike creatures that inhabit the ancient basins of the continental shelf. As they grow, they take the shape of adults and the colour of blued gun-steel. By the time they reach 30 centimetres in length, they are formidable predators. They have been reported to "leap out of the water when the depth was 40 fathoms (73 metres), and…[to be caught] on the hand-line when…only halfway down. Several have followed the bait to the surface, and one even followed the thermometer up twice in succession. Feathers were pulled out of the mouth of one, and the skeleton of a gull, *Larus tridactylus*, was found in the stomach of another."[140] More typically, Greenland halibut feed on capelin, other small fishes, and shrimp. They mature at five to seven years of age, and migrate either back to their northern spawning grounds or to smaller southern grounds. Details of these movements are not well known. Greenland halibut can grow to over a metre in length and to masses of over 10 kilograms. Most of the largest fish are females.[141]

Yellowtail flounder. There are five genera of plaice and flounder that have important species in Newfoundland and Labrador waters. Most species of plaice and flounder inhabit the North Pacific, where the group likely originated. However, there is some opinion that these species originated in the northeast Atlantic.[142] In northern oceans, there are six flounder species of the genus *Limanda*; of these, four occur only in the North Pacific. The so-called "dab" (*Limanda limanda*) occurs in the northeast Atlantic

2.29
Atlantic halibut and Greenland halibut distribution in Newfoundland and Labrador waters.

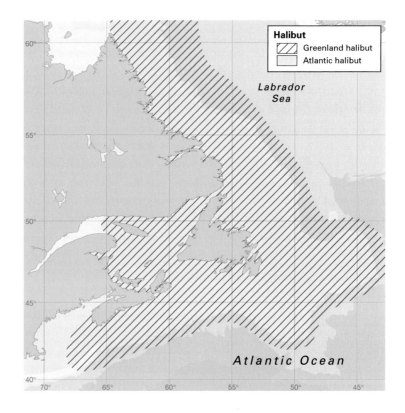

2.30
Flounder and American plaice distribution in Newfoundland and Labrador waters.

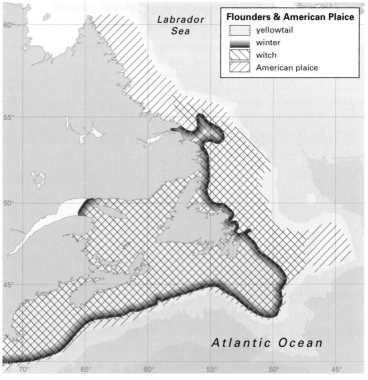

from the Bay of Biscay to Iceland, while the yellowtail flounder (*Limanda ferruginea*) ranges from southern Labrador and the Grand Banks southward to Chesapeake Bay. Yellowtail flounder are particularly abundant in shallow waters (less than 90 metres) on the southern Grand Banks and on Georges Bank.

Like all flounder species, yellowtail flounder spend most of their lives on or near the seafloor.[143] They spawn in the spring, releasing hundreds of thousands of buoyant eggs that float to the surface and drift with the currents. On the southern Grand Banks and on Georges Bank, the eggs and larvae are likely caught in the oceanographic gyres which retain and concentrate the young fish on the banks.[144] Yellowtail adults have small mouths and feed mainly on worms and other small invertebrates found on or near the bottom. They may also take the occasional sand lance.

Winter flounder. There are three species of *Pseudopleuronectes* flounder. Two occur in Japanese waters of the North Pacific and one occurs in the northwest Atlantic. The Atlantic species is the winter flounder (*Pseudopleuronectes americanus*), known as the "blackback" in Newfoundland and Labrador. These are small coastal flatfish that inhabit shallow waters mostly near shore. Winter flounder are true bottom fish, and their backs provide exceptional camouflage: they are difficult to detect unless they move. Although much about the life history of this species is unknown, they can be commonly seen feeding on the eggs of demersal-spawning capelin and herring. Winter flounder are known to develop very high levels of blood antifreeze that allow them to live in the coldest waters, as cold as -1.7°C, apparently without ill effects.

American plaice. There are four species of *Hippoglossoides* plaice. Three of these species occur only in the North Pacific, while the fourth, the American plaice (*Hippoglossoides platessoides*), is widespread in the North Atlantic. Interestingly, American plaice occur in both the northwest and northeast Atlantic, as well as in Icelandic waters. There is evidence that American plaice may be undergoing speciation at present, and two or more subspecies are currently recognized.[145] American plaice are known by different names in different places. In Newfoundland and Labrador, they are simply called plaice. In the United Kingdom, they are called the sand dab. The subspecies off Iceland and north to the Barents Sea is called the long rough dab or rough dab. In the United Kingdom and northeast Atlantic, the term "plaice" refers to a different species of flatfish entirely.[146] American plaice occur from the Arctic to Georges Bank, with the centre of distribution falling on the Grand Banks. American plaice inhabit a wide range of depths and temperatures, but will shift their distributions to avoid particularly cold waters.[147] The American plaice is a slow-growing species. In Newfoundland water, females do not mature until they reach eight to ten years of age, while males mature at four to five years of age. Plaice spawn near the seafloor and have relatively large, buoyant eggs that drift in surface currents before hatching.

Arctic flounder. The Arctic flounder belongs to the genus *Liopsetta*, and there may be four species or only one, depending on the classification system being used. Arctic flounder are confined to Arctic and Alaskan waters. A very similar fish has been reported in Newfoundland and Labrador waters, and is called the smooth or eel-back flounder.[148] Little is known about this species, other than it lives in very cold waters.

Witch flounder. There are three species of *Glyptocephalus* flounders, two of which occur only in the Pacific. Witch flounder (*Glyptocephalus cynoglossus*) inhabit both the northeast and northwest Atlantic, where they range from southern Labrador to the Carolinas. In some areas, they are called pole flounder, but are known simply as "witch" in Newfoundland and Labrador. Witch flounder are wafer-thin fish – the ultimate in the flattened design – and a bottom dweller. Witch flounder inhabit channels and basins with muddy bottoms, and waters with temperatures in the 2-6°C range. They are a slow-growing fish that may reach ages of over 25 years; the largest individuals may be 75 centimetres long and weigh 4-5 kilograms. Witch flounder have small mouths and feed mainly on benthic invertebrates such as worms and other small invertebrates.

In Newfoundland waters, witch flounder never became very numerous, perhaps because of the low temperatures. However, they prosper in some of the deep, silty trenches of the Grand Banks. Witch flounder concentrate to spawn in the main channels (from the Hawke Channel off Labrador and the trenches around Funk Island Bank to the Halibut Channel south of the island of Newfoundland) over an extended period from January to March of each year.[149] They spawn near the seafloor, but have floating eggs that superficially resemble those of cod. The pelagic juvenile stage is prolonged, lasting up to a year, and juvenile witch flounder may inhabit deeper waters than the adults.

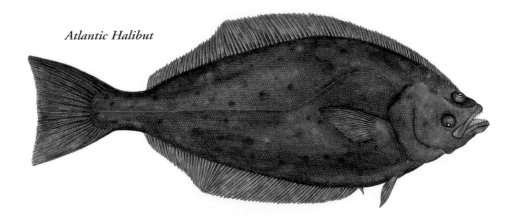

Atlantic Halibut

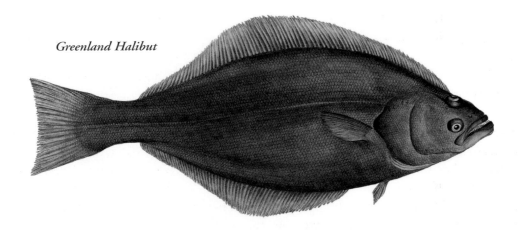

Greenland Halibut

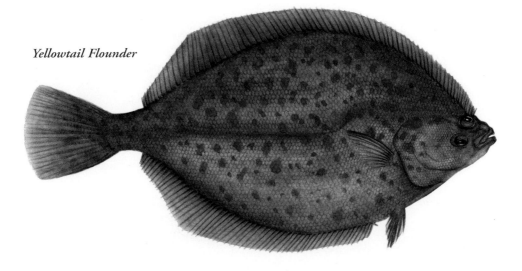

Yellowtail Flounder

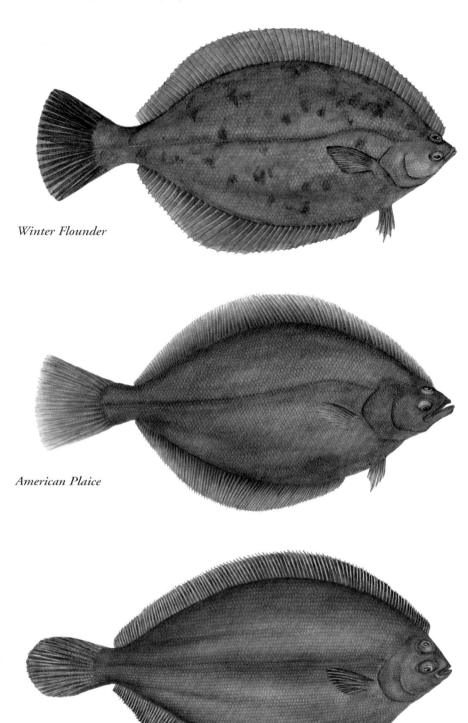

Winter Flounder

American Plaice

Witch Flounder

Other Species

Green sea urchin. Sea urchins are ancient animals called echinoderms and are distributed in all the oceans of the world. Most species of sea urchins initially evolved and currently occur in the tropics. The green sea urchin (*Strongylocentrotus droebachiensis*) is widely distributed in Newfoundland waters, lives on the ocean floor, and feeds mainly on kelp. As kelp beds exist only in shallow coastal waters that were frozen during the ice ages, sea urchins repopulated the coastal zones during the last 10,000 years. Adult sea urchins are not mobile animals; recolonization was likely accomplished by the more mobile early life stages. Sea urchins cannot escape predators, so they are armed against them – many have venomous or sharp spines. The green urchin has sharp spines that provide some protection from lobsters and crabs, although they are not venomous like the spines of some urchins.

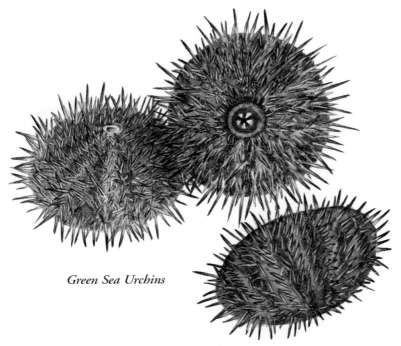

Green Sea Urchins

Sea butterflies. Sea butterflies include about 70 species of what are called tunicates, so-called because of the soft tunic surrounding the body. They are found in the surface waters of the world's oceans, sometimes in great numbers – one report from British Columbia indicated nearly 26,000 animals in one cubic metre of seawater.[150] The adults resemble a tadpole, but the tunic of some species can be a metre wide. Tunicate species in the waters of Newfoundland and Labrador are much smaller, but at times their numbers can be enormous. The discarded tunics of sea butterflies are known as "slub" in Newfoundland and Labrador, and "slime" in Nova Scotia. There are years and times of great slub and slime,[151] well known to fishermen because it fouls fishing gear.

PACIFIC INVADERS

Capelin

Capelin (sometimes spelled caplin or capelan) are members of the widespread smelt family of fishes. Capelin (*Mallotus villosus*) are small, silvery, pelagic fish that are high in oil content and have a unique smell resembling cucumbers. They inhabit the cool waters of the northern Pacific and Atlantic oceans and, although they are not an Arctic species, venture into Arctic waters regularly. Although capelin are the most important small pelagic fish in the North Atlantic, they originated in the Pacific, where they are of lesser ecological importance. In the Atlantic, capelin are the chief food and the main source of fat for many predatory fishes, marine mammals, and seabirds.[152] The importance of capelin to the food webs in North Atlantic ecosystems cannot be overstated. This is particularly true in Newfoundland and Labrador. Before their arrival, the large abundances of capelin-consuming species could not have been supported.

Capelin

Capelin migrated from the North Pacific through the Canadian Arctic to the North Atlantic, likely within the past few million years. Arctic populations of capelin still mark the route from Coronation Gulf–Bathurst Inlet to Hudson and Ungava bays. After reaching the northwest Atlantic, capelin spread to the northeast Atlantic. There, they developed specialized behaviours, in particular relating to reproductive abilities: in the northeast Atlantic and in Icelandic waters, capelin have given up their ancient habit of spawning on beaches, as they do in most areas of the North Pacific and the northwest Atlantic, in favour of spawning in shallow coastal waters.[153] There has been speculation that another incursion of capelin from the Pacific came very recently, since the last ice age, but recent genetic work suggests that all North Atlantic capelin are more similar than are Atlantic and Pacific capelin, so that theory seems unfounded.[154]

In Newfoundland and Labrador waters during the last ice age, capelin were restricted to spawning on the ice-free beaches of the southern Grand Banks. They likely extended their range southward on the Nova Scotia shelf and to Georges Bank until the ice retreated, then recolonized the waters to the north and west. Spawning beaches had been improved and new ones created by the gravel depositions of the

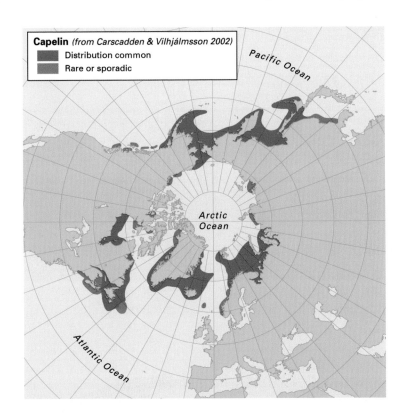

2.31
Capelin distribution worldwide.

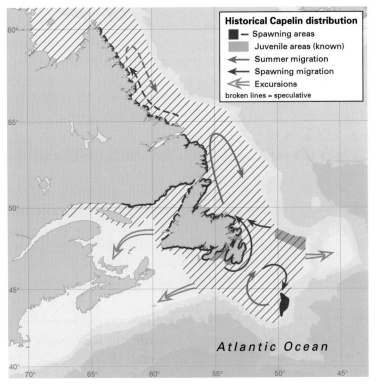

2.32
Historical capelin distribution in Newfoundland and Labrador waters.

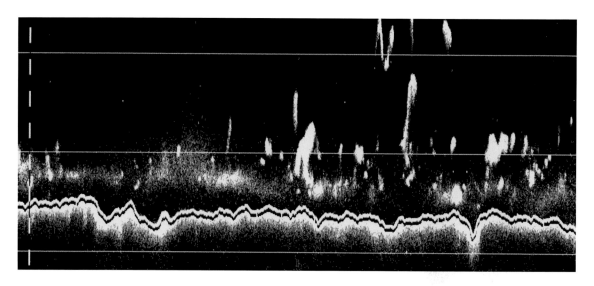

2.33
Echogram of migrating adult capelin in the Bonavista Corridor in 1992. View spans about 10 kilometres, total depth is 320 metres, and horizontal lines are 50 metres apart (only the bottom portion of echogram shown).

glaciers, and capelin spread north to Labrador and west into the Gulf of St. Lawrence and St. Lawrence River. As sea levels rose, the relic beaches of the southern Grand Banks sank under the waves, but capelin still spawn there, underwater.

Capelin typically choose spawning beaches with specific gravel sizes, but spawning in 10-20 metres of water just off the beaches is also common.[155] Spawning occurs from Groswater Bay in Labrador to above Trois-Rivières in Quebec, as well as around the island of Newfoundland. Offshore spawning occurs only on the Southeast Shoal of the Grand Banks. Schools of males and females approach the spawning grounds separately. Mating occurs in an instant, typically with two males embracing a female, one on each side, with special spawning ridges that form along their bodies.[156] The eggs are sticky and adhere to the gravel of the grounds. Most capelin die after a single spawning at ages of three to four years, but some females may survive. Larvae emerge from eggs in the gravel and, if conditions are right, are dispersed seaward.[157]

Capelin have a short life cycle. At one year of age, capelin are nearly translucent and look very little like adults. At two years of age, they are known in Newfoundland and Labrador as "whitefish" because of their colour.[158] By age three, they achieve their more familiar form, and then it is time to spawn and die so that the cycle may be repeated again. Juvenile capelin can be found both inshore in bays and on the offshore banks, especially on the North Cape of the Grand Banks. But they are hardly sedentary – a bay full of juvenile capelin one week may hold few the next.[159] The exact migration patterns of these fish are not well known, but cycles have likely developed that take them to feeding areas until they are ready to spawn and then inshore to beaches as they

2.34 Brador beach near Blanc Sablon, just inside the Strait of Belle Isle, is one of the largest capelin spawning beaches in Newfoundland and Labrador. Photograph by the author.

2.35 Capelin about to spawn at the beach at Petley, Trinity Bay, in late June 2006. Water depths about 20 centimetres. Photograph by the author.

mature.[160] Capelin feed on increasingly large zooplankton as they grow larger themselves.[161] Historically, large schools moved north to the rich feeding grounds around the Labrador Banks in the late summer and fall, moving southward again in early winter.

Many species of fish, marine mammals, and seabirds eat capelin. When cod stocks were abundant, they were by far the greatest consumer,[162] ingesting some 2-3 million tonnes of capelin annually. Capelin made up one-third to one-half of the energy consumed by cod each year, and all Newfoundland and Labrador cod stocks came to depend on them. During the spring and early summer, capelin comprise nearly 100% of the diet of cod in some regions, and they are particularly important as cod regain condition after spawning.[163] Whales and seals likely consume at least half a million tonnes of capelin annually and seabirds another few hundred thousand tonnes. Other fishes, such as Greenland halibut, Atlantic salmon, and Arctic char, take another few hundred thousand tonnes. In addition, winter flounder, eelpout, and other species feast on capelin eggs on the spawning beaches, and gulls and eagles feed on dying adults.

Wolffish

Wolffish are large, highly specialized, bottom-dwelling predators commonly called "catfish" in Newfoundland and Labrador. They occur only in northern oceans and there are only five species in total. There are three species of wolffish in the North Atlantic, all close relatives of the Asiatic Pacific wolfish: the northern or broadhead wolfish (*Anarhichas denticulatus*), the striped wolffish (*Anarhichas lupus*), and spotted wolffish (*Anarhichas minor*). The ranges of all three species overlap. The spotted and striped wolffish are very similar, while the northern wolffish is most like its northwest Pacific cousin.

The origin of the wolffish is controversial.[164] Early work suggested an Atlantic origin like the cod and herring, based on the existence of more species in the Atlantic. However, a later view suggests the opposite – the most primitive wolffish species occurs in the Pacific, and the northern wolffish may have a common ancestor with its cousin in the Asiatic North Pacific. If this is true, then wolffish would likely have penetrated the northeast Atlantic about 10 million years ago. Fossil wolffish teeth of this age have been

Striped Wolffish

found in England, supporting this theory.¹⁶⁵ According to this view, the striped and spotted wolffish of the North Atlantic evolved later from an original invader that migrated from the Asiatic Pacific to the northeast Atlantic. The invader's closest living relative is the northern wolffish.

Spotted Wolffish

Atlantic wolffish have colonized deeper waters than their Pacific cousins.¹⁶⁶ This shift in habitat likely occurred during the recessions of the ice ages. Nevertheless, the wolffish habit of preying on sea urchins and other echinoderms, molluscs, and crustaceans persists. The wolffish are not particularly fast or strong swimmers – they are burly with large sharp teeth specialized to prey on bottom-dwelling organisms with tough exteriors. Wolffish are solitary fishes and do not form schools. In fact, they appear not to like each other much – even very small wolffish put into a tank will set up territories, with the largest usually holding a central position and attacking any that approach its site. How they reproduce is not known, but clutches of sticky demersal eggs are produced in summer. In the northeast Atlantic, egg clutches attributed to the striped wolffish have been found in coastal waters 9-85 metres deep. Striped and northern wolffish may spawn in deeper waters. Young striped wolffish are known to inhabit waters as deep as 450 metres off the northeast coast of Newfoundland and Labrador.¹⁶⁷

Redfish

Redfish are the North Atlantic version of the Pacific rockfishes. All belong to the genus *Sebastes*. There are at least 100 species of rockfish known worldwide; almost all of these occur only in the North Pacific.¹⁶⁸ There are four species of redfish in the North Atlantic that are found nowhere else. The rock- and redfish are close relatives of and likely descended from the seafloor-dwelling scorpion fishes, best known for their poisonous barbs. Over time, some rockfish became semipelagic, with various species adopting habits that took them well up into the ocean waters. Still, most species of modern *Sebastes* are found in close association with a seafloor habitat, and they are not typically long-distance migrators.¹⁶⁹

2.36
Wolffish distribution in Newfoundland and Labrador waters.

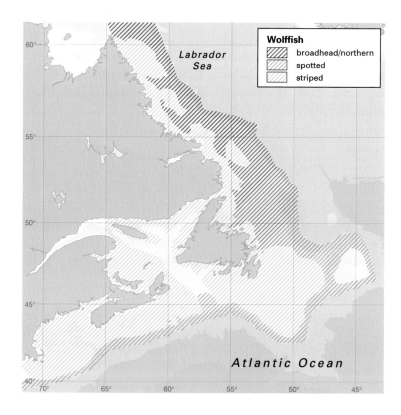

2.37
Redfish distribution in Newfoundland and Labrador waters.

Rockfish evolved in the North Pacific about 10 million years ago. Fossil rockfish have been found in California and Japan dating from the upper Miocene epoch. They have colonized many diverse habitats since then, from deepwater slopes off Alaska to the Gulf of California islands. Many of the species are so recently evolved that it is very difficult to tell them apart. However, DNA analyses have helped a great deal in sorting out species and distributions.[170] Present-day distributions of species reflect a southerly evolution, with the number of species highest off California and becoming progressively lower to the north. There are only a few species in the Bering Sea, similar to the number in the North Atlantic.

Rockfish have some unusual life history characteristics. They are very long-lived – up to 100 years – and grow very slowly. Redfish in Newfoundland and Labrador waters take more than 10 years to reach 20 centimetres in length. They do not spawn eggs, as do most other fish, but have a form of internal fertilization and hatching of eggs, and give birth to well-formed larvae.

The redfish of the North Atlantic are true rockfishes all descended from a single North Pacific migrant. This adventurous fish was likely the Pacific ocean perch (*Sebastes alutus*) or its immediate ancestor. The Pacific ocean perch is a northern rockfish, most numerous off Alaska and in the Bering Sea. It likely migrated into the Bering Sea and then to the North Atlantic very recently, within the past 3.5 million years, after the sinking of the Bering land bridge. From this initial migration, the four Atlantic species of redfish have evolved: the Acadian redfish (*Sebastes fasciatus*), the deepwater redfish, (*Sebastes mentella*), the golden redfish or rosefish (*Sebastes marinus*), and the Norway redfish (*Sebastes viviparous*). The Acadian, deepwater, and golden redfish are found both in the northwest and northeast Atlantic, while the Norway redfish occurs only in the northeast.[171] Given this pattern, it is likely that the three most common species of redfish evolved in the northwest Atlantic and then moved across to Iceland and Norwegian waters, with the Norway redfish evolving later.

Redfish

*2.38
Echogram of redfish on the southern Grand Bank in 1998. Redfish schools appear as red-blue vertical structures above the bottom, which is the blue band. Water depths were about 350 metres. Panel spans about 5 kilometres.*

Invading rockfish found an abundance of food in the North Atlantic, notably euphasiids, and populations developed along the edges of all the major banks from Labrador to the Gulf of St. Lawrence to Georges Bank. In turn, young redfish provided a source of food for the largest of the Atlantic cod – those grown too big to feed profitably on capelin or sand lance.

In the North Atlantic, the redfish are all deepwater species and share most of the habits of their cousin, the Pacific ocean perch. They have become one of the best examples of Thienemann's Law. None of the rockfishes of the North Pacific reach huge abundance levels and the school sizes are modest, with dimensions on the order of hundreds of metres or less.[172] In contrast, historical redfish aggregations off the Grand Banks and the northeast Newfoundland and Labrador shelves measured many kilometres across, although individual schools of fish within the aggregations were similar in size to those of the Pacific rockfishes. Distribution differences may relate to habitat: the bank and sloping bathymetry that forms the habitat of rockfishes is far more extensive and connected in the North Atlantic than in the North Pacific, which tends to have smaller banks and isolated seamounts.

Redfish reproduce and grow slowly. Like their ancestors in the Pacific, their populations may only produce a large number of young once in many years. Long-lived fish may reproduce up to 100 times over their lifetime, each time producing thousands of young, but only rarely will many survive.[173] A low reproductive rate is necessary for stability in long-lived species.[174]

Eelpout

There are about 50 species of eelpout in the northern oceans, most of which are cold-water fishes. Eelpout are all quite similar, although they vary in length from the 20-centimetre Newfoundland eelpout (*Lycodes terraenovae*), also known as the fish doctor, to the large ocean pout (*Macrozoarces americanus*), which has been reported at lengths of over a metre.[175] Newfoundland and Labrador waters are home to only nine of the 50 species of eelpout. Like wolffish, eelpout are sometimes called catfish. In some places, the large ocean pout is called the "mother of eels."

Eelpout likely originated in the Pacific: there are more species of eelpout there and they are not restricted to the far north as they are in the Atlantic. However, some species have adapted to the coldest ocean waters and at least some have an Arctic origin. The blood of northern species contains antifreeze that prevents the fish from freezing in sub-zero waters. Their antifreeze proteins are the same as those in cod, indicating that these fishes may have shared a common evolutionary background.[176]

The life histories of the eelpout are not well known. Most are thought to inhabit rocky bottoms over a wide range of depths from a few to thousands of metres. The larger species appear to migrate from coastal to deeper waters where they overwinter. Eelpout are almost exclusively bottom feeders, taking mostly invertebrates – including worms, sea urchins, brittle stars, sand dollars, shrimp, and molluscs – and an occasional small fish. Eelpout either have live births or deposit a few thousand very large eggs in a gelatinous mass. The ocean pout spawns in autumn, with the eggs hatching in two to three months. They generally grow slowly and are not a very productive fish.[177]

Eelpout

Seasnails

Seasnails are small and soft little fishes that originated in the North Pacific and migrated to the North Atlantic. Shallow-water species came directly across the opened northern passage, but at least one deepwater group of species apparently came the long way, via Antarctica: their ancestors migrated south in the Pacific, then back north along the mid-Atlantic ridge and all the way to the Arctic. Some snails![178]

Atlantic Walrus

The walrus is the survivor of a family of marine mammals that evolved from primitive meat-eating bears in the North Pacific. It is thought that the ancestor of the modern walrus migrated from the Pacific to the Atlantic about 5-8 million years ago, taking a southern route through the then-open Panama strait, then became extinct in the Pacific. The modern North Atlantic walrus migrated back to the North Pacific within the past million years, this time taking the northern route through the Arctic. The walrus in the two northern oceans are very similar, although they now are considered to be separate species.[179]

The walrus depends on haul-out sites (resting places where the less-than-graceful walrus – on land – can haul itself out of the water) on ice and land. The rocky shores of Newfoundland and Labrador provide few sites that have sea access sufficiently ramped for use by these large creatures. However, the walrus was once distributed as far south as Cape Breton and Sable Island, and was relatively abundant in the Gulf of St. Lawrence, with breeding colonies at the gently sloping Magdalene Islands, Sable Island, and perhaps at some sites off Labrador. Feeding walrus probably once coursed the coasts of Newfoundland, raking the seafloor with their tusks to gather food, which consists primarily of molluscs that are sometimes shelled before eating.[180]

The size of the walrus population that was present in the early marine ecosystems of Newfoundland and Labrador is difficult to determine, but, based on surviving populations, their numbers were likely modest. In the North Atlantic, current estimates are about 20,000 animals, of which about half are in Canadian waters. Pacific populations are much larger – perhaps 200,000 animals. There may have been thousands to tens of thousands of walrus in Newfoundland and Labrador waters in ancient times. Accounts of vast numbers of walrus being harvested are likely far-fetched.[181]

Atlantic Walrus

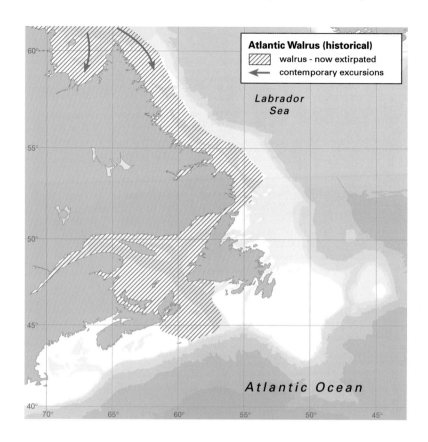

2.39
Historical walrus distribution in Newfoundland and Labrador waters.

SOUTHERN MIGRANTS

Atlantic Mackerel

The Atlantic mackerel (*Scomber scombrus*) is a member of the tuna family, which includes 49 species worldwide.[182] Tuna are pelagic fishes of the tropical and temperate oceans. There are no Arctic or cold-water tuna, but some related species, like the Atlantic mackerel, invade northern waters on both sides of the Atlantic during the summer when water temperatures warm. They typically remain in surface waters with temperatures above 7°C.[183] Tuna species may be found in all of the world's oceans, but the Atlantic mackerel occurs only in the Atlantic. On the western side of the North Atlantic, mackerel inhabit the coasts from the Carolinas to the Strait of Belle Isle; on the European side, they are found in the Mediterranean, on the Norwegian and Icelandic coasts, and are abundant off the British Isles.[184] Like all members of the tuna family, Atlantic mackerel are highly migratory; however, the cold waters of the north prevented any ventures to the Pacific. Atlantic mackerel spawn well to the south of Newfoundland and Labrador. They migrate north in summer to feed in the Gulf of St. Lawrence, and may venture farther through the Strait of Belle Isle and then down the northeast coast of

Newfoundland, especially during warm years. In cold years, no mackerel will reach Newfoundland. As late as the early 1900s, no one knew where mackerel came from: they just showed up in the summer months. One theory held that mackerel ventured only a short ways offshore to deeper water and stuck their heads in the mud for the winter.[185]

Atlantic Mackerel

Mackerel are the speedsters of the smaller pelagic fishes. They migrate thousands of kilometres between overwintering, spawning, and feeding grounds, travelling in very dense schools and at great speed. During the overwintering and spawning periods, they do not feed much, if at all, and inhabit areas with water as deep as 200 metres. In keeping with their tropical heritage, they spawn pelagic eggs in the spring in waters with temperatures greater than 9°C.[186] As spawning ends, the dense schools migrate north, prowling the upper layers of the warming ocean in search of prey. Mackerel are constantly on the move and feed voraciously on small fish and large zooplankton. If their numbers are great, the survival rate of young cod may decline.[187] Mackerel do not waste time in areas with poor feeding. In the northern Gulf of St. Lawrence, marauding schools of mackerel can be found devouring post-spawning capelin near the shore of the Strait of Belle Isle. When the capelin are gone, so too will be the mackerel.

Atlantic Saury

The Atlantic saury (*Scomberesox saurus*) is another small-sized, subtropical migrant. The saury has a long beak formed by both jaws and is known as the billfish in Newfoundland and Labrador. It is a very active fish, sometimes called the "skipper" because of its habit of leaping from the water in great schools to avoid predators. Atlantic saury are most closely related to the flying fishes of the tropics and inhabit warm surface waters with

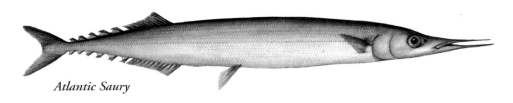

Atlantic Saury

temperatures of about 8-25°C. Spawning occurs well south of Newfoundland, where temperatures are greater than 17°C, and may be rare north of the Carolinas.[188] In these southerly climes, Atlantic saury are fed upon by the dominant predators, including tunas, mackerel, and pollock. Atlantic saury migrate north to feed on zooplankton in summer. It is common on the Nova Scotian banks and in the Cape Breton Island region in summer and early fall. In the warmest years, Atlantic saury may migrate north to the Grand Banks and coastal Newfoundland waters.[189]

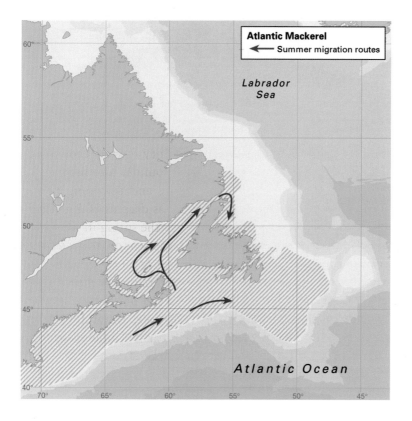

2.40
Atlantic mackerel distribution in Newfoundland and Labrador waters.

Bluefin Tuna

The bluefin tuna (*Thunnus thynnus*) is the largest member of the tuna family. It occurs in oceans worldwide and its ancestors predate the North Atlantic. As the North Atlantic opened, new and rich feeding areas became available, and the migration patterns of many fish species, including the bluefin tuna, evolved to exploit northern waters each summer. This large fish is one of a few tuna species that have specially adapted internal heaters that enable them to venture into cold waters to feed.[190] The advantage to the fish is clear: northern waters contain high levels of potential food, much more than in the tropics. Bluefin tuna are built for speed and possess a specialized, sideways, quick-tail-flick swimming motion. Despite their large size (three-metre-long fish are common), they are quick, although not as fast as their smaller cousin the mackerel.[191] Bluefin tuna are capable of consuming all types of fish, even cod – as a research crew tagging cod on the south coast found out. As each cod was released, the crew heard a slap on the water. The cod were being picked off one by one by a large bluefin tuna. Needless to say, these were tags that would never be returned. Despite the bounty of summer food, bluefin tuna do not stay long in northern waters as they require very high water temperatures, such as those found in the Caribbean, for successful reproduction.

Bluefin Tuna

Dogfish

Dogfish (*Squalus acanthius*) are small sharks – the most numerous of all shark species. The ancestors of these boneless fish with non-movable gills have inhabited the oceans for hundreds of millions of years and predate all the bony fishes. In the northwest Atlantic, the centre of dogfish distribution is well south of Newfoundland in the coastal waters from New England to the Carolinas. Spawning takes place from fall to winter. No births take place in the coastal waters of Newfoundland and Labrador.[192] However, in some years huge schools migrate to Newfoundland waters to feed in summer on capelin and herring, and occasionally venture as far north as southern Labrador, but seldom if ever north of Cape Harrison. They can move quickly – up to 13 kilometres per day.[193] In years when they are abundant, they appear first on the south coast in mid- to late June, then on the west coast in early July. By late July to early August, dogfish may move around the northeast coast and by September be off Labrador, perhaps following the capelin. By October, they will be scarce in all areas, although a few may stay year round.[194]

Dogfish

Cod: The Ecological History of the North Atlantic Fisheries 121

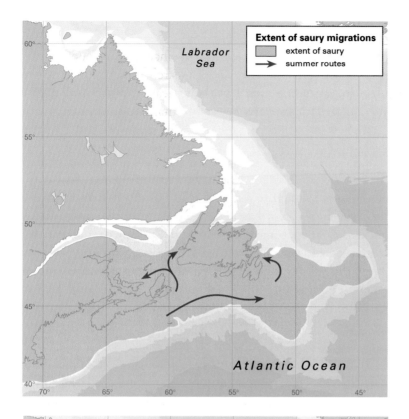

2.41
Atlantic saury distribution in Newfoundland and Labrador waters.

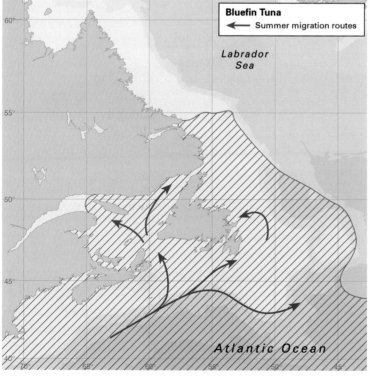

2.42
Bluefin tuna distribution in Newfoundland and Labrador waters.

ANADROMOUS FISH

Anadromous fish spawn in freshwater but live most of their lives in the ocean. Salmon and several trout and char are anadromous species.

Atlantic Salmon

Salmon and related trout have an ancient lineage in both the North Atlantic and Pacific oceans. Fossils have been found that are similar to modern Atlantic salmon that date back 10-20 million years. Although the fossil record is too incomplete to allow any direct tracking of salmon evolution, it is clear that a protosalmon was present well before the earliest ice ages. The classification of salmon and trout has changed greatly over the past 200 years. In the early 1800s, all the Atlantic and Pacific salmon species were thought to be close relatives,[195] but that view was modified substantially in the mid-1800s. Modification of classifications continues to this day: the rainbow and steelhead trout (*Oncorhynchus gairdneri*) were recently transferred from the Atlantic salmon group to the Pacific group. More genetic research has been done on salmon than on any other fish group, and has clarified some of these relationships, but the record is still far from clear. It was thought that the anadromous salmon and trout evolved from a purely marine form, but that may not be the case; the debate over the marine or freshwater origin of salmon has yet to be concluded.[196]

Atlantic Salmon

Although the ancient history of the salmon remains a matter of controversy, there is little doubt that the modern distribution came about very recently, since the Wisconsin retreat, and that the current populations were established within the past 10,000 years.[197] During the ice ages, glaciers would have completely blocked the spawning rivers of the Atlantic salmon (*Salmo salar*). The species apparently took refuge in the rivers of New England and the mid-Atlantic American states, south of the leading edge of the glaciers. As the ice receded, Atlantic salmon recolonized northern rivers in Newfoundland and Labrador. Atlantic salmon, and perhaps most salmonids, were able to colonize new habitats quite effectively and quickly. Salmon species have been transplanted all over the world to habitats as remote as mountainous regions of Kenya, rivers of New Zealand and South America, the North American Great Lakes, and streams of the Avalon Peninsula in Newfoundland. This ability to adapt to a wide variety of habitats appears to be part of their genetic make-up, forged by the necessities of their history.

Atlantic salmon spawn in hundreds of large and small rivers in Newfoundland and Labrador. Salmon young, or fry, feed mainly on insects and small crustaceans. During their first few months at sea, they gradually switch their diet to fish – primarily capelin and sand lance. Their survival is likely related to climatic conditions in coastal waters and in the Labrador Sea, with noted declines during cold periods.[198] In the northeast Atlantic, lanternfish (myctophids) and euphasiids appear to be important in the diet of the adults at sea. Like cod and many other species, adult salmon in the northwest Atlantic depend heavily on capelin for food.

The migrations of Atlantic salmon take many of the fish to the Labrador Sea and to the south and west of Greenland.[199] They follow a counter-clockwise route with the gyre of the Labrador Sea. In the area of west Greenland, salmon from both North America and Europe are present. Fish from many rivers, both American and European, utilize the same feeding grounds in the Labrador Sea and then somehow know how to find their way home. The migrations of Atlantic salmon in the open ocean are most impressive. Fish tagged at sea north of the Faroe Islands in the northeast Atlantic have been caught as far away as Atlantic Canada, and presumably would have made it back to their home stream there.[200] They swim near the surface most of the time, but take some excursions to deeper waters. It is not entirely clear why Atlantic salmon undertake such long migrations or what advantage it offers them, although it is almost certainly related to temperature conditions and feeding. The migration may have ancient roots that date back to the opening of the North Atlantic. Most Newfoundland and Labrador salmon overwinter in the Labrador Sea and across the eastern and southeastern edge of the Grand Banks. This distribution parallels the ice edge during the maximum glaciation of the Wisconsin and may be a ghost of former restrictions to their range.[201] Extensions northward to west Greenland in summer are likely responses to favourable temperatures and food.

Arctic Char

The Arctic char (*Salvelinus alpinus*) is related to the brook and lake char or trout (*Salvelinus fontinaliis* and *S. namaycush*, respectively). Char are basically freshwater species that like cold waters, but they utilize the ocean in some places. The Arctic char is the most cold-loving, and also the most marine. They are plentiful in the rivers of northern Labrador and the Arctic, all the way to the Mackenzie River delta in the Northwest Territories of Canada. Farther west, the related Dolly Varden char dominates and may have taken over from earlier distributions of Arctic char.[202] It is likely that the char evolved

2.43
Atlantic salmon distribution and migrations in Newfoundland and Labrador waters (data from Reddin and Friedland 1993 and other works by these authors).

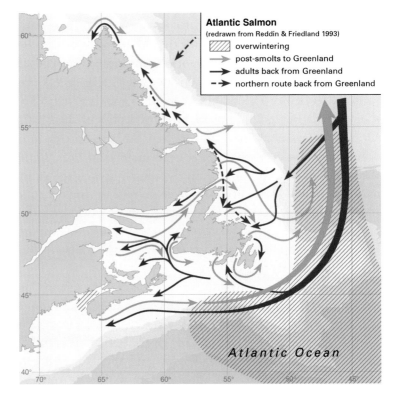

2.44
Arctic char distribution in Newfoundland and Labrador waters.

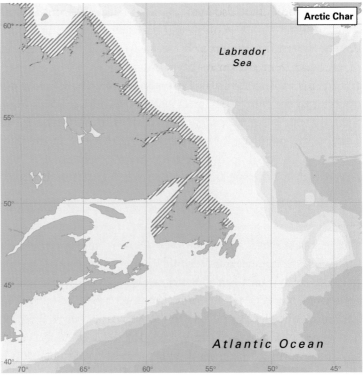

from a common ancestor with the Pacific salmon.[203] Being primarily freshwater forms, they would have been driven south during the ice ages and recolonized the northern regions in the wake of receding Wisconsin ice in the past 10,000 years.

Like the salmon, the Arctic char migrates to sea to feed; unlike the salmon, it does not undertake prolonged journeys. The Arctic char migrates within coastal waters, seldom venturing far from its home river.[204] It feeds on pelagic fish, primarily capelin off Labrador, and other organisms. Change in capelin distribution has been shown to influence char feeding,[205] so it is likely that migrations in this region are a response to capelin availability in the coastal oceans.

Arctic Char

The brook char, also known as the speckled trout or mud trout (in Newfoundland), will run to sea to feed from many rivers and streams in Newfoundland and Labrador, and is often caught by anglers at river mouths as sea-run trout. Sea-run trout grow to a much greater size than stream or pond fish. On the Avalon Peninsula, and some other areas, there are introduced rainbow (*Onchorhynchus mykiss*) and brown trout (*Salmo trutta*) that also run to sea from some streams.

CATADROMOUS FISH

Catadromous fish spawn in the ocean but spend much of their life in fresh water. This type of life history is generally uncommon: in Newfoundland and Labrador ecosystems the only catadromous species is the Atlantic eel (*Anguilla rostrata*) that inhabit the rivers and ponds of northeastern North America and Europe as adults. Their origin confused generations of investigators;[206] it was not until the 1920s that it was realized that all of these eels spawn in the Sargasso Sea in the mid-Atlantic Ocean, and make long distance migrations both in their early life to the rivers of both continents and as adults back to the spawning grounds.[207] In the words of Léon Bertin, "…there can hardly in the whole of natural history be a more remarkable example of response to environment –

to temperature, to salinity, to light, and to current."[208] Tectonics might also be added to this list. The remarkable life history of the eel depends on the Gulf Stream and North Atlantic Drift, which carry the young eels from the spawning grounds to their destinations in the rivers of North America and Europe. The Gulf Stream developed only within the past 12 million years and its flows may not have been fully developed until much later. Prior to this time, the eel life history as it is today likely did not exist.

Many rivers, creeks, and ponds of Newfoundland that have access to the ocean have been home to adult eels. These habitats were obliterated during the ice ages; adult eels recolonized these areas after the retreat of the Wisconsin ice.

CRUSTACEANS

The modern crustaceans, including the copepods, shrimp, lobsters and crabs, represent some of the most important species in the oceans. There are over 30,000 freshwater and marine species of crustaceans.[209] Crustaceans evolved long ago from a primitive marine hard-bodied, trilobite-like, bottom-dwelling animal.[210] Fossils of this ancient species are numerous in shale deposited in British Columbia over 400 million years ago.[211] Crustacean evolution has produced a variety of forms, from parasitic copepods (*Laernia* spp.) that suck blood from the gills of cod to free-living copepods (*Calanus* spp.) that comprise the most important food for many fishes in the North Atlantic to commercial species such as shrimp, snow crab, and lobster.

An interesting feature of crustacean evolution is the development of planktonic and widely dispersive life stages in many species whose adults are relatively sedentary bottom dwellers. This evolution has enabled some crustaceans to colonize the seas almost without limit.[212] Nevertheless, there are many more endemic species of crustaceans in the North Pacific than in the North Atlantic. For example, there are 89 endemic species of shrimp in the North Pacific, but only 6 in the North Atlantic; 14 shrimp species occur in both, including the all-important Pandalid northern or ice shrimps.[213] There are 13-15 endemic species of near-shore crabs of the genera *Cancer* and *Lithodes* in the North Pacific, but only 2-4 in the North Atlantic. No species of crab occurs in both.

Copepods

Copepods are small crustaceans[214] that are numerous in all oceans; there are 8000 species in marine and fresh waters worldwide. Most species of copepods are marine and planktonic – that is, they float in the upper waters of the sea. The Atlantic has fewer shallow-water (less than 100 m) species than the Pacific; endemic species account for about 20%-25% of the total number of species in both regions. Species numbers are greatest at depths from 100 metres to about 1000 metres. Deepwater copepods are generally not widely distributed – about 94% of the North Atlantic species are endemic.

Fossilized copepods are rare, but it is likely that they are as ancient as most crustaceans. Among the copepods with the most primitive features are the Calanoids; they include the most important species in the North Atlantic. *Calanus finmarchicus* is one of the dominant copepods of the North Atlantic, at times comprising over 90% of the copepod biomass in Icelandic waters.[215] This is a relatively large copepod, with a length of two to three millimetres. It has a bullet-shaped body with the first appendages longer than the body and the second shorter. Like most free-living copepods, it is a vegetarian; it feeds on phytoplankton, the tiny plants of the ocean. Most copepods are older than the Atlantic, but *C. finmarchicus* may not be, given the high rate of endemism.

Not all copepods are free-living – some are parasitic and feed on fish. In the waters of Newfoundland and Labrador, the parasitic copepod *Lernaeocera branchialis*, or "lernea," feeds on cod and flounder. Its early life stages are much like other copepods, but the parasitic adult female attaches itself to the gill arch of the fish to suck blood.[216] Not content with taking blood from the gill surface, it grows a long hose-like tentacle that penetrates the main artery of the fish and may extend all the way to its heart. The lernea ends up resembling a bloated and grotesque blood-filled worm with attached egg sacs. Infection rates are much higher in coastal than bank cod.

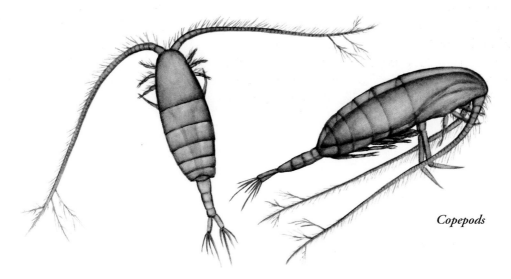

Copepods

Euphasiids

Euphasiids are often called krill, a Norwegian word for these small shrimp-like animals. Krill look similar to shrimp, but are smaller and more closely related to lobsters. They have no fossil record, but their larger relatives have been observed as far back as the Age of Fishes in the Devonian period. There are only 85 species of euphasiids, all of which are marine species. They are mostly tropical in distribution. Sixty-two species occur in the Atlantic, but only 14 of these are found north of 40° north latitude, and only six breed there. Seventy-five of the 85 species are found in the Pacific, which has twice as many endemic species of euphasiids as does the Atlantic. Euphasiid-like animals are an old and numerous life form: they have been present on earth for more than 300 million years. Although there are a relatively few species, it is likely that euphasiids or their ancestors were present before the cod ecosystems of the North Atlantic developed. They may have been among the first arrivals as the North Atlantic opened.

Thienemann's Law applies well to euphasiids. The huge abundances of krill that developed in the Antarctic Ocean, and to a lesser extent in the Arctic and around Newfoundland and Labrador, were achieved by only a handful of species, and in some cases by only one. There are many more species of euphasiids in subtropical and tropical waters than in cooler areas, but their abundance in warmer waters may only be 25% that of the polar regions.[217] Ecologically important euphasiids in the North Atlantic include *Thysanoessa inermis* and *Thysanoessa raschii*; these are a primary food for many whales and fishes. *T. inermis* and *T. raschii* also occur in the North Pacific and may have originated there.[218] Along with copepods, euphasiids provided the basis for the burgeoning pelagic fish biomass that developed in the North Atlantic.

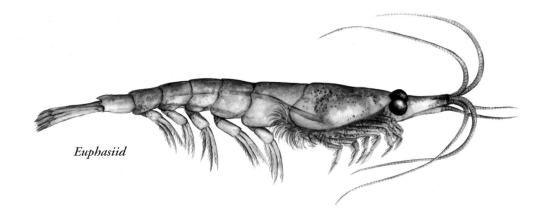

Euphasiid

Decapods

Decapods are a large and diverse group of crustaceans that include shrimps, crayfish, crabs, and lobsters. Decapods are named for their legs – decapod means "ten feet." Decapods have ten legs, five on each side, but in some species they may not all be functional. The first three pairs are used for feeding.[219] There are approximately 10,000 species of decapods in marine, freshwater, and terrestrial habitats. Decapods are an ancient order: a fossil decapod that looks much like a modern crayfish or small lobster was present in the Age of Fishes, 350 million years ago.[220] Fifty-six percent of all decapods occur only in the tropics; the greatest number of decapod species occurs just north of the equator in all oceans.

There are two types of decapods: swimmers and crawlers. The swimmers (e.g., shrimp) use large, sometimes feathered abdominal paddles for propulsion. The crawlers don't have such specialized paddles, but some manage to swim, if awkwardly, using their tails. The crawlers include the really big crustaceans, such as the Alaskan king crab, the snow crab, and the lobster, and can reach masses over 15 kilograms. The giant Japanese spider crab (*Macrocheira kaempferi*), the titan of this body form, has legs spanning two metres in length. Both types of decapod feed on dead animals, but are capable of securing live food as well. In particular, shrimp are efficient predators that migrate well up into the water column to feed on planktonic organisms. Only a relatively few species of decapods occur in Newfoundland and Labrador waters. These include the Pandalid shrimps, the snow crab, and the lobster. The first two at least have achieved widespread abundance, and again demonstrate Thienemann's Law.

Pandalid shrimp. In the North Atlantic, there are two species of Pandalid shrimp: *Pandalus borealis*, the most numerous, also known as the northern or ice shrimp, and *Pandalus montagui*, sometimes called the striped shrimp. Both species have very close relatives[221] in the North Pacific, where at least a dozen other species of Pandalid shrimp are also found. The most likely origin of Pandalid shrimp is the Pacific Ocean where they evolved perhaps 10-20 million years ago and then radiated to their present distribution.[222] A common ancestor to the northern and striped shrimps probably entered the North Atlantic from the North Pacific within the past 3.5 million years. The earliest immigrants were likely larvae, which may drift for months prior to settling to a bottom-oriented life.

Most Pandalid shrimp start out life as males. They moult their exterior shells in order to grow, and do so often, especially in the early years. In mid-life they turn into females.[223] Changing sex is not simple, nor does it appear to be a fixed phenomenon. Some shrimp may be female right from the start, and a shrimp that breeds as a male at the age of one to two years may or may not then change to being female. Both the environment and the state of the shrimp population may influence these features of shrimp life.[224]

Northern shrimp spawn in spring to early summer in Newfoundland and Labrador waters. Spawning occurs earlier in the south (March-April) and perhaps a month later off Labrador.[225] The timing of the reproductive cycle varies over the range and depends upon temperatures during the egg-bearing period and at spawning. The female shrimp carries the fertilized eggs attached to her body.[226] The hatched larvae are small and not well developed, and must pass through five moults before becoming adults. During the larval period, they may drift long distances with ocean currents or be retained within oceanic gyres. They are able to control their vertical position in the water as they grow. At the age of one year, most northern shrimp will have settled to the seafloor as juveniles. They settle over a wide range of bottom types and may vary their behaviour depending on whether the bottom is sandy or rocky. Most settled juveniles will become males and will reproduce in the coming season. After one or more years as a male, the majority will undergo a sex-change moult and become functioning females. Their maximum life expectancy may be five to six years.

Although shrimp are considered to be bottom dwellers, this is far from the whole story of their behaviour. Most shrimp migrate vertically at night and sometimes during the day. At times, these vertical movements may take them hundreds of metres off the seafloor. The purpose of these movements is to feed: when they ascend, they gain access to many small planktonic organisms. Shrimp are aggressive predators and do not wait for food to come to them: they make quick darts to capture food, using their tails to flip quickly and their swimming legs for longer excursions. They grow to large sizes in Newfoundland and Labrador waters, although growth rates are slower than in some other areas.[227] Shrimp aggregate in large swarms that may use currents for movement, but can stay in the same areas for many days, despite their vertical migrations.[228]

The arrival of shrimp into the ecosystems of the Grand Banks and surrounding areas brought a new source of food for many species, including Atlantic cod. Next to capelin, shrimp became the second most important prey for cod north of the Grand Banks, especially for juveniles, and provided a buffer against years when capelin were not abundant.

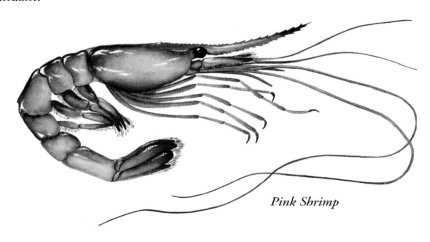

Pink Shrimp

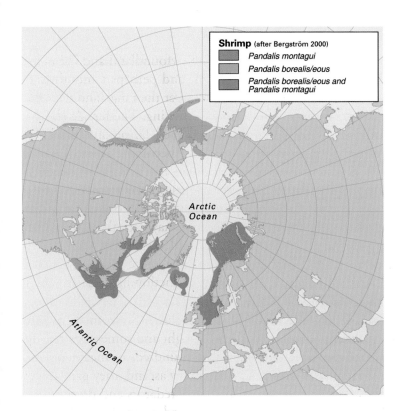

2.45
Northern and striped shrimp (Pandalis borealis and montagui) distribution in northern hemisphere.

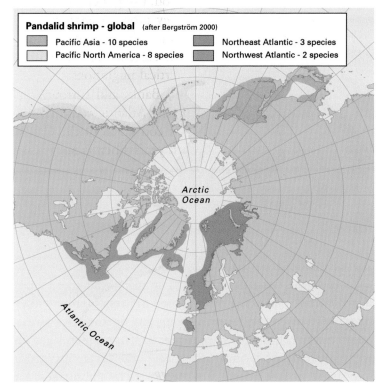

2.46
Pandalid shrimp distribution worldwide.

Snow Crab. Crabs are the most numerous of the crawling decapods. There are approximately 4500 species throughout the world.[229] As with many other types of marine animals, the North Atlantic has many fewer crabs than does the North Pacific, and most originated in the Pacific.[230]

The *Chionoecete* crabs are a small group that originated in the North Pacific. There are four species there, including the giant Alaskan red king and snow crab (in Alaska called the tanner or queen).[231] The snow crab migrated to the North Atlantic relatively recently, after the Bering land bridge collapsed and perhaps later still. The species occurs only in the northwest Atlantic, so it is a fair assumption that the range expansion from the Pacific to Atlantic hugged the North American side of the Arctic. There is little difference between the North Pacific and North Atlantic snow crab today.

Male snow crabs become very large, but the females are small. A large male may achieve a shell width of over 16 centimetres, a leg span of nearly a metre, and a mass of over 1.25 kilograms. Females seldom reach 9 centimetres in shell width or leg spans of over 35 centimetres, and have masses only one-third those of males.[232] Snow crabs occupy muddy or sandy ocean floor at depths of about 70 metres to 400 metres. Juveniles may be found in shallow waters with rougher bottoms. Snow crabs are cold-water animals and seem to thrive at temperatures ranging from -0.5°C up to about 4.5°C.

Snow Crab

Crabs, like all crustaceans, have hard protective shells and must moult in order to grow or reproduce. Mating occurs in late winter or spring. The male holds the female until she moults and then deposits sperm into the female's sperm sacs. A large female may have 150,000 eggs that can either be released soon after mating or stored for a year. The eggs are not released into the water, but rather onto special hairy appendages on the abdomen. Hatching generally occurs in spring. The young crab larvae are about three millimetres long and float to the surface, where they are carried by the currents for three to five months before settling to the ocean floor in September and October.[233] The range expansion of snow crab from the Pacific likely occurred during this highly mobile life stage. Snow crabs go through several developmental stages before becoming adults with the frequency of moulting slowing with age. Between each moult, the animal grows new muscles and shell under the old one. A moulting crab literally backs out of its old shell, and the new shell will be soft and wrinkled for up to two months. With each moult, mature males increase by about 20% in width and 60% in mass. Snow crabs mature at ages of five to eight years.

Snow crabs are remarkably mobile for a bottom-dwelling animal: tagging experiments have shown that males may travel up to 50 kilometres in a year, with most crabs staying within an area of about 2000 square kilometres. The snow crab diet includes mostly bottom-dwelling animals such as seastars, sea urchins, and worms. They will also feed on dead fish when available – herring is often used as bait in snow crab traps.[234]

Snow crab numbers vary greatly over time: there are episodes of very high abundance that are then typically followed by a period of scarcity. These episodes may occur in eight-year cycles in the Gulf of St. Lawrence. The cause of these cycles is not well known, but appears to be related to the biology of the species and changes in the environment.[235] Two species of so-called toad crabs (*Hyas araneus* and *H. coarctatus*) occur in coastal waters of Newfoundland and Labrador. Toad crabs are similar to the snow crab but generally occur in shallower waters. Little is known of their biology.

Lobsters. Lobsters are ancient crustaceans. A "living fossil" was recently recorded off the Philippines with features similar to those of 200-million-year-old ancestors previously thought to be extinct.[236] However, most existing lobsters likely evolved much more recently in concert with changing climates and ocean conditions. Migration, then isolation, may underlie the rather strict divisions of lobsters into specialized habitats. The four major groupings of spiny lobsters that occur in the Indo-West Pacific, east Pacific, west Atlantic and east Atlantic regions may have developed as a consequence of the glaciations of past million years.[237] There is little doubt that the Newfoundland distribution of the American lobster (*Homarus americanus*) in the shallow coastal waters has developed since the last Wisconsin ice retreat, and that the lobsters came from their preferred range to the south. The American lobster likely evolved much earlier from the European lobster (*Homarus gammarus*) or a shared Atlantic ancestor.

Lobsters are widely distributed in the world's oceans in almost all habitats from the shoreline to depths of 3000 metres.[238] There are about 160 species worldwide. Some are long-lived (to at least 50 years of age),[239] and achieve large size. Commercially important species include the American lobster, which occurs on the continental shelf on the North American side of the North Atlantic from North Carolina to Newfoundland. American lobsters are most abundant from New England to the southern Gulf of St. Lawrence, where local densities can be very high, with up to three lobsters per square

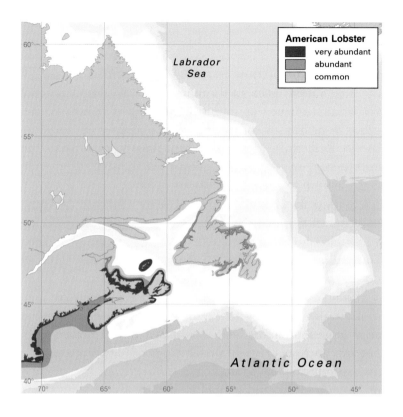

2.47
American lobster distribution in eastern North American waters.

metre. In Newfoundland waters, lobster densities are much lower, with perhaps one lobster per ten square metres in the better habitats.[240] Two similar clawed lobsters occur in the northeast Atlantic: the European lobster (*Homarus gammarus*) and the Norway lobster (*Nephrops norvegicus*). In other oceans, spiny and other clawless lobsters are more common.

Each lobster species has specific habitat requirements, including depth, water temperature, and bottom type. The American lobster lives primarily in the shallow waters of the continental shelf, but may also occur at depths of up to 700 metres. The typical habitat for this species has a mix of sandy and rocky ocean floor.[241] Adults tolerate water temperatures from -1°C to 28°C, but younger life stages require much higher temperatures (10-25°C) and saltier water.[242] This limits lobster distribution in Newfoundland and Labrador and explains why they are at their northern range limit there.

The American lobster in its typical nearshore habitat lives a relatively sedentary life once it has settled in a shelter site. Daily movements may be as small as a few metres. Storms that cause physical disturbances and temperature changes will induce lobsters to move farther from shore into deeper water, but usually not more than a few hundred metres. In Newfoundland waters, lobsters move to deeper water in the fall and return to the shallows in spring.[243] Longer movements and migrations have also been observed: new-shelled lobsters tagged in Port-au-Port and St. George's bays in August 1938 were caught an average of 2 miles (3.3 kilometres) from their release point within a year, but a few moved from 10 to 15 miles (16 to 24 kilometres).[244] Offshore lobsters tend to be more mobile than nearshore individuals. Tagging studies of offshore lobsters in the New England area have shown that some lobsters move as much as 11 kilometres in a day, with one tagged lobster moving 335 kilometres in 71 days. These extensive movements were both to the north and towards shore. It is not known for certain if lobsters return to their point of departure (a migration) or if these are one-way movements.[245]

Female lobsters mate while still in the soft-shell condition after moulting. They carry 5000 to 60,000 developing eggs underneath their tail and are referred to as "berried" lobsters. Lobster eggs hatch into larvae that drift in the currents. Young lobsters are highly cannibalistic – a trait that restricted the success of early attempts to raise them in captivity. The "free-swimming young may kill and suck the blood from as many as four or five young lobsters of its own size in a single day in the crowded condition existing in a hatchery."[246] In turn, small lobsters are preyed upon by fishes, including dogfish, cod, and skates. The large barn door skate is reputed to take even quite large lobsters.[247] Once settled on bottom, lobsters eat mostly less mobile bottom animals, such as crabs and seastars, and will scavenge dead fish and invertebrates. In Newfoundland waters, juvenile crabs make up the bulk of the lobster diet.[248]

CETACEANS

Cetaceans – whales, dolphins, and porpoises – are mammals, not fish, and are most closely related to horses. There are two basic types of whales: toothed whales, such as the sperm whale and all species of dolphins and porpoises, and baleen whales, such as the minke, humpback, fin, and great blue. Whales have a complex and relatively recent evolutionary history that is clearly derived from land animals, but many of the "missing links" have yet to be found.

The most impressive thing about whales is their size.[249] Consider this: elephants seem large, but elephants are mere miniatures compared to some of the extinct dinosaurs. We marvel at the size of the dinosaurs, but the largest animal that has ever lived still lives: this is the blue whale, which can grow to 30 metres length and weigh 300 tonnes. Even the largest dinosaur would be dwarfed by a blue whale, or even by the more modestly sized fin or sei whale. Most of the large whales are baleen whales that feed on plankton and small fish.

The evolution of the whales occurred relatively recently. The earliest whales, which first appeared in the fossil record about 30-45 million years ago, no longer exist. They were smaller than modern baleen whales and more eel-like in body form. Even then, the division between toothed and non-toothed whales was evident. The non-toothed whales feed mainly on plankton, and therefore don't need teeth. But even the non-toothed whales descended from animals like horses with teeth in both upper and lower jaws. During foetal development the long-lost tooth buds on both jaws are clearly evident.[250] Several modern species of whales date from the Pliocene epoch (1-7 million years ago).

Both toothed and baleen whales occur universally in the oceans of the world. Many of the same species occur in the Atlantic, Pacific, and Antarctic regions. In general, their distribution is vast and often worldwide. However, there are some species that occur only in certain oceans and some areas sustain larger populations than others. The most common baleen whales in the North Atlantic are the minke (*Balaenoptera acutorostrata*), humpback (*Megaptera novaeangliae*), sei (*Balaenoptera borealis*), fin (*Balaenoptera physalus*), and blue (*Balaenoptera musculus*) whales. These are all closely related species that evolved within the past 15 million years. Also present are two species of right whale – the bowhead, or Greenland whale (*Balaena mysticetus*), and the Atlantic right whale (*Eubalaena glacialis*). The bowhead whale occurs only in the Arctic Ocean and the Atlantic right whale only in the North Atlantic.

Minke Whale

Toothed whales are more numerous in species and less universal in distribution than the baleen whales. Species that can be found in the North Atlantic, including the waters of Newfoundland and Labrador, are the sperm whale (*Physeter macrocephalus*), the northern bottlenose whale (*Hyperoodon ampullatus*), the long-finned pilot whale (*Globicephalus melas*), the killer whale (*Orcinus orca*), and several species of dolphins and porpoises, including the Atlantic white-sided dolphin, (*Lagenorhynchus acutus*), known as "jumpers" in Newfoundland and Labrador, the white-beaked dolphin, (*L. albirostris*), known as "squidhounds," and the harbour porpoise (*Phocoena phocoena*). Two small Arctic whales, the beluga, or white whale (*Delphinapterus leucas*), and the narwhal (*Monodon monoceros*), may also be found to the north.[251]

Baleen whales feed mostly on euphasiids (krill) and small fish, such as capelin. In the Antarctic Ocean, krill consist mainly of a single species, *Euphausia superba*, and form virtually 100% of the diet of humpback, fin, and blue whales. These large whales transcend a principal organizing feature of food chains: that larger animals eat larger prey. The fact of the matter for whales is that no other food is sufficiently productive to feed them. There are billions of tonnes of krill in the oceans, and each year they produce hundreds of millions of tonnes more. Only such high levels of production could nourish the large whales. If an animal the size of a blue whale exclusively ate larger prey, either there could be only a very few whales or they would exterminate their prey, and hence themselves, very quickly.

In Newfoundland and Labrador waters, where krill are not as abundant as in the seas surrounding Antarctica, some species of whales developed more diverse tastes. Right whales and blue whales are perhaps too slow to feed on anything that moves quickly and feed on krill almost exclusively. The sei whale is a specialized feeder on even smaller *Calanus finmarchicus* zooplankton. However, minke whales and humpback whales have become fish-eaters in the North Atlantic, and even more so around the coasts of Newfoundland and Labrador.[252] Here, these whales feed heavily on capelin, herring, and mackerel. Minke whales will even take cod.[253] That whales can catch these fast-swimming and elusive fish is a testimony to their hunting abilities. Fin whales eat so much herring that the Norwegians call them herring whales (*sillhval*). Humpback whales will occasionally take seabirds,[254] although this may be by accident.

Whales likely invaded the North Atlantic and Arctic oceans as they opened. The attraction was, no doubt, the abundant food found in these waters, particularly around Newfoundland and Labrador. Even after the Wisconsin ice retreated, freezing and icing conditions arrived each winter, and that, combined with the cold waters of the Labrador Current, sent most whales south for the winter. Although there is little food in the tropical oceans – not enough to sustain such large creatures – they provide warm environments suitable for calving, provided that the mother has sufficient northern fat

Humpback Whale

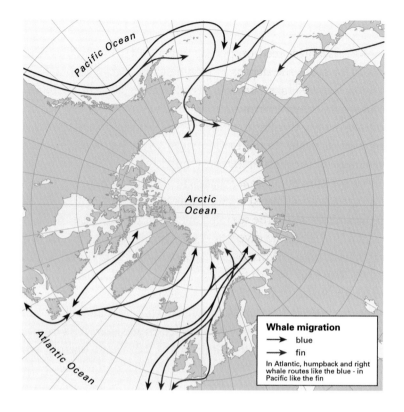

2.48
Blue and fin whale migrations in the North Atlantic and Pacific (data from Slijper (1979)).

reserves from the previous summer to produce milk. So this migration pattern, based on securing huge reserves of fat in northern summers to be used to nourish a tropically born calf, has taken firm hold, and with that pattern these whales have increased dramatically in size over the past millions of years. These giants for the most part cannot live in cold waters year-round. Mostly smaller-sized and specialized Arctic whales, that have evolved very recently, are capable of living year-round in the icy waters.

SEABIRDS

The Newfoundland and Labrador marine ecosystems are home to prolific seabird populations. The productivity of the Grand Banks and adjacent continental shelves resulted in abundant food for surface-feeding birds, and the huge schools of capelin, sand lance and herring provided the basis for large colonies of diving birds. Many seabirds have particular breeding requirements, including close access to feeding and isolation from predators. The rough and isolated rocky headlands and islands of Newfoundland and Labrador, close by the migratory routes of capelin and other species, proved to be ideal for seabirds.

Most seabirds feed on pelagic fishes, while a few scavenge at the surface on what has floated up from the depths. The various species of diving birds have different specialties. The "top guns" are the northern gannets (*Morus bassanus*) that target unsuspecting pelagic fish and dive arrow-like from one hundred metres in the air straight into the water, making barely a splash, with better form than any Olympian. Arctic terns (*Sterna paradisaea*), using several breeding islands in Newfoundland during their epic migrations from the Arctic to Antarctic seas, make similar if less spectacular dives from lesser heights. Common murres (*Uria aalge*), thick-billed murres (*Uria lomvia*) – (both known locally in Newfoundland as turrs) – and Atlantic puffins (*Fratercula arctica*) also dive for fish, but start their dive from the sea surface and literally swim down to their prey. They often catch several capelin in one dive and return to the surface with a mouth stuffed with fish. These diving birds have short paddle-like wings which limit their flying abilities. A puffin or murre fully loaded with fresh capelin may not be able to take off at all from the surface of the water. At best, they use a long runway.

Surface-feeding seabirds such as the gulls and shearwaters are designed differently: they cruise the up-and-down air drafts, barely skimming over the waves in search of food. Their wings are large and provide a lot of lift, rather like gliders. Many of these species are such agile and acrobatic flyers that they can course a gale within metres of a raging sea and pick a bit of food off the surface as easily as a human picks a pea from a plate.

In general, seabirds are never very abundant relative to the many species of invertebrates and fishes that live in the ocean. However, on the Grand Banks and the coastal breeding sites in Newfoundland and Labrador, seabirds are decidedly, if not uniquely, abundant. Despite their numbers and voracious feeding, there are not enough seabirds to have a negative impact on the trillions of capelin and sand lance. On the other hand, the success of seabird breeding depends upon the abundance and distribution of these small fishes. If prey is scarce near the colonies, it will have a direct effect on feeding of chicks and, ultimately, on their survival – these birds cannot simply pick up and move.

The four main species of seabirds that characterized the Newfoundland and Labrador oceans are the great auk (*Pinguinus impennis*), the Atlantic puffin, the common murre, and the northern gannet. The great auk, now extinct (see Chapter 6) was a large flightless bird – the penguin of the North Atlantic. Early observers indeed called it a penguin, which it superficially resembled. The largest colony was at Funk Island and numbers may have reached 100,000 breeding pairs.[255] There were also colonies at the Penguin Islands off the south coast, and perhaps farther south at Cape Breton and Cape Cod. Great auks could be seen over the Grand Banks and continental shelf during the non-breeding season. Very little is known of their biology. It is thought that the auk came to its colonies in mid-May and probably laid only one large egg, which was incubated for a relatively short period of four to five weeks. Chicks took to the sea as soon as possible, with reports that auk chicks would ride on their mothers' backs if threatened. Seals may have been their major predator. The food of the great auk was likely predominantly capelin, but perhaps such a large bird would also have fed on larger fish, possibly even relatively rare species such as lumpfish.[256] How a lumpfish, other than a very small one, would have been swallowed, however, requires some imagination.

Murres

Murres are the most common seabird of the Newfoundland oceans. There are two species: the common and the thick-billed murre. The ancestors of these birds occupied the North Pacific; the two modern murres evolved during the recent glacial periods, likely from a common ancestor. Perhaps a single species became separated, one group to the far north, which became the thick-billed species, and the other farther south, becoming the common variety.[257] Since the last Wisconsin glaciation, their distributions have overlapped, with both species occurring in the waters of Newfoundland and Labrador. They often occupy the same colonies, but do not interbreed.

Murres are true seabirds – they spend their lives in the open sea, coming ashore only to breed. Both species of murres are colonial nesters. They use the same sites as did the great auk, but others as well. Funk Island, once perhaps the largest great auk colony, is now largely a murre colony. Other major murre colonies are found at Cape St. Mary's and the Witless Bay islands. All murre colonies are located near capelin migration

Murres

routes. The murres make nests in crevices and on steep rocky islands or mainland capes. Funk Island is flat, an exception to the cliff colonies of most murres, perhaps because of its unique location 50 kilometres from land and most predators. When the great auk dominated the Funk Island colony, murre numbers were likely much smaller. Like the auk, the murre breeds in spring, but spends more time on land than the auk did and does not take its young to sea until they are half-grown. Murres are long-lived and do not breed until they are five to six years old.[258] They build no nests: a single egg per breeding female is laid on a bare rocky ledge. The eggs are almost pear-shaped, an apparent anti-roll adaptation to being deposited on a narrow ledge. Falling eggs are a prime source of early death.[259] Murre chicks are fed by both parents, largely on capelin and sand lance in the south and on Arctic cod farther north.[260] Parent birds fly 5-80 kilometres from the colony[261] and may dive to depths of 180 metres to find food.[262] When the chick is ready to leave the colony, it dives or jumps into the sea from its rocky ledge. If it survives, it will not come back to land for two years when it is ready to breed.

Murre numbers fluctuate in response to the abundance of their prey and other factors. In the Newfoundland and Labrador region, there were large increases in murre numbers over a 20-year warming period from 1940 to 1960: the population of murres grew from 50,000 birds in 1936 to 500,000 birds in 1960. At least 10 million murres inhabited the seas between Newfoundland and the eastern Arctic-Greenland area in the 1960s. By the early 1990s, the numbers had declined somewhat, but still measured almost 7 million.

Atlantic Puffins

The Atlantic puffin occurs only in the North Atlantic. It has three close relatives in the North Pacific, where the original puffin likely evolved. About 15 million puffins inhabit the North Atlantic, with 60% at Icelandic colonies. There may be a million puffins at Newfoundland and Labrador colonies.[263] Puffin colonies require isolated islands away from predators, but puffins use burrows to breed so the islands cannot be entirely composed of rock. The few sites in Newfoundland and Labrador that are suitable have been colonized since ancient times and puffins are remarkably loyal to their colony. As with the murres, puffin colonies are strategically located adjacent to major capelin spawning migration routes. The same islands may support breeding colonies of both puffins and murres. Puffins are long-lived, and may not attempt to breed until they are six to seven years old. In all instances, after spending much of the year away from land, each pair returns to the colony to breed. The female lays one egg which is incubated by both parents. The chick is tended for nearly two months and fed a protein- and fat-rich diet of capelin or sand lance brought back to the nest by its parents. During their hunting forays, puffins may dive to depths of 60 metres.[264] Like all alcids (the family of northern diving seabirds which includes the auks, puffins, and murres),[265] puffins are fast but not agile flyers, their wings being a compromise between a flying wing and a paddle.

They can achieve airspeeds of up to 80 kilometres per hour, but their landings are often ungainly.[266] In winter, puffins range far out to sea, farther from land than murres. The shallow and productive Grand Banks are an important winter area for these birds.

Atlantic Puffins

Northern Gannets

Northern gannets are colonial breeders that nest on remote islands or cliff edges. They are spectacular divers and make picture-perfect dives into the sea from great heights to capture pelagic fishes. In 1834, their numbers in the North Atlantic were about 334,000 breeding birds; two-thirds of these nested on the Bird Rocks in the Gulf of St. Lawrence. These are the same Bird Rocks first described by Jacques Cartier in 1534 as being covered with an infinite number of great white birds.[267]

Gannets may not nest within the same colony from which they originated. Unlike the more site-keeping alcids, gannets may establish new colonies or abandon old ones if conditions are not suitable. In Newfoundland waters, gannet colonies have been successfully and rapidly established at Cape St. Mary's, Baccalieu Island, and Funk Island. The most recent colonization of Cape St. Mary's occurred in 1879, when the first three breeding pairs were observed. By 1883 there were eight to ten breeding pairs and by 1990 the birds were well established. Gannets colonized Baccalieu Island in 1901; by 1941, there were 200 breeding pairs present. At Funk Island, there were no breeding birds in 1934, but a population began in 1936 with seven breeding pairs.[268] Gannets are thus fast invaders, and although the specific reasons for leaving one colony or establishing another may not be known, they are likely related to changes in the state of the ecosystem and its available pelagic fish.

Of the other species that inhabited the marine ecosystems of Newfoundland and Labrador prior to the coming of humans, we know little. There are many species of plankton, crustaceans, echinoderms (e.g., starfishes, sea urchins, and sand dollars), small invertebrates, and fishes that are known only to ichthyologists (scientists who study fish) or other specialists, and few are of commercial importance. Many are rare and form only a trivial fraction of the living matter in the ecosystem. Their ecological niches and the role they play in marine ecosystems are, for the most part, unknown. Even less is known about the benthic species that live on and in the sediments at the bottom of the sea. These invertebrates and fishes have diverse forms and habits, with colourful names such as sea butterfly, pumpkinhead, and sea cucumber.[269] Little-known fishes in Newfoundland and Labrador waters include the wrymouth, pimpled lumpsucker, sea raven, inquilline snailfish, glacier lanternfish, northern alligatorfish, stoplight loosejaw, and the plainchin dreamarm. Little is known about these species beyond the fact of their existence.

Northern Gannets

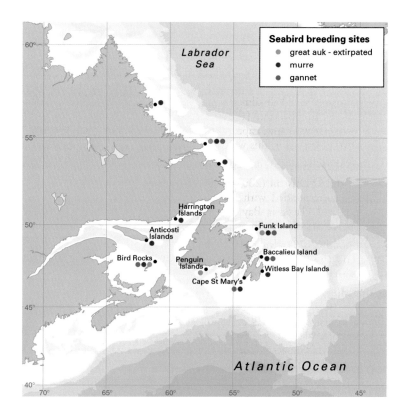

2.49
Major seabird breeding colonies in Newfoundland and Labrador waters.

We can conclude that these ecosystems were occupied and formed by an assemblage of species from several sources, including ancient species, those homegrown in the North Atlantic, and migrants from the Arctic and especially the North Pacific. These ecosystems became productive although never rich in species. A few species achieved great abundance, following Thienemann's Law, including species at all levels of the food chain, from plankton, such as *Calanus*, to benthic invertebrates such as shrimp and pelagic fishes such as capelin, to fishes that feed on the invertebrates and pelagics such as cod and redfish, and top predators such as harp seals. Since the ice retreated from the Wisconsin glaciers about 10,000 years ago, the marine ecosystems of Newfoundland and Labrador burgeoned to carrying capacity of about 50 million tonnes of fish, seabirds and marine mammals among 15 groups of species. It was a spectacular display of marine life.

The coming of humans to these systems began as a slow trickle, with limited impact, but soon became a flood, eventually resulting in the devastation of these ecosystems in the second half of the 1900s. That story will be told in the remainder of this book.

NOTES

1. Astthorsson and Vilhjálmsson (2002) gave a detailed discussion of the simple food chain of the dominant zooplankton (*Calanus finmarchicus*), capelin, and cod, and how one affects the other. When *Calanus* is low, capelin do not do well, and when capelin is low, cod do not grow. There are important lessons in this well-studied Icelandic ecosystem for Newfoundland and Labrador.
2. Vilhjálmsson (2002) showed that cod growth in Iceland is directly related to capelin biomass.
3. Rose and O'Driscoll (2002) showed that the lack of capelin in the diet of northern cod in the late 1990s corresponded with poorer condition and reproductive capacity, compared to cod in Smith Sound and Placentia Bay in more southerly Newfoundland waters that had more ample feeding opportunities on capelin.
4. Diamond (1975) advanced the theory that there are assembly rules of communities that give guidance for how an ecosystem will respond to perturbations such as overfishing. An assembly rule might be that capelin has to surpass some threshold level of abundance to allow cod to achieve large populations.
5. Longhurst (1998) used plankton responses to physical oceanography to describe the biomes of the world's oceans. He did not consider animals above the plankton in the food web. He showed that there were characteristic rates of plankton processes that can be used to characterize these biomes. The Labrador ecosystem was part of his Boreal Polar Province that included the Arctic ocean all the way to Alaska and Siberia. South of the Strait of Belle Isle, the Newfoundland ecosystems were part of the Northwest Atlantic Shelves Province. This characterization was not so different from those described many years ago by Ekman (1953) and Briggs (1974), based on species occurrence and endemism.
6. See Goddard et al. (1999).
7. Rose and Leggett (1989), Rose (1993), and Deblois and Rose (1996) showed that capelin being pursued by migrating and feeding cod move into waters with sub-zero temperatures where cod are less likely to follow them.
8. This phenomenon, with widely ranging harp and hooded seals preying on depressed local stocks of capelin and cod, would come back to haunt the Newfoundland and Labrador marine ecosystems thousands of years later.
9. Vilhjálmsson (2002) showed that growth in Icelandic cod was directly related to capelin consumption, and Gjøsæter (1998) showed similar results for the Barents Sea. Rose and O'Driscoll (2002) illustrated how Newfoundland and Labrador cod show reduced condition and reproductive potential when their diet lacks capelin.
10. Robichaud and Rose (2004) concluded that over 40% of the 174 known cod groups were sedentary and that most of these were small groups. The few large populations were all migratory and supported the main historical fisheries for cod in both the northeast and northwest Atlantic and in Iceland and Greenland.
11. Bundy et al. (2000) estimated that 1 million tonnes of cod would consume about 100,000 tonnes of shrimp annually. This is within the range measured in the West Greenland ecosystem by Hvingel (2003, 3), who noted the effect of cod on shrimp and stated that "when the cod stock declined in the late-1960s, and predation pressure was lifted, shrimp stock biomass increased."
12. See Myers et al. (2001) and Robichaud and Rose (2004) for modelling and empirical studies, respectively.
13. See Robichaud and Rose (2004, 209).
14. Abundance and biomass are like capital or money in the bank, whereas production is the interest earned on the capital. Sustainability implies extracting an amount equal to or less than the interest. If more than the interest, or production, is extracted, the capital, or population, will move towards zero. Fisheries management attempts to set the extraction at an optimum sustainable level.

15 Most observers of the cod-capelin relationship have concluded that the cod follow capelin in the Barents Sea, Iceland, and Newfoundland (Hjort 1914; Hansen 1948; Templeman 1966; Rose 1993; Nakken 1994; Vilhjálmsson 2002). In addition to these studies, there are numerous historical anecdotes of cod following capelin. Spanish fishing reports from 1956 on the southern Grand Bank after a hurricane state: "The preferred food of the cod at this time of year is the capelin…The only species in the stomach from 12th September and onwards was the capelin, and in very large quantities. The effect of the hurricane on the capelin was to make it seek deeper waters, and the cod followed…observed through the fishery" (Rojo 1956).

16 Bergstrom (2000, 215) states: "It seems plausible that more or less localized densities of predators may seriously affect the density of 'settling' juveniles, possibly creating the aggregated distribution patterns often observed in these shrimps. If these speculations hold true, one may postulate that, for example, the recent 'collapse' of the cod stock in large parts of the North Atlantic ought to give rise first to a general increase in abundance of *P. borealis* and second to a less patchy distribution pattern."

17 There is no fixed balance in nature. Ecosystems are constantly changing. Species are subject to modifying influences and may or may not evolve over time within a changing environment. However, given a relatively stable community of species, a dynamic balance will be attained, at least over short time periods.

18 Longhurst (1998, 106) states that "we may expect that the MIZS [marginal ice zones] are a focus of biological activity. Indeed, this is seen in the CZCS [Coastal Zone Colour Scan] imagery for the Greenland and the Labrador currents, both characterized by a receding ice edge in summer and each of which tends to be defined in satellite chlorophyll fields as an area of enhanced chlorophyll."

19 Carscadden et al. (1989) argue that capelin continued to spawn on the beaches of the southeast shoal of the Grand Banks throughout the Wisconsin ice age. With the area being well above sea level during the Wisconsin, and ice not extending that far southeast, this is likely correct. A population of spawning capelin still exists on the southeast shoal.

20 See Gjøsæter (2004) for Barents Sea capelin, Vilhjálmsson (2002) for Icelandic capelin, and also Carscadden et al. (2001). All major stocks of capelin are known to respond quickly to environmental changes.

21 Hansen (1949, 40) wrote of Greenland: "Among the fishes the capelan (*Mallotus villosus*) is the most important. In the early summer, when the capelan in large shoals is spawning near the shore, it is eagerly pursued by the cod, which then becomes choke-full of this small fish. Both when capelan before spawning swims in shoals over the deeper parts of the fjords and coastal waters and when it gathers near the shore to spawn, one can follow the cod hunting after them right up to the surface; frequently the cod spring right out of the water." I have observed the same jumping of cod in the mid-1980s off Blanc-Sablon (from a distance they were thought to be salmon). Cod were clearing the water by a metre, coming up from beneath capelin schools, with the capelin literally spraying in all directions. I have also observed cod breaching the surface in pursuit of capelin near the beaches of Smith Sound in Trinity Bay in June, 2006.

22 Rose (2005) summarized historical data, primarily from Iceland and Norway between 1920 and 1940, on capelin distribution shifts and the shifts in distribution of cod that followed. In those years, capelin moved their spawning grounds north more than 500 kilometres in a warm period. Cod followed. Many other species also moved north; some were new immigrants from the south, while others were local fish that had extended their range.

23 An ecological niche is hard to define, but describes how a species fits into an ecosystem. It describes its "place" in the system, which is thought to be unique, and implies a function. Some ecosystems appear to have more niches than others, with more functions to fulfill and/or more places to fill. Northern ecosystems have relatively few niches.

24 Dragesund et al. (1980) told the story of the decline of this large herring stock by overfishing in the 1960s and the beginnings of its rebuilding.

[25] Robichaud and Rose (2004) estimated the total maximum historical biomass of Newfoundland and Labrador cod stocks at approximately 6.2 million tonnes.

[26] Regarding Gadidae, the family of fishes to which cod belong, Marshall and Cohen (1973, 491) state: "Gadidae proper is virtually unique among fishes in having its headquarters on the continental shelf of the temperate North Atlantic, a region with a fish fauna comprised chiefly of the tag ends of groups having their main centers of diversity farther to the south or in the temperate North Pacific. Apparently much of the evolution of the gadids has taken place in the North Atlantic, the Arctic or in Tertiary locations of the Atlantic which covered parts of Europe. Such diversions as those of the Lota [Burbot] into fresh water, and a few species of *Gadus, Microgadus, Theregra,* and *Eleginus* into the Pacific derive from Atlantic or Arctic sources. Only a few gadids live in the southern Hemisphere."

[27] Svetovidov (1963) gave a full history of herring species, although the book is centred on the fishes of the former U.S.S.R.

[28] Genetic studies of northern gadoid species indicated an evolution within the past few million years (Steve Carr, Memorial University, personal communication), and the more southerly distributed gadoid hakes show their origin to be similarly recent (2-3 million years) and their diversity and disjunct distribution based on evolution in different environments (Grant and Leslie 2001).

[29] Meso-pelagic fishes occur in the midwaters of the open ocean.

[30] Carr et al. (1999, 19) demonstrated the close relationships between gadids in the North Atlantic and Pacific oceans using molecular genetics. They state that: "The Greenland cod (*Gadus ogac*) and the Pacific cod have essentially identical mtDNA sequences; differences between them are less than found within *G. morhua* [Atlantic cod]. The Greenland cod appears to represent a contemporary northward and eastward range extension of the Pacific cod, and should be synonymized with it." It could be the other way around – Pacific cod may be a range extension of the Greenland cod. In either case, the two so-called species appear to be the same.

[31] Steve Carr, Memorial University of Newfoundland, personal communication.

[32] See Carr et al. (1999).

[33] Some related hakes have colonized coastal waters of South America, Southern Africa, and New Zealand.

[34] Svetovidov (1948) relates that although the exact number of species and the grouping of species may be argued by taxonomists, all taxonomists agree that the centre of the cod is in the North Atlantic and pinpoint their origin to the eastern side of that ocean. This matches the tectonic descriptions of the opening of the North Atlantic, with the eastern side opening prior to the western side.

[35] Robichaud and Rose (2004) described 174 cod groups across the North Atlantic, ranging in biomass from a few thousand to 7 million tonnes.

[36] Historically, most Newfoundland and Labrador cod commenced spawning at ages of six to seven years. Rideout et al. (2000) showed that over 30% of adult female cod in Smith Sound, Trinity Bay, may not spawn each year. The proportion of "skipped spawners" may vary between stocks and years for unknown reasons. Stocks with females that do not spawn each year will obviously have lower reproductive output.

[37] Measures of overwintering densities from the Banks are not available, but Rose (1993) provides measures of prespawning densities, which are likely to be somewhat lower, of about one fish per cubic metre. Rose (2003b) showed that overwintering densities of coastal cod in Smith Sound, Trinity Bay, exceed one fish per cubic metre and may approach two to three fish per cubic metre.

[38] See Rose (1993).

[39] Brawn (1961) first observed the courtship rituals of cod in the laboratory. She observed that male cod performed several repeated gyrations to attract a female, and that the final mating and actual spawning occurred pelagically with the female on top and the male lining up the genital pore as best he could from beneath. Hutchings et al. (1999) confirmed these observations and added the notion

that male cod formed leks, or breeding territories, and that females came in and out of the territories to spawn. Recent observations at sea suggest that some aspects of a lek do occur, but that they appear to be movable over the spawning grounds and are not tied to a fixed piece of ground (Windle and Rose, 2006). Hutchings et al. (2000) reported that captive cod spawn near bottom of tanks, but observations at sea do not support this contention for wild cod. In all areas of Newfoundland where spawning has been studied, movements off bottom are the rule, as they are in Norway. Rose (1993) showed evidence of pelagic columns of fish formed during the spawning period in the Bonavista Corridor. Recent evidence that columns are formed by rising pairs of fish has been recorded in Placentia and Trinity bays (unpublished data).

40 Morgan and Tripple (1996) noticed that many trawl catches on the Grand Banks in springtime had high proportions of either males or females. They speculated that males and females behaved differently prior to spawning and were located in different areas. Robichaud and Rose (2002) used acoustic telemetry to record the position of males and females at the Bar Haven spawning ground in Placentia Bay. They found that males did occupy the ground more consistently than did females, with females moving in and out to spawn. Further study at Bar Haven has indicated that the spawning ground may not be fixed in space (Windle and Rose 2005), but that males do form groups which females approach in batches when ready to spawn. Prior to spawning, the ground may be held by groups that are composed nearly entirely of males.

41 Studies of fecundity (the number of eggs) have been conducted in Newfoundland and Labrador waters by May (1966), Pinhorn (1984), and Fudge (unpublished).

42 See Tripple (1998).

43 Windle and Rose (2005) showed a consistent segregation of cod on the Bar Haven spawning ground by size, irrespective of sex. Large males and large females were found together, as were small males and small females. This division was stronger than any male-female divisions attributable to a lek-type behaviour.

44 There are many early descriptions of the size-segregation of cod. Goode (1887, 139), describing the cod in the Gulf of St. Lawrence, stated: "The large cod appear, both in size and habits, to be a distinct school of fish from the smaller kind. Among the latter it is rare to find an individual more than 35 inches long, so that there seems to be no intermediate sizes between the day school of small cod and the night school of large fish. Of the latter, one in rarely taken that would weigh less than 45 to 50 pounds. This is all the more remarkable, since, on all outer fishing banks, there appears to be a regular gradation from the smallest to the largest cod." Rose (2003b) showed that in the aggregation of cod in Smith Sound, there were areas occupied by large fish, nearer the centre, with smaller fish near the edges. When cod migrate, the larger fish tend to lead the way and have been termed "scouts" (Rose, 1993).

45 In southern areas, such as Placentia Bay, St. Pierre and the southern Grand Banks, and as far north as Trinity Bay, cod feed through the spawning season and do not become as exhausted. This occurs because of greater food availability in southern waters.

46 See Rose (1993).

47 Cod have been observed spawning in sub-zero waters in both Trinity and Placentia bays.

48 See Hutchings et al. (1993), Rose (1993), and Lawson and Rose (2000).

49 Rose and Leggett (1988) reported that cod in spring and summer off the Strait of Belle Isle inhabited temperatures from -0.5°C to 8.5°C. Cod on the northeast coast overwinter in waters with temperatures of -1.5°C (Rose 2003b). In Placentia Bay, cod may also overwinter in sub-zero waters. However, the large northern and Grand Banks cod stocks spent most of the year in waters with temperatures of 2-4°C.

50 See Anderson and Dalley (2000).

51 See Pepin et al. (1997).

52 Gotcietas et al. (1997) discussed the use of eelgrass by juvenile cod on the northeast coast of Newfoundland.

53 Anderson and Gregory (2000) discuss the factors regulating survival of juvenile northern cod as well as the potential for cannibalism, especially when numbers of cod are high.

54 Methven et al. (2003) showed the development of shoaling behaviour in cod on the northeast coast of Newfoundland. Juvenile shoals of fish of one to two years of age were reported.

55 Dalley and Anderson (1997) documented the seaward range expansion of young cod from the coastal nursery areas based on an examination of research catches over many years. On the northeast coast of Newfoundland, they move seaward. At age three they move farther offshore, and by age four are ready to join with adult aggregations.

56 Harden-Jones (1968) first indicated that juvenile cod would undertake "dummy runs" to the spawning grounds in the accompaniment of adults. In a study on the northeast coast of Newfoundland, Rose (1993) reported that juveniles might be learning the migration behaviour from the adults they joined at four years of age. DeBlois and Rose (1996) showed that, during migration, larger "scouts" were followed by smaller fish during feeding forays after capelin. McQuinn (1997) developed a similar learning-based theory for Atlantic herring migrations and spawning sites. There is a large literature on the ability of fishes to learn spatial orientation (Odling-Smee and Braithwaite 2003).

57 The juveniles that join the adults and either become adults themselves or are available to be caught by fishermen are recruits; the process is called recruitment by fisheries science. A full treatment of cod recruitment can be found in Cushing (1995).

58 The carrying capacity of an ecosystem is the upper limit to the total number or mass of a species that can exist within that system.

59 Robichaud and Rose (2004) showed that historical maximum cod stock abundance is a function of the range of the stock, regardless of the type of migratory behaviour.

60 Dalley and Anderson (1997) showed that the distribution of juvenile cod of two to four years of age spreads from the coastal region seaward with each advance in age. By age four, many juveniles join up with adult aggregations and adapt adultlike feeding and migration patterns. They may even join the spawning aggregations, although they stay at the periphery of them (Rose 1993). Since the decline of the cod stocks in the 1980s, the age at maturity has dropped so that many fish now mature at ages as young as 3-4 years off Labrador, and 4-5 years farther south.

61 deYoung and Rose (1993) showed that cod are located farther north in warm years and farther south in cold years, and that warm waters are a necessary condition, but no guarantee, of successful reproduction.

62 See Frost (1938a).

63 Larger cod came from the more southerly stocks, where growth rates are higher. Innis (1940, 4) states: "One caught off the Massachusetts coast in May, 1895, on a line trawl was over six feet in length and weighed 211 pounds. A fish which dressed 138 pounds, and probably weighed 180 when alive, was caught on Georges Bank in 1838; and various weights from 100 to 160 pounds have been recorded. In the Gulf of Maine the average 'large' fish caught near shore weighs about 35 pounds, and those on the Bank about 25. 'Shore fish' in Newfoundland waters average 10 to 12 pounds. In the Gulf of St. Lawrence and on the east coast of Labrador, the fish are of smaller size, while those to the north of Labrador are shorter and thinner than those taken at the Straits of Belle Isle. Grenfell gives as the largest fish caught on the Labrador coast one that weighed 102 pounds and was five feet six inches long. According to him 'The average Labrador cod taken in the trap-net is about twenty inches long and weighs between three and four pounds. Those caught with hook and line in the autumn are much larger and heavier.' Throughout the general range along the Atlantic coast from New York and New Jersey, but particularly from the Nantucket shoals to northern Labrador, the weight tends to grow less as one goes north."

64 Goode and Collins (1887) comment on the feeding habits of Grand Banks cod. On page 168, they state that "the fish is liable to swallow almost anything that comes in its way, so that stories are by no means uncommon of jack-knives lost overboard returning to their owners again when the day's catch of fish was dressed."

65 Bundy et al. (2000) summarized much of the early diet work on cod and other species in preparing a mass balance model. According to these data and the model, capelin historically accounted for 50%-60% of the diet of cod in Newfoundland and Labrador stocks, and for over 90% seasonally in some areas. No other diet source came even close to this level of importance. On the southern Grand Banks, sand lance replaced capelin as the main food. On St. Pierre Bank this may also be true, but few data exist. All other cod stocks in the Newfoundland and Labrador region depend on capelin.

66 Regarding herring movements on the south coast of Newfoundland, Winters (1977, 2401) states: "The reason for such high activity at this time of year is uncertain but may be related to coincidental migrations of cod. In late fall there is a differential emigration of large cod (> 80 cm) from the northern Gulf of St. Lawrence to the southwest coast of Newfoundland, where overwintering occurs. During this period these large cod feed very heavily (85%-90%) on herring (Minet and Perodou 1977). Thus, the high degree of activity characterizing overwintering herring along the southwest Newfoundland may be associated with avoidance behavior induced by the presence of abundant concentrations of predatory cod." More recent observations in Placentia Bay suggest that large cod may target herring during late fall and winter.

67 A genetic cline is a gradual change with distance in genetic properties within a species – it contrasts with an abrupt change in properties between two or more distinct genetic entities.

68 Juvenile cod are more cold tolerant than older fish because they produce higher levels of blood antifreeze. Goddard et al. (1999) showed that cod from the Great Northern Peninsula of Newfoundland had superior antifreeze producing abilities than those from the more southerly Trinity Bay.

69 Robichaud and Rose (2004) provided a description of the cod groups that have been identified in Newfoundland and Labrador waters.

70 Jeffers (1931) and Munn (1922) described the movements of the northern Gulf of St. Lawrence cod stock through the Strait of Belle Isle and around the tip of the Great Northern Peninsula. The likely reason for this migration was feeding on capelin.

71 Rose (1993) described the large aggregations (35-50 kilometres across) of spawning and migrating cod that occurred in the Bonavista Corridor in the early 1990s.

72 Ironically, the waters off Labrador where cod overwinter are much warmer (3-4°C) than those in the southern bays in winter, which typically have sub-zero temperatures.

73 The importance of capelin to the distribution of cod in many stocks, including most of those in the waters of Newfoundland and Labrador, has been known since the first observations of cod. An early written report by Whitbourne (1622), repeated in Innes (1940, 5), stated that "[Cod] take their fills of smaller fishes which they follow there are commonly above thirty dayes together before they hale off from the shore again, and in such manner there come there severall shoales of the cod-fishes in the summer time. The one of them followes on the herrings, the others follow the capling…and third follows the squid." Among the later and more systematic examinations of this phenomenon are those of Templeman (1966), Lilly (1987), Rose and Leggett (1989), and Rose (1993). The dependence of cod on capelin is well recognized in Norway and Iceland.

74 Hutchings and Myers (1993) discuss spawning timing and its variability.

75 Helbig et al. (1997) conducted a detailed exercise on potential drift patterns of cod eggs and larvae on the northeast shelf using information from ocean drifters. There was a southward drift in all cases, and at the temperatures and flow speeds of this region, the drift would cover hundreds of kilometres.

76 Robichaud and Rose (2004) provided full details of these and other cod stocks in the North Atlantic.

77 Innis (1940, 3) summarized earlier knowledge and states that "the cod usually spawns in water less than 30 fathoms deep and apparently in fairly restricted areas." He was partially right in his second point but the first shows an ignorance of the true migratory range of the cod that was widespread in that era. Later work by the Russians, based largely on egg studies, indicated most spawning occurred near the shelf break (Serebryakov 1967). This was likely not the case either. If it were true, most eggs would end up lost in the North Atlantic and the large settlement of juveniles in coastal waters would be impossible.

78 Hutchings et al. (1993) summarized trawl data that indicated more widespread spawning on the shelf and Banks. Rose (1993) observed spawning in large aggregations on the shelf in the Bonavista Corridor in the early 1990s.

79 ICNAF (1964, 39 and 1965, 34).

80 See Rose (1993).

81 Cod will only freeze if they come into contact with ice crystals in the water. Otherwise they can withstand sub-zero sea temperatures even with limited amounts of antifreeze in their blood.

82 Morris and Green (2002) reported the existence of a genetically separate population of cod in Gilbert Bay, Labrador, a small bay to the south of the Hamilton Bank.

83 Some fish have eggs that sink to bottom and are termed "demersal." Other species have eggs that float and are called "pelagic." Some demersal eggs have a glue on them which makes them stick to rocks, gravel or vegetation (e.g. herring and capelin).

84 The easiest way to tell Atlantic from Greenland cod is to look at the lateral line, the sensory organ that runs the length of the body on the side of the fish: in the Atlantic cod the line stands out and is invariably light coloured, whereas in the Greenland cod it is dark and hidden by the more mottled colouration of the fish.

85 It has been assumed by many authors that the ancestor of the Pacific cod was the Atlantic cod or its ancestor (Meek 1916; Ekman 1953; Svetovidov 1963). This could be true, but the Greenland cod might also be the progenitor. In either case, the Greenland cod and the Pacific cod are closely related, perhaps even the same species as indicated by mtDNA characteristics (Carr et al., 1999). The Greenland cod could simply be a Pacific cod that moved to the Atlantic recently.

86 Haddock prefer warmer temperatures than do cod by several degrees. Most haddock are found where temperatures are in the range 4-15°C (Frost 1938; Scott and Scott 1988).

87 Carr et al. (1999) presented genetic data to suggest that haddock separated from Atlantic cod about 8 million years ago, and Atlantic pollock about 1 million years before that.

88 See Frost (1938a, 23-24).

89 See Frost (1938a, 24-25).

90 Meek (1916) gave the range of Arctic cod as "at Greenland, Iceland, White Sea, Spitzbergen, Northern Asia, Bering Sea (rare), and young examples have been recorded from Japan." He does not mention Labrador where this species is also common, as it is in the Baffin Island region of northern Canada.

91 Hop et al. (1997, 71) reported juvenile schools of Arctic cod that feed on Calanoid copepods, and schools of larger adults that feed less often and on larger amphipods and copepods. In general, Arctic cod are slow growing, reaching 22 centimetres at about seven years of age. Fish in adult schools were mostly over 15 centimetres in length.

92 Craig et al. (1982) summarized life history traits of Arctic cod in the Beaufort Sea. They emphasized that this species exhibited "small body size, relatively short life span, early maturity, rapid growth, and large numbers of offspring." They also stressed that varying population levels of Arctic cod may influence their predators, which include several species of marine mammals and seabirds. Gillispie et al. (1997), in another study from Alaska, confirmed that result.

93 Svetovidov (1963, 32) gave a full account of herring distribution and evolution in the world's oceans.

He states: "Of 50 genera and about 190 species, nearly 150 species belonging to 37 genera are restricted to the tropics, with only a number of species belonging to several genera…penetrating the southern regions of subtropical areas of the northern hemisphere. A minor part of the herrings are characteristic of temperate latitudes. These include the genera distributed in the boreal region and in the major part of the subarctic region. About 20 species of 7 genera are encountered here, including species of a number of genera of forms of a number of species of which the main concentration is in more southerly subtropical regions. There is not one species or genus of the herrings that is truly arctic. Only a few temperate-arctic forms of *Clupea harengus* are found in the Arctic."

[94] Svetovidov (1963, 33-34) reported about 25 genera of 100 marine species, 15 genera of 30 freshwater species, and 7 genera of 30 anadromous species. Most of the anadromous species are found in temperate regions, such as the genus *Alosa* (Alewife, Shad).

[95] Some classifications considered the Pacific herring to be a separate species, *Clupea pallasi*. Whatever the classification, there was no doubt that the two groups are close relatives and separated only recently. McAllister (1960, 9) considered the Atlantic and Pacific herring to be subspecies. He stated that "the herring of the Pacific has been regarded as a separate species, *Clupea pallasii* Valenciennes 1847." However, he dismissed this classification in favour of a later grouping of the Atlantic and Pacific herring as a single species, based on work by Svetovidov (1952).

[96] Svetovidov (1963, 52), in comparing the herring to the cod (the true Atlantic fishes), stated: "Of the considerably more warm-loving herrings, one of the most cold-loving species (*Clupea harengus*) has penetrated only into the temperate parts of the Arctic-circumpolar subregion. Among the anadromous herrings there were no cold-loving forms that could penetrate as far north as *Clupea harengus*. This is the reason for the absence of anadromous herrings in the northern part of the Pacific, since boreal herrings could not pass around northern Asia and North America, whereas for anadromous subtropical herrings the appropriate thermal conditions were lacking. The absence of anadromous herrings in the northern part of the Pacific is one of its outstanding characteristics."

[97] Svetovidov (1963, 50) stated: "At present, it is difficult to say when *Clupea harengus* penetrated into the Pacific – whether in the Pliocene or, as assumed by Deryugin (1928 in Russian), later, in the post-glacial 'transgression.' All the factors noted by me as bearing on this problem for *Gadus morhua* are also wholly applicable to *Clupea harengus*, i.e., it is most probable that this species too arrived in the Pacific during the post-glacial transgression."

[98] The schooling behaviour of herring has long been known. Meek (1916, 87) reported that: "We have to remember, in the first place, the strong impulse of the herring to seek protection at all stages by crowding together. They instinctively congregate together for safety and form the shoals so well known in many localities."

[99] Winter observations of herring in Bloody Reach, Bonavista Bay, have shown fish of various sizes in close proximity, with no obvious school structure. During the day, herring were in close contact with bottom, but at night they spread out throughout the water column and almost to the surface.

[100] Winters and Wheeler (1996) provided information on herring reproduction in Newfoundland waters. The timing of herring spawning was not invariant but depended on environmental conditions, especially sea temperatures, in the previous year. Peak spawning occured over at least eight weeks in May and June.

[101] Corten (2002) provided evidence that herring in the North Sea learn their migration routes and spawning areas as they approached maturity and then kept them over their lifetime. McQuinn (1997) provided similar evidence primarily for Gulf of St. Lawrence herring.

[102] Winters and Wheeler (1987, 882) discussed the recruitment dynamics of herring in Newfoundland waters. They stated that recruitment patterns "were found to be determined largely by annual variations in over-wintering temperatures and salinities associated with the Labrador Current." Herring populations that declined to below a critical level will suffer more from the variable survival of young and had fewer recruits to the adult population.

103 Stobo et al. (1982) summarized the stock structure of herring in the northwest Atlantic. Wheeler and Winters (1984) and Moores and Winters (1984) reported on tagging experiments from White Bay to St. Mary's Bay and on the west coast. All herring groups showed a high rate of homing, ranging from 65%-92%, averaging 81%. Wheeler and Winters (1984) indicated that post-spawning feeding migrations take herring northwards from most areas, resulting in mixing of fish from the various bays on northern feeding grounds.

104 Sergeant (1991, 1) cited earlier work and described how the hair seals have been separate from the eared seals for most if not all of their evolutionary history. McLaren (1960) reviewed their history and suggested that the Phocidae likely evolved from an otterlike ancestor in Asia (again suggesting a Pacific origin to the group) that invaded the seas through marine transgressions in the Miocene epoch. McLaren also thought that eared seals and walruses arose from a bearlike ancestor in the North Pacific.

105 Sergeant (1991, 1) stated that "the ancestral harp seal probably evolved during the Pliocene in the North Atlantic," but offered no supporting data for this claim.

106 Sergeant (1991, 1-4) described fossil and anthropological evidence that harp seals occurred much farther south during the Wisconsin ice age and after its retreat until a few thousand years ago. The reasons for the demise of the southern distributions were not clear, but may be as simple as an unacceptably warm climate.

107 Whitecoats are young that have striking white fur prior to their first moult. Raggedy-jackets are moulting young, beaters are moulted young, bedlamers are immature one-year-old seals, and dogs and bitches are adult males and females, respectively.

108 Scott and Scott (1988) reported that only 14 of 177 species of Cyclopteridae (lumpfish and snailfish) occurred in the North Atlantic.

109 Cox and Anderson (1924, 4) first reported on the biology of lumpfish in Canadian-Newfoundland waters. They stated: "The lumpfish is found in the Atlantic littoral of North America, from Greenland to New Jersey, penetrating bays and channels, and even extending its range into the great Canadian inland sea of Hudson Bay."

110 Stevenson and Baird (1988, 5), citing J. Carscadden, reported that: "The pelagic existence of lumpfish in Newfoundland waters is substantiated by the occurrence of the lumpfish, at times in large numbers, in midwater trawl catches during offshore capelin surveys."

111 Blackwood (1983) reported on a tagging study of spawning lumpfish in Newfoundland waters. Most of the tags were recovered close to the release site in subsequent years; 86% of the tags were recovered within 35 kilometres. Studies in Norway and Iceland also suggest return repeatedly to the same coastal spawning areas (homing).

112 Collins (1976, 65) provided some description of the spawning of lumpfish. He stated that: "Observations by divers seem to indicate that egg masses are laid by the females when the water warms up to 4°C, even though they arrive a week or two before in coastal waters."

113 Stevenson and Baird (1988) cited earlier Icelandic work by Thorsteinsson indicating that spawning female lumpfish reach a length of about 45 centimetres by the age of ten years.

114 See Cox and Anderson (1924) and Collins (1976) for more details on the diet of lumpfish.

115 Marshall and Iwamoto (1973, 513) stated that "most of the 300-odd kinds of macrourids lived just above the continental slopes (they are benthopelagic) at depths between 200 and 2000 m." They also indicated that some benthopelagic species occured at depths greater than 2000 metres. Some rattails were more bathypelagic (that is, they live more in the water column) at depths to 5300 metres. One of the best-studied rattails is the hoki (*Macruronus navaezelandiae*) of New Zealand, which supports a large commercial fishery (O'Driscoll et al. 2003).

116 Marshall and Iwamoto (1973, 513) stated: "One outstanding feature of this fauna is that about a half (32 species) lives in the Gulf of Mexico and the Caribbean Sea. Indeed, three out of four of the western Atlantic rattails occur in these subtropical pockets of the ocean."

117 See Katsarou and Naevdal (2001).

118 Marshall and Iwamoto (1973, 515) advanced ideas about present rattail distributions in the North Atlantic. They stated that "at least 33% of the Atlantic macrourids occur on both sides of the ocean." They then posed an interesting question: "Why is this proportion of amphi-Atlantic species less for crabs (8%) and echinoderms (15%)?" Crabs in particular have a very dispersive larval life stage.

119 See Murua et al. (2005) for a review of grenadier biology on Grand Banks.

120 Marshall and Iwamoto (1973) indicated that the onion-eye grenadier has eggs of 3.4-3.85 millimetres in diameter. These are large eggs. By comparison, cod eggs are less than one millimetre in diameter. Newly hatched larvae of this grenadier may be 10 millimetres in length.

121 Marshall and Iwamoto (1973, 508), citing earlier Russian work, gave the diet of the roundnose grenadier in Newfoundland and Labrador as shrimps, amphipods, and cumaceans, with lesser quantities of cephalopods and "luminous anchovies" (presumably myctophids). In a Norwegian fjord, the same species fed on similar items, but also on capelin. It is unlikely that capelin are a major part of the diet over most the range because the species do not generally overlap.

122 Robards et al. (1999, 3) stated: "Although 23 nominal species of the genus have been described, current literature consistently recognizes only six based on a combination of genetic, meristic, ecological, and biological parameters."

123 Winters (1983, 409), citing earlier sources, stated: "*Ammodytes dubius* is the single most important prey species of American plaice (*Hippoglossoidees platessoides*) comprising nearly 80% of the food spectrum of that species in the southern Grand Banks area. Sand lance also contribute to the diet of the yellow-tail flounder (*Limand ferruginea*) on the Grand Banks although not to such a substantial degree as in American plaice. In addition sand lance form the dominant prey species of cod in the southern Grand Banks area and are also important in the diet of cod in the northern Grand Banks area."

124 Winters (1989) described how, in coastal Newfoundland, the American lance has a slightly earlier spawning season and maturity than the northern lance, and there were evident differences in the growth rates between species. Winters and Dalley (1988) also argued that the American and northern lances were reproductively isolated in Newfoundland waters, but offered some evidence that the American lance and the lance in coastal Greenland and southwards from Georges Bank were the same species as the lance which occurs in the northeast Atlantic. This must be regarded as speculative.

125 Winters (1983, 410) summarized bottom trawl surveys conducted from 1959 to 1980. He stated: "The highest densities of sand lance from 1959 to 1980 occurred on top of the Bank[s] with abundance levels decreasing towards the slopes...Sand lance appear to be most abundant in the eastern half of the Grand Bank[s], particularly in the northeast portion and around the Southeast Shoal area. These areas...are predominantly comprised of sand and fine gravel." Winters mentioned that sand lance appear to inhabit colder waters on the Grand Banks than in other areas. He stated that "they also tend to prefer cold water, being abundant mainly in the temperature range of -1°C to +2°C."

126 Winters (1983) reported that Grand Banks lance mature at ages of three to four and may live up to ten years of age, which is old for sand lance but not inconsistent with lance having a lower productivity in these cold waters.

127 Springer and Speckman (1997, 774) used this term. Their work focused on the North Pacific, where lance may be more important than capelin to many seabirds, at least at times. They stated that: "Sand lance is a quintessential forage fish. As a group of very closely related species, it is possibly the single most important taxon of forage fish in the Northern Hemisphere."

128 Nafpaktitis et al. (1977) provided detailed descriptions of the myctophids and their lanterns (or photophores), in addition to providing some aspects of their biology.

129 See Briggs (2003). His conclusions were based on earlier Russian work.

130 Backus et al. (1977, 271-279) showed 17 species in their Atlantic sub-Arctic region, which included

Newfoundland and Labrador waters, but state that: "The most striking aspect of the sub-Arctic region is its very low diversity. Midwater-fish catches in this region are dominated by *Benthosema glaciale*, which comprised 96% of all myctophid specimens taken in the Woods Hole Oceanographic Institute shallow (< 200 m) nighttime collections." Most of the fish were caught at night because of the vertical migration behaviour of most myctophids.

[131] Backus et al. (1977, 271), in a summary of the Woods Hole midwater trawl data for the period of 1961-1974, which covered much of the Atlantic ocean, reported 17 species of myctophids in the Atlantic subarctic, 213 species in the North Atlantic temperate region (south of New England), and 169 species in the North Atlantic subtropical region. However, the volume of myctophids per hour of fishing was much higher in the subarctic, at 554, than in any other region. The second highest was in the West Mediterranean Sea of the temperate region at less than 200 measurements of specimens per hour of fishing were more equal, with 529 in the subarctic, 519 in the northern gyre of the temperate region, 438 in the West Mediterranean, and lesser but substantial numbers elsewhere. This indicated that northern myctophids may not always be more numerous than in other regions, but they grow bigger.

[132] Vast layers of myctophids and their apparent associates, the euphasiids, can be observed off the edge of the Newfoundland and Labrador shelf. The layer typically starts at about 350-450 metres, and extends downward at times to near 1000 metres or more. Layers over 500 metres deep can be commonly observed.

[133] In an attempt to describe the feeding relationships of species within the range of the northern cod, Bundy et al. (2000) put myctophids into a broad group that includes herring and other species such as squid and saury. They state that "in the absence of further information, a biomass range is used, from 352,500-587,500 tonnes, or 0.712-1.187 tonnes per square kilometre." This is really a guess and may be far too low, given the extensive distribution of myctophids along the shelf edge.

[134] The origin of these interesting fishes has long been questioned. Meek (1916) stated: "The family appeared much earlier than the Eocene, and was, as will be seen, generally distributed in the Cretaceous period. The evidence of primitive forms and of distribution indicates that the family originated in the East Indian region."

[135] The classification of the flatfishes is not simple. The one suggested in Norman (1966) is used here.

[136] Meek (1916, 258) stated: "It is interesting to observe that in this case, as in the case of other families which predominate in the Pacific, there is an absence of fossil remains. The Atlantic genera appear, therefore, to have come from the Pacific. The Atlantic species are all northern in habitat. In the meantime, in the Pacific region the sub-family has had time to become modified, and has the primitive Atheresthes and Psettodes of the tropical waters, and Paralichthys, which gained the Atlantic before the Isthmus of Panama was formed."

[137] Norman (1966, 293) stated: "In Europe this species attains a weight of at least 500 [pounds]. There is in the British Museum a cast of a specimen which weighted about 456 [pounds] when ungutted. The length of this fish was nearly 8 feet." I have seen specimens in the northern Gulf of St. Lawrence that were more than two metres in length and weighed more than 150 kilograms.

[138] These are the arrow-tooth halibut or flounder (*Atheresthes stomias*), and similar species in Japanese waters.

[139] Meek (1916, 257) stated: "In the Atlantic it extends from Greenland, Iceland, and the Murman Sea south to Sandy Hook (New York) on the west, and to the Channel (English Channel) on the east. It is plentiful at Iceland, Faroes, north and west of Scotland, northern North Sea, the Norwegian coast, and the Skagerak."

[140] See Scudder (1887, 119).

[141] Bowering (1983) reported that males and females grow at the same rate up to five to seven years of age, but after that females grow faster and live longer. All fish over 90 centimetres are females. Fish of 110 centimetres and up to 20 years of age have been recorded.

[142] Meek (1916, 260) stated that "in the absence of a full knowledge of the relationships of the species [it is difficult to know] that the sub-family was present in both the Atlantic and Pacific during the

Tertiary era but the facts of distribution may be said to indicate that Pleuronectes passed from the Atlantic to the Pacific, probably north of Asia, during the period of the post-Glacial submergence." There was evidence against this, and no universal agreement.

143 Walsh (1992) reported on the shallow depths and cold temperatures at which juvenile yellowtail are found on the Grand Banks. Despite being a bottom fish, Walsh and Morgan (2004) used data-storage tags to show that yellowtail will at times swim well off bottom, likely in search of more pelagic food.

144 Simpson and Walsh (2004) reported centres of distribution for yellowtail within the oceanographic gyre on the southern Grand Banks. As the population grew, additional grounds around the area of concentration were utilized.

145 Norman (1966) listed three subspecies of American plaice from the North Pacific and two in the North Atlantic. Earlier classifications differed from this; the exact relationships among both the Pacific and Atlantic species are uncertain.

146 The common plaice of the North Sea and northeast Atlantic is not the American plaice, which is known as the sand dab, but another flatfish, *Pleuronectes platessa*, which does not occur in the northwest Atlantic.

147 See Morgan and Colbourne (1999).

148 McAllister (1960) called the Atlantic variety the smooth flounder, but added that "this species is questionably distinct from *glacialis*" (the Arctic flounder). Norman (1966) termed the Atlantic fish the eel-back flounder.

149 See Bowering (1989, 1990).

150 See Ruppert and Barnes (1994, 896).

151 Mahoney and Buggeln (1983, 1) spoke of the "years of great slime" in Nova Scotia. In Newfoundland and Labrador, local knowledge tells of years and periods of slub related to local ocean conditions. Taggart and Frank (1987) showed that slub occurrence in coastal waters was related to wind-driven upwelling that washed the tunics ashore and onto fishing gear.

152 Capelin are rich in oils and lipids (Montevecchi and Piatt 1984). Lipids are the limiting factor to marine fish growth and reproduction (Bell and Sargent 1996).

153 Carscadden et al. (1989) provided a detailed description of the spawning modes of capelin and their geographical description.

154 Carscadden et al. (1989) described earlier work suggesting a more recent incursion of beach spawning capelin from the Pacific, as well as genetic work that does not support that contention. They provided a theory that beach spawning persisted through the ice ages in Newfoundland waters, but has been lost in capelin in the northeast Atlantic.

155 See Nakashima and Wheeler (2002). I have observed capelin spawning in 5-15 metres of water off Blanc-Sablon, and in Trinity and Conception bays, as well as on the beaches in those areas.

156 George Jeffers was the first to study capelin in Newfoundland. His doctoral thesis at the University of Toronto (1931) is a classic reference. Regarding the spawning process, Jeffers states: "The school swam gracefully back and forth just beyond the crests of the small waves that were breaking on the beach...The males were more active in seeking a mate...When a female was found the two became attached side by side and quite often a second male attached itself to the free side of the female, and the three together rushed up the beach...they go up the beach as far as they possibly can get in this way and then settle in one spot as the wave recedes, all the time using their fins and tails with great rapidity. In this way they scoop out a slight hollow in the soft sand as if trying to bury the eggs as far as possible, and the vigorous action of the fish in the tiny puddle of water thus formed can be distinctly heard. After separating the capelin lie still on the sand for a second as if exhausted before starting to paddle furiously in an attempt to regain the water, for by this time another wave has rolled up the beach."

157 During the 1980s, extensive work on capelin spawning and reproductive success was conducted at Bryant's Cove in Conception Bay by the research group of W.C. Leggett at McGill University. See Leggett et al. (1984) and other papers by these authors.

158 I have had many interesting discussions with fishermen in Newfoundland and Labrador who did not realize that the familiar whitefish were juvenile capelin.

159 Year-round acoustic observations in Smith Sound, Trinity Bay, have shown that at times in spring juvenile (and adult) capelin will fill the Sound with schools of all shapes and sizes. A month later, however, not a school can be found (unpublished research).

160 Nakashima (1992, 2423) presented evidence on movements of adult capelin along the coast. He states that: "Capelin released in a particular bay were recaptured from the same bay [at] locations farther north. Upstream migration using the Labrador Current was hypothesized as a directional clue to the prespawning migration." Observations with echo sounders have tended to confirm Nakashima's hypothesis about the use of the Labrador Current by capelin in their migrations.

161 O'Driscoll et al. (2001) studied the diet of capelin in the Newfoundland region in the late 1990s.

162 Bundy et al. (2000) summarized information on the amount of capelin consumed by various predatory fishes and marine mammals in the northern cod ecosystem. In brief, cod were by far the most important predator and up to 50% of the annual energy budget of cod came directly from capelin.

163 Thompson (1943, 84) stated that "[t]here is no doubt that cod in the region mentioned [Newfoundland and Labrador] depend upon caplin for the period of gross feeding and recuperation which succeeds the spawning season." More recent research has shown that Labrador cod do not regain condition after spawning without capelin as a major part of the diet (Sherwood at al., in press).

164 Barsukov (1959, 97) provided a discussion of wolffish distribution and evolution. He suggested a Pacific origin and a migration into the North Atlantic during the Pliocene epoch. He stated that: "the original species which erected the family has an eel-like body form and a weakly differentiated dentition. Like *Anarhichthys ocellatus* it was a temperate coastal Pacific species. Another species which erected the genus *Anarhichas* separated itself at some point of time from the original species in the northern part of their range. In the body form and the number of vertebrae this species was close to *A. orientalis*, and in the teeth structure to *A. latifrons*. It penetrated into the Atlantic Ocean probably not later than the Lower Pliocene." This must be regarded as speculative. In any event, the similarity of the species does give some guidance to the recent evolution of the north Atlantic wolffishes.

165 Newton (1891) reported wolffish teeth, which are quite distinctive, in Lower Pliocene deposits in the U.K.

166 Barsukov (1959, 106) reported data showing the striped wolffish inhabiting depths to 435 metres across its range in the North Atlantic, the spotted wolffish somewhat deeper to 550 metres, and the northern wolffish still deeper from 65-940 metres.

167 Bottom trawls made for cod in 450 metres of water off the Hawke Channel in 2000 encountered one catch of approximately 20 striped wolffish about 5 centimetres in length (unpublished data).

168 Love et al. (2002) provided almost everything you might want to know about Pacific rockfish and a bit about Atlantic redfish too.

169 Although redfish and all *Sebastes* are not known to be long-distance migrators, the Pacific ocean perch or its ancestor did make it as far as the North Atlantic and then across it. Also, south of Greenland there are pelagic redfish which migrate in midwater. In the late 1990s and early part of the 21st century, these fish moved southwest towards Labrador. Large redfish have appeared on the Labrador shelf in recent years that are unlikely to have originated or grown up there, because no midsize redfish have been found there for many years (unpublished data).

170 Rockfish are evolution in action. Many recent divisions among species have been postulated. For example, Love et al. (2002, 21-22) explained the existence of species in the Gulf of California: "Preliminary evidence from mitochondrial DNA suggests that spiny-eye and gulf rockfishes evolved

from an ancestor that probably entered the Gulf between 817,000 and 400,000 years ago." They also explained the few species in the southern oceans: "Assuming a range of rates of DNA evolution, this equatorial crossing probably happened between 141,000 and 80,000 years ago." This is very recent in geological time.

171 Love et al. (2002, 22) stated: "the invasion of the Atlantic probably occurred during a relatively abbreviated warming period between 3.5 million and 3 million years ago. During this time, termed the "Great Transarctic Biotic Interchange", opening of the Bering land bridge and warming water in the Arctic facilitated species movements between the Pacific and the Atlantic oceans. Evidently, most of the ultimately successful fish dispersals were from the Pacific to the Atlantic as various poachers, smelts, eelpout, and sculpins made the journey eastward. Only a few eastern fish species, notably members of the cod family, appear to have successfully colonized the west."

172 Stanley et al. (2000) estimated the widow rockfish shoals to be on the order of several hundred metres across off Vancouver Island. This is a commercial species and relatively abundant in the North Pacific. Note that most rockfish populations have been overfished and are at lowered abundance levels.

173 Love et al. (2002, 31) stated: "[The] longevity of many rockfish species is in part a response to the extremely variable oceanographic conditions found in the northeast Pacific. Successful reproduction depends on pelagic larvae (the most sensitive life stage) surviving to become benthic juveniles (a process known as recruitment). Many years will pass before there is the right combination of water temperature, food supply, and upwelling intensity that will encourage larval survival. Thus, many species have successful juvenile recruitment on relatively rare occasions. For instance, large numbers of Bocaccio larvae survive to juveniles only once in about 20 years." In the deepwater redfish (*Sebastes mentella*) off the south coast of Newfoundland, good year classes are produced every decade or so, but there has not been a good year class in over 25 years, the last being 1980.

174 Longevity coincides with low growth and reproductive rate, otherwise, individuals and populations would become too large to be supported. The other side of the coin is that high growth and reproductive rates go with short life cycles, the best example from the Newfoundland and Labrador marine ecosystems being capelin. The differences in life histories are very important not only to ecosystem functions but also to fisheries management. Sustainable harvest levels in fisheries depend on productivity, the sum of growth and reproduction, not the numbers or biomass of the harvested species. Long lived species with low productivity, such as redfish, cannot sustain harvests at as high rates as can species with high productivity.

175 Liem and Scott (1966, 327) reported the size of the ocean pout as "up to a length of about 3.5 feet and weight of 12 pounds, but specimens over 2.5 feet in length are rare."

176 See Scott et al. (1986).

177 Liem and Scott (1966, 327-328), regarding the ocean pout, stated: "The eggs are very large, about one-quarter inch in diameter and yellow in colour. Each female produces from 1300 to 4200 eggs." In terms of growth, "studies of ocean pout in the Bay of Fundy indicate slow growth, viz. a length of 12 inches at 5 years; 24 inches at 12-13 years; and 27-28.5 inches at 16-18 years." No doubt the growth in Newfoundland and Labrador waters was even slower, as a result of the colder water conditions.

178 See Briggs (2003).

179 Richard and Campbell (1988, 1) gave the evolutionary history of the walrus.

180 Loughrey (1959, 40), citing earlier reports, stated that from a single walrus stomach was taken "more than a bushel of crushed clams in their shells…it is in digging this shellfish food that the services rendered by the enormous tusks become apparent." But there were also reports of walrus shelling their food before eating by crushing the shells in their mouths and spitting them out. Probably both types of feeding occur.

181 Richard and Campbell (1988, 347), in response to earlier reports that 175,000 walrus had been exported from Labrador between 1925 and 1931 by the Hudson's Bay Company, calculated that "the 165,963 pounds of hides represent approximately 592 walruses; that is, a catch of little more than 100

walruses per year: a number considerably less than some authors would have us believe."

[182] Readers are referred to Lockwood (1988) for details of the life history of the Atlantic mackerel, although most all details pertain to the northeast Atlantic, especially the area around the British Isles. General information of the systematics of tunas and mackerels can be found in Collette et al. (2001).

[183] Castonguay et al. (1992) showed that Atlantic mackerel will go to colder temperatures temporarily to feed, but generally migrate with the warmer temperatures.

[184] Lockwood (1988, 27) gave a complete description of the distribution of the Atlantic mackerel. He states: "While the total range of the mackerel's distribution is from North America and across the North Atlantic to the Black Sea coast of the Soviet Union, the seasonal migrations are localized by comparison. Because these movements are restricted to a limited geographical area, they help to separate the species into smaller units or stocks. Each stock reacts, relatively independently of adjacent stocks, to those factors which affect all fish stocks, such as environmental change and exploitation by commercial fishing fleets."

[185] Meek (1916, 322-323) told the tale of debates of the early 20th century on the migrations of mackerel in the northwest Atlantic. He stated that "the mackerel…could be seen in the clear waters of the bays of Greenland and in other similar situations in spring, hibernating with their heads buried in the mud." Mackerel are not normally found north of the Strait of Belle Isle. Meek also reports that: "The migrations of the mackerel have in modern times assumed a place of importance, in the first place, because of the many disputes on the subject relating to the rights of fishery in the region of the Gulf of St. Lawrence. The Canadian view of the migrations, as expressed by Hind and Whitchet, is that the mackerel spend the winter at the bottom in a quiescent condition, and that the migrations are confined to moving from the shore into deep water in the immediate vicinity, and to a return migration from the deep water to the surface and the coast." The American view was more correct in that it argued for a much longer migration, likely wrong in its extreme position. Meek related that: "In the second place, the mackerel catches on the American coast during the last decade have declined to an alarming extent, while on the European side the catches have been steadily improving, and the suggestion has been made that the mackerel are leaving America and migrating, evidently permanently, to the other side of the Atlantic."

[186] Lockwood (1998, 69) showed egg distributions over the spawning season. He stated: "Early in the spawning season (April), while the temperature is more or less equal (10-11°C) from the surface to depths of 200 metres of more, eggs may be found throughout the water column. From May onwards the surface waters get warmer than the deeper layers and a thermocline forms. In the second half of the spawning season eggs are most abundant above the thermocline."

[187] See Swain and Sinclair (2000).

[188] Leim and Scott (1966) stated that "fry have not been caught in northern waters but are abundant between 11° and 40°N latitude." However, Scott and Scott (1988) reported that "dead sauries cast up on beaches near Canso, NS, have been reported to occur in fall (November), a result of post-spawning mortality." This infers that more northerly spawning is possible.

[189] Dudnik et al. (1981) provided details of the distribution and biology of the Atlantic saury.

[190] Graham and Dickson (2001) provided details of the specializations for endothermy in tunas.

[191] Altringham and Shadwick (2001, 319) gave data on the swimming speeds of several species of tunas. The Atlantic mackerel was capable of swimming 5-18 body lengths per second, whereas the bluefin can manage only about one. This equated to a maximum speed of over 8 metres per second for mackerel, which is blazing fast, and a respectable 3.5 metres per second for a large bluefin. Most fish cannot sustain speeds above one body length per second. All tunas are quick.

[192] See Templeman (1944, 99). Many sharks, including the dogfish, have live birth.

[193] One tagged dogfish migrated from St. John's to Massachusetts, at least 1600 kilometres in 132 days (Templeman 1944, 99).

[194] See Templeman (1944).

[195] Meek (1916, 115) summarized the state of knowledge about salmon evolution after the first 100 years of study. By then, fossil salmon had been found in both Europe and western North America that dated back 10-20 million years. It was not known if these fossils represented one or several ancestors of modern salmon. It was thought that the Atlantic variety was free to enter the Arctic, while the Pacific variety was not. However, Meek says that: "It is not improbable either, considering the differences in habit and structure of the Pacific and Atlantic salmon and the close correspondence in both these respects between the Pacific and Atlantic trouts, that the former were separated in the Cretaceous, and the Pacific trouts were derived from the Atlantic during the post-glacial submergence." Recent genetic study has shown that this is likely not true. Kinnison and Hendry (2004) summarized recent knowledge based on genetic studies and the reclassification of the Pacific trout as belonging to the Pacific salmon group, and not to the Atlantic salmon or char group. According to this new information, the former marine origin of salmon was not supported, but it did support the notion that protosalmon were present in both the North Atlantic and North Pacific before the cooling and ice ages of the past 10 million years.

[196] Hendry et al. (2004, 93) in a recent study on anadromy and nonanadromy in salmonids avoid the issue. They stated that: "Whether salmonids had a freshwater or marine origin has long been debated…with some evidence suggesting an origin in fresh water (e.g., only some extant populations are anadromous whereas all breed in fresh water) and other evidence suggesting the opposite (e.g., non-anadromous populations seem to be the derived state in many species)." There was no doubt that anadromous forms can become nonanadromous, based on transplants all over the world. And some anadromous forms have become migratory under certain circumstances. It may be that the evolution of these behaviours, and the marine versus freshwater crossings that are implied, have occurred more than once in history and even in different directions.

[197] Kinnison and Hendry (2004, 220) stated: "Most modern populations of salmon likely became established in the last 8000-15,000 years. Glaciation prior to this time restricted most salmonids to a limited number of isolated refugia. Following the recession of the ice sheets, these refugial populations acted as sources for dispersal, leading to colonization of their current ranges in North America, Asia, and Europe. Although the refugial lineages may have undergone substantial evolutionary diversification prior to and during the most recent glaciation, a significant component of extant inter-population variation has clearly arisen post-glacially."

[198] Reddin et al. (2000, 89) showed smolt-to-adult survival rates at west and south coast rivers in Newfoundland which appeared to correlate with declining sea temperatures and productivity.

[199] Reddin and Friedland (1993) summarized the marine life of Atlantic salmon in the northwest Atlantic, which included those from Newfoundland and Labrador. They emphasized the importance of sea temperatures in determining the distribution of migrating salmon.

[200] Hansen and Jacobsen (2000, 80) showed about 3.5% of returned tags from Atlantic salmon tagged at sea north of the Faroe Islands were returned from Atlantic Canada (where presumably they would have spawned). Most returns (> 40%) came from Norway.

[201] See Reddin and Friedland (1993).

[202] Reist et al. (1997) reported that Arctic char prefer colder waters than do the Dolly Varden, and the boundary between these species appears to be the Mackenzie River in the Northwest Territories, Canada, with Arctic char to the east and north, and Dolly Varden to the west and south.

[203] Stearns and Hendry (2004, 19) showed recent conclusions that the char evolved from a common ancestor with the Pacific salmon, and not with the Atlantic salmon as most earlier works reported.

[204] Dempson (1984) studied the Arctic char of northern Labrador and concluded there were a number of discrete stocks along the coast. Based on tagging experiments, he concluded that char do not migrate great distances from their home river.

205 Dempson et al. (2002) showed that Arctic char off northern Labrador changed their diet and their distribution as capelin declined in the 1990s.

206 Aristotle and Pliny believed the eels self-generated somehow from the mud or from rotting vegetation.

207 Details of the life history of eels are given in Bertin (1956).

208 See Bertin (1956).

209 Bowman and Abele (1982, 4-6) gave details of the crustaceans, including the approximate numbers of species and their environment and habitat.

210 Cisne (1982, 74-75) described the characteristics of the ancestral crustacean. He concludes that "such an arthropod would probably be called a trilobitoid if known…which is to say that it would have a trilobitelike but basically unclassifiable marine arthropod." On the issue of whether crustaceans and other organisms evolved from a single ancestor, or there were multiple lines, he states that "there is no general agreement as to which among several possible solutions is best, and no agreement as to whether Crustacea share a common ancestry with other groups."

211 Green (1961, 1) stated that "the Burgess Shales…are remarkable for the fossils which they contain. Some of these fossils look very similar to Crustacea which are alive today." However, many of the details of these ancient animals have not been preserved, so most classifications put the Burgess Shale so most classifications put the Burgess Shale animals of British Columbia in a group called Pseudocrustacea.

212 Cisne (1982, 85-86) stated: "Many evolutionary changes involved in the origin of the Crustacea probably centered around the evolution of adaptations for increased dispersal and colonizing abilities through rather drastic reorganization of the ancestor's simpler development pattern and life history design. Diverse facts about primitive Crustacea fall into place when viewed in terms of this organizing theme, which is the main reason for giving any credence to the following scenario: the evolution of the ancestral crustacean's developmental pattern represented the evolution of adaptations for maintaining high intrinsic rates of natural increase (r) in populations – prime characteristic for species variously called colonizing, fugitive, or ecologically opportunistic."

213 See Ekman (1953, 158).

214 Green (1961, 130) indicated that there are eight species of Cancer crabs in the North Pacific, with three in the North Atlantic; Ekman (1953) suggested six and one, respectively. Taxonomists are known for quibbling, but the message is the same.

215 See Astthorsson and Gislason (1999).

216 Green (1961), page 112, related the life history of *Lernaeocera*. He stated: "The adult female of *L. branchialis* lives on cod, whiting and Pollack. Its body is large, swollen and deformed, with two long coiled egg strings. The head is buried deeply in the tissues of the host and penetrates the main artery leaving the heart. The copepod feeds on the host's blood; it takes its meals infrequently, but digestion is complete, and there is no anus for the expulsion of undigested material." The eggs hatch and go through a nauplius stage before moulting into a copepod that can swim and seek out a flounder. It attaches to the gills of the flounder and at its next moult produces an attachment thread that penetrates the host. It is then known as a chalimus, which has nonswimming legs. The life cycle is complex. Green continues: "Four moults are passed in the chalimus stage, then the males become sexually mature and produce sperms which they transfer to the females. Both males and females develop swimming setae on their legs and can swim actively. The males die after mating, but the females swim away." At this point the females search for a cod. If one is found, the female parasite attaches to the fourth gill arch, sucks blood, and grows to be a bloated blood-filled worm. It then releases eggs and dies. It is thought that the whole life cycle from egg to dead female takes about eight to ten weeks and occurs year round, so that at any one time there may be *Lernae* is various parts of the life cycle.

217 Ponomareva (1966, 44-45), relating data from the Pacific, stated that "euphasiid biomass often reaches 1000 mg/m3 and even more, while in tropical waters it rarely exceeds 200-300 mg/m3." He also stated that "the tropical euphasiid fauna, quantitatively poor in the study area, was distinguished by the diversity of its composition."

218 Ponomareva (1966, 42) was not clear on the origin of North Atlantic euphasiids, but stated: "If the period at which related forms became isolated is determined by the extent of their taxonomic divergence, then the gap (if any) in the ranges of the eastern and western *T. inermis* and *T. raschii* occurred not so long ago, probably in the Miocene or Pliocene." The comments of the earlier work of Einarsson (1945, 52) are particularly relevant. In response to a noted similarity in these two euphasiids in the North Atlantic and North Pacific, he stated: "To account for this phenomenon, two questions may hereby be brought forward for further consideration. On the one hand we may have to do with parallelism in the development. If we assume that northern waters have been colonized by Euphasiids from the south, the inermis type may have developed independently from two different species of the genus *Thysanoessa*, perhaps as the species colonized the northern coastal waters. On the other hand there may well be a closer relationship between the species of the inermis type, i.e. they may have a common ancestor; in that case the Atlantic stock of *T. inermis* must have its origin in the Pacific area."

219 Green (1961, 15) stated: "The name Decapoda implies the presence of ten legs, and this is true of most decapods, though the legs are not always large and functional. The first three pairs of thoracic limbs are modified as mouthparts (maxillipeds) and help in the feeding mechanism, while the remaining five pairs of thoracic limbs are leg-like. The first pair of these legs usually bears a pair of pincers or chelae which are used to catch food and for defence." In some species of crabs and lobsters, the pincers have reached large proportions.

220 Abele (1982, 288) stated that "the earliest record of a decapod is…known from the late Devonian." Schram (1982) contained a fine line drawing of this fossil.

221 The classification of the Pandalid shrimps is controversial. Earlier reports indicated that *P. borealis* occurred in both the North Atlantic and North Pacific. However, Squires (1992) suggested that what had been called *P. borealis* in the North Pacific was actually a different species, named *P. eous*. The same has occurred with *P. montagui*, earlier thought to also inhabit the North Pacific, but later the North Pacific variety was recognized as a separate species, *P. tridens*. Recent genetic studies have shed some light on this, but also muddied the waters, naming new species of shrimp in the Pacific and finding differences between shrimp in adjacent ocean basins. As to where Pandalids originated, and how they spread, there is more controversy. The presence of many more species in the Pacific and their broader distribution there strongly suggests a Pacific origin, and the similarity of the *P. borealis/eous* shrimp and the *P. montagui/tridens* shrimp suggests a recent split (whether or not they are separate species). This argument was reinforced by Baldwin et al. (1998). However, Komai (1999) has suggested the opposite, but the evidence is less convincing. A summary of these arguments may be found in Bergstrom (2000).

222 Bergstrom (2000, 111), stated: "In Penaeus the greatest species diversity today is found in the Indo-Pacific Ocean, which contains about five times more species than the Atlantic Ocean. In Pandalus two or maybe three species are found in the Atlantic Ocean while six to eight times as many are found in the Pacific."

223 Pandalid shrimps are protandric hermaphrodites (midlife sex changers).

224 Bergstrom (2000, 12) reviewed the theory and information on reproduction and sex change in shrimp in some detail. To stress the complexity of this phenomenon, for northern shrimp, he states: "For example, although the overwhelming majority of post-larval shrimps in boreal populations of *P. borealis* show male external morphology during their first year of life, some may develop directly into females without passing through the male stage. Most one-to-two-year-old shrimp in these populations breed as males, but a small proportion change sex during their first year and reproduce as females at about 1.5 years of age. Between subsequent reproductive periods, varying proportions of males change sex."

225 See Bergstrom (2000, 136-138). For *P. borealis*, the period of egg bearing is reported to be seven to ten months at an average temperature of 3°C.

226 Parsons and Tucker (1986) reported on the fecundity of *P. borealis*.

227 Bergstrom (2000, 188-189) provided information on shrimp growth. Parsons et al. (1986) showed that shrimp in Newfoundland and Labrador waters grow to a substantial size, so much so that *P. borealis* is sometimes considered a marine "prawn." However, they corrected earlier misinterpretations of shrimp age that inferred shrimp were far more productive than they really are. This is consistent with the coldwater environment of that region.

228 Shumway et al. (1985) expressed the view that *P. borealis* is a schooling species. Crawford et al. (1992) showed how *P. montagui* in Hudson Strait formed dense aggregations that held the same ground for many days despite strong vertical movements.

229 Green (1961, 16-17) gave the history and number of decapod species.

230 Green (1961, 130-131) stated: "One surprising feature of the crustacean fauna of the North Atlantic shelf is that a large part of it is derived from the Pacific, and it can in some ways be regarded as a poorer version of the North Pacific fauna. For instance, the genus *Spirontocaris* has 64 species in the two oceans; 51 of these are purely N. Pacific, 10 are common to the N. Pacific and N. Atlantic, and only 3 are purely N. Atlantic. Another example is given by the edible crab, Cancer, which has only three species in the N. Atlantic while the N. Pacific has eight species of the genus. These facts, together with information about the fossils in the two areas, indicate that the waters of the N. Pacific have been temperate for a longer period than those of the N. Atlantic. When, during the Eocene, Britain had a tropical climate the warm water in the N. Atlantic reached farther North than did the warm water in the Pacific, so that a temperate water fauna developed first in the N. Pacific, and only later in the N. Atlantic."

231 The snow crab was often known as the spider or queen crab, but this changed after the US dictated that all *Chionoecete* crabs to be marketed there must bear the name snow crab. The name does fit the cold-loving nature of this species, and is now universally accepted.

232 See Jamieson and Bailey (undated), *The Atlantic Snow Crab*, Communications Branch, DFO, Ottawa.

233 See Sainte-Marie et al. (1996) for additional references.

234 See Jamieson and Bailey (undated), *The Atlantic Snow Crab*, Communications Branch, DFO, Ottawa.

235 Sainte-Marie et al. (1996, 451) stated: "The cycle is apparently endogenous and may result primarily from negative, density-dependent interactions between a settling year class and age classes established in the preceding 6 years. Cycles cause striking changes in population features." Conan et al. (1996, 60) stated that "strong natural fluctuations of recruitment prevail in the Bonne Bay snow crab stock in the absence of any important fishing activity. Series of 3 to 4 successive years of good recruitment are followed by series of 5 to 6 years of poor recruitment." Also, "most of the recruitment variability observed…results from intra-population effects involving selective cannibalism by predation of older males on early benthic recruits and molting individuals."

236 See Forest and Saint-Laurent (1976).

237 See references in Phillips et al. (1980).

238 Phillips et al. (1980, 7) stated that: "In almost every marine habitat one is likely to find a lobster of one sort or another. Some have been trawled at great ocean depths of 3000 metres or more. Others live in holes on shallow tidal reefs." Details of the biology of lobsters, and many additional references, are given in this work.

239 See Cooper and Uzmann (1980, 120) for growth up to age 100 years.

240 See Cooper and Uzmann (1980, 113) and the references therein.

241 See Cooper and Uzmann (1980, 98) and the references therein.

242 Phillips et al. (1980, 10) cited earlier work by Huntsman and Templeman.

243 Ennis (1983) showed lobster movements from about 5-10 metres depth in spring and summer to about 15 metres in winter.

244 See Templeman (1940).

245 See Cooper and Uzmann (1980, 122-124) for additional references.

246 See Templeman (1940, 31).

247 See Templeman (1940).

248 See Ennis (1973).

249 Slijper (1979, 254) related how a blue whale measuring 100 feet and 280 tonnes weighs as much as 4 brontosaurs, 30 elephants, 200 cows, or 1600 men. This whale contained as much fat as the combined production of 275 cows over a full year!

250 Slijper (1979) provided a broad description of the evolution of whales and of their jaws. On page 72, he states "even modern Mysticetes (non-toothed whales) cannot be said to be entirely devoid of teeth, which still occur in their fetuses. This has been known for the past 150 years, for foetal teeth were first discovered in the lower jaw of a Greenland whale by Geoffroy St. Hilaire in 1807. If an incision is made into the soft mucous membrane of the upper or lower jaw of a Fin Whale foetus aged 4-8 months (ca. 4 feet 3 inches – ca. 10 feet long) a row of conical tooth buds will appear. The beginnings of the baleen, a row of cornified transverse ridges, first appears in fetuses when they exceed a length of 10 feet; at this stage the rudimentary teeth of both the lower and upper jaw disappear without a trace. However, their presence during the early stages of foetal development is clear evidence that the Mysticetes are descended from a line of ancestors with teeth in both jaws."

251 A complete classification of cetaceans (whales, dolphins, and porpoises) was given in Slijper (1979, 433).

252 There may have been 150,000 humpbacks in the North Atlantic historically. There were likely even more minkes (Jon Lein, personal communication, 2005).

253 See Haug et al. (1995).

254 Slijper (1979, 259) stated "a Humpback whale caught in the North Atlantic was found to have six cormorants in its stomach, with a seventh stuck in its throat."

255 See Montevecchi and Tuck (1987).

256 Gaskell (2000, 155, 163) summarized what is known of the diet of the great auk. He mentioned lumpfish and sculpins as being important, which seems unlikely as they are relatively rare species. Hobson and Montevecchi (1991) mentioned capelin and euphasiids, based on isotope analyses of bones from Funk Island, which seems far more likely as these species are abundant in that locale. Some penguins, the southern hemisphere's large flightless seabirds, feed euphasiids to their chicks. It is interesting that these data suggest the great auk fed lower on the food chain than do the murres, which opposes the general rule that larger animals eat larger prey. But then whales are also an exception. Regrettably, we will never know for sure.

257 There are 20 species of Alcid seabirds, all in the northern hemisphere. The North Pacific has 16 species (14 endemic) and the North Atlantic 7 (5 endemic). The common and thick-billed murres occur in both oceans. There are several races of each species spread across their range. A summary description of the evolution of murres was given in Tuck (1960, 19-26).

258 See Harris et al. (1994).

259 Tuck (1960, 25) stated: "Fewer eggs fall toward the end of the incubation period than at its beginning. This may partly be explained by increasingly assiduous incubation, but also important are changes taking place within the egg which aids its stability. As incubation proceeds, the radius of the curve described by the egg when disturbed is diminished. Consequently, the chance of rolling off the ledge is similarly reduced."

260 See Rowe et al. (2000) and Gaston (1985).

261 Cairns et al. (1987) summarized the foraging patterns of common murres in Newfoundland waters. They stated that middle "potential ranges were 37.8 kilometres for incubating birds and 5.4 kilometres for chick-rearing birds."

262 See Piatt and Nettleship (1985) and Burger and Simpson (1986). Cairns et al. (1987) summarized the foraging patterns of common murres in Newfoundland waters. They stated that middle "potential ranges were 37.8 km for incubating birds and 5.4 km for chick-rearing birds."

263 Boag and Alexander (1986) gave "guesstimates" of the numbers of puffins. They stated that: "In the western Atlantic, there are many large and well-established sites. In excess of 330,000 pairs breed in over fifty different sites with the majority on the coast of Newfoundland…The Labrador coast also contains good-sized colonies."

264 See Burger and Simpson (1986).

265 See de L. Brooke (2002). Alcids include 17 species confined to the Pacific, 2 to the Atlantic, and 4 which overlap.

266 Boag and Alexander (1986, 31) put it this way: "There is nothing graceful about a puffin landing on the water; it simply begins to lose height until it gets caught by the first wave, when it crashes into the wave in a most undignified fashion. In a calm sea, the puffin will lose height until only a few centimetres above the water and then it will simply stop flying; the resulting belly flop is as amusing as the wave crash." Take off is no more graceful, and a puffin loaded with capelin may not make it off the water.

267 Hakluyt reported of Cartier's journey in 1534 that "these ilands were as full of birds as any medow is of grasse, which there do make their nestes; and in the greatest of them there was a great and infinite number of those that we call margaulz, that are white and bigger than any geese." Cited in Fisher and Vevers (1954, 202).

268 Details of gannet numbers and their increases were given in Fisher and Vevers (1954).

269 Sea butterflies are floating invertebrates that have a large "wing" resembling a butterfly. Pumpkinheads are large jellyfish that look like floating pumpkins. Sea cucumbers are benthic invertebrates that look a bit like their namesake.

THE ARRIVAL OF HUMANS
6000 BC TO AD 1450

A tradition of old times told of the first white men that came over the great lake were from the good spirit, and that those who came next were sent by the bad spirit; and that if the Boeothics [sic] made peace and talked with the white men which belonged to the bad spirit, or with the Mic-maks, who also belonged to the bad spirit, that they would not, after they died, go to the happy island, nor hunt, nor fish, nor feast in the country of the good spirit, which was far away, where the sun went down behind the mountains...

SHANADITHIT,
the last of the Beothuk people, in response to a query by W.E. Cormack in the late 1820s

Archaic Maritime Indians . . . 168
Palaeo-Eskimos . . . 168
Recent Indians . . . 169
Europeans . . . 172

ARCHAIC MARITIME INDIANS

About 7000-8000 years ago (6000-5000 BC), Maritime Archaic Indians followed the retreating ice north to the Labrador coast. Approximately 5000 years ago (3000 BC), as the last of the ice melted, they found their way across the Strait of Belle Isle to the island of Newfoundland. The island was an uninhabited and hospitable place with a milder climate than the Labrador coast. Maritime Archaic Indian villages soon appeared along the shoreline around most of the island. Archaeological relics indicate that they were marine hunters: in spring and summer they fished for salmon and harpooned whales, seals, and other marine mammals. There is little evidence that they fished for cod – salmon was likely preferred, being a fattier fish and easily caught in the rivers. In winter, the people supplemented their diet and clothing by hunting caribou, bear, and beaver, but whether or not they vacated the coast on a seasonal basis remains uncertain. Over time, these people became well established in Newfoundland. Using the typically low population densities of other hunting cultures as a guide, their population was likely fewer than 2000-3000 people, and remained relatively stable after the initial period of colonization. Ancient cemeteries have been found at Port au Choix and on Twillingate Island.[1] The Maritime Archaic Indians were the sole occupants of the island of Newfoundland for 2000 years. Their influence on the marine ecosystems was likely slight, although some local reductions in salmon may have occurred. About 3000 years ago (1000 BC), these people disappeared; what happened to them is not known.[2]

PALAEO-ESKIMOS

Shortly after the Maritime Archaic Indians disappeared from the archaeological record, about 2800 years ago (800 BC), a group of northern people often referred to as the Early Palaeo-Eskimos arrived on the island of Newfoundland. Like the Indians before them, they came from Labrador; their tools have been found on the west coast of the island and in Groswater Bay, Labrador. They occupied summer camps in the outer reaches of the bays or on the headlands, always close to the sea, hunting marine mammals and spearing fish. In winter they may have hunted caribou. Their tools were more refined in design than those of the Maritime Archaic Indians. Only a few Palaeo-Eskimo camps have been found, the most notable near Cow Head on the west coast, and their population numbers were almost certainly very low. They did not stay long on the island, only 700 years.[3]

Two independent groups of people arrived in Newfoundland about 2000 years ago (AD 0). The first were the Late Palaeo- or Middle Dorset Eskimos, who may have usurped the Early Palaeo-Eskimos. The Late Palaeo-Eskimos dispersed and occupied much of the island, especially the northeast and west coasts.⁴ Like the earlier peoples, they were sea hunters; the Late Palaeo-Eskimos, however, specialized in hunting seals, which explains their major camp locations near the harp seal migration routes and the haul-out sites of harbour seals. They took salmon from the rivers in summer and pursued caribou in winter, but their culture was essentially tied to the ocean. They dressed in seal fur and used seal hides to build kayaks, which they used expertly. They also carved soapstone into various vessels, which the earlier Eskimos did not do, and carved bone and ivory. The Late Palaeo-Eskimos reached their peak about 1400 to 1700 years ago (AD 300-600), after which they declined, and there are no records of them later than 1000 years ago (AD 1000).⁵ Their decline coincided with the warming climate of the Middle Ages.

The ancestors of the modern Inuit people, the Thule Eskimos, migrated south from the Arctic about 700 years ago. They reached the mid-Labrador coast around AD 1550, at the beginning of the Little Ice Age in the northern hemisphere. Like the aboriginal peoples before them, they were sea hunters and they exploited whales, harp and ringed seals, and walrus. In the late 1500s, Thule Eskimos made excursions farther south to the Strait of Belle Isle and perhaps onto the east coast of the Great Northern Peninsula of the island of Newfoundland. Here they came into conflict with the Beothuks.

RECENT INDIANS

After the decline of the Maritime Archaic Indians about 3000 years ago (1000 BC), no Indian people inhabited Newfoundland for 1000 years. However, about 2000 years ago (AD 0), at the same time as the Late Palaeo-Eskimos were exploring the island, another group of aboriginal Indians from Labrador came to Newfoundland. These people are often referred to as Recent Indians in order to distinguish them from the Archaic group. The earliest groups camped near Cow Head on the Great Northern Peninsula (and are therefore referred to as the Cow Head Indians), venturing east and south at least as far as Bonavista Bay and the southwest coast.⁶ Their tools indicate that they were marine hunters, but not so tied to the sea as the Eskimos. They likely also hunted caribou and ventured inland in winter. Their numbers are unknown, but were probably few.

The Cow Head Indians were succeeded by another group known as the Beaches Indians, named after a key campsite in Bonavista Bay. The Beaches Indians inhabited the northeast coast and some other areas, and like the Cow Head people, were marine hunters who also pursued caribou.⁷ A third group of Recent Indians appeared between 750 and 850 years ago. These were the Beothuk people, whose ancestors are sometimes called Little Passage Indians after the site in Hermitage Bay where their first remains were

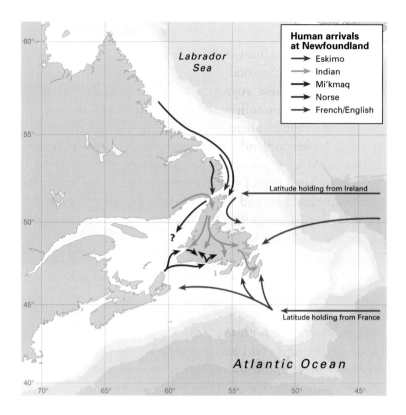

3.1
The arrival of a succession of peoples to southern Labrador and Newfoundland. The southern route of the French and others may have been somewhat farther south across the Atlantic, depending upon season.

uncovered. The Beothuks inhabited most of the island and became well established, with camps ranging from the Avalon Peninsula to the west coast and Great Northern Peninsula.[8] In keeping with their hunting and gathering ways, they lived in small groups, which limited their impact on food resources. Their summer camps on the coast were located close to salmon rivers and to the haunts of seals and seabirds.[9] Fall and winter camps exploited caribou migration paths. The numbers of Beothuk people are difficult to determine: estimates range from fewer than 500 to an unrealistic 50,000. Most scholars believe that there were fewer than 1000 Beothuks around AD 1500 when the first lasting contact with Europeans occurred.[10]

There is diverse opinion on the origin of the Beothuk people. Some early investigators, including the noted W.E. Cormack, held the opinion that the Beothuks were descendants of the Norse colonists intermixed with the Cow Head Indians.[11] They based this on clues from the Beothuk language, which was thought to be distinct from all others. But these early speculations have little foundation. The Beothuks were North American Indians, most likely descended from the Beaches people.

In AD 1450, there were likely on the order of 1000 Beothuks harvesting the oceans around Newfoundland. They were accomplished marine hunters; salmon, seals, seabirds, and eggs were important items in their summer diets. They undertook sea journeys in uniquely constructed birchbark canoes,[12] travelling along the coast and as far offshore as the Funk Island to hunt seabirds and collect eggs. They must have been accomplished canoeists who understood the quick-changing sea conditions of their home shores to make such extensive voyages.[13] Caribou made up their main winter meat, perhaps supplemented by ptarmigan, beaver, and other animals. They also consumed berries and some edible roots.[14] There is no evidence that they fished for cod. They had the island world of Newfoundland to themselves, apart from an occasional visit from the Inuit to the north, Mi'kmaq who crossed from Cape Breton Island, and Innu (Montagnais and Nascapi Indians) from Labrador. The Beothuk took from the ecosystems only what they could carry and eat, and by all accounts they lived well.[15]

Of the ecosystems themselves, little is known with certainty, but it is likely that the species most utilized by the Beothuks, and also species that were not harvested, such as capelin and cod, were highly abundant. Seals were numerous, perhaps 5-6 million or more, with migratory habits that limited their contact with the Beothuks to a few weeks per year. Caribou may have numbered about 200,000 in several herds.[16] Their harvests, and those of all native peoples before them, were almost certainly well within the productivity of both the marine and terrestrial ecosystems.

If the Beothuks had any effect on the marine ecosystems, it would most likely have been on salmon and seabirds, especially the great auk. There is evidence that they killed considerable numbers of these species.[17] However, it is unlikely that the Beothuks impacted these species in any significant way. The abundance of both seabirds and salmon at the time of arrival of the Europeans after AD 1450 argues that their influence was minor. If there were 1000 Beothuk people living on the island of Newfoundland, each accounting for one three-kilogram salmon per day over a peak season of six weeks,[18] then the total salmon harvest would have been 42,000 fish weighing 126 tonnes. This is modest in comparison to later commercial salmon catches of the English settlers. The same logic may be applied to the effect of the Beothuks on seabirds, which is thought to have been similarly limited.

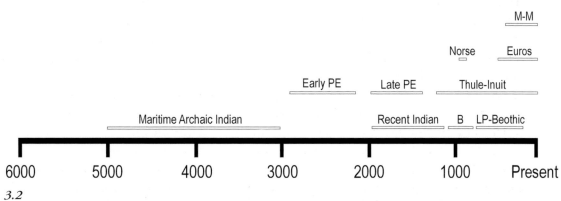

3.2
Chronology of human occupancy of Newfoundland. PE means Palaeo-Eskimo, B means Beaches, LP means Little Passage, Euros are Europeans, and M-M are Mi'kmaq.

EUROPEANS

Iceland played a key role in the earliest explorations of the North Atlantic. Legend has it that in the Middle Ages, Irish monks sailing in boats made of animal skins, likely oxhide, searched the North Atlantic for uninhabited islands where monasteries could be established away from worldly influences. Around AD 800, the Irish writer Dicuil described an uninhabited island to the northwest of Ireland, an island the monks called Thule (Iceland) where the sun shone all night in summer. There were also rumours of lands farther west. The Irish monks became the first settlers in Thule (Iceland), but left when the Norse arrived around AD 870 because they did not wish to live with heathens.[19] Christianity did not come to the Norse until much later.

After AD 870, Iceland became the stepping stone to the New World for the adventurous Norse. First Greenland was located and settled; then, about 950 years ago (AD 1050), the Norse came to Newfoundland. The Norse sagas relate that the first explorer to set foot in North America was Leifr Eiriksson (spelling as in the Norse sagas), son of Eirik the Red, who had fled Iceland to become a farmer in West Greenland. Exploring west from Greenland, Eiriksson likely followed the currents south down the Baffin Island and Labrador coast to the Strait of Belle Isle. Along the way the Norse encountered people who used skin boats – these were likely either Late Palaeo-Eskimos or Recent Indians.[20] Norse sagas tell of conflict with the resident peoples, and these difficulties must have discouraged the Norse from staying in the newly discovered land, if not driven them out entirely. During one encounter, "early one morning in spring, they [the Norse] saw a great horde of skin-boats approaching from the south round the headland, so dense that it looked as if the estuary were strewn with charcoal. [At first there was trade, but the skin-boat people returned three weeks later and]…there was a fierce battle [during which two Norsemen and four skin-boat people were killed. Just as the skin-boat people were about to overrun the Norse, a brave young woman named Freydis, who was pregnant], pulled one of her breasts out of her bodice and slapped it with a sword." [The skin-boat people fled at the sight].[21]

Norse legends tell of the fabled Vinland where wild grapes grew and where it seldom snowed. A Norse village existed about 1000 years ago at L'Anse aux Meadows at the tip of the Great Northern Peninsula; both Cow Head Indian and Late Palaeo-Eskimo artifacts have also been found there. Whether or not this was Vinland remains uncertain. The Norse were born explorers and they likely ventured farther south through the Gulf of St. Lawrence and perhaps beyond, returning with stories of grapes and fine weather.[22] The Norse had no lasting influence on Newfoundland.[23] They hunted and fished only for subsistence and their numbers were few. By 800 years ago (AD 1200), the Norse had vacated Newfoundland, and their settlements in West Greenland were abandoned a century later as the climate cooled.

The Newfoundland and Labrador marine ecosystems were flourishing in the still warm climate of the late 1400s AD (see chapter 1 and the ecosystem model) when a new wave of people arrived from Europe, taking a more direct route across the North Atlantic, followed by increasing incursions of Mi'kmaq from Nova Scotia. For the first time in several hundred years, the Beothuks no longer had Newfoundland to themselves, and the ecosystems would never be the same.

NOTES

[1] Marshall (1996, 254) described the history of the Maritime Archaic Indians in Newfoundland and noted that "the location of their coastal sites and the presence of slate lances (or bayonets) suggest that these Indians pursued sea mammals such as porpoise, small whale, perhaps even walrus; toggling and barbed harpoons indicate seal hunting, and fish spearpoints and plummets (small ovate stone weights) imply the spearing and netting of fish."

[2] The Maritime Archaic Indians "are believed either to have left the island, to have become extinct, or to have become 'archaeologically invisible' before the arrival of Early Palaeo-Eskimo groups from Labrador" (Marshall 1996, 254).

[3] Marshall (1996, 256) noted that "carbon dates from Early Palaeo-Eskimo sites indicate that they lived in Newfoundland from about 2800 [BP] to 2100 BP [150 BC to 85 BC]."

[4] The larger campsites of the Late Palaeo-Eskimos included those at Port au Choix on the west coast, at the Pittman Site in White Bay, at Cape Ray on the southwest coast, and at Stock Cove in Bull Arm, Trinity Bay (see Marshall 1996, 256).

[5] Marshall (1996, 257) noted that "the most recent carbon date from a small site in Placentia Bay is close to 1000 BP (AD 950)."

6. Marshall (1996, 257) indicated that "the Cow Head population represents the earliest of the Recent Indian groups." Their tools and camps have been excavated at "Port au Choix, L'Anse aux Meadows, Cape Freels, and on several smaller sites on the northeast and southwest coast."
7. Marshall (1996, 258) stated that "remains from the Beaches population have largely been excavated in Bonavista and Notre Dame bays, though some sites have also been found at Port au Choix and on the south coast."
8. Marshall (1996, 265) gave a listing of all known sites of the Recent Indians in Newfoundland.
9. The first reliable description of the Beothuks was made by Jacques Cartier at Quirpon at the tip of the Great Northern Peninsula in AD 1534. According to English maritime historian Hakluyt (see Howley 1915, 10), Cartier claimed: "These are men of indifferente good stature and bigness, but wild and unruly. They wear their hair on the top like a wreath of bay, and put a wooden pin in it, or other such thing instead of a nail, and with them they bind certain bird's feathers. They are clothed with wild beasts' skins, as well the men as the women, but the women go somewhat straighter and closer in their garments than the men do, with their waists girded. They paint themselves with certain roan colours. Their boats are made of the bark of birch trees, with which they fish, and take great store of seals, and as far as we could understand, since coming hither, that is not their habitation, but they come from the mainland out of hotter countries, to catch the said seals and all necessaries for their living." Cartier was likely correct in discerning that the Beothuks were summering at the tip of the Great Northern Peninsula to hunt seals and take fish (likely salmon for the most part) and can be forgiven for not appreciating that the Beothuks did not come from any mainland, but from somewhere south in the caribou country where they overwintered.
10. Marshall (1996, 283) reviews most estimates of Beothuk population size. She concluded that "it should be clear from this discussion that any attempt at estimating the size of the Beothuk population is fraught with uncertainty." However, Newfoundland archaeologists suggest that the numbers would not have greater than 1000, and these are likely the most reliable estimates.
11. Regarding theories on the origin of the Beothuks, Howley (1915, 251) stated that "W.E. Cormack... conceived the idea that the Beothuks might possibly have derived their origin from the Norsemen." Howley states that he had evidence from a friend of Cormack's in New Westminster, British Columbia, where Cormack lived the later part of his life, that he held onto this belief until his death in 1868. Cormack was likely incorrect about this, but his interest in the Beothuks never left him. As a young man, he searched the island from east to west in the 1820s, but by that time there was likely only a single Beothuk survivor, the young woman Shanadithit who by then was at Twillingate. Shanadithit died in St. John's a few years later, after a stay in Cormack's residence, and was buried on the south side of St. John's harbour.
12. Howley (1915, 17) cited John Guy: "Their canoes are about twenty-foote long, and foure foot and a half broad in the middle aloft, and for their keele and timbers, they haue thin light peeces of dry firre, rended as it were lathes: and instead of boards, they use the outer burch barke, which is thin and hath many folds, sowed together with a thred made of a small root quartered. They will carry four persons well, and weight not one hundred weight." Further details of canoe construction can be found in Marshall (1996).
13. Howley (1915, 17) cited John Guy: "For euery place is to them a harborough: where they can goe ashore themselues, they take aland with them their canoa: and will neuer put to sea but in a calm, or very faire weather."
14. Marshall (1996, figure 19 2) described the Beothuks' seasonal exploitation of food and what was likely to have been available to them.
15. Howley (1915, 11) cites Hakluyt: "Touching their victuals, they eate good meat, but all unsalted; but they dry it and afterward they broil it as well fish as flesh. They have no certain dwelling place, but they go from place to place as they think they can best find food, and they live very well for they take care for nothing else. They drink seal oil, but this is at their feasts."

16 See Millais (1907).

17 Marshall (1996, 295), citing earlier accounts, stated: "It was widely known that the Beothuk stocked up on seabirds and eggs; on one island to the north, so it was said, they drove 'penn gwynes' into their boats from the shore. This may have been Funk Island, about sixty kilometres north-east of Bonavista Bay, where the Great Auk nested in large colonies."

18 Marshall (1996, figure 19 2) indicated a peak harvest period of six weeks during July and early August.

19 Karlsson (2000, 9-10) related: "In the 8th century AD Irish monks, who used to drift along on the ocean in search of solitary islands, came upon a country which seemed to fit Pytheas' description. This is first mentioned by the English historian Bede around [AD] 730, although the Thule that he describes could be the western coast of Norway. On the other hand the Irish author Dicuil, writing around [AD] 825, can hardly be referring to any other country than Iceland when he says that some priests had told him, thirty years previously, that they had come to the uninhabited Thule, where it was so bright at midnight in summer that 'a man could do whatever he wished as though the sun were there, even remove lice from his shirt.'" At around this time, the Age of the Vikings started, and the Norse attacked present-day England, Scotland, Ireland, and many other places, and colonized Iceland, later Greenland, and later still Newfoundland and Vinland. If we take 800 AD as the approximate starting point of the Viking Age, it lasted 500 years until about 1300 AD, when settlements were confined to the Scandinavian countries and Iceland.

20 Magnusson and Palsson (1965, 27) dismissed the skin-boat evidence and concluded that it is "quite clear from the Vinland Sagas that the Skraelings of North America were not the same as the Eskimos of Greenland, however, but Red Indians." They may well be right, but it was an overstatement to say it was quite clear. Nevertheless, recent scholars (e.g., Marshall 1996) tended to agree that although both Late Palaeo-Eskimos and Cow Head Indians may have been encountered by the Norse, the Indians were the most likely to have driven the Norse out.

21 Eirik the Red had three sons, Leif "the lucky," Thorvald (who was killed on one of the Vinland voyages), and Thorstein "the black." Eirik also had a daughter, Freydis. Freydis is described in the Grænlendinga Saga as "an arrogant, overbearing woman" with a feeble husband who had been married for his money. She is reported to have murdered several other Norse with an axe, both men and women, who had displeased her. (See Magnusson and Pálsson 1965, 52). Prior to the breast incident in Vinland, under attack by indigenous peoples, she is reported to have shouted at her comrades: "Why do you flee from such pitiful wretches, brave men like you? You should be able to slaughter them like cattle. If I had weapons, I am sure I could fight better than any of you" (from Eiriks Saga, see Magnusson and Pálsson 1965, 100).

22 Karlsson (2000) believed that the sagas of wild grapes and finding of butternuts at L'Anse aux Meadows suggested that the Norse traveled much farther south from their base camp in northern Newfoundland.

23 Marshall (1996, 263) stated that "there is no indication that they [the Newfoundland Indians] adopted Norse methods of extracting bog iron to fashion tools or other Norse techniques."

THE BEGINNINGS OF EXPORT FISHERIES
AD 1451 TO 1576

Terra Nova, New-founde-land, Terre Neuve, Terra do Bacalhau – the land of cod.

Latin, English, French, and Portuguese names
for the new world of the northwest Atlantic in the early 1500s

Terra Nova of the codfish is a cold place.

From a map of Terra Nova drawn by P.A. Mattioli, Ptolemy, 1547-48

Early Explorations . . .	178
John Cabot . . .	180
Early Politics . . .	181
Early Fisheries . . .	183
Portuguese . . .	183
English . . .	184
French . . .	184
Basques . . .	189
Effects of the Fisheries . . .	191
French Explorations . . .	191
Giovanni da Verrazzano . . .	191
Jacques Cartier . . .	192
Early Conflicts . . .	193
The Ecosystems . . .	194

EARLY EXPLORATIONS

By 1100, the Norse had explored much of the North Atlantic, sailing their longboats to Greenland and North America, and relaying their knowledge and stories of Vinland to the courts of Scandinavia. Adam of Bremen, a German priest, specifically mentioned the islands of the north and the fabled Vinland in his writings on the peoples of Scandinavia. Bremen reported that, around 1075, King Svein of Denmark "recounted that there was another island in that ocean which had been discovered by many and was called Vinland, because vines grow wild there and yield excellent wine, and moreover, self-sown grain grows there in abundance; it is not any fanciful imaginings that we have learned this, but from reliable reports of the Danes."[1] By 1300 – the time of the demise of the Norse settlements in Greenland – the Scandinavian courts were replete with stories of a rich new world farther to the west.[2] Sailors and adventurers from the major maritime nations of Spain, Portugal, and England may have heard these stories, as well as the details of Norse settlements at Greenland and the explorations farther west to Vinland.

After the introspection of the Middle Ages, the 1400s became a century of European exploration and adventure that led to the first detailed descriptions and maps of the world beyond Europe and Asia. Portugal and Spain, often aided by Italian mariners and navigators, led these adventures. Exploration focused on finding a sea route to the Orient and China – Marco Polo's land route across the Middle East and Asia to the Orient, charted in the 1200s, was a tortuous and dangerous journey. The development of wooden sailing ships displacing 50 to 100 tonnes made ocean travel much safer (although still dangerous, as accident rates attested) than it had been during previous centuries. In the early 1400s, Henry the Navigator, prince of Portugal, financed voyages south along the African continent. Later in the 1400s, the Portuguese adventurers Vasco da Gama and Bartholomeu Diaz sailed around the Horn of Africa into the Indian Ocean, reaching the towns on the east coast of Africa and India. West met East for the first time then, forever changing the history of those regions. By the late 1400s, relatively accurate maps of the African continent and Indian Ocean were widely circulated in Europe.[3] The world had taken on smaller dimensions, but more was to come.

The American continents were unknown to the courts of southern Europe in the mid-1400s. Heinrich Hammer, known by his Latinized name Henricus Martellus Germanus, was a German mapmaker working in Florence, Italy, then the European centre of mapmaking, in the late 1400s.[4] Martellus's maps of the known world in 1489 did not show the New World. At best, based on ancient stories that could be traced back to southern European writings from around 400 BC by the Greek explorer Pytheas and the more recent Scandinavian reports of Norse exploration, the New World was thought to be a collection of small islands. By the late 1400s, the monarchies of Spain, Portugal, and England, interested in finding a quicker route to the Orient and having long shrugged off the ancient notion of a flat earth, supported voyages that would travel east by sailing west. Young navigators, many from the Italian navigation tradition, were eager to use their superior compasses and navigational instruments to attempt the methods of dead reckoning and even latitude holding across the open oceans.[5] There was little consideration that a major landmass might block the way.

Thus was the stage set for rediscovery of the New World. The Italian navigators were not the only ones pushing the limits of geographical knowledge: French and Spanish Basques fishing for whales and cod off Iceland may have heard the legends of lands and fishing grounds farther to the west. Icelandic records show English vessels in Iceland in the early 1400s:[6] English fishermen from Bristol, Devon, and Cornwall carried shiploads of stockfish (air-dried cod that had been lightly salted) from Iceland to a growing market in England. By 1500, English fishermen from Devon were fishing and drying cod on the west coast of Ireland, which would become the jumping off point to the *New-founde-land*.[7] It is likely that many of the early Atlantic navigators, including Columbus, Cabot, and Corte Real, cut their teeth in the Icelandic or Irish Sea cod trade. Some may have heard the legends of "Brasil"[8] and the islands of the North Atlantic[9] that supposedly lay farther to the west.

Between 1450 and 1500, Europeans from several countries arrived at Newfoundland. Whether adventurers or fishermen found it first is not known. At some time prior to 1500, fishing vessels from Bristol, England, or the Basque region of France or Spain, perhaps blown off course in a voyage to Iceland, may have found major concentrations of whales and cod off the coast of Labrador and northern Newfoundland, and perhaps even the Grand Banks themselves.[10] The records of this era are vague, but Bristol fish merchants may have attempted trips west as early as 1480.[11] Basque fisheries may predate these events.[12] It may never be known who sighted Newfoundland first, but it hardly matters. Its discovery was inevitable in this age of exploration.[13]

Of greater importance was the motivation behind the discovery of Newfoundland. The Norse voyages were motivated by an urge to explore the North Atlantic, a wish to settle and live in new lands, and perhaps an element of desperation – some of the Norsemen left their homeland after various "troubles."[14] They found Iceland and Greenland largely unoccupied and much to their liking. The Norse had few religious goals and were less driven by prospects of riches and commerce than the later Spanish, Portuguese, French, and English colonizers.[15] The Norse tended to settle and stay in their new lands, although this did not happen in Newfoundland, and their numbers were few. Although there is some indication of logging in Labrador for use in Greenland, and further evidence that the Norse traded with their home Scandinavian countries in a two-way exchange, the dealings were too limited to be called an export industry. Perhaps this is why the ancient legends of the Beothuks, as cited by Shanadithit, referred to the first wave of white men as good spirits – they took little. However, this was destined to change: the next Europeans to visit the New World were driven by prospects of commerce, profit, and expanding national spheres of interest. By 1500, England and Portugal had both claimed the *New-founde-land*, and Spain was scrambling to claim its piece of the action.

JOHN CABOT

The Italian explorer, Zuan Caboto – or, as most know him, John Cabot[16] – likely used the method of latitude keeping when he sailed west for the Orient in 1497 and officially discovered the *New-founde-land* for the King of England.[17] Whether or not Cabot knew of the Icelandic sagas and other stories of a land to the west has been hotly debated.[18] However, he did know that his countryman, Columbus, had sailed west from Spain in 1492 and struck land in the Caribbean, claiming it for Spain. Cabot was an experienced navigator and traveller: he had been to Mecca, where his interest in the Orient no doubt was peaked, and to Iceland, apparently to conduct negotiations with the King of Denmark regarding English trade for Icelandic cod.[19] With Columbus to the south and the Norse tradition of island hopping to the north, Cabot sailed down the middle.

Cabot first attempted a crossing from Bristol in 1496, but turned back.[20] On May 2, 1497, he sailed from Bristol once again with a crew of 16 local merchant mariners and a Burgundian mate and mapmaker.[21] Cabot left the Irish coast, perhaps from Dursey Head,[22] in his little 50-tonne vessel, the *Matthew*, sailing first north and then due west across the North Atlantic, certain that he would strike land. A few weeks out, he did. The location of Cabot's landfall has been debated at length. His son, Sebastien, depicted a landfall on Cape Breton Island,[23] but there is local tradition that the landing was at Cape Bonavista in Newfoundland. The truth will never be known, but the weight of the evidence suggests that the most likely landfall of Cabot's first voyage was either Cape Degrat at the tip of the Great Northern Peninsula or farther north on the Labrador coast.[24] Cabot made a brief exploration of this new place, either along its northeast coast or into the Gulf of St. Lawrence, and was back in England by August 6, 1497. King Henry VII barely noticed the achievement, but gave Cabot a small reward nevertheless. On August 10, 1497, it was recorded that "to hym that found the new isle, 10*l* [pounds sterling]." Later that same year, Cabot was granted a pension of 20 pounds sterling per year, and on February 3, 1498, he received a royal warrant to explore the new isle again with six English ships.

Cabot's discoveries were not well documented: he was a dreamer, not a note-taker, and left few personal records of his voyages. Prowse considered that "the discovery of

Newfoundland and North America, as told by the Cabots [referring to John and Sebastien], is as dull as the log of a dredge-boat."[25] Nevertheless, the Cabot voyages left their mark. Sebastien's map contained notes reporting that the sea "abounds in fishes, and those very large, as sea-wolves [seals], and salmon; there are soles of a yard in length [halibut]; but, above all, there is a great abundance of those fishes which we call *bacualaos* [various spellings, all referring to cod]."[26] Merchants living in England and Italy wrote about Cabot, leaving the most important records of interest from this voyage. A letter written on December 18, 1497, by Raimondo di Soncino, an Italian merchant living in London, to the Duke of Milan became the first and longest-lasting advertisement of the marine riches of Newfoundland. In this letter was recorded the now-famous description of fishing with baskets: "The sea is full of fish which are taken not only with the net but also with a basket in which a stone is put so that the basket may plunge into water." The fish that di Soncino described undoubtedly were cod.[27]

In the spring of 1498, armed with his royal warrant and a fleet of five vessels replete with merchants and trade goods, Cabot sailed west again. One vessel returned to port almost immediately, apparently in trouble. There are few records from the rest of the voyage, and neither Cabot nor any of the vessels would be heard from again. They may have overwintered in North America, and there were rumours that one vessel limped back to England in 1499 with a cargo of furs and fish. There were also unconfirmed notions that Cabot survived this voyage and later visited Lisbon and Seville seeking further backing.[28] It is more likely, however, that he was lost at sea.

Cabot did not find a route to the Orient, but he did find the *New-founde-land* and an ecosystem teeming with marine life, particularly cod. Cabot's explorations and the potential richness of the fish stocks did not go unnoticed in Spain – the geography of his first, and perhaps second, trip appears on Spanish charts as early as 1500.[29] The race was on, albeit with many false starts, for the fisheries of the *New-founde-land*.

EARLY POLITICS

Political intrigues were part of the fishery from the earliest times. In March 1501, Cabot's warrant for the new isle appears to have been cancelled, with a new royal charter granted to three English merchants and three Portuguese merchants from the Azores. Later the same year, another royal charter was issued to an English merchant with two Portuguese partners giving them large powers and a virtual trade monopoly. Southern Europe was also engaged in political intrigues over Terra Nova. By the late 1400s, both Spanish and Portuguese navigators and explorers were ranging widely over the seas of the world, although they tended to keep to the tropics. Competition between these Catholic countries was so intense that, in 1493, the year after the Columbus voyage, Pope Alexander VI intervened by drawing an imaginary line 100 leagues west of the point midway between the Azores and Cape Verde Islands, and by papal decree gave all of the western area to Spain and all of the eastern to Portugal.[30] However, by 1500, Portugal and England had both claimed the western-lying *New-founde-land*.[31] Spain, concerned that the Portuguese were flagging Spanish lands by right of the papal decree, licensed their own expedition to Terra Nova around 1506, but had to rely on French pilots to guide them there. On early Portuguese-influenced maps, Terra Nova is located suspiciously too far to the east, within the area decreed to be Portuguese.

The Portuguese were the first to explore and name the eastern coast of Newfoundland and Labrador. The Portuguese explorer Gaspar Corte Real sailed northwest on May 10, 1501, after previous but unsuccessful attempts to cross the North Atlantic.[32] This time he succeeded. Although his exact course remains uncertain, it is likely that he cruised the coasts of Labrador and perhaps the northeast coast of Newfoundland. Corte Real was probably looking for a route to the Orient, but the motivation for his voyage is far from clear. He had a charter from the King of Portugal that gave him title to the land. His relatives claimed the title "Governor of Terra Nova" long after his death, but at best this was a paper designation. Corte Real's convoy arrived back in Portugal on October 8, 1501, without their leader. Corte Real had been lost somewhere off Labrador, either in a storm or in a conflict with the aboriginal peoples, possibly the Labrador Indians. In the returning ships were 50-60 aboriginal people who were regarded by the exuberant king as having great potential as slaves. The fate of the aboriginals who were taken to Portugal is unknown, but presumably they did not undertake this voyage voluntarily, thus conflict in Labrador was likely. Corte Real's brother, Miguel, organized a follow-up voyage in the next spring to search for the missing Gaspar. Miguel sailed west in early May, but like his brother, failed to return.

The Portuguese influence in Newfoundland and Labrador was fleeting, but their explorations gave names to prominent capes and bays, and material for a new generation of world maps produced in Europe in the early 1500s.[33] Portuguese names (sometimes Latinized) appeared on most of the early maps and many anglicized versions of the names are still used today.[34] Among the best known are Cabo Raso (Cape Race), Baccalaurus (Baccalieu Island), Cabo Verde (Cape Verde), Cabo de Boaventura (Cape Bonavista), Cabo de S. Maria (Cape St. Mary's), and Rio do Sam Joham (St. John's). There are also at least two outports named Portugal Cove.[35]

EARLY FISHERIES

Portuguese

There are several views of the Portuguese participation in the earliest fisheries in Newfoundland. One assumes a dominant Portuguese presence shortly after the exploratory Corte Real voyages,[36] while another view claims a more fleeting Portuguese presence – a paper ownership of Terra Nova with little sustained or substantial involvement in the development of the Newfoundland fishery.[37] As stated by Lorenzo Sabine, "Portugal like Spain soon abandoned all attention to the claims derived from the voyages of her navigators to the northern parts of our continent [North America], and devoted her energies and resources to colonization in South America."[38] In support of this second view, Portuguese fishermen had few vessels capable of the trans-Atlantic voyage[39] and records of Portuguese fisheries are sparse.[40]

Another line of evidence comes from maps of the era. By the mid-1500s, both English and French maps recognized Newfoundland as an island, although its shape was only a very rough approximation of reality. Most landmarks, especially those facing the Atlantic, bore Portuguese names,[41] suggesting a dominant Portuguese influence in at least parts of Newfoundland in the early 1500s. The highly detailed Spanish map produced by Diego Gutiérrez in 1562, based on contemporary Spanish and Portuguese geographical knowledge, provided realistic descriptions of the New World from the Carolinas to South America, but still shows Newfoundland as a bud on the mainland.[42] This indicates that Portuguese and Spanish knowledge of Newfoundland had declined by the mid-1500s and had been eclipsed by more detailed descriptions from French navigators, particularly Jacques Cartier and fishermen from Brittany and Normandy. In any event, there is little doubt that there was substantial Portuguese involvement in the very earliest fisheries, but these fisheries were small, lasted only a few decades, and would eventually be dwarfed by the fisheries of the English and French.[43]

Despite their small fisheries, Portugal was a large market for both English and French salt cod. In the 1400s, the fish came primarily from Iceland, but would later come from Newfoundland. The Portuguese were, and still are, substantial consumers of dried fish and quite particular about the product they buy. Salt cod sold in Portugal in the 1500s and 1600s was graded as *vento* (air-dried), *pasta* (wet or green), or *refugo* (poor quality) fish. Early trading records indicated that *vento* was supplied primarily by English ships, whereas *pasta* was supplied only by the French (the French kept most of their air-dried cod for their own consumption, but sold their *pasta* and *refugo* on the international market). *Vento* was most expensive, reflecting the greater investment in making it.[44]

English

English merchants from Bristol and their Azorean partners may have brought cod from Newfoundland waters to Europe as early as the 1480s, prior to Cabot's first voyage in 1497.[45] After Cabot's voyage, a few English vessels may have sailed to Newfoundland to fish, but records are sparse. The first concrete record is from 1531, in which a Bristol-Portuguese company led by Bristol financiers and supporters of John Cabot landed cod from Newfoundland worth 180 pounds sterling. In 1536, London merchant Richard Hore chartered two fishing ships for a Newfoundland voyage; on board were thirty gentlemen who set sail in an early example of ecotourism analogous to tourism in space in the twenty-first century. Although one of the vessels made it to Newfoundland, the cruise ended in starvation, cannibalism, and the pirating of a French fishing vessel. Many of the gentlemen died, but some fish were apparently caught and returned to England, the shares of which resulted in a lawsuit.[46] These sporadic and ill-organized voyages are the only evidence of English fisheries in Newfoundland waters prior to the 1570s.[47]

On these early voyages, it is thought that English ships sailed northwest from Ireland between January and April to take advantage of prevailing easterly winds and westward currents present at that time of year, coursing south in the Arctic Current south of Greenland to Newfoundland or Labrador.[48] There was also an alternative route that took ships south of the North Atlantic Drift at about 40° north latitude, then northwest across the Grand Banks. The southern route was well known by at least 1580, but favourable winds were rare this far south until early summer.[49]

Most of the capes, headlands, coves, and harbours of the Avalon Peninsula had Portuguese names prior to the arrival of a substantial English presence. The English retained and anglicized most of the Portuguese and French names, sometimes keeping the original meaning, but sometimes not. For example, Cabo de la Spera (Cape of Waiting, perhaps referring to this most eastward point of North America as a rallying point for convoys across the north Atlantic) became Cape Spear, while Baie d'Espoir (Bay of Hope) became Bay Despair. A few original English names were also adopted, including Bay of Bulls (later Bay Bulls), which may have referred to the bull bird, or dovekie (*Plautus alle*). Why this bay had not been named earlier, or if it had been, why its original name was not retained is not known.[50]

French

Sixteenth-century France was a federation of ancient provinces that included the maritime regions of Normandy and Brittany and the Protestant city state of La Rochelle. French fisheries of the time depended mostly on the large herring stocks in the North Sea and English Channel. There is little evidence of French involvement in the early Icelandic fisheries,[51] but within a few decades of Cabot's discovery, the French came to dominate the Newfoundland and Labrador fisheries.[52]

Many Normans were descendants of the Norse and, similar to their ancestors, were fine seamen with adventuresome spirits. In the early 1500s, a rapid development of maritime technology took place in Normandy, centred at Dieppe, using new navigation instruments and mapping techniques.[53] It is not surprising that Norman fishing vessels on the Grand Banks were equipped with printed charts, magnetic compasses first made in Italy, and astrolabes as early as 1504. The first authentic records of a French ship at Newfoundland are those of Jean Denys of Honfleur, Normandy, a few years later. Denys apparently fished between Bonavista and the Strait of Belle Isle, which suggests that he had navigated the same northern route from Ireland as Cabot. The merchant Jean Ango of Dieppe entered the fishery soon after. Ango became wealthy and very powerful, and his crescent-shaped house flag is emblazoned on many early charts and maps of the New World.[54] Ango sponsored Captain Thomas Aubert in 1506 to further explore the fishing grounds of Newfoundland.[55]

The Bretons and Rochellais were never far behind the Normans. The Bretons had the advantage of ready and local availability of cheap sea salt.[56] In 1515, the Breton merchant Michel Le Bail sold 17,500 salt cod in Rouen; there are more or less continuous records of cod sales in Brittany after that.[57] In addition, the more southern port of La Rochelle sent many vessels to Newfoundland in the early 1500s: five vessels from La Rochelle sailed for Newfoundland waters in 1523 and 49 in 1559.[58] In 1534, the Norman navigator Jacques Cartier recorded the French fishing harbour of Brest (a Breton name) inside the Strait of Belle Isle on the north shore of the Gulf of St. Lawrence and noted a vessel from La Rochelle fishing farther to the west.

> The first French fishing capital was located at Brest, inside the Strait of Belle Isle on what is now the Lower North Shore of Quebec, close to the original landing sites of the earliest trans-Atlantic navigators. Brest was already well established when Jacques Cartier cruised the Gulf of St. Lawrence in 1534. The exact location of this fishing capital has not been determined, but was likely near the present-day outports of Bradore, Bonne Espérance, and Old Fort, just west of the entrance of the Gulf of St. Lawrence. The *Dictionary of Commerce* by Lewis Roberts, printed in London in 1638, referred to Brest as being the chief town in New France. Brest appeared on maps drawn in the 1500s and early 1600s, but was absent from the many maps drawn in the 1700s. Although its fate is not known for certain, it presumably crumbled into obscurity after the wars of the late 1600s.[59]

The quickly expanding French fishery of the early 1500s had a large internal market and an abundance of salt – both local and imported from Portugal and Spain. The French developed two types of fisheries: the first was a land-based fishery, especially in the northern Gulf of St. Lawrence, north of Cape Bonavista, and, later, on the south coast of Newfoundland, while the second was a "green" fishery, based mainly on the Grand Banks. The land-based cod fishery was established in the early 1500s and produced a hard and dry salted cod product of good quality. This fishery used large cargo ships to come and go from France, but the fishing was conducted from small *chaloupes*, each needing five men (three to fish and two to dry the fish on land). As many as 30-40 *chaloupes* would fish from a harbour, including special boats that caught bait (capelin, mackerel, and herring) for the others.[60]

By 1504, French fishermen voyaged to and fished the area around the Strait of Belle Isle, then called the Baie des Chasteaux, or Bay of Castles.[61] Staying to the north allowed the French to avoid competition with the Portuguese who fished farther to the south.[62] A short sail to the west from the Bay of Castles brought fishing vessels to Blanc Sablon (an original French place name that was never anglicized in print, but often has been in speech) at the mouth of the Gulf of St. Lawrence and, only slightly farther on, the protected waters of the north shore fishing grounds at Bonne Espérance (originally the area of Brest). The fishing there proved to be exceptionally good, with both the northern cod and the northern Gulf of St. Lawrence cod stocks making migrations to this area to feed on the abundant capelin that crowded these waters in summer.[63] One of the largest capelin spawning beaches in the northwest Atlantic is at Brador Bay between Blanc Sablon and Bonne Espérance, as the seabird colonies at nearby Greenly and Perroquet Islands attest.

As the numbers of French vessels increased in the 1520s, the range of their fisheries expanded. Some French vessels likely made landfall south of the Strait of Belle Isle and prosecuted fisheries to Cape Bonavista.[64] More southerly areas had been explored since the early 1500s: in 1506, the Norman captain Jean Denys, sailing under the auspices of Jean Ango of Dieppe, explored the northeast coast of *Terre Neuvfe* (old French spelling of Terre Neuve) from the Bay of Castles. Captain Denys was apparently familiar with the coast as far south as the Avalon Peninsula – the town of Renews is called Havre du Jehan Denys on some early Dieppe maps – and it would not be long before French fishermen were on the Avalon and south coasts.

The Avalon fishery offered many advantages over fisheries farther north. The Avalon shore was the end point for migration of the two largest cod stocks – the northern and Grand Banks stocks – and on a seasonal migration route of a third stock on the south coast. There were also small inshore and less migratory groups of cod on the southern shore.[65] The reason that this region was such a crossroad for cod was undoubtedly the presence of capelin, which spawned along the Avalon Peninsula and migrated north past it in early summer.⁶⁶ Whales and seabird colonies were similarly numerous in this region. With each cod stock exhibiting varying migration tracks from year to year, the Avalon crossroad was less likely than other sites to suffer a failed fishery. Modern fishermen at Petty Harbour, one of the earliest fishing ports, know this well, and still refer to fish coming from the north (northern cod), the south (south coast cod), or from the east (off the Grand Banks).⁶⁷

The south coast fishing grounds had further advantages over the Strait of Belle Isle region and even the eastern Avalon Peninsula, both of which relied for the most part on migrating fish whose movements were always somewhat uncertain. The Straits area was iced over from February to May in most years, and ice touched the Avalon Peninsula in most late winters and early springs. The south coast, on the other hand, was free of ice year round and cod were present inshore at nearly all times of year. Placentia Bay, which would become the heart of the French south coast fisheries, had large schools of cod that overwintered and spawned in its shallow waters and migrated along its eastern shore seasonally to feed on capelin in its outer reaches.⁶⁸ In addition, fish migrated into outer Placentia Bay from the west and from St. Pierre Bank in spring, typically along the western Burin Peninsula.⁶⁹ The early French fishing grounds in Placentia Bay reflected these seasonal migration patterns, as did their many fishing settlements in and around Placentia Bay. The French claimed the islands of St. Pierre and Miquelon, the nearest land to the rich grounds of St. Pierre Bank, to further exploit a fishery that had a much longer fishing season compared to the Strait of Belle Isle and southern Labrador.⁷⁰

By 1520, France dominated the fishery in the *Terra Neuvfe*.⁷¹ Records from this era are incomplete, but at least 100 vessels were sent from France each spring to fish for cod beginning in about 1520.⁷² By contrast, the Portuguese may have had a dozen vessels and the English only a few.⁷³ English explorer John Rut reported finding eleven Norman, one Breton, and two Portuguese vessels in St. John's harbour on August 3, 1527.⁷⁴ The size of the vessels varied, but may have averaged less than 100 tonnes. It is likely that total landings of French, English, and Portuguese vessels from 1500 to about 1577 were sporadic and relatively modest, seldom exceeding 35,000 tonnes of cod and a few thousand tonnes of salmon.⁷⁵

The early French fishermen named many sites around the Strait of Belle Isle (e.g., L'ance au Loup, Forteau, Blanc Sablon, Isle aux Bois (often pronounced "Isle a boys" today), Belles Amour, Bonne Espérance, Quirpon, and Port aux Choix). Later, many French names took root on the south coast (e.g., Tasse d'argent (which became Tortello John and then just Tortello), Petite Forte, Baie Bois (pronounced "bay boys" today), Merisheen [the current use may come from Mer à Chine or Sea to China], and Baie d'Espoir). On the Avalon Peninsula, the earlier Portuguese and Basque names were shifted to a French form (e.g., Plaisance from an earlier Basque form and Carbonear from Carbonier). Some of the original French names survived nearly intact, such as Harbour

Grace (Havre de Grâce), Port de Grave (based on grève – a shingle used for drying cod), and Petty Harbour (Petit Havre), while others were radically anglicized (e.g., Cap Dégrat, which became Grates Point).[76]

By the 1540s, the Dieppe school of mapping rivalled that of the Italians, and drew cartographers from Britain and France who produced beautifully illustrated maps of the New World. The Dieppe cartographers incorporated earlier Portuguese knowledge as well as the records of contemporary French explorations, especially those of Jacques Cartier. The Dieppe maps showed Newfoundland correctly as an island, although the outline was crude, with improved representation of the Great Northern Peninsula and Strait of Belle Isle areas, but poor representation of the Avalon Peninsula (shown as a separate island) and the south coast (the Burin Peninsula was also shown as an island). These differences reflected the better knowledge of the northern areas at this time. The Dieppe maps also showed that most of the place names in southern Newfoundland were of Portuguese origin, with French names occurring farther to the north; however even these trends are not constant from map to map from this period.[77]

The navigation techniques and routes across the North Atlantic used by the earliest fishing vessels are not well known, but most likely involved a combination of dead reckoning and latitude keeping, and are instructive as to the likely sea routes taken by early navigators from Europe to Newfoundland.[78] If we back-calculate from fishing positions centred in northern Newfoundland and the Strait of Belle Isle, it is most likely that the vessels took a northern route starting from southern Ireland. However, a more southerly route that skirted south of the Gulf Stream and then north across the Grand Banks was adopted sometime later.[79] The Grand Banks are so large and the presence of seabirds and whales is so pronounced that no experienced sailor could have missed them. The sea rides very differently on the Banks than in the open ocean, and soundings with lead lines, as was the practice of the times, would have shown depths of less than 50-100 metres far from sight of land.

In addition to their land-based salt fishery, the French also developed a "green" fishery on the Grand Banks. The origins of this fishery were not recorded. Perhaps the discovery of the Grand Banks led a fisherman to try his luck in this area, which would have proved to be substantial. Another possibility is that south shore fishermen first found the nearby and shallow St. Pierre Bank, a key spawning and feeding ground for the south coast cod stock, and from there found the Grand Banks to the east. The precise date that the Grand Banks fishery began is not known, but it was likely underway by the mid-1500s[80] and well-established by the early 1600s. Maps from 1550s onward show how knowledge of Newfoundland and the Grand Banks accumulated as the fisheries developed. The first maps of the New World in the early 1500s did not show the Grand Banks, likely reflecting the use of the northern trans-Atlantic route which did not cross the Banks. The Dieppe maps of the 1540s still did not show the Grand Banks. The first maps to show the Grand Banks were drawn in the 1550s, but the Banks appear as an unrealistic thin formation – the mapmakers were probably told of the existence of a large bank, but were given few other details. As knowledge and soundings accumulated through the 1600s and 1700s, the bathymetry and outline of the Grand Banks became more accurate. In the early 1600s, Richard Whitbourne wrote "the Banke, which lieth within 25 leagues from South-cape of that Countrey [Newfoundland], where the French use to fish Winter and Summer, usually making two voyages every year thither: to which places, and to the Coast of Canady [likely Nova Scotia and Cape Breton Island], which lieth neere unto it, are yearly sent from those Countries, more than 400 saile of ships."[81]

The French developed a new technique to fish the Grand Banks. They drifted with the wind, with little sail, never anchoring except at night. Cod were caught with single hooks rigged on weighted lines, first baited with salt herring and then with cod gut contents.[82] The fish were split and heavily salted in barrels, and the product taken back to France and marketed in a wet condition known as green fish. This was the first "load and go" fishery.[83] Wet fish were not as durable as hard, air-dried fish, but if they were sold quickly in France, Portugal, or Spain, it didn't matter. Once this type of fishery was established, it may not have been necessary for these ships to land in *Terre Neuvfe* – or even see it, as the main fishing grounds on the Grand Banks are out of sight of land. Fishermen sailed from French ports as early as possible in spring in small, fast vessels in order to arrive on the Banks by April when the cod were most aggregated. The first cod brought back to France in early summer commanded the highest prices. With a swift vessel, a second or even third trip to the Banks could be made later in summer and fall.[84]

Basques

Fishermen from the Basque region of Spain joined the Newfoundland fishery around 1530-1540. French Basque merchants, following their Norman and Breton confederates, chartered Spanish Basque fishing vessels for voyages to Newfoundland in 1541, and by 1547, the Spanish Basque whaling industry in Terra Nova was documented in barrel factories in Spain.[85] The Basque fishery tended to concentrate in the western Strait of Belle Isle and Gulf of St. Lawrence, perhaps partly to avoid the Bretons and Normans who fished farther east. The Straits were a prime area for capelin in spring and early summer, and for herring and mackerel later in the season; all these species attracted both cod and whales. The Basques fished and dried cod, but as whale oil was in high demand in Europe, they became whale specialists. It was a dangerous business: catching whales, even the docile right whales, from small wooden boats with only a hand-thrown harpoon required more skill and daring than did catching and curing cod.[86] Perhaps that was part of the attraction – like bullfighting, whaling may have appealed to the Spanish machismo – or perhaps it was because whaling was the most lucrative of the fisheries. Whatever the reason, the Spanish Basques excelled at it.

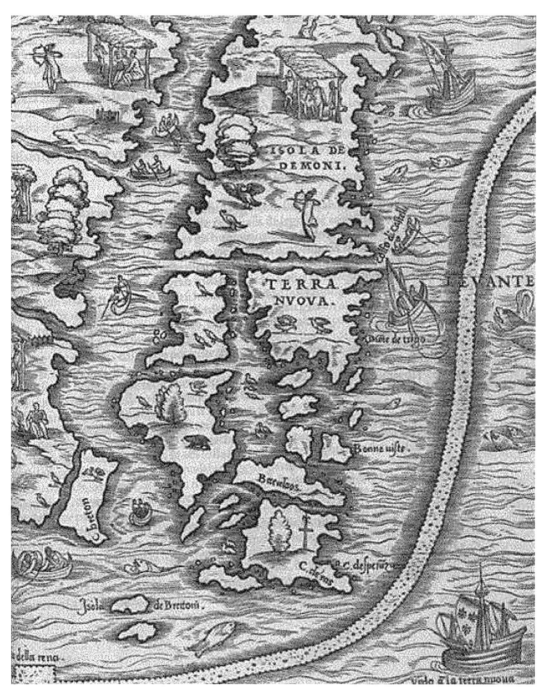

4.1 Italian map of La Nvova Francia (New France) by Giacomodi Gastaldi published by Giovanni Battista Ramusio in 1556. Details are likely based on reports by Cartier and Verrazzano. Features on this map show the Isola dela Renna (Island of Sand, likely Sable Island), C. de Ras (Cape Race), Bacalaos (Baccalieu Island?), C. desper... (Cape Spear) and Bonna Vista (Bonavista) and Golfo di Costell (Strait of Belle Isle). The long feature offshore may be the earliest representation of the Grand Banks. Courtesy of the Centre for Newfoundland Studies, Memorial University of Newfoundland.

Effects of the Fisheries

The period from the late 1400s to mid-1550s was a learning stage in the fishery. By 1577, Newfoundland was known to all the fishing nations of Europe, in particular to Portugal, France, Spanish Biscay (Basques), and England. Its geography had been roughed out, if not known in detail. Knowledge of the many kinds of fish, invertebrates, and marine mammals that were in great abundance in these oceans began to accumulate,[87] and rapid increases in the fishery would follow.

All of the cod stocks, with exception of those in northern Labrador, were being fished by the mid-1500s. Catch levels were poorly documented until the late 1500s, but they are thought to have been modest.[88] Catches were not high enough to impact any but the smallest coastal cod stocks. It is curious that many early harbours were made in Conception Bay, which in later centuries was not known to be a good fishing bay and is not on the main migration routes of the large cod stocks. It is possible that much of the early fishery in Conception Bay was based on a local coastal stock that was depleted early on, beginning in the mid-1500s by Portuguese fisheries.

Most other species of fish were not exploited at all at this time. The Europeans did, however, decimate seabird populations – especially the great auk, which was slaughtered for food in large numbers by the earliest French vessels.[89] Although accurate population estimates for these birds are unavailable for this period, it is likely that significant reductions in their numbers began early in the 1500s.

FRENCH EXPLORATIONS

Giovanni da Verrazzano

Similar to several European monarchies, the French "House of Ango" had great ambitions to find a sea route to the Orient. It helped sponsor the voyage of yet another Italian adventurer under the flag of France, and Giovanni da Verrazzano sailed from Dieppe in 1523 with a fleet of four ships. Two were lost early in the journey, while another returned after looting in Spain. Verrazzano's remaining ship took a southerly route along 32° 30' north latitude, then veered northwest after a storm and set along 34° north latitude. He made land at or near Cape Fear, North Carolina, then cruised south to Florida and redoubled north to land near Kitty Hawk, North Carolina. As seemed to be the custom of the times, the adventurers kidnapped an aboriginal child, and tried but failed to make off with a beautiful young woman.[90] Verrazzano then sailed north again to the waters off Maine, coursing past present-day Nova Scotia and the Avalon Peninsula of Newfoundland. His most northerly measured latitude was 49° 50' north latitude, which put him in the vicinity of Funk Island or Fogo Island. He used

Portuguese names for all recognized landmarks in Newfoundland, but he made no new landfalls and claimed no new discoveries. His failure to find a route to the Orient did not inspire his political and commercial backers.

Jacques Cartier

The fishing grounds of the *Terre Neuve* were about to get another boost, albeit unintentionally, from the master mariner Jacques Cartier. Cartier was born in St. Malo, Brittany, in 1491, into a family of respected mariners. Cartier first sailed local boats and learned both the theory and practice of his craft well. Unlike the limited successes and scanty records of his predecessors, Cartier made three extensive voyages to the New World without losing a ship, entering many new harbours and keeping accurate and well-documented records of his discoveries. Although unconfirmed, Cartier may have been a crewman on Breton ships prior to his own voyages that began in 1534 – his ease of navigation along the northeast coast of Newfoundland and entry into the Strait of Belle Isle suggested that this was not his first time in these waters.

Cartier left St. Malo on April 20, 1534, and struck due west along 48° 39' north latitude. He made land at Cape Bonavista (48° 42' north latitude; Cartier recorded it as 48° 30' north latitude). From this record, we can surmise that Cartier was confident in measuring latitude and that he knew that holding west from St. Malo would lead to a known landfall on the *Terra Neuve*. Ice blocked his way off the northeast coast, so he slipped into Trinity Bay to Havre Ste. Katherine (Catalina), as would any modern sailor. He then sailed west and north to the L'Isle des Ouaisseaulx (Funk Island) and loaded up with fresh meat. Cartier referred to the great auk as the *apponatz*,[91] its Beothuk name, indicating that he may have met the Beothuks. He then cruised north to Cape Degrat and to the harbour at Karpont (Quirpon), setting new standards in navigation and accuracy. He recorded the latitude of Karpont as 51° 31' north latitude – this measurement was as accurate as any until the invention of satellite-based global positioning systems.

Cartier then crossed the Strait of Belle Isle and reached the beaches at Blanc Sablon and the French fishing post at Brest. He noted large numbers of seabirds here and feasted on the eggs of eider ducks. Somewhat further on, around St. Augustine, he encountered a fishing vessel from La Rochelle, France. Just offshore, he was reported to have fixed a cross on a small round island as a landmark for future voyages. French fishing enterprises were also active in the area near Natashquan; Cartier reported taking aboard a party of Montagnais Indians fishing for Captain Thiennot there. Cartier may have known the captain, as he named Cape Thiennot after him. Cartier then doubled back to Blanc Sablon and took only three weeks to return to St. Malo on September 5, 1534. Cartier's explorations in the Gulf led early mapmakers to label the north shore as "Nouvelle France." In a later voyage in 1536, Cartier claimed the islands of St. Pierre and Miquelon for France, although they received little French attention until the early 1600s.[92] Three hundred years of conflict between English and French interests in Newfoundland followed.

Cartier stands alone among New World explorers of the early 1500s. His record of seamanship, navigation, record keeping, and exploration is unrivalled among his peers. Although he is better known for his discovery of the River of Canada (the St. Lawrence) and the native settlements at Stadacona (Québec) and Hochelaga (Montreal), Cartier provided the first descriptions of the west coast of the island of Newfoundland. He established that Newfoundland was an island, not an extension of the mainland: Dieppe cartographers from the early 1540s onward clearly depicted Newfoundland as an island or a collection of islands, while Spanish maps based on Portuguese information continued to show Newfoundland as an extension of the mainland until decades later.[93] In addition, Cartier used the name *Terra Nova* to refer to these islands alone, whereas it had previously described all of the New World. Cape Anguille was Cartier's last view of Newfoundland, spotted through the mist before he headed home. In spite of all his discoveries, however, he seemed disappointed – for on the west coast of Newfoundland, he had found nothing but fish.

EARLY CONFLICTS

The fishery would likely have developed more quickly during the 1500s were it not for the sporadic outbreak of wars between the European fishing nations. One of the first and most brutal of these actions occurred in the 1550s. In 1554, French Basque whalers attacked the Spanish Basques at their stronghold in the Strait of Belle Isle and the Spanish returned the favour by attacking the French cod fisheries on the Avalon Peninsula. There are various accounts of the amount of damage inflicted, but there are indications that at least 300 French vessels were either captured or sunk, and that the French Newfoundland fleet was effectively destroyed.[94] This is a realistic number given that that there were commonly 500 French vessels fishing at Newfoundland from March to August in the 1570s.[95] The French grip on the Avalon fishery was loosened during this conflict, but was not replaced by the Spanish or Portuguese as they were more interested in whales and bank fishing. These events encouraged the English to assert themselves in the Avalon fishery, which they did over the next two decades.

THE ECOSYSTEMS

During the early years of the export fisheries, most aspects of the vast marine ecosystems of Newfoundland and Labrador continued to function as they had since the last ice age. There are scant records from this era, but all existing documentation points to an abundance of many species, sufficient to evoke considerable economic interest in Europe. A work published in 1552 described "a great stretch of land and coast which they call Bacallaos, extending to forty eight and a half degrees [north latitude]. Those living there call a large fish *bacallao*, of which there are so many that they impede ships in sailing."[96] This was undoubtedly an overly enthusiastic description, likely based on reports given by the mischievous Sebastien Cabot or the performance of a very poor ship. Exaggerations aside, however, there can be little doubt that large and dense schools of cod were encountered, and other species were almost certainly in considerable abundance. The fishery had very modest impacts during this period as evident in the marginal influence on the outputs of the ecosystem model. As the northern hemisphere began to cool in the late 1500s (see Chapter 1), climate would remain the dominant influence on fish productivity and biomass in the marine ecosystems of Newfoundland and Labrador.

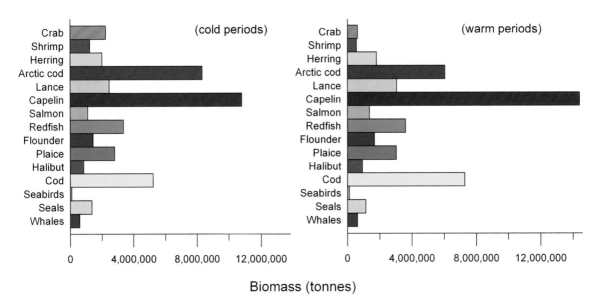

4.2
Newfoundland and Labrador ecosystem species abundance from AD 1450-1576 (warm and cold periods).

NOTES

1. See Magnusson and Pálsson (1965, 24-25).

2. Icelandic geographers and poets regaled northern European society with stories of a new world in the tenth and later centuries. An Icelandic geographical treatise dating from around 1300 indicated that "to the north of Norway lies Finnmark [Lapland]; from there the land sweeps north-east and east to Bjarmaland [Permia], which renders tribute to the King of Russia. From Permia there is uninhabited land stretching all the way to the north until Greenland begins. To the south of Greenland lies Helluland [Baffin Island?] and then Markland [Labrador?]; and from there it is not far to Vinland [America-Newfoundland]" (Magnusson and Pálsson 1965, 15). An Icelandic map produced in 1570 showed Vinland in a location that could only be Newfoundland.

3. See Ehrenberg (2006, 56).

4. See Ehrenberg (2006). The compass was perfected in Italy in the 1300s and led to a cartographic revolution in Europe and the Middle East.

5. Dead reckoning is a navigation method whereby successive positions are determined from previous positions by knowing direction and speed. It requires a compass and a means to measure speed. Latitude holding or keeping requires celestial observation and instruments to measure the position of stars (astrolabes and other inventions) that became available in the late 1400s. Latitude keeping was far from foolproof, but a good navigator should have been able to fix his latitude within one degree (60 nautical miles). Longitude was much more difficult, if not impossible, to determine during this era.

6. Prowse (1895, 24), citing earlier sources, stated that "in 1400 England had fishery rights around the Island, but even in 1360 it must have been a regular voyage, for one Nicolas of Lynn had been five times there, on one occasion making the voyage in a fortnight." The *Icelandic Annals* confirmed the presence of English fishermen in Icelandic waters by the early 1400s, and reported the frequent conflicts between the English and the Norse. In 1336, "Arne Abbot of Lysa [was] beheaded off England and all his ship's crew [slain] by the Duggars and all cast overboard." In 1467, "Sir Bjorn was slain by the English at Rif…and in vengeance thereof afterward in the summer she (the Lady Olaf) had all the English slain and she had twelve of them bound with a rope and all their heads cut off."

7. Pope (2004, 12) mentioned that, in the late 1400s, Devon fishermen were "involved in a new seasonal fishery on the west coast of Ireland, where they produced salt fish at shore camps, as they would do a century later in Newfoundland."

8. A letter from John Day, an English merchant with dealings in Spain, written sometime between December 1497 and January 1498 to the Lord Grand Admiral of Spain, who was almost certainly Christopher Columbus, was uncovered in the mid-twentieth century, published in Vigneris (1956), and now resides in the Spanish Archives. Cabot's voyage was described in the letter: "It is considered certain that the cape of the said land was found and discovered in the past by the men from Bristol who found 'Brasil' as your Lordship well knows. It was called the Island of Brasil, and it is assumed and believed to be the mainland that the men from Bristol found." This translation was taken from Cumming et al. (1971).

9. Cuff (1997, 61) described the early legends of Atlantic islands west of Ireland: "Aside from St. Brendan's islands, two other mythical islands might be mentioned as having particular significance. Brasil, which by popular belief either floated or appeared every seven years, was frequently depicted off the coast of Ireland. Farther to the south was Antillia (or the Isle of the Seven Cities). According to Portuguese legend, the Seven Cities were populated by Christians who had been forced by the Moors to leave Portugal in the eighth century." It is evident from the letter of John Day after Cabot's first voyage that these legends were widespread and not regarded as total fiction.

10. The route of the early discoverers is not known, but if they were blown off course from Iceland, they may

[11] Prowse (1895, 8) cited early documents that hinted of trips west prior to 1497, but "if they bore any results they were kept secret."

[12] See Rowe (1980, 46).

[13] Prowse (1895, 5) gave background on the Age of Discovery in Europe as the ignorance of the Middle Ages was shaken off: "Many remarkable events combined to make this age illustrious; the invention of printing, the general use of gunpowder and artillery in war, and the capture of Constantinople by the Turks. This last event put to flight numerous Greek scholars, who bore their precious manuscripts to Italy and thus inaugurated the new learning commenced in England by an Oxford scholar, Grocyn, bringing Greek writings to his ancient university. But of all the causes, great and powerful…undoubtedly the greatest influence of all was the daring maritime enterprise which simultaneously discovered New Continents in both East and West…all wonderful things seemed possible."

[14] The Grænlendinga Saga included the account of a man named Thorvald who, along with his son Eirik the Red "left their home in Jæderen, in Norway, because of some killings and went to Iceland, which had been extensively settled by then; so to begin with they made their home in Drangar, in Hornstrands. Thorvald died there, and Eirik the Red then married Thjodhild and moved south to make his home at Eirikstead, near Vatnshorn. They had a son called Leif. Eirik was banished…after killing Eyjolf Saur and Hrafn the Dueller, so he went west to Breidafjord and settled at Oxen Island…[further troubles led to Eirik being] sentenced to outlawry at the Thorsness Assembly…[after which] Eirik prepared his ship…for a sea voyage…and told [his friends] that he was going to search for the land that Gunnbjorn, the son of Ulf Crow, had sighted when he was driven westwards off course." Eirik located, explored and settled in West Greenland (Magnusson and Pálsson 1965, 49-50).

[15] Karlsson (2000, 11) stated that the Norse discovered Iceland and continued west as the result of an "insatiable lust for travel." While this is undoubtedly true, many of the original Icelandic settlers were likely fleeing various troubles in Norway or the Celtic countries. This was a very different motivation than that of the later European exploiters and colonizers, which was purely commercial and designed to enrich the home country.

[16] Pope (1997, 13-14) discussed Cabot's oft-miswritten name.

[17] See Pope (1997, 16-23).

[18] Cabot and other navigators made a geographical deduction that if they sailed west, they would end up east in China or Japan. Cabot may have believed that a northern route would be the shortest route to get there. Stories of a large island and Vinland to the west fit into his worldview, but there is no clear historical record of exactly how. An early map indicated a belief that there were islands between Europe and China; Vinland was perhaps thought to be one of these. Some accounts indicated that the Bristol merchants had sent out earlier voyages to islands they called "the Seven Cities" or "Brasil," but these must be regarded as legends. Two recent accounts of the Cabot story are interesting to compare. Williams (1996, 9) stated that "there is nevertheless consensus that neither Columbus nor Cabot knew about the Norse voyages, even though Columbus may have visited Iceland." It is likely that Cabot had also visited Iceland. In contrast, Fardy (1994) related how the Bristol merchants, based on a sure knowledge of the Norse voyages gained in their ample trade with Iceland, attempted westward voyages of exploration as early as 1480. By this account, the Cabot voyages occurred two decades after the first attempts. Unfortunately, few records survived from these earlier voyages. Whichever is the truth, there is consensus that no European knew of the extent of the American continents.

[19] Anspach (1819) related the background of Cabot's early dealings in Bristol. It is not possible to confirm the initial reason why Cabot was in Bristol, but given his background and the fact that Bristol was the leading trading area with Iceland, it likely concerned trade in cod.

[20] John Day's letter to the Lord Grand Admiral is the only surviving document that describes the course of Cabot's first voyage, his landfall in the New World, and Newfoundland: "They left England towards the

end of May, and must been on the way 35 days before sighting land; the wind was east-north-east and the sea calm going and coming back, except for…a storm two or three days before finding land; and going so far out, his compass needle failed to point north and marked two rhumbs below. They spent a month discovering the coast and from the above mentioned cape of the mainland which is nearest to Ireland, they returned to the coast of Europe in 15 days. They had the wind behind them, and he reached Brittany because the sailors confused him, saying that he was heading too far north. From there he came to Bristol, and he went to see the King" (Vigneris 1956).

[21] The first map to clearly show Newfoundland was made by Ruysch, who may have been Cabot's Burgundian mapmaker. His map was published in the *Ptolemy* at Rome in 1508.

[22] Fardy (1994, 54) showed a map of the North Atlantic and the course from Dursey Head, Ireland, to the tip of the Great Northern Peninsula along latitude 51° 35' north latitude.

[23] Sebastien (also spelled Sebastian) Cabot was a son of John Cabot and did little to shed light on his father's explorations for which he claimed some credit. In much of the earlier literature, there is confusion about the trips and accomplishments, fed by the misinformation of Sebastien. Overall, he appears to have been more armchair courtier than explorer. A voyage he supposedly led in 1508 produced no lasting record. Cuff (1997, 81) stated that "in short, Sebastien Cabot was a successful, lifelong blowhard – and the older he got, the harder he blew."

[24] John Day's letter to the Lord Grand Admiral of Spain stated: "I am sending…a copy of the land which he has found. I do not send the map because I am not satisfied with it…from the said copy your Lordship will learn what you wish to know, for in it are named the capes of the mainland and the islands, and thus you will see where land was first sighted, since most of the land was discovered after turning back…the cape nearest to Ireland is 1800 miles west of Dursey Head which is in Ireland, and the southernmost part of the Island of the Seven Cities is west of Bordeaux River…he landed at only one spot of the mainland, near the place where land was first sighted…with a crucifix and raised banners." Also: "Thus your Lordship will know that the cape referred to nearest to Ireland is 1800 miles west of Dursey Head which is in Ireland, and the southernmost part of the Island of the Seven Cities is west of Bordeaux River, and your Lordship will know that he landed at only one spot of the mainland, near the place where land was first sighted, and they disembarked there…" From this description, landfall was made at Cape Degrat on the Great Northern Peninsula or in southern Labrador. It is highly unlikely that a skilled navigator could have traversed the Grand Banks – which rise to less than 50 metres depth in several places, break the seas at the Virgin Rocks, and are alive with seabirds (a sure sign of nearby land) – and missed all of southern Newfoundland to reach landfall in Cape Breton. A route to the south of the Grand Banks and Newfoundland to Cape Breton is far-fetched. On the other hand, Day's description in his letter and a direct latitude-holding route (51° 33' north latitude) across the North Atlantic from Cape Dursey, Ireland, brings landfall at Cape Degrat, a prominent cape rising over 150 metres out of the sea and visible for over 24 kilometres (15 miles) from sea. From this account, it seems inescapable that the "Island of the Seven Cities" is Newfoundland, and that landfall most likely was made on the mainland in southern Labrador. In the same letter, Day confirmed that Cabot made an earlier attempt from Bristol in 1496, but returned to port after various difficulties. The discovery of an already-named island begs the question of who named it and when. Various royal documents – for example, Cabot's pension – referred to Newfoundland as the New Isle. Sometime after this, the name New-founde-land came into use, but early on it applied to all of the northern new world. The Portuguese and Spanish used various forms of the Latin name Terra Nova, meaning "new land," for the same region. Details of the Day letter and Cabot's landfall can be found in many works on the subject – see, e.g., Fardy (1994), Pope (1997), and Cuff (1997).

[25] See Prowse (1895, 6).

[26] See Murray (1829, 64).

[27] di Soncino's letter was published in 1865 in the *Annuario Scientifico* – the original is housed in the State Archives of Milan, Italy. di Soncini was apparently an acquaintance of Cabot. He wrote: "The king has

gained a great part of Asia without a stroke of the sword. In this kingdom is a popular Venetian called Zoanne Caboto, a man of considerable ability, most skilful in navigation, who having seen the serene kings, first him of Portugal, then him of Spain, that they had occupied unknown islands, thought to make a similar acquisition for His Majesty [Henry VII]. And having obtained the royal privileges which gave him the use of the land found by him, provided the right of possession was reserved to the Crown, he departed in a little ship from Bristol with 18 persons, who placed their fortunes with him. Passing Ibernia [Ireland] more to the west, and then ascending towards the north, he began to navigate the eastern part of the ocean, leaving for some days the north to the right hand, and having wandered enough he came at last to firm land, where he planted the royal banners, took possession for his Highness, made certain marks, and returned. The said Messer Zoanne, as he is a foreigner and poor, would not be believed if his partners who are all Englishmen, and from Bristol, did not testify to the truth of what he tells. This Messer Zoanne has the description of the world in a chart, and also in a solid globe which he has made, and he shows where he landed; and that going toward the east he passed considerably beyond the country of the Tanais… The sea is full of fish which are taken not only with the net but also with a basket in which a stone is put so that the basket may plunge into water…And the Englishmen, his partners, say that they can bring so many fish that the kingdom will have no more business with Islanda [Iceland], and that from this country there will be a very great trade in the fish they call stock fish. They say, now they know where to go, the voyage will not take more than 15 days if fortune favours them after leaving Ibernia…The Admiral, as Messer Zoanne is already styled, has given his companion, a Burgundian, an island, and has also given another to his barber, a Genoese, and they regard themselves as Counts, and my lord the Admiral as a Prince. And I believe that some poor Italian friars will go on the voyage, who have the promise of being Bishops. And I, being friend of the Admiral, if I wished to go, could have an Archbishoprick."

[28] Prowse (1895, 13) stated that, "according to Spanish writers, John Cabot was a kind of second-class Columbus, and had been both to Lisbon and Seville trying to get aid for his scheme of Transatlantic discovery. They admit, however, that there is no very reliable authority for this statement." The Spanish view of Cabot is undoubtedly biased and political. Cabot was an extraordinary navigator and explorer, and tackled much more dangerous waters than did Columbus. If Cabot is to be faulted, it is for his lack of systematic record keeping. However, the letter of John Day confirmed that Cabot did have his own maps, and perhaps records, but these have never been located. It is a great loss to history.

[29] Juan de la Cosa was a well known Biscayan navigator and geographer, and the sailing master on Columbus's *Santa Maria* on the famous voyage of 1492. His chart of the Atlantic in 1500 contained information on the Columbus voyages, but also from much farther north than Columbus ventured. The chart included flags and marked capes and islands that have been interpreted as derived from Cabot's 1498-1499 voyage. Alas, the map is difficult to interpret, and raises more questions than answers. See Wilson (1991) for further information.

[30] Pope Alexander VI wrote a Papal Bull dividing the New World between Portugal and Spain. The Portuguese area was to the east (e.g., Brazil) and the Spanish to the west (e.g., Argentina, Chile, Mexico). Newfoundland was contested, and France, a Catholic country, later had to deal with this Bull. England, on the other hand, paid little attention to it.

[31] Prowse (1895, 14) stated that "Spain and Portugal claimed the new world as their own; they looked upon the English as poachers on their preserves…certainly by his charters Henry VII repudiated any such interpretation." Perhaps Pope Alexander was unaware of English ambitions in the new world, or maybe regarded the Spanish and Portuguese as more loyal Catholics (foreshadowing Henry VIII's break with Rome).

[32] Prowse (1895, 4) gave the chronology of quickly unfolding events of European exploration in the last decade of the 1400s and the first decade of the 1500s. Although the exact routes of the Corte Real voyages, like those of the Cabots, are not known, there has been a considerable amount of speculation. Prowse concluded that "he sailed along the Greenland coast and made this island and Labrador," although he also indicated that the accounts of the voyages were "shadowy" (Prowse 1895, 14-15). As with the

voyages of Cabot, without additional information it will be impossible to know with any certainty where these early explorers went.

[33] Around 1506 or 1507, maps with detailed descriptions of Africa and Asia and the first sketches of the New World were drawn in Europe by mapmakers Giovanni Contarini, Marin Waldseemüller, and Johannes Ruysch. None of these maps depicted Newfoundland correctly. Ruysch's map actually showed North America as part of Asia, although Waldseemüller's used the name America. Ruysch's map clearly showed Newfoundland, identified as Terra Nova, with markings for C. de Portogesi (Cape Race?), N. Baccalavras (Baccaleu Island?), R. Grado (Gander River?), Baia de Rockas (Notre Dame Bay or Bay of Rocks), and C. Glaciato (which could either be Cape Bauld or Cape Degrat covered with snow). These landmarks were likely drawn from a combination of Cabot's findings and early Portuguese sources, but their exact locations could not have been clear to Ruysch, as the names were only generally placed on his map. See Morison (1978, 69) for further details.

[34] Morison (1978, 90) compared Portuguese names on maps from the 1520s with modern landmarks.

[35] See Seary (1971) for the origin of place names of Newfoundland.

[36] Most of this view can be attributed to various poorly documented Portuguese sources, identified by Abreu-Ferreira (1995), which were repeated in English by Patterson (1890), Prowse (1895), and Innis (1940). Repetition, however, does not measure accuracy, and there is little hard evidence to support these claims. Perhaps they are as thin as the paper claims of the family Corte Real to be "Governors of Newfoundland" long after the establishment of the French and English fisheries.

[37] See Abreu-Ferreira (1995).

[38] See Sabine (1853, 34).

[39] Sauer (1971, 12) stated: "In 1506 the Portuguese Crown imposed a tax in its northern ports on cod brought from Terra Nova, the name used for the land found by the Corte Real voyages of 1501 and 1502. It does not follow that the Portuguese commerce in cod resulted from these later voyages, which did not sail from nor return to those north Portuguese ports. Fishing in Portuguese home waters was a simple matter of small boats and nets, going out overnight, as anchovies and tunny still are fished. Cod are found in waters distant from Portugal, those of Iceland being of early fame. This mode of fishing is still characteristic of the Portuguese ports named in 1506. The *bacalhao* fishery, as it is known in Portugal, required cargo ships that stayed at sea for months, carried small rowboats for fishing by hook and line, and had room where the catch could be dressed, salted and stored. It is unlikely that such a different kind of fishing, gear, and ship, the necessary working capital, and the establishment of a market could have taken place in the short time after the Corte Real voyages."

[40] See Abreu-Ferreira (1995).

[41] The Harleyan map of about 1542 was drafted by John Rotz with English titles, but mostly Portuguese-derived landmarks, and is shown in Prowse (1895, 34, 41). The French map drawn by P. Descellier from 1553 is shown in Prowse (1895, 40). Both maps were typical of the era and depicted Newfoundland as a disjointed group of islands. The French map showed new landmarks, including Forillon (Ferryland), B. de conception (Conception Bay), Ys. des oiseaulx (Funk Island), and Bellisle (Belle Isle). The French name for Conception Bay was also used on the English map.

[42] See Ehrenberg (2006, 88).

[43] Patterson (1890, 145) stated: "Immediately after Gaspar Cortereal's first voyage – 1500 or 1501 – fishing companies were formed in Viana, Aveiro, and Terceira, Portugal, for the purpose of founding establishments in Terra Nova. In 1506 the King of Portugal gave orders that all fishermen returning from Newfoundland should pay a tenth part of their profits at the Custom House. At different times Aveiro alone had 60 vessels sailing to Newfoundland, and in 1550 150 vessels. Equal numbers sailing from Oporto and other ports, gave a large increase of revenue." However, there is little evidence that this fishery was prosecuted consistently. A search of the Portuguese archives for records of cod from Terra Nova showed a very modest involvement of Portuguese vessels in the Newfoundland fishery in the early

44 Abreu-Ferreira (1995, 40-42) provided descriptions of early Portuguese cod grading systems and the likely sources of the fish, and stated that "except for the *refugo*, which was clearly a refuse or inferior quality of the other two, little is known about what exactly was *vento* or *pasta* cod. The *vento* name suggests that the cod in question was wind dried…while the *pasta* was probably wet or green cod …almost all *pasta* entering Porto arrived in French ships, while *vento* was brought in primarily by the English."

45 Cell (1969, 3) summarized arguments that Bristol merchants discovered Newfoundland prior to Cabot's voyage. The principal of the royal charter issued in December 1502 to the company of adventurers into the *New-founde-land* was Hugh Elyot of Bristol. Some historians have suggested that Bristol sent one or two or more vessels to Newfoundland each year after 1502, but records supporting this idea do not exist.

46 See Morison (1978, 103-105).

47 Prowse (1895) argued that English fisheries were dominant and substantial after Cabot's voyage, but there is little evidence to support that view.

48 See Cell (1969, 4).

49 Cell (1969, 4) related how Humphrey Gilbert took this southern route.

50 See Seary (1971).

51 See Morison (1978, 124).

52 Pope (2004 ,15) stated that "French and Iberian records of the middle decades of the sixteenth century suggest a scale of effort by Bretons, Normans, and Basques unmatched, until the 1570s, by the few English ports that occasionally took part in the early industry [in Newfoundland]."

53 The Dieppe maps were some of the finest of this era. They were coloured works of art and represented the state of geographical knowledge in Europe in the early 1500s. See Ehrenberg 2006 for further details.

54 Morison (1978, 115) indicated that "in twenty years, 1520-40, his ships captured prizes worth a million ducats. The Portuguese condemned several Ango ships for trading with their possessions, and he retaliated by blockading Lisbon on his own account, with the approval of Admiral Chabot de Brion."

55 See Morison (1978, 124).

56 See Morison (1978, 118).

57 See Morison (1978, 126).

58 Morison (1978, 120-121) also stated that other information existed which suggested that there were one hundred departures from La Rochelle for Newfoundland in the years 1534-1565.

59 The *Dictionary of Commerce* indicated that Brest "was the chief town of New France, the residence of the Governor, Almoner, and other public officers; the French draw from thence large quantities of *baccalao*, whale fins, and train (oil), together with castor (beaver) and other valuable furs" (Prowse 1895, 596). There may have been as many as 200 houses and 1000 winter residents in Brest, indicating that it was likely the largest European settlement in the New World in the 1500s. It is no coincidence that most place names from Bonne Espérance to the Labrador Straits are French. Brest declined after 1600, for reasons that not are entirely clear. French interests shifted markedly towards the Quebec and Nova Scotia areas around this time.

60 Cushing (1988, 58), based on de la Morandière (1962).

61 The Baie des Chasteaux was named by early sailors for the castlelike features of an island in what is now Henley Harbour, Labrador. The castle rocks were distinct and easily recognized from sea, and provided a landmark for early sailors as well as a base for fishing.

62 Competition with the Portuguese is suggested as a reason for the northern French presence by Cuff (1997).

63 See Rose and Leggett (1988, 1989, 1990).
64 Cushing (1988, 61), based on de la Morandière (1962).
65 Conclusions reached by Templeman and Fleming (1962), based on tagging studies.
66 See Nakashima (1992).
67 Doug Howlett and Tom Best (commercial fishermen, Petty Harbour, Newfoundland and Labrador), personal communication, 1992.
68 See Lawson and Rose (2000).
69 See Mello and Rose (2005).
70 Sabine (1853, 32) after describing the French shore and bank fisheries, stated that "the third fishery, at St. Pierre and Miquelon, is similar in some respects to that between Cape Ray and Cape St. John [the French Shore], on the coast of Newfoundland…As this is the only business of the islands nearly all the men, women and children are engaged in catching or curing. The season opens in April, and closes usually in October."
71 In a letter to Richard Hakluyt in 1578, the Bristol merchant Anthony Parkhurst wrote that the Newfoundland fishery in the 1570s consisted of approximately 50 boats from England (dry cod from land), 20-30 Basque whalers, 100 Spanish wet-cod vessels, 50 Portuguese wet-cod vessels, and 150 French vessels (mostly wet cod, but some dry cod from land) (Quinn 1979, 7-10). These numbers were estimates, but gave some idea of the composition of the fishery prior to the English expansion in the late 1570s.
72 Prowse (1895, 49) stated that "the French fishing fleet was by far the largest during the early period from 1504 to 1580," and "French fishing in Newfoundland was so flourishing in 1540 that the authorities of St. Malo had to stay ships in order to get Cartier a crew. In 1541 and 1542 no less than sixty French ships were fishing in Newfoundland."
73 The exact number of vessels that prosecuted the Newfoundland fishery during the 1500s is uncertain, but there were not many. Regarding English vessels, Cell (1969, 22), reviewing earlier letters and reports, wrote that "port book evidence suggests that only a handful of vessels went annually to Newfoundland during the greater part of the sixteenth century and that the trade did not yet have a significant part of the commerce of the west country. A Bristol man, Anthony Parkhurst, who had considerable personal experience of the fishery, reported that in the early 1570s the English fleet consisted of only four small barkes. While this figure may well be something of an underestimate in support of Parkhurst's special pleading for greater English involvement in the whole St. Lawrence area, the fact remains that no Newfoundlanders are recorded as entering Plymouth in 1568-69 or in 1570-71, and that only three came into Dartmouth in 1574 and none in 1575-76. The returns to Exeter are no more impressive." There were early assumptions that the Portuguese fleet was large in the early 1500s. Chadwick (1967, 4) wrote that "by 1530 close on 150 English vessels had been diverted from Icelandic waters. The Portuguese were even more fully involved, with fishermen from the northern and north-eastern ports of France not far behind." However, Chadwick provided no evidence to support this statement. In another example, Prowse (1895, 40) attempted to refute statements that only 50 vessels were fishing in Newfoundland in 1517, largely by declaring that Portugal had more than that. Prowse also provided no evidence of this. A recent examination of Portuguese fishery import records (Abreu-Ferreira 1995) yielded no support for Prowse's contention. The evidence indicates that the French fishery was the largest, but it is unlikely that it exceeded 100 vessels averaging 100 tons (102 tonnes) until the late 1700s (de la Morandière 1962).
74 The text of the Rut letter is given in Prowse (1895, 40-41).
75 The catch records of this early era are incomplete and it is possible that catches were considerably higher. Rose (2004) provided a summary of the best estimates. For comparison, consider Whitbourne's estimates that a 100-ton (102-tonne) vessel could carry the equivalent of 360 tonnes of live fish (see Whitbourne 1620). Most vessels were likely smaller than this. If there were 120 vessels from all countries, which may not be far off given early French records, with average vessel sizes ranging from 50

to 100 tons (51 to 102 tonnes), the capacity of the fleet would have been between 22,000 and 43,000 tonnes of harvested live fish.

76 See Seary (1971, 34-41).

77 See Ehrenberg (2006).

78 Brière (1997, 50), citing earlier French memoirs that described later crossings to the Grand Banks, stated: "The ships, packed with salt, went to take up their positions between the 43rd and 45th parallels, in the same latitude as the southern edge of the Grand Bank[s]. In case of a departure very late in the season, they would position themselves between the 45th and 47th parallels, as the fishery then shifted toward the north. They would then run straight toward the west. The Atlantic crossing lasted about a month, often a bit more, but never less than three weeks."

79 See Whidborne (2005, 49).

80 Morison (1978, 126) gave an earlier date for the beginnings of the Grand Banks fishery, perhaps the 1510s: "We have indubitable evidence that these intrepid fishermen were accustomed to make two fishing trips annually to the Grand Bank[s]. The first set out in late January or early February and, braving the winter westerlies of the North Atlantic, returned as soon as their holds were full; then discharged, started off again in April or May, and were home in September. This, of course, was the 'wet' fishing, which required no call at a Newfoundland harbour. Saint-Malo set aside a rocky section of the shore called Le Sillon for curing the fish, as early as 1519." Morison did not say what his indubitable evidence was.

81 See Whitbourne (1620, 11).

82 Cushing (1988, 57) based on de la Morandière (1962).

83 Trawler fisheries in the later half of the 1900s were referred to as "load and go" when the overwintering and prespawning aggregations of cod were located. The implication is that loading took little time; they filled their holds with fish in short order and returned to port.

84 See Morison (1978, 126).

85 Quinn (1979, 91-93) described a contract for the charter of Spanish Basque ships to French Basque merchants from La Rochelle, France, to fish in Newfoundland in 1541, and another contract for several hundred barrels to be used in the Spanish whale fishery in Labrador in 1547. Anthony Parkhurst related in his letter of 1577 that, in addition to cod, "nowe shal yowe understand what other commodytes may growe by that cuntrye more then hitherto we have had. Chefely above all other the kyllynge of wale, which woulde be one of the rytchest trades in the worlde, as the bascons knowe right well, that use that trade only" (Quinn 1979, 6).

86 Quinn (1979, 84) stated that "whalers were the aristocrats of the fisheries and the Spanish Basques rigorously defended their bays and camps along the north shore of the Strait of Belle Isle." The whaling station that operated at Red Bay in the mid-1500s is now the Red Bay National Historic Site of Canada. Quinn (1979, 83) also related that the "Spanish Basques, uniquely, made the best harpooners."

87 Some of the earliest biological reports came from Anthony Parkhurst. In a letter to Richard Hakluyt in 1578 (Quinn 1979, 7-10), he wrote that the seas around Newfoundland had many types of fishes in addition to cod: "As touching the kindes of Fish beside Cod, there are Herrings, Salmons, Thornebacke, Plase, or rather wee should call them Flounders, Dog fish, and another most excellent of taste called of us a Cat, Oisters, and Muskles, in which I have found pearles…There are also other kinds of Shel-fish, as limpets, cockles, wilkes, lobsters, and crabs, also a fish like a Smelt [called by the Spaniards Anchovas, and by the Portugals Capelinas] which commeth on shore, and another that hath the like propertie, called a Squid: these be the fishes, which (when I please to bee merie with my old companions) I say, doe come on shore when I commaund them in the name of the 5 ports, and conjure them by such like wordes: These also bee the fishes which I may sweepe with broomes on a heape, and never wet my foote, onely pronouncing two or three wordes whatsoever they be appointed by any man, so they heare my voice: the vertue of the wordes by small, but the nature of the fish great and strange. For the Squid, whose nature is

to come by night as well as by day, I tell them, I set him a candle to see his way, with which he is much delighted, or els commeth to wonder at it as doe our fresh water fish, the other commeth also in the night, but chiefly in the day, being forced by the Cod that would devour him, and therefore for feare coming so neere the shore, is driven drie by the surge of the Sea on the pibble and sands. On these being as good as a Smelt you may take up with a shove-net as plentifully as you do Wheate in a shovel, sufficient in three or foure houres for a whole Citie…I may take up in lesse than halfe a day Lobsters sufficient to finde three hundred men for a dayes meate." Parkhurst's sense of humour may have clouded his biological reports (he later described how he commanded his dog to fish), but the descriptions of the various species appear sound, and he provided one of the earliest descriptions of beaching capelin.

88. Hutchings (1995) presented a series of catches for the 1500s that were much lower than those provided by Rose (2004). Pope (1995) also questioned Hutchings's low estimates for the later 1600s.

89. Parkhurst's letter of 1578 (Quinn 1979, 7-10) stated that: "There are Sea Guls, Murres, Duckes, wild Geese, and many other kinds of birdes store, too long to write, especially at one Island named Penguin, where wee may drive them on a planke into our ship as many as shall lade her. These birds are also called Penguins, and cannot flie, there is more meate in one of these then in a goose: the Frenchmen that fish neere the grand baie [Gulf of St. Lawrence], doe bring small store of flesh with them, but victuall themselves always with these birdes." Apparently they also took caribou, as Parkhurst went on to write "nowe againe, for Venison plenty, especially to the North about the grand baie, and in the South neere Cape Race, and Pleasance."

90. See Morison (1978, 149-150) for a translation of the original text from Verrazzano.

91. Cuff (1997, 121) stated that "Cartier's note that the Breton fishermen called the great auks *apponats* is of interest, for this would appear to be the Beothuk word for these now-extinct, flightless seabirds." Cartier's second voyage made landfall at the Funk Island. Cuff cites translations of Cartier's record indicating that "this island is so exceedingly full of birds that all the ships of France might load a cargo of them without one perceiving that any had been removed." Apparently, perceptions can be misleading: the last great auks were killed on Funk Island by 1800, filling someone's boat.

92. See Breummer (1971).

93. See Ehrenberg (2006, 88) for the Spanish map of the Americas from 1562.

94. In 1555, an inquiry was held in Spain to investigate the conflicts between Spanish and French interests in Newfoundland; parts of the text of the resulting report are given in Quinn (1979, 95-98). The damage inflicted on French fisheries varied in the accounts of different witnesses, but there was consistent mention of at least 500 men being killed, many more wounded, and 300 vessels captured (most loaded with cod) or destroyed. One witness estimated 1000 dead and "ships captured from the enemy, large and small, at over a thousand, including more than four hundred of two hundred tons and upwards, with more than five thousand artillery pieces of iron and bronze, and 12 to 15 thousand prisoners, not counting the dead." This last account may have been exaggerated, but if the vessels included the small fishing shallops, then it is not entirely unrealistic. It is clear that the French fishery was severely impacted for a number of years, at least up to 1560.

95. A narrative in 1580 by Robert Hitchcock, a proponent of an English fishery at Newfoundland, stated that "there goeth out of Fraunce commonly five hundreth saile of shippes yearely in March to Newfoundlande, to fishe for Newland fishe, and comes home againe in August" (Quinn 1979, 105). If this record is accurate, then (using Whitbourne's estimates of French vessel size and capacity) it suggests that French catches of cod ranged from 90,000 to 180,000 tonnes of live fish in the 1570s.

96. de Gómara (1552), translated in Sauer (1971, 24).

MIGRANT FISHERIES AND LIVEYERS
AD 1577 TO 1800

If efver you looke for money agayne in this country, you must send fisher men.

T. WILLOUGHBY, in a letter to his father from Cupids, Newfoundland, August 4, 1616

The iland of New-founde-land is large, temperate and fruitefull... the Seas are so rich, as they are able to advance a great Trade of Fishing; which, with Gods blessing, will become very serviceable to the Navie.

RICHARD WHITBOURNE, *Discourse and Discovery of New-found-land*, 1620

Liveyers: people who "lived here."

Dictionary of Newfoundland English, 2006

Iceland . . .	207
English Expansion in Newfoundland AD 1577-1600 . . .	207
English-Spanish War . . .	211
The Newfoundland Fishing Station . . .	211
Was Settlement Discouraged? . . .	215
The Real Issue: Fish . . .	216
The Settlers . . .	217

The Origins of Prejudice . . .	218
Early Settlements . . .	218
Outports . . .	224
St. John's . . .	229
English and French Newfoundland . . .	230
Early Laws and Fishing Admirals on the English Shore . . .	235
A Fishing Station Again . . .	238
The Fishery . . .	240
Cod . . .	240
Salmon . . .	245
Walrus . . .	246
Seals . . .	246
The Ecosystem . . .	247

ICELAND

The 1400s are known as the "English Century" in Icelandic history. Icelandic sources first indicated the presence of an English fishing vessel near southern Iceland in 1412, with a rapid increase in the fishery thereafter.[1] In 1419, at least 25 English vessels perished in a storm, and some 100 or more English fishing boats voyaged to Iceland annually by the late 1400s.[2] There are few records of French or Portuguese fisheries in Icelandic waters during this century or the next.[3]

The rise of the German Hanseatic League[4] in the late 1400s changed English dominance in Iceland. In the late 1400s, German vessels brought salt cod to England, and Hanseatic merchants began to displace the English from the fishery at Iceland. Treaties and edicts were put forth by the Danish Crown, the Hamburg Council in Germany, the English Crown, and the Icelandic Althing (parliament) in the early 1500s to try to settle disputed fishing and trading rights in Iceland, but they did not prevent deadly battles between English and German mariners. The Althing passed an edict in 1533 banning all foreign fishing, particularly by the English, but it was summarily ignored. Nevertheless, a temporary alliance of German merchants from Hamburg, the Danish royal administration, and a relatively small group of wealthy Icelanders continued to try to expel the English from Icelandic waters throughout most of the 1500s.[5]

By 1580, Denmark had regained control of the Icelandic state and its fishery, passed measures to remove the influence of the Germans, and imposed licence fees that discouraged expansion of the English fleets. By this time, the Newfoundland fisheries of the French and Portuguese were well known to the English and the timing was right to switch much of their North Atlantic cod fishery to the grounds of the New World. The Icelandic fishery had provided the training ground for English expansionism; in the words of Icelandic historian Gunnar Karlsson, it was "a rehearsal for sailing to North America in the sixteenth century and the subsequent conquest of the world."[6]

ENGLISH EXPANSION IN NEWFOUNDLAND: AD 1577-1600

The last two decades of the 1500s brought major increases in the English fisheries in Newfoundland. Expansion of the English fisheries was driven by rising populations in Europe and a depletion of local fish stocks, which caused a sudden increase in the demand for salt fish, particularly in Spain and Portugal.[7] English merchants increasingly used salt cod as a commodity to trade for goods with their counterparts in Spain, Portugal, Italy, and France. Increasing production of salt cod demanded more salt – although the

English had been making salt since 600 BC,[8] natural desalination did not occur in the cold and misty British Isles, and salt had to be purchased from France or Spain. Thus, the air-dried variety of salt cod (the Portuguese *vento*), which required only a modest amount of salt, suited the English enterprise well, and was in high demand in Spain, Portugal, and Italy, countries with which the English wanted to trade.[9] In Newfoundland, this dictated a shore fishery with seasons limited by the timing of the seasonal migrations of cod and the drying weather of the short summers.[10]

The fishery was a training ground for navy sailors, and as war with Spain loomed in the late 1500s, this role of the fisheries became increasingly important. The English monarchy wanted to expand the fisheries for this reason alone and stimulated the domestic market for salt cod by doubling the number of dietary "fish days" – in which eating meat was illegal – to two per week.[11] With the "rehearsal" in Iceland behind them and conflicts with the Danes and Germans increasing in Icelandic waters, it would be the Newfoundland fisheries that were expanded.

In 1577, Bristol merchant Anthony Parkhurst thought the English fishery at Newfoundland might be doubled or tripled, and that salt might be made there more cheaply than in England.[12] He was right on the first count, but not on the second. Salt was never produced in Newfoundland, and the lightly salted cod product became the standard of the English fisheries. Air-dried curing required "rooms" on land (beaches or raised wooden platforms, usually built over the sea and called stages, on which the cod could be dried) and took all summer to produce the best quality fish,[13] thus requiring large ships that could hold a full season's worth of dried cod. Richard Whitbourne, drawing on more than forty years of experience, gave detailed instructions for the Newfoundland fisheries. He recommended vessels of about 100 tonnes, each with a crew of 36 men, that deployed eight small fishing boats manned by three fishermen apiece.[14] Each vessel would make about 2000 quintals[15] of dry fish (equivalent to 102 tonnes of dry fish or about 360 tonnes of live fish)[16] and 100 quintals (5 tonnes) of wet fish. The wet fish was the last taken and heavily salted for the trip home.[17] There was little advantage to vessels being small and quick, as were many of the earliest English pinnaces (vessels as small as 15-20 tonnes), or the Basque and French *caravelles* that ranged much more widely to the north and in the Gulf of St. Lawrence.

As the English fishery developed, vessels returned year after year to the bays and coves of the Avalon Peninsula, but seldom ventured farther north than southern Bonavista Bay. The English concentrated on the key cod grounds of the Avalon Peninsula, particularly around St. John's, and fishing harbours were occupied from Trepassey on the south coast to Conception Bay, and, later in the 1600s, farther north to Trinity Bay. By the mid-1570s, English fishing ships had begun to make major inroads in the French, Portuguese, and Spanish Basque fisheries that had preceded them, and English merchants began to act like the owners of Newfoundland. English ships would typically arrive with fishermen and fishing gear in May and leave with their sealed holds filled with salted dried cod in late August to early October. Three to four weeks later, they would be back in England where foreign buyers were waiting for them. The trade was initially dominated by fish merchants from the West Country ports of Plymouth, Dartmouth, Exeter, Bristol, Barnstaple, Southampton, Weymouth, and Poole. In the late 1500s,[18] competing London merchants entered the trade. The most notorious of these were the Kirkes (later of Ferryland), who perfected a three-way trade between England, Europe, and Newfoundland in large freighters referred to as "sack ships."[19] Sack ships carried cod directly to European markets from Newfoundland and were named for the exported *vino de sacca* that they often carried from Europe to England. Many sack ships were of Dutch origin in the early 1600s, but most were from London and other English ports by 1700.

The English method for curing cod produced a high-quality product that found favour in the best European markets. The method required rooms on shore, and it was an advantage to have these rooms built and ready to use each spring when fishermen arrived. Leaving rooms unattended over winter increased the risk of having them pilfered by other crews or by the Beothuks. English merchants therefore began leaving winter crews in Newfoundland to look after their fishing stages, boats, and equipment. Winter crews gave English ownership of Newfoundland credibility, under the creed that "possession is nine-tenths of the law," and gave their fisheries an advantage. Anthony Parkhurst travelled to Newfoundland several times during this era and reported that although the English were well outnumbered in terms of boats and men by Spain and France, they were commonly lords of the harbours.[20] Whether this was true or simply jingoism is uncertain, but the tradition of giving authority to those who were "first there" became the rule in Newfoundland.

Any Portuguese and French aspirations of ownership of Newfoundland were soon challenged. The English adventurer Sir Humphrey Gilbert sailed to St. John's in 1583 and read aloud to a gathered throng of fishermen and merchants his decree that Newfoundland belonged to England and perhaps to him.[21] Sir Humphrey has been described as a romantic but not very practical man.[22] His seamanship was certainly questionable: while making his grand entrance into St. John's harbour, his ship struck a rock in the Narrows.[23] Despite this rather humbling start, he claimed Newfoundland for England and temporarily subjugated the many French and Portuguese fishermen already there. British expansionism began on that day in the harbour of St. John's, resulting centuries later in an Empire upon which the sun never set.

Sir Humphrey did not entirely ingratiate himself to anyone, including the English sailors and merchants already present in St. John's. He offered to extend land grants; however, the grants meant that the land was to be kept for fishing ships, with no permanent buildings. In any event, after leaving St. John's that same summer, his ship was swamped on the voyage back to England in a storm off Sable Island, with all hands drowned.[24] A dreamer to the end, Sir Humphrey's last reported words were, "We are as near Heaven by sea as by land."[25] Gilbert's claim of Newfoundland for England marked the beginning of several attempts to set up a permanent colony. In the same year, Sir George Peckham, who had been part of the Gilbert adventure, wrote the first pamphlet on Newfoundland, extolling its benefits to England and the need to fortify St. John's.[26] Thus, Sir Humphrey's efforts signalled the beginning of the English dominance of the Newfoundland fishery that would endure for several centuries.[27]

English exploration in the northwest Atlantic also spurred their cod fishery and claims on the region. Captain John Davis, perhaps the most skilled navigator since Cartier, made several voyages in the late 1500s in search of a passage to the Orient; while he failed to find one, he located and described the northern regions of Newfoundland and Labrador. He also found cod. In August 1586, Davis discovered a "great store of cod" and "marveilous great store of birds, guls and mewes" off northern Labrador, and put into an unnamed harbour on the Labrador coast at 54° 30' north latitude, near the northern entrance to Hamilton Inlet. On September 1, 1585, his crew "found great abundance of cod, so that the hooke was no sooner overboard, but presently a fish was taken. It was the largest and the best fed fish that ever I sawe, and divers fisher men that were with me sayd that they never saw a more suavle or better skull of fish in their lives: yet they has seene great abundance."[28] Davis returned to the New World in 1587, this time sailing north to explore Baffin Island. In an early show of entrepreneurial spirit, he brought salt and diverted his two best vessels to the Labrador to fish for cod.[29]

CAPTAIN JOHN DAVIS

John Davis was born of humble parentage near Dartmouth, West Country England, in 1550. His boyhood friends included Sir Humphrey Gilbert and Sir Walter Raleigh. Davis's early sailing career remains obscure, but he became a master mariner and was one of the first scientific seamen. He was also a character with a "strong sense of humour"[30] and he took several hunting hounds with him on his vessel. The dogs became so fat and lazy from lack of exercise and being fed to excess that they could barely run. In the summer of 1587, an attempt at land sport led to an English-style fox hunt. Much to the chagrin of the sailors, however, the dogs were no match for the fox and quickly became exhausted.[31] Davis was such an accomplished captain that sailors lined up to sign on board with him. He was the first to accurately describe the coast of Labrador and the Arctic waters of the northwest Atlantic. Among the practical results of this great seaman's work were the opening of lucrative whale and seal fisheries in the Davis Strait, the extension of the cod fishery to the coast of Labrador, and the eventual recolonization of Greenland.[32] Davis fought alongside Admiral Nelson, Sir Francis Drake, and several vessels from the Newfoundland fishery in the war with Spain in the late 1580s. He was killed by Japanese pirates in the Pacific Ocean in 1605. The Davis Strait is named in his honour.

ENGLISH-SPANISH WAR

England and Spain sparred on the oceans during the mid-1580s, and war between the two countries became inevitable. England was a relatively small and insignificant country at the time, an underdog that challenged the rich and powerful Spain with its famed and feared warships – the supposedly invincible Spanish Armada.[33] The war reached Newfoundland in the mid-1580s; ships were seized and various atrocities were committed. War profiteering in fish products was rampant, with merchants ever on their toes for a profit. In 1585, six hundred Spaniards (likely Basques) were taken prisoner in St. John's and most of their fish confiscated; much of it was then sold illegally by English merchants in Spain and France, perhaps to the very military with which their country was at war.

In 1588, with a war at sea looming, the English fishing fleet mostly stayed home to become part of the challenge to the Armada. Among the sailors was a young Richard Whitbourne, back from Newfoundland with his own ship; he was soon joined by many other English West Country vessels.[34] In his own words, Whitbourne relates how "in the year 1588, I served under the then Lord Admiral [Nelson], as Captaine in a ship of my owne, set foorth at my charge against the Spanish Armada."[35] The defeat of the great Spanish Armada by the ragtag Englishmen, some of whose skills had been honed in the bays and coves of Newfoundland, effectively ended Spain's maritime supremacy and its role in the Grand Banks fishery. It was the first, but not the last, contribution of Newfoundland to the resolution of wars in Europe. Spanish claims to fishing rights would resurface periodically over the next 400 years, but the English declared that Spain's right to fish in Newfoundland would never be recognized, even if they "were in possession of the Tower of London."[36]

THE NEWFOUNDLAND FISHING STATION

> Ai Roca [meaning, literally, "woe to them rocks"]
> – from a Portuguese map in 1602, may be the Virgin Rocks

From the late 1500s to 1799, Newfoundland was a fishing station for English and French merchants and the migratory fishery they controlled. Attitudes towards settlement varied. On the one hand, permanent settlements were strongly discouraged – first by mercantile anarchism and then by colonial decree – and only essential winter crews were tolerated. Exactly when winter crews first stayed in Newfoundland is uncertain. One report claimed that there were 40 or 50 houses – more accurately described as shacks or tilts – in Newfoundland by the 1520s.[43] Whenever overwintering first occurred, it was not true settlement – winter crews were mostly men who served in

CAPTAIN RICHARD WHITBOURNE

Richard Whitbourne was born in 1561 in Devonshire, England, and spent much of his adult life as one of the first literate fishing skippers in Newfoundland.[37] His *Discovrse and Discovery of New-Found-Land*, published in 1620, was the first substantial pamphlet on Newfoundland. As historian Gillian Cell pointed out, Whitbourne's descriptions stood out from those of others because he knew, through first-hand experience in the Newfoundland fishery, of what he wrote. His was no armchair account and "authority inform[ed]…every page."[38] Whitbourne made many fishing voyages to Newfoundland from at least 1583, when he was at St. John's to witness the arrival of Sir Humphrey Gilbert, until at least the late 1620s, when he joined the navy. His writings on seamanship are reserved and modest, but his skills must have been excellent. He survived many trips without major mishap and entered many new harbours, including the notoriously rocky Bay of Flowers (Bonavista Bay). Whitbourne was accosted by pirates – including Peter Eason, Newfoundland's nemesis in the early 1600s, as well as brigands from La Rochelle – at least twice and experienced failure in early attempts at settlement, but he never lost faith in Newfoundland or its potential. In 1615, Whitbourne was charged by Governor John Mason with instituting law and order.[39] He travelled extensively along the coast and to the north, where he recorded the name Trinity Bay, which had formerly been called by its Portuguese name of Baya de Santa Cyria.[40] He held the first court in North America at Trinity in 1615, attempting to exert some control over destructive fishing practices. Richard Whitbourne was a strong proponent of colonization in Newfoundland and wrote, "Nay, what can the world yeeld to the sustentation of man, which is not in her [Newfoundland] to be gotten? Desire you wholesome ayre? (the very food of life) It is there. Shall and Land powre in abundant heapes of nourishments and necessaries before you? There you have them. What Seas so abounding with fish? What shores so replenished with fresh and sweet waters? The wants of other Kingdomes are not felt here."[41] Whitbourne's credibility was recognized in his time and has survived the ages. Robert Hayman, a seventeenth-century English poet and "Sometimes Gouernour of the Plantation" at "Harbor-Grace in Britaniola, anciently called Newfound-Land," wrote from Harbour Grace in 1628 to "Sir Richard Whitbourne, Knight, my deare friend, Sometime Lieutenant to Doctor Vaughan for his Plantation in Newfound-Land, who hath since published a worthy booke of that most hopeful Country":[42]

Who preaching well, doth doe, and liue as well,
His doing makes his preaching to excell:
For your wise, well-pend Booke this Land's your debter;
Doe as you write, you'le be beleeu'd the better.

Newfoundland for a few years before returning to England or moving on to other locations, such as New England. On the other hand, official settlements that included a few women were sanctioned by the early 1600s and were for a time the most modern and advanced colonies in the New World. Their marine facilities and trading positions were unparalleled in the North Atlantic.

Knowledge of the geography of Newfoundland and the continental shelf was limited prior to the settlements of the early 1600s. The earliest maps that followed the Cabot and Verrazzano voyages showed little detail and were not very accurate in their depictions of the New World. However, after the voyages of Cartier and other navigators in the early to mid-1500s, a more complete picture of Newfoundland and the coast of North America emerged, particularly on French and Portuguese maps. A Portuguese map drawn in 1602 had the general configuration and coast of the region correct, but gave little detail. The Grand Banks were apparently known by then, but were shown as a wall or barrier. A landmark named the Ai Roca may be the Virgin Rocks. Many other landmarks and features were also depicted, including C. Raso (Cape Race), Farilhon (Ferryland Head), C. de Spera (Cape Spear), Isla dos Bakalhaos (Baccalieu Island), I. de Aucs (Funk Island), C. de Pena (Cape Pine), C. d. S. Maria (Cape St. Marys), I. de S. Paulo (St. Pierre Island), and the French fishing harbour of Brest on the Labrador coast.

5.1 *Portuguese map from around 1602, possibly drawn by Bertius. Courtesy of the Centre for Newfoundland Studies, Memorial University of Newfoundland.*

5.2
Portion of French map drawn by Marcus L'Escarbot in 1609 showing the Newfoundland and Grand Banks region. Courtesy of the Centre for Newfoundland Studies, Memorial University of Newfoundland.

By this time, the cod stocks of the Grand Banks were well known to at least the French fishing fleets. A French map dating from 1609 shows Le Grand Banc aux Morues (Grand Bank of the Cod) and St. Pierre Bank (not named) more accurately than earlier Portuguese maps. The French depiction of Newfoundland is highly inaccurate, perhaps reflecting their lack of interest in the island itself, while landmasses farther to the south, such as Cape Breton and the mainland, are more accurately drawn. This map recognizes many of the same landmarks as the Portuguese map and adds a few more, including Campseau (Canso), B. de Chaleur (Bay of Chaleur), C. Breton (Cape Breton Island), I. S. Pierre (St. Pierre Island), I. aux Oiseaux (Funk Island), and G. des Chasteaux (Gulf of St. Lawrence – named after the castlelike rock feature at Henley Harbour, Labrador). Such was the state of knowledge when settlement by the English began.

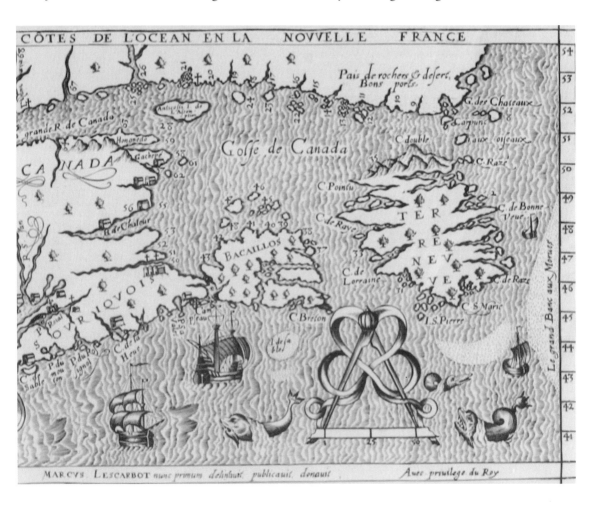

Was Settlement Discouraged?

The settlement of Newfoundland was a contentious issue for almost 200 years and remains so amongst historians. Nineteenth-century opinion emphasized the antisettlement campaigns of some merchants.[44] In the early 1600s, several petitions were put by English merchants to the Crown arguing that settlement should be curtailed or stopped altogether. According to these merchants, colonists usurped their fishing places, either harboured pirates or were pirates themselves,[45] and hindered migratory fishermen in their operations, including preventing them from taking seabirds at Baccalieu for fish bait. The Western Charter of 1634 accused settlers of being lawless and of committing murder against aboriginals (which was likely true in some cases).[46] For their part, early colonists typically denied all charges. Pushing harder, the merchants declared from their premises in England that they knew best how to manage the fishery, and that they would not be ordered about by any loutish would-be colonists.[47] It was an early expression of what would become a common attitude towards the Newfoundland fishery and the people who chose to live there and prosecute it.

As late as 1671, His Majesty's Council for Foreign Plantations recommended that the English migratory fishery needed protection from resident Newfoundlanders, that "masters of ships [should] be required to bring back all seamen, fishermen, and others, and none...to remain in Newfoundland," and that would-be settlers should be removed by force and sent back to England or to New England or the Caribbean.[48] These types of policies, and the attitudes underpinning them, were put forth for another 100 years in various forms.

Antisettlement attitudes were far from universal. Richard Whitbourne wrote eloquently in 1620 about the benefits to England and to the cod fishery that would be realized by settlement. He argued that to have facilities ready for fishing each year would save time and money, and that "if such Pinnaces [small fishing boats] and such Stages and Houses may be maintained and kept in such readinesse yeerely, it would be the most pleasant, profitable, and commodious trade of fishing, that is at this time in any part of the world."[49] Sir John Berry, who became Naval Governor in the 1670s, was another noted prosettlement advocate. Berry refused to evict settlers after the edicts of 1671 and rebuked merchant opinion against the settlers. He also expressed the view, as Whitbourne had decades earlier, that settlement would aid the fishery rather than harm it. His forceful rebuff of the antisettlement campaign in England seemed to turn the tide and ameliorated much of the discord about settlement by the end of the 1600s.[50]

Recent historical reviews of the earliest English settlements have offered a more balanced view of the settlement/antisettlement issues established by Prowse.[51] These reviews do not discount that attempts were made to restrict settlement, but they disagree about their effects on Newfoundland, and the degree to which settlement was impeded by decree or by the actions of merchants. Early settlements were no doubt part of a logical extension of the North Atlantic trading pattern and, in fact, were supported by influential merchant enterprises, especially the Kirke family.[52]

Reconstructing history is almost as difficult as reconstructing fish stocks. We know for certain that some merchants organized petitions to restrict or evict settlers and that the Western Charter stated that settlers were to be removed. However, we also know that other influential persons in early Newfoundland, including Whitbourne and Berry, strongly supported settlement and that the draconian antisettlement edicts were seldom if ever enforced. The views that antisettlement actions dominated Newfoundland in the 1600s and 1700s are likely to have been overstated, if not subject to some level of invention.

The Real Issue: Fish

The real issues for the merchants were fish, fishing rights, and economic gain – everything in Newfoundland and its historical development has depended on fish and fisheries: "The fishery first drew men to Newfoundland; the fishery shaped the policies of the nations concerned with it; the fishery both created and limited the way of life of the colonists; and the fishery, through its fluctuating prosperity, its assumed value to Europe and the conflicts it caused, determined when, where, in what numbers, and under what conditions the colonists should settle."[53] There was no other reason to come to Newfoundland. Moreover, the fishery did not require large numbers of settlers – as stated previously, some people believed that it might even be hampered by them. There were, of course, many different views on and different strategies for the exploitation of these very lucrative fisheries.[54]

Some merchants regarded Newfoundland as their fishing station, pure and simple, and not as a potential colony – they would oppose settlement under any circumstances. Under this creed, the only reason to stay in Newfoundland was to look after merchant property. Additional settlers would add nothing, with only minor labour required once the fishing season was over for the year. The antisettlement argument was predicated on the right of open access to fishing.[55] Settlement would hinder open access because settled fishing communities would inevitably take possession of the best grounds. The principle of adjacency – that proximity to a fishery implies primary rights of use – was not recognized in the 1600s. Hence, open access was fiercely protected by the English, even at the expense of giving privileges to potential or real enemies – such as the French, who were several times guaranteed freedom of access to the Newfoundland fishery and the shoreline for drying fish.[56]

Other people saw settlement in Newfoundland as offering opportunities, as Richard Whitbourne argued in his writings. In the long run, this would prove to be correct, but not all opportunities would prove to be fruitful: the first attempts to install colonies that intended, somehow, to eschew the fishery inevitably failed. Opportunities based on the fishery did well, however: the Kirke family fishing enterprise at Ferryland in the 1600s developed an integrated trade that involved Newfoundland salt fish and the much larger European trade between England, continental Europe, and, later, with the New England colonies. In this triangular trade, Newfoundland settlement supported the salt fish apex of the triangle in addition to providing merchants with a market for European goods.[57] In the end, settlement would displace the migratory fishery, and some merchants, such as the Kirkes, became settlers themselves.

The Settlers

The first settlers were of two types. A minority were well-off gentry who became "planters"[58] – middle-class property owners who had their own boats and fishing premises. Some planters lived out their lives in Newfoundland, while others kept homes in England that they visited regularly.[59] The majority of residents, however, were either servants of the planters or bye-boat keepers – men who used boats owned by planters or merchants and fished for wages. The majority of these servants could not be considered as settlers – most were poor men and women who did not stay long, and either moved back to England or on to more southerly colonies. For them, Newfoundland was solely a place to work and to leave as soon as possible. Nevertheless, some servants stayed on for many years and a few accumulated sufficient capital to become planters themselves. Bye-boat men were generally in a better position to acquire the necessary capital to settle. Some men no doubt found wives in the servant girls who came to the plantations,[60] and went on to have families and settle permanently. Several prominent early planters were female, most often widows, including Lady Kirke of Ferryland.

These early pioneers found freedom in Newfoundland, whatever their station in life. The rocky shores and coves of the coast held a certain charm for independent-minded people, then as well as now – some may even have come to love the place. Unfortunately, there are few records of the early fishermen;[61] they were not considered important to the class-conscious English and crew lists were not kept. Prowse, in his typically colourful way, concluded that "it is no argument that history does not inform us about these first residents; their presence here would not be mentioned. No lists were kept of fishing vessels or their crews – at any rate not in Devonshire – all through the Tudor period; and no notice would be taken of the doings of those obscure fishermen in the far-off land."[62]

THE ORIGINS OF PREJUDICE

The beginnings of a pejorative view of the fishery and those who partook in it – and therefore, by extrapolation, all Newfoundlanders – were laid down in the 1600s. It began in England: long before the first settlers had any notion of who they were, English society judged Newfoundlanders and those who engaged in the fishery as disorderly and debauched primitives living in miserable and barely subsistent conditions. The stereotypes were based on an "underlying, profoundly negative cultural attitude to Newfoundland"[63] on the part of middle-class England.[64] According to this silk-hankied view, Newfoundland was unfit for decent folk. At best, a young man should make his pay and get out as quickly as possible after having learned his seamanship, preferably to the English Navy. These views may have impeded the settlement of Newfoundland and become a self-fulfilling prophecy.

Norwegians had a similarly negative view of Iceland. Iceland was supposed to be "so cold, unattractive, infertile, and wicked that it could not be seriously considered as a possible place to emigrate."[65] Icelanders initially acceded to this view and discouraged their young men from engaging in the fishery, preferring to keep them home on the farm. In the early 1900s, Icelanders threw off such attitudes, but Newfoundlanders have been slower to do so and have, at times, internalized these imposed negative stereotypes of their culture and the fishery.

EARLY SETTLEMENTS

The commercial prospects of the New World beckoned in the early 1600s, overriding any negative views of Newfoundland. The Newfoundland Company was formed in England, with the influential Sir Francis Bacon as a principal and with backing from King James. For settlement and commercial exploitation of Newfoundland, the Company was granted all known lands of the Avalon Peninsula from Placentia Bay to Bonavista. The first official settlement was established in 1610 at Cuper's Cove (now Cupids), a "little sequestered nook"[66] in Bay de Grave, Conception Bay, with Sir John Guy as the first governor. The small and unknown cove in Conception Bay was a curious choice, but perhaps a deliberate one: Guy's colonists were not fishermen.[67] A cove that was well removed from the main fishing areas would prevent confrontations with the migrant fishermen and agents who seasonally occupied St. John's, Fermeuse, Bay Bulls, and the nearby harbours of Carbonear and Harbour Grace. In retrospect, Guy chose a lovely spot, with plenty of fresh water, trees, and reasonable soil close at hand.[68]

Governor Guy did not wait long to exercise his authority and issued Newfoundland's first proclamation of laws on August 30, 1611. There had been no laws in Newfoundland before this – only convention, power, and piracy. Guy's proclamation became the first effort to divert some of the destructive practices of the fishery. Actions that were declared illegal included polluting the harbours with ballast or "anything hurtful" (with a penalty of five pounds sterling), vandalizing stages or flakes (ten pounds sterling), defacing or robbing another person's boat (five pounds sterling), setting fire in the woods (ten pounds sterling), or transporting any settler away from the colony without written permission of the Governor.

At the time, deforestation was becoming common around most fishing harbours. The rules of the fishery promoted the practice of destroying all premises at the end of each fishing season because the first ships to arrive in the next spring would appropriate the best places, including any buildings left from previous years. Rather than leave any facilities for others, crews burnt them or disassembled them and took the wood back to England.[69] The resultant annual rebuilding and "rinding" (debarking) of trees was denuding the countryside. In addition, the rock ballast of ships was often hurled into already shallow harbours. Guy's laws were not appreciated by many of the migrant fishermen and various attempts were made to destroy the Cuper's Cove colony.

These attempts, however, did not succeed. In 1612, Guy brought 16 women,[70] mostly from Bristol, to the colony, and at least one child was born at Cuper's Cove in 1613. Guy quarrelled with the Newfoundland Company soon after and resigned in 1615. He may have taken some of the colonists a short distance out Conception Bay to Bristol's Hope in Harbour Grace,[71] thereby undertaking what was perhaps the first internal migration to a better fishing harbour and beginning a local tradition of giving colourful and, some would argue, overly optimistic names to the rugged coves.[72] Robert Hayman, an Oxford-educated self-proclaimed poet, became governor at Bristol's Hope in 1618, and continued the tradition of colourful language by writing various verses to all and sundry, including an antidote for drunkards: "If that your heads would ake before you drinke, As afterwards, you'd ne'r be drunke, I thinke."[73] Who succeeded Hayman as governor is unknown, although the colony apparently was still in existence in 1631. Hayman died in an expedition to Guyana in 1629.[74]

Despite the loss of Guy and his supporters, the Newfoundland Company did not give up on the Cuper's Cove colony. John Mason became second governor of the colony, and of Newfoundland, in 1615. Like Hayman, Mason was Oxford-educated, but unlike him, a navy man and geographer. With Mason came his wife; neither was cut from fishermen's cloth, but their accomplishments were several. In the six years that they stayed, Mason produced the first accurate map of the island of Newfoundland and wrote his *Briefe Discourse of the New-found-land*, a short book published in 1620.[75] After leaving Newfoundland in 1621, Mason went on to found the colony of New Hampshire in New England. The colony's fate after Mason's departure is uncertain, but it likely survived for at least several decades. It did not succeed as a fishing station, nor was that its intention. Cuper's Cove is not well located with respect to cod migration routes and many of the would-be colonists either disdained or were incompetent at fishing.⁷⁶ Nevertheless, it would be simplistic to judge the Cuper's Cove colony as a failure – if nothing else, it showed unequivocally that settlement without fishing was not going to work. In addition, although the settlement was not profitable to the Newfoundland Company, some of the earliest colonists resettled in different locations, had families, and became the first generation of Newfoundlanders. The Guy name lives on to this day in descendants in the towns of western Conception Bay.⁷⁷

Settlements sprang up near Cupids in several coves of Conception Bay, as well as at Ferryland, Trepassey, Fermeuse, and St. John's on the southern and eastern shores of the Avalon Peninsula in the 1620s. In 1616, the Welsh poet William Vaughan purchased the rights to the southern Avalon Peninsula from the Newfoundland Company and

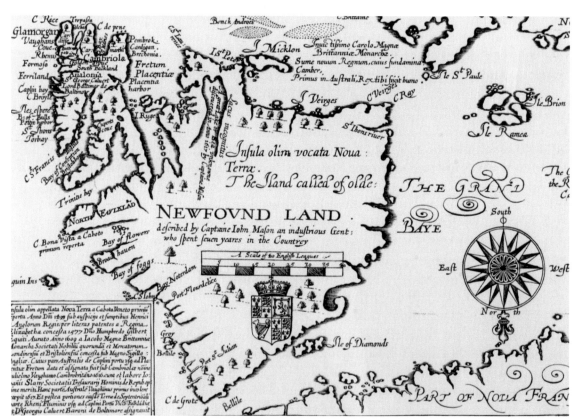

5.3 *English map of Newfoundland "described by Captaine John Mason an industrious gent: who spent seven yeares in the Countrey," published in 1625. This map was the first realistic depiction of the island of Newfoundland. It shows the colonies of the time, including the original Cuperts Cove [Cupids], Bristols Hope, the Bay of Flowers [Bonavista Bay], and the Penguin Ins [Funk Island]. Cape Bonavista has a notation C. BonaVista a Caboto primum reperta [Cape Bonavista first discovered by Cabot] – not likely true. Courtesy of the Centre for Newfoundland Studies, Memorial University of Newfoundland.*

> ## NEWFOUNDLAND'S FIRST WOMEN SETTLERS
>
> Settlement cannot properly be claimed or brought about without women. Robert Hayman wrote an ode to the early women settlers of Newfoundland from Harbour Grace in 1628, with a particular nod to Lady Mason. He wrote "to all those worthy women who have any desire to live in Newfound-Land, specially to the modest and discreet Gentle-woman Mistris Mason, wife to Captaine Mason who liued there divers yeeres":[78]
>
> > *Sweet creatures, did you truly understand,*
> > *The pleasant life you'd live in Newfound-land,*
> > *You would with teares desire to be brought thither,*
> > *I wish you, when you go, faire wind, faire weather:*
> > *For if you with the passage can dispence,*
> > *When you are there, I know you'll ne'r come thence.*

attempted to start a settlement at Aquaforte. Vaughan hired Richard Whitbourne to transport and organize the settlers, but despite moving the settlement to a better harbour at Renews, it was abandoned within a year. These harbours were well known to fishermen by this time, with Richard Whitbourne reporting that "there vsually come euery years in the fishing trade, vnto the harbours of Formosa [Fermeuse], and Renowze [Renews], aboue eight hundred English men," and "there is yeerely all the Summer time, great fishing neere vnto both the said harbours mouthes," and in the harbours "there is store of Lobsters, Crabbes, Muscles, and other Shell-fish."[79]

In 1623, Whitbourne, who had assisted with most of the earliest colonies, left his position as magistrate and returned to England. That same year, Lord Baltimore was issued a royal charter for the Province of Avalon, which led to a substantial settlement being constructed at Ferryland. Attitudes in England towards Newfoundland were equivocal, but never totally disinterested. Baltimore's charter aimed to set up a profitable colony, with the profit coming from fish, yet the charter clearly stated that settlers had no prejudicial rights to the migratory fisheries. In spite of this handicap, Ferryland became a wealthy settlement, was described as the "pleasantest place in the whole Island,"[80] and ranks among the longest continuously occupied settlements in North America. Its eventual owners, the Kirke family, led by Sir David and Lady Sara, were given control of four previous grants encompassing the entire Avalon Peninsula in 1638.[81] With their three sons and Sara's sister, the Kirkes formed Newfoundland's first dynasty, established trade links to Europe, Canada, and New England, and were the "preeminent planter gentry on the English Shore through the better part of a century."[82] The success and wealth of the Kirkes[83] were not appreciated by the English merchants or in London. Sir David was a

5.4
Ferryland Harbour from the north side with Ferryland Head in the background and the "pool" in the middle distance. The wharf in the foreground is of recent construction. Photograph by the author in 2004.

noted royalist, and was branded by anti-royalist sentiment in parliament as a malcontent and "an inveterate enemye to this present state and government."[84] He died in prison in 1654. In the years that followed, English officials conducted censuses of the resident population in Newfoundland "to assess the numbers of settlers to be removed, either to England or to other colonies, with the intention of discouraging further colonization in Newfoundland."[85]

Not all plantations were economically successful. In most cases, few or no profits were returned to their owners or backers. Failures were attributable almost entirely to incompetence in the fisheries and farming was likely undervalued.[86] The Cuper's Cove colony could not pull its weight in the fishery and failed to develop an alternative and sustainable economy. In John Guy's first letter home, he revealed that they only had eight fishermen and one fish splitter, and he requested from his backers a "ship of an hundred and fiftie tunns [to be] sent

hether, with only thirtie fishermen and foure Spilters."[87] Ignoring the singular importance of the fishery was a phenomenon that would be played out many times over the next 400 years in Newfoundland.

The on-and-off attempts in the 1600s and 1700s by some merchants to curtail settlement and to control trade in produce and other commodities were ultimately doomed to failure – there was simply far too much resistance to these ideas by influential people. Captain Mason of Cuper's Cove was a senior officer in the Royal Navy, Sir David Kirke of Ferryland a wealthy merchant, and many settlers in the 1700s were English naval officers who curtly thumbed their noses at antisettlement notions. Lieutenant Griffith Williams cleared one of the first farms at Quidi Vidi, near St. John's, and lived there for many years. Major's Path, near the modern St. John's airport, led to the farm of Major Brady of the Royal Navy in 1787. Bally Haly golf course nearer to town was the farm of Colonel Haly around the same time. West Country merchants may have run roughshod over poor servants, but naval officers were a different matter.

Outports

> The Aire, in Newfound-Land is wholesome, good;
> The Fire, as sweet as any made of wood;
> The Waters, very rich, both salt and fresh;
> The Earth more rich, you know it is no lesse.
> Where all are good, Fire, Water, Earth, and Aire,
> What man made of these foure would not live there?
> — ROBERT HAYMAN, *The Second Booke of Qvodlibets*, 1628

Unofficial settlements – the outports of Newfoundland – were located in scattered and awesomely rugged places as close to the cod migration routes as possible, with many houses and fishing stages barely clinging to the rocks. It was in these outports, some home to only one or two families, that Newfoundland's independent character took shape and gained purpose. Most of the grand and more effete sponsored colonies did not survive, at least not in their initial forms, but the outports did.

Outport settlement was a risky proposition – fish migrations varied from year to year, and the fishery could fail at any locale even when the stocks were strong. Migrant fishermen could pack up and move if the fish did not show up, but settlers could not. It was never entirely clear that settlement was advantageous for the fishery or that fishing alone could support settlement.[88] In addition, the best places for fish were often inhospitable and poor in soil, but rich in wind and ice, which made growing crops difficult. Perhaps there were other considerations for the settlers. Outport life may have been a struggle for survival, but it offered to many a sense of freedom that was better than a life of certain poverty in England.[89] Servitude was hard to escape, however, and over time most fishing families[90] became indentured to English merchants with whom they traded their salt fish for imported goods. Despite the hardships, and in part because of them, the outport settlers learned about the country and how to survive in it. They cut wood, picked the luxuriant berries, and learned how to grow vegetables in the rocky and boggy soils. They kept livestock, cattle, pigs, and sheep. They hunted caribou and seals for meat and figured out a recipe for heavy seal meat that became Newfoundland's infamous flipper pie.[91] Most of all, they learned about cod.

By 1626, there were perhaps 200-350 English settlers between Cape Race and Bonavista.[92] At the same time, about 250 English vessels were fishing in Newfoundland and given every advantage by the English government. One decree indicated that settlers were not to build homes or other buildings within six miles of shore or permitted to fish before English vessels arrived in spring. Even gardens were supposedly prohibited by decree as late as 1676.[93] These high-handed dictates were not followed and were not supported by Sir John Berry, the English naval commodore of the day.[94]

By 1675, settlements existed at St. John's, Carbonear, New Perlican, and Bonavista, with many smaller outports scattered between Trepassey on the south coast and Salvage in Bonavista Bay. An informal census of Newfoundland settlers in 1660 suggested that there were about 1500 winter residents.[95] Nevertheless, disagreements regarding settlements persisted. In 1670, William Hinton, a royal hanger-on with ambitions to govern Newfoundland and little interest in the settlers, stated that "the inhabitants of St. John's build houses and make gardens and orchards in places fitt for cureing and drying fish, which is a great hinderance and not to be suffered."[96] On the other hand, St. John's resident Thomas Oxford complained to the authorities about the depredations of seasonal fishermen in 1675. In 1677, England officially rescinded the nonfunctional and unenforced six mile rule, but the underlying sentiments against settlements did not change for another 100 years.[97] As late as 1799, Governor Waldegrave proposed a solution to the settler problem: "Could half the inhabitants of this island be sent either to Nova Scotia or Upper Canada, it would be, by no means, an undesirable measure."[98]

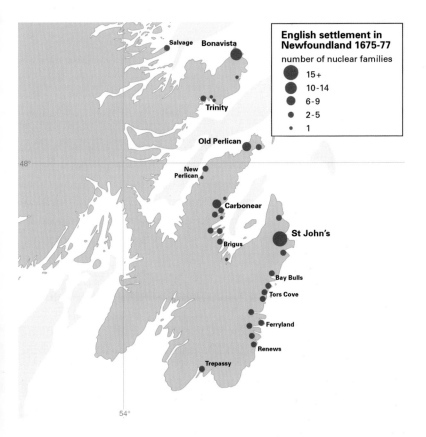

5.5
English settlements around 1675 (data from Handcock 1977).

The most successful early settlements (e.g., those located in Placentia Bay, on the Avalon shore, and on the northeast coast headlands) were invariably located where fish were available from multiple stocks, including local coastal stocks and the more abundant migratory fish from the banks. On the Avalon Peninsula, where the first fixed settlements were concentrated, the

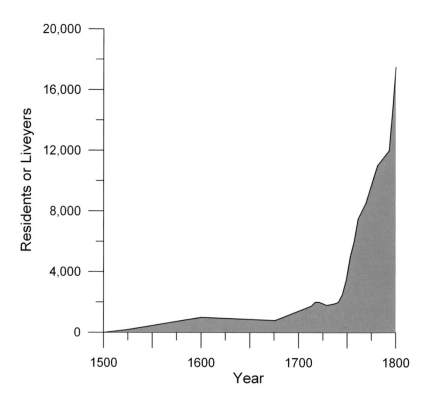

5.6 Resident population of Newfoundland, 1500-1800 (data from various sources, including Handcock 1977).

likelihood of failure in the fishery was low because this region received fish from three stocks: northern cod from the north, south coast cod from the south, and Grand Banks cod from the east. The same principle applied to seasonal fishing stations, such as the harbours in the Strait of Belle Isle located at the crossroads of the northern and northern Gulf cod stocks. Locations that received fish from only one stock, as in many areas of the northeast coast, the southwest coast, and Labrador, were far more prone to failure, making settlement there less sustainable.

Despite the problems of fixed settlements, by the late 1600s Newfoundland had taken on at least some of the trappings of a permanent fishing colony and less resembled a "great ship moored near the banks."[99] Most of the early residents were immigrants from England. As time went by, an increasing number were Newfoundland born, although complex family interactions between Newfoundland and England were maintained for many generations.[100]

By the mid-1700s, settlements in Newfoundland had grown considerably and, with them, a culture based almost entirely on the cod fishery. By then, there were a few justices and constables, a supreme court, numerous churches, and even a few schools. Settlement and development occurred as a trickle, not a flood: 200 years after the first winter crews fought the cold and the winds of a Newfoundland winter in the 1500s, the total English "liveyer" (those who lived here)[101] population on the east coast outside of St. John's probably did not exceed 2000 people.[102] However, it would soon grow at a much more rapid pace.

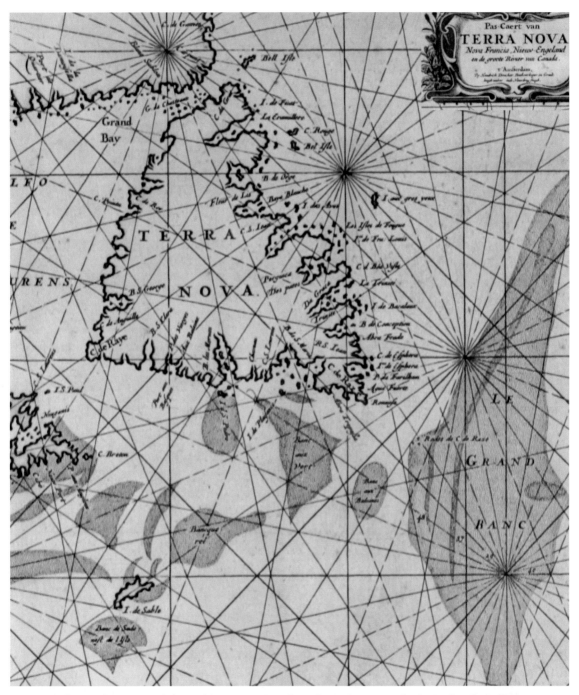

5.7 *Dutch map from around 1660 showing Le Grand Bank, St. Pierre Bank, the Roches de C. de Raso [Virgin Rocks], and Blanc Sablon. Courtesy of the Centre for Newfoundland Studies, Memorial University of Newfoundland.*

5.8 *Placentia Bay outport in the early 1900s (Gordon King collection).*

5.9 *Drying of fish on wooden flakes at an unknown outport, likely on the Avalon Peninsula in the early 1900s (Gordon King collection).*

St. John's

St. John's was a central fishing harbour from the earliest times and became an important commercial settlement. The harbour itself is uniquely protected on an Avalon coast that is, for the most part, exposed to the North Atlantic. It was also well placed for fishing, on a crossroads of cod migratory routes from the north, east, and south. It was no accident that Sir Humphrey Gilbert came to St. John's harbour to make his declaration for England. St. John's had a few settlers early in the 1600s – the first houses built around 1610-1620. St. John's would not be left unsettled much longer – the location, security, and military significance of its harbour could not be ignored. The wars between Old World nations resulted in fortifications being built by the English in St. John's as early as the 1620s: earth embankments were placed on the south side of the harbour entrance, construction of a major fort was begun, and batteries were built on both sides of the Narrows.

The English military presence in St. John's had a shaky start. Lieutenant Thomas Lloyd, a man more than mildly interested in wine, women, and song, took command in 1703 after serving for three years at Fort William. In 1705, the Reverend John Jackson of the local Church of England accused Lloyd of preventing people from going to church and holding sex orgies, complete with naked dancing girls, at the Fort.[103] Lloyd was subsequently recalled to England to face charges. Lieutenant John Moody, Lloyd's replacement, didn't fare much better. Moody ordered that a young servant girl be whipped and thrown in the streets for being saucy with Reverend Jackson's daughter. The servant girl died and Moody was sent to London for trial. Both Lloyd and Moody were cleared of all charges.

By the mid-1700s, St. John's had about 1000 permanent residents, mostly involved in the fishery and related trades. The growing town became a hive of activity for visiting sailors. During this era, Newfoundland had over 120 taverns; 50 of these were located in St. John's, resulting in a ratio of one tavern for every 20 residents. St. John's also became the English administrative and trade centre of Newfoundland. A customs house was established in 1762, more in theory than in practice under the direction of the Colonial Department in Boston, the capital of the British North American Colonies until the American Revolution.[104]

> ### LIEUTENANT GRIFFITH WILLIAMS
>
> In 1765, a published account of the Newfoundland colony urged that better statistics be kept on the fishery. Navy Lieutenant Griffith Williams, who first lived on Carbonear Island then cleared a farm near Quidi Vidi outside St. John's, had a distinguished military career. After 20 years in Newfoundland, realizing the lack of adequate statistics on the fisheries, he implemented his own estimates of landings, effort, and economic returns. Lieutenant Williams may qualify as the first fisheries scientist in Newfoundland.

ENGLISH AND FRENCH NEWFOUNDLAND

Although the English and French shared Newfoundland and its fishery in the early 1600s, French history in Newfoundland is often overlooked. The English shore stretched from Trepassey on the south coast to Salvage in Bonavista Bay, with the new settlements lending the fishery a sense of permanency. The French fishery occupied St. Mary's and Placentia bays on a seasonal basis – overlapping with the English in Trepassey Bay – in addition to areas to the north and around the Strait of Belle Isle.

The French were slower to settle on the island than the English, and this, more than anything else, led to a decline in their fishery on what became the English shore in the early 1600s. By 1635, France was supposed to obtain permission from the English to fish and dry cod on the northern and southern coasts, even well outside the English shore, and to pay a duty of five percent of landed value to Sir David Kirke at Ferryland.[106] Whether this duty was ever collected is uncertain. By this time, the French fisheries had expanded out onto the Grand Banks and St. Pierre Bank, which were portrayed with increasing accuracy on French and Dutch charts.[107]

There were no year-round French settlements in Newfoundland until the 1660s, when a naval captain from the French fishing port of La Rochelle was granted land in Placentia Bay and named governor by Louis XIV. The new Compte de Plaisance built the fortified town of Plaisance (now Placentia) in 1662. Other small settlements were developed on the island of St. Pierre and in Placentia and St. Mary's bays. The French attitude towards their settlers seemed to be similar to that of the English, and in 1686, it was ruled that vessels from France had preference for fishing grounds over settlers.[108] Nevertheless, the French population had increased to about 650-1000 persons by 1687.[109]

Although there was harmonious coexistence for a time between English and French fishermen, and many reports of mutually beneficial trade, the peace did not last.[110] In 1689, France declared war on Britain; within a year English privateers had taken Placentia and it had been looted by settlers from Ferryland. The French then retook Placentia and looted Ferryland and Bay Bulls in the mid-1690s.[111] By 1696, the French at Placentia were positioned to challenge all the English settlements to the east and north. In the dead of winter, a French force of regular soldiers and Canadian aboriginals led by Captain d'Iberville from Québec marched overland from Placentia and burned every English village, including St. John's, to the ground, leaving nothing standing and committing terrible atrocities.[112] Of the 25-30 English settlements attacked both by land and sea, only Carbonear survived, its settlers having successfully defended the island at the mouth of the bay.

In the end, however, the French were defeated and the Treaty of Utrecht ceded Newfoundland to England in 1713. The French retained the right to catch and cure fish on the French Shore – the stretch of coast from Cape Bonavista on the northeast coast to Pointe Riche in the Gulf of St. Lawrence. The French also retained Cape Breton Island. After the Treaty of Utrecht, relative peace returned to the fishery. English settlements spread north to Fogo, Twillingate, and Notre Dame Bay, and south to Placentia Bay and the southern coastal areas vacated by the French. For fifty years, from 1713 to 1763, the islands of St. Pierre and Miquelon became St. Peter's and Miquelon and were in the possession of the English.

Forty-odd years after the Treaty of Utrecht, the Seven Years War once again pitted England against France, and in particular against the French empire in North America. British troops took Louisburg on Cape Breton and General Wolfe captured the heavily fortified citadel at Québec in 1759.[113] In 1762, near the war's end, French troops took and occupied St. John's. In September 1762, at the Battle of Signal Hill, English troops ousted the French and retook St. John's in the last battle of the Seven Years War in North America. This final defeat dampened French ambitions of establishing an empire in the New World, but did little to dissuade them from participating in the Newfoundland fishery.

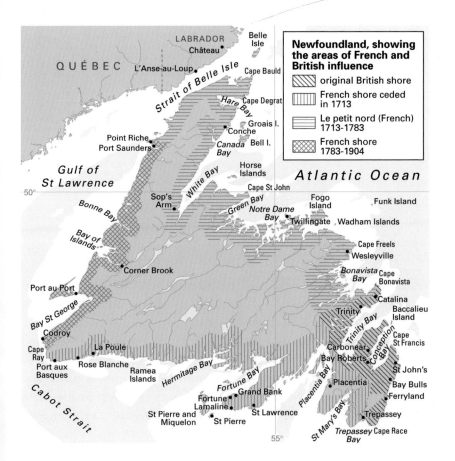

5.10

Areas of British and French influence in Newfoundland until 1904.

The Seven Years War ended in 1763 with the Treaty of Paris,[114] which succeeded the Treaty of Utrecht. The Treaty of Paris not only reaffirmed the French fishing rights from the Treaty of Utrecht, but conceded to France the island of St. Pierre, albeit without rights to erect any permanent buildings. English authorities in Newfoundland were instructed to show French fishermen every courtesy, but to tolerate no fortifications or settlements.[115] The French did not want St. Pierre alone as it was deemed to be too small, so the English threw in the neighbouring island of Miquelon. The French were disgruntled with the loss of their fishing rights under the Paris Treaty:[116]

> Let us begin with what relates to the very valuable fishery in the North American seas…France cedes to Great Britain, besides Cape Breton, all the other Islands in the Gulf and River of St. Lawrence without restriction: and…we are excluded from fishing within three leagues of any of their coasts…We have nothing left us but a *precarious right*, subject to cavil and insult, to the "*morue verte*" [literally, green fish], a commodity not marketable in Portugal, Spain or Italy, but only fit for our own home consumption. At the breaking out of this war we had in the Bay of Fundy, in Acadie, in Cape Breton, in St. John's [Prince Edward Island], Great Gaspé, and other places in the Gulf, above 16,000 fishermen, who carried on most successfully in shoal water the *pêche sedentaire* [shore fishery]. Now all this is in the hands of the British; all our settlements are unpeopled. From the single island of St. John's [Prince Edward Island] Admiral Boscawen removed 5,000 inhabitants. What, then, is left to France? Nothing but the North coast from Cape Riche to Cape Bonavista, with liberty to land and erect stages for a short season, so that we must carry and recarry both our fish and fishermen; whilst the British settled on the spot, and carrying on the *pêche sedentaire*, will forestall us and undersell us in every market in the Mediterranean. Miquelon and St. Peter's, two barren rocks indeed, are to be ours yet; even for them we have pledged the Royal Word, engaging not to erect in them any fortifications, so that even they, with their guard of fifty men for the police, will always lie at the mercy of the British.

The Spanish Basques were excluded from any participation in the Newfoundland fishery under the Treaty of Paris: Article XVIII states that Spain "for ever relinquished all claims and pretensions to a right of fishing on the Island."[117]

One can be forgiven for wondering why England did not exclude the defeated French entirely from the Newfoundland fishery, as they did in Prince Edward Island and the maritime settlements in Cape Breton and Nova Scotia. The logical rationale was that the maritime settlements, in particular Prince Edward Island and Nova Scotia, were thought to have colonial potential, whereas Newfoundland was of lesser consideration

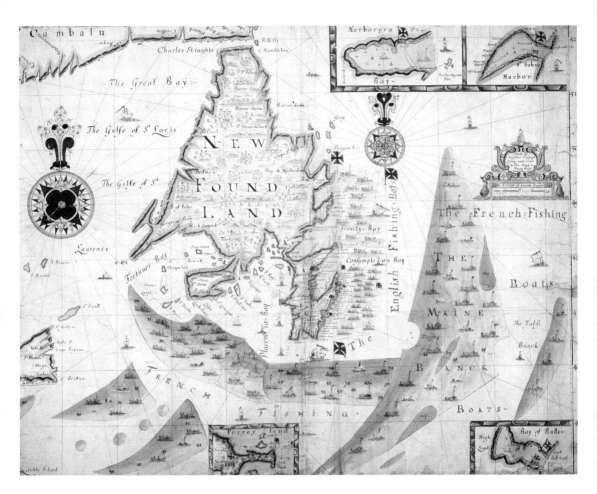

and more suited to remain a fishing station. Hence, a concession of limited fishing rights in Newfoundland to the French was more than a fair trade for considerations that benefited colonial expansion by Britain, but did little for the development of Newfoundland. Some in England argued that this was a clear giveaway of the Newfoundland fishery, but commercial interests in England were already looking beyond Newfoundland to the riches of Canada. In 1774, as if to signal the shift of English interest from Newfoundland to the potentially rich and much larger Upper and Lower Canada, the Quebec Act placed Labrador under Canadian rule, largely as a result of opposition to Newfoundland jurisdiction over the fisheries.[118] The peace following the Treaty of Paris did not last long – the unrest in the New England colonies was about to spill over.

5.11

English map drawn by Augustine FitzHugh in 1693, showing the Maine Banck [Grand Banks], the distribution of French fishing boats on the banks, and English fishing boats near shore. Note that the island is portrayed fairly well from the Avalon Peninsula to Trinity Bay, the area the English knew best in the 1600s, while the remainder is less accurately portrayed. The English were well aware of the French bank fishery. Courtesy and permission of the British Library, London.

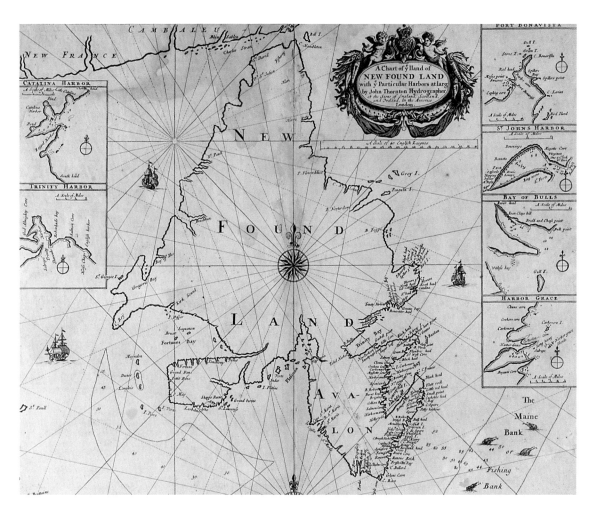

5.12
English map drawn by John Thornton in 1698, showing the Maine Bank [Grand Banks], the Virgins [Virgin Rocks], and details of the major English fishing harbours in the late 1600s. Courtesy of the Centre for Newfoundland Studies, Memorial University of Newfoundland.

In 1783, the newly independent United States of America, France, and a defeated England signed the Treaty of Versailles and rescinded the Treaty of Paris. The Treaty of Versailles was intended to reconcile once again fishing rights in Newfoundland. The boundaries of the French Shore changed: France gave up the shore east of Cape St. John in exchange for rights on the west coast of Newfoundland (to Cape Ray) and regaining St. Pierre and Miquelon.[119] The French, perhaps buoyed by the success of the American Revolution, argued for exclusive rights to fish their shore, as there had been some confusion about this under the Utrecht and Paris treaties. The English did not openly concede to this demand, but made amendments that effectively allowed a new form of French control over part of Newfoundland. For their part, the French interpreted the Treaty of Versailles as granting them more-or-less full authority over the fishery and, therefore, the right to exclude and remove Newfoundlanders from their coast and their fishery.[120]

Farther north, Labrador was under the rule of the Canadian colonies. The issue of Labrador ownership would not be settled until 1927, and overtones of Quebec and Canadian imperialism with respect to the resources of Labrador haunts Newfoundland and Labrador to the present day.

Near the closing of the 1700s, Napoleon Bonaparte ruled France, and aimed to rule much of Europe and the British Isles. Britain had been at war with Napoleon's France on and off for a decade, and the conflict inevitably spilled over once again to Newfoundland. The French attempted to take St. John's again in the summer of 1797: a fleet of ten warships approached the Narrows, but was dissuaded from attacking by a show of force, mostly bluff, at Fort Amherst and on Signal Hill. The French ransacked unfortified Bay Bulls instead, abandoning their scheme to take St. John's and retreating to their base in St. Pierre and Miquelon. Apparently, Bay Bulls settlers purposefully exaggerated the strength of the St. John's garrison, its fortifications, and the size of the chain that had been stretched across the Narrows from Chain to Pancake rocks in 1770,[121] and in so doing, very likely saved St. John's.

Disputes between the French and English over the fishery continued for another 200 years, pivoting on interpretation of treaties drafted in Paris and London. French fisheries may have declined in relative importance after their successive defeats in the 1700s, but they would remain a key part of the Newfoundland and Labrador fisheries until the 1970s.

EARLY LAWS AND FISHING ADMIRALS ON THE ENGLISH SHORE

In the early 1600s, English governors were appointed for Newfoundland, beginning with Sir John Guy and later Sir David Kirke, but they had little real authority over the fishery, the only important industry. From its earliest days, however, the fishery was not without order, and a system developed whereby the first captain to arrive at each cove in spring would be its fishing admiral for that season, with rights to direct the fishery. The admiral had his choice of fishing rooms. The second and third captains to enter the harbour might be appointed as vice-admiral and rear-admiral, and they would assist the admiral. West Country fishermen, showing a mix of deference and cheek, often referred to the admiral as the lord and the vice-admiral as the lady, which explains the many Lord's, Lady's and Admiral's coves on the English shore. The authority of the admirals was informally recognized by the British government under the Western Charter of 1634.[122]

Prowse painted a disparaging and comical picture of the admirals as illiterate and self-serving brigands, some of whom gave themselves grandiose titles. The admiral of St. John's around 1700, for example, was "Commander-in-Chief and Generalissimo"

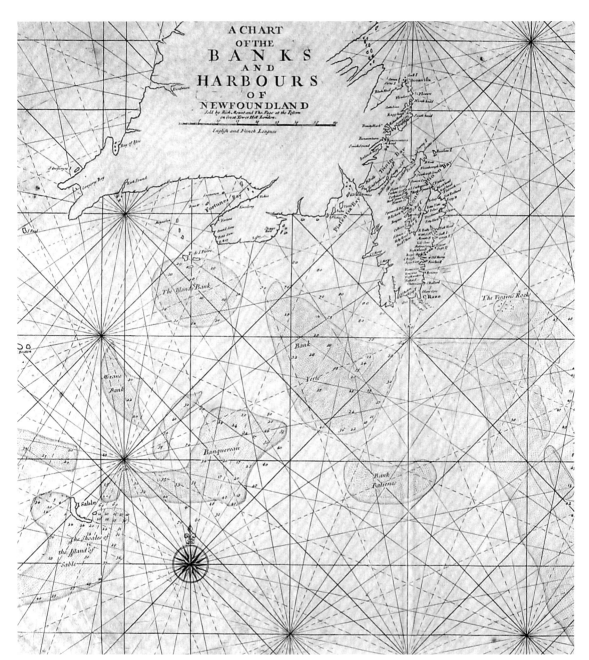

5.13 English map from 1702, showing detailed depth soundings of the Great Banks of Newfoundland, the Virgin Rocks, Bank Verte [Green Bank], and the Islands Bank [St. Pierre Bank]. Only the English shore is well detailed and accurate. Courtesy of the Centre for Newfoundland Studies, Memorial University of Newfoundland.

Arthur Holdsworth, who was accused of hoarding the best fishing sites and planting his own fishermen to the exclusion of all others.[123] Nearby admirals were purported to include the "Governor of Petty Harbour" and the "King of Quidi Vitty."[124] Prowse's depictions of these admirals were likely based on earlier negative commentary beginning with Whitbourne's stating that the admirals did "great mischeife" and maliciously pulled down stages and gear so that later arrivals would be hampered and delayed in setting up their fishery.[125]

A more recent description of the admirals casts doubts on many of Prowse's claims (but not Whitbourne's), insisting that "there is remarkably little hard evidence that the admirals were the corrupt despots described by Judge Prowse."[126] Few records of the admirals exist, which in itself tends to support Prowse's view that the admirals were at least illiterate, if not corrupt, as some governors reported.[127] On the other hand, many admirals could sign their own names.[128] Overall, it seems likely that at least some admirals did abuse their powers, given the nature of the times and competition within the fishery, and as Bannister points out, "we cannot say with absolute certainty that they never acted like the admirals described by Judge Prowse."[129] But in the end there is little proof of this, and Prowse's near-Falstaffian portrait of the admirals must be viewed as colourful but of questionable accuracy.

King William's Act of 1699 formalized the concept of fishing admirals. Under the Act, all fishing rooms used before 1683 were to remain as open access property forever. Both settlers and migrants, however, wanted rooms they could use exclusively year after year.[130] In an attempt to preserve open access, the Act gave the admirals the legal right to regulate the conduct of the cod fishery and, most importantly, the allocation of shore rooms where the fish were cured. There is disagreement amongst historians on exactly how much power the admirals had. According to one interpretation, they had no jurisdiction beyond the fishery for civil or criminal cases, or even to settle the common disputes between planters or ships' masters and servants, while another attributes a much broader authority, including "power to decide all cases, both civil and criminal, except those which carried the death penalty."[131] Overall, it appears that by 1700, at least, the admirals did little but try to further their own fishing interests, which in practice meant keeping the best rooms for themselves as quasi-private property. It would not be long before their role in the fishery was displaced by military and civil authorities.

Although the admirals may be the colourful villains of Newfoundland history,[132] they provided the first fisheries management and some level of order to the inevitable conflicts over access to a limited resource. The bottleneck in the fishery was access to curing and drying sites on shore, not the fish themselves.[133] It is likely that the allocation of shore sites by the admirals, as self-interested or even criminal as they may have been at times, provided some sort of limit to local fishing efforts, and forced new entrants to seek new grounds and hence spread out the fishery.

Newfoundland received its first true governor, Captain Henry Osborne, in 1729, although there was no requirement for him to live year round in the colony.[134] Governor Osborne's term marked the beginning of the end for the fishing admirals.[135] Naval officers replaced the admirals and were commissioned to hold court in summer. Resident Constables and Justices of the Peace were appointed in six districts from Bonavista to Placentia, and assumed authority in winter after the fishery was over for the year. A system of courts and law was established in Newfoundland, largely to govern the fishery, although it would be another 20 years before criminal cases could be tried.[136] Nevertheless, by the 1730s, there were Justices and Constables in most major settlements.[137] The role that the admirals played in limiting fishing effort was not transferred to the new authorities. With their decline, and with settlement rates rising quickly in the late 1700s, local competition for the fishery inevitably increased.

A Fishing Station Again

By the mid-1700s, the admirals were no longer a major factor in the Newfoundland fishery, but arguments for and against settlement had not disappeared. The wild colonial boys of Newfoundland were clearly getting under the skin of a new breed of haughty and proper English governors. In 1764, Sir Hugh Palliser, a Royal Navy officer, came to St. John's as governor and wrote a scathing and reactionary piece about the Newfoundland settlers:[138]

> Of these people full 9/10ths of them are of no use to that country and are lost to this during six months of the year; for during that time they are perfectly idle, abandoned to every sort of debauchery and wickedness, become perfect savages, are strangers to all good order, government and religion, by habitual idleness and debaucheries they are averse to and unfit for labour, never becoming either industrious fishermen or usefull seamen; or if they were either they are never of use for manning our fleets or for defense of the mother country, have not attachment to it and are always out of reach of it they are subsisted with the produce of the plantations and use a great deal of foreign manufactorys; they as all inhabitants of Newfoundland ever did, always will carry on a trade prejudicial to the mother country; they claim and hold as property all the old and best fishing conveniencys which by law belongs to ship fishers; by such claims a great deal lies waste and on such as are occupyd they do not employ half so many or so good men as shipfishers would; in my humble opinion such inhabitants instead of being a benefit or security to the country and the fisherys are dangerous to both, for they always did and always will join an invading enemy.

Palliser saw Newfoundland as an English fishing station rather than as a colony,[139] but despite his tirade against the settlers and "his determination to crush [them],"[140] there is no evidence to support his claim that they would be traitors. In fact, settlers at Bay Bulls, many of them Irish, risked their lives to save the British garrison at St. John's from a French assault a few decades later. However, there is evidence that his main concerns about settlers taking the fishery away from the English merchants proved correct, as did his assertion that the settlers would carry on a trade whose main beneficiary might not be England (and here exposed is the main reason for colonialism). Already, the same fires were burning in New England to the south.

Despite an undisguised disdain for Newfoundland, Governor Palliser did much to assist its development. He was particularly interested in advancing the Labrador fishery and curtailing American and French interests there and in Newfoundland.[141] The renowned Captain James Cook, who served under Palliser in the Navy, charted the oceans around Newfoundland.[142] Although Palliser was no longer governor at the time, the first Colonial Statute Fisheries Act of 1776, issued under King George III, became known as Palliser's Act. The Act had several objectives, the foremost of which was to secure the Newfoundland fisheries for England.[143] Palliser's naval background made him well aware that the migrant fishery was the main training ground for sailors, and in his view, this was reason enough for settlement to be further curtailed.[144] The Act banned passengers to Newfoundland, because they might stay, and the 10,000-odd migrant fishermen that made the voyage to Newfoundland each year were not to be paid their full wages until they returned to England. The migrant fisheries, however, were given all encouragement. The newly developing English Grand Banks fishery was subsidized with a bounty, and the salmon fisheries and extensions of the migrant cod fisheries north to the Labrador and Magdalen Islands were supported. A problem arose when it was realized that some of the new Grand Banks fishermen were Irish, but in a rare moment of bureaucratic efficiency, they were considered to be Englishmen for fisheries purposes and given the bounty too.

In summary, Newfoundland and Labrador in the 1600s and 1700s functioned largely as an oversized fishing station, with a large seasonal population of English and French fishing crews and a minor resident population who derived their livelihoods almost exclusively from the fishery.[145] A unique legal and administrative system developed as per the customs of the fishery and the local developing culture. Fishing admirals were the main authority over the fishery until the 1730s when they were displaced by naval commodores. Local customs and developments were often, although not always, at odds with the British administration and the governors of Newfoundland, and the interests of the Newfoundland colony were at times traded away to English merchants and other countries for greater political gain. Fishing privileges in Newfoundland were traded with France, with the newly formed United

States of America, and even with Bermuda.[146] Local objections to these privileges were based on competition in the world markets in which Newfoundland fish would be sold and also based on some early nationalism, but there is little evidence that they were ever based on concerns about the fish stocks.

THE FISHERY

The period from 1577 to 1800 witnessed the development of a fishery from its earliest and pristine condition through periods of a largely migrant fishery from Europe to the beginnings of rudimentary settlement and a permanent fishing culture. By the end of the 1700s, export fisheries for cod, salmon, lobsters, and marine mammals were already 200-300 years old. Other fishes, especially capelin and herring, were taken for local consumption and to supply fertilizers for gardens, but the resident population was less than 20,000 people, so those catches were inconsequential to the stocks.

Cod

During the 1600s, cod catches averaged about 100,000 tonnes per year and in some years may have approached 200,000 tonnes, with the largest proportions taken by the migrant fisheries of England and France.[147] Basque and New England vessels were also involved in the fishery.[148] Catches varied from year to year and from decade to decade, mostly as a result of market changes caused by wars between England, France, and Spain, and the American War of Independence. Despite these substantial landings, however, the fisheries likely reduced the essentially untouched cod stocks by less than ten percent and the cod stocks remained strong.

 The English fishery continued until the mid-1700s much as it had from the 1500s, with only a few minor changes in methods or strategies. Migrant fishermen, most of whom were servants of merchants or bye-boat men, came to Newfoundland in spring, fished for the summer, and returned to England or moved on to New England or Jamaica after the season was over. Growing settlements in the second half of the 1700s meant that the proportion of the catch taken by settlers increased quickly. Settlers used the same fishing methods as the migrants and, by 1675-1680, accounted for one-third of the total British catch.[149]

 From the earliest times, year-to-year variations in the fishery were observed to relate to environmental changes, particularly with extremes in weather. A decrease in temperatures began in 1660 and, after some respite, resumed in 1683 and persisted into the first decades of the 1700s. The fishery declined badly in 1684 when "participation in the migratory fishery dropped by more than half from the level of roughly eight hundred boats, typical of the early 1680s, to about three hundred. Even the planters [settlers], who had no real alternative to participation in the fishery, reduced their commitment of boats by 10 or 20 percent at this time."[150]

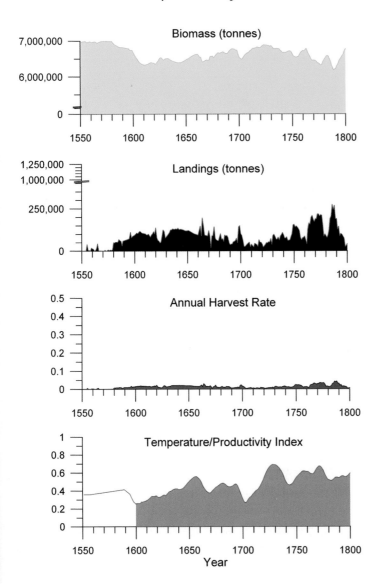

5.14
Cod population biomass, landings, annual harvest rate (the proportion of the stock caught each year) (data from Rose 2004), and index of temperature and productivity from tree-ring analyses – from 1550 to 1600 derived from qualitative descriptions (index ranges from 0 to 1 with data from D'Arrigo and Jacoby 1992).

The worst was yet to come. Temperatures rebounded slightly in the 1690s and catches increased,[151] but the fishery failed again during the cold and ice-ridden years from 1713 to 1724,[152] and few fish were caught in the shore fisheries north of Cape St. Francis. These were tough times – English "merchants and fishermen were cruelly affected…[even though]…as far as we can see, between 1713 and 1740 the markets seem to have been more or less stable, prices changed little and they absorbed all the fish that was exported."[153] The French fisheries were similarly affected. French Shore fishermen occupied mostly northern locations and largely abandoned Newfoundland during these years, moving south to Cape Breton Island, the southern Gulf of St. Lawrence, and – beginning in 1720 – the Gaspé region. Although there are no direct records of what happened to the fish during this period, the cold conditions likely caused distribution and migration changes that did not bring the fish to shore.[154] About a decade later, after 1725, ocean conditions apparently warmed considerably and catches improved on the Newfoundland coast, but continued to decline in Cape Breton and on the mainland until 1750. The Grand Banks fishery also declined during this period.[155]

The collapse of coastal fishing in the early 1700s motivated the English to make their first attempts to fish the Grand Banks, as the French had already been doing for over 100 years. Lacking an abundant salt supply, they developed a shore cure method for Banks-caught fish that suited their markets. Although the south coast cure initially received a lower price than the "merchantable" air-dried fish that sold as *vento* in Portugal, it would become Newfoundland's best in later centuries.[156]

By the 1760s, great changes were coming in the prosecution of the fishery. Most of the coves and fishing grounds south of Cape St. John were home to planters, and the production of cod was shifting quickly from migrant fishing ships to the settlements. A rapid expansion in trade freighters, or sack ships, occurred to take dried fish back to the markets in Europe and the Caribbean in exchange for trade goods needed in the fishery. Many fishing ships were converted to sack ships; the number of sack ships from England grew from about 30 in 1716 to more than 120 by 1750.[157]

By 1783, with the French now ousted from the "petit nord," English merchants were extending their trade northward to the Labrador coast. At the same time, Jersey Island firms established businesses in the former French strongholds in the Strait of Belle Isle. Migrant fishermen, followed by traders and then settlers, pushed farther north in waves. These first fisheries were modest – large fisheries did not develop in northern Labrador until the mid-1900s. The large northern stocks, with the incredible abundance reported by John Davis some 200 years earlier, were still largely untouched.

CARTWRIGHT'S PREDICTION

Expanding settlement in Newfoundland led to inevitable conflict with the Beothuk people. The Exploits and Notre Dame Bay regions initially provided protection for the few hundred remaining Beothuks because they were not areas of intensive cod fishing. However, in the 1760s, settlers moved north to fish for salmon and trap fur-bearing animals, and both of these activities led to numerous atrocities against the remaining Beothuks. By the late 1700s, the Beothuks had retreated from most of their former lands and were confined to small areas near Trinity Bay and the Exploits River valley and headwaters. In the 1760s, under the direction of Governor Palliser, brothers John and George Cartwright made investigations as to the fate of Beothuks in Newfoundland. For almost 150 years, since the initial friendly interchanges with John Guy's party, the Beothuks had shunned contact with Europeans – understandably, as Beothuk petty thievery had been followed by murderous acts on the part of a few settlers. By 1760, these atrocities and the reduced state of the Beothuks were well known. With the salmon fishery in Notre Dame Bay and the numbers of furriers growing quickly, George Cartwright offered the ominous prediction that, "I fear that the race will be totally extinct in a few years."[158]

The early decades of the 1800s brought about the final demise of the Beothuk people. By 1800, the remaining Beothuks were concentrated in the area between Gander Bay and the Bay of Exploits. They overwintered in the upper reaches of these watersheds and especially favoured Red Indian Lake. Several Beothuk women were captured and brought to St. John's: the first (name unknown) in 1803, followed by Demasduit (Mary March) in 1819 and Shanadithit, the last known Beothuk, in 1823. Shanadithit died in St John's in 1829 and was buried there. Several belated attempts were made to engage the remaining Beothuks during these years, but rewards offered to bring Beothuks to St. John's resulted in men being killed so that their women could be captured. In 1810, Governor Duckworth engaged Lieutenant David Buchan to make contact with the Beothuk people. Buchan overwintered in the Bay of Exploits and travelled to Red Indian Lake where he encountered the few remaining Beothuks. This well-intended attempt to reach out to the Beothuks ended in the deaths of two of Buchan's party. There was no trust left. Before her death, Shanadithit provided a great deal of information on Beothuk habits, culture, and language. William Epps Cormack attempted to locate the Beothuks in his epic tramp across the island in 1822 to catalogue its flora and fauna and geology, employing a Mi'kmaq guide. He found none. In 1827, Cormack traversed the country from the Exploits River to Red Indian Lake, specifically to find the Beothuks, but managed to locate only the burrial lodge of Demasduit. The following year his Indian guides made one final attempt to locate living Beothuks, but could only confirm Cartwright's prediction.[159]

The French fisheries on the banks and shore continued with little change during the 1600s and 1700s. The French fished the southern Grand Banks, as well as the Green and St. Pierre banks close to St. Pierre and Miquelon. The French were keen strategists: their fleet left France during winter and arrived on the banks in April in time to strike the large schools of cod that formed prior to and during spawning. They fished until cod aggregations dispersed in June and fished again in the fall when the overwintering aggregations began to form. Fishermen were well aware that food and temperature were key to the distribution of cod, and knew that cod forsook the banks in summer to feed on migrating capelin. In summer, the shore fisheries were prosecuted, much like the English fisheries. As a consequence of their observations and interest, the French were aware very early on that too much fishing could result in reduced catches.[160]

French fishermen also noted the differences between cod of the Grand Banks and those from the Avalon Peninsula and French Shore. Then, as now, the biggest fish came from the Grand Banks, while fish from the northeast coast – from Notre Dame Bay to Cape Degrat – were typically smaller. Larger fish were noted at the tip of the Great Northern Peninsula at Quirpon later in the season. Smaller cod were associated with an early run of capelin up to the end of July, with larger fish associated with herring after that. These are some of the first observations of the migrations of the northern cod to the northeast coast, the arrival of what may have been larger northern Gulf of St. Lawrence fish from the south in August, and the dependence of all these movements on the forage fish capelin and herring.[161]

Overall, the size of cod caught in the French fisheries remained large. In 1740, it "was said that the length distributions had remained the same for a hundred years." The fish were big: the top two market categories – "gaffe cod" and "officer's cod" – contained fish ranging from one to two metres in length, and fish less than 50 cm fell below any of the main recognized categories.[162]

The more northerly French fisheries were reported to be less certain and more risky than those on the Grand Banks and the Avalon Peninsula. The northern fish were very abundant in most years, but there were periods of several years when the fisheries failed entirely from the northern Gulf of St. Lawrence to the Labrador Straits on the northeast coast.[163] The late 1780s and early 1790s were poor years: fishing harbours were deserted, boats were stashed or tied up, and desperate attempts were made to find fish and make a "saving voyage." In contrast, the southern fisheries did well in those years. As a result of their more variable nature, the northern fisheries became less sedentary and schooners were used to search the coast for fish when they failed to show up at the primary harbours. This practice was used at Blanc Sablon, Bonne Espérance, and other main harbours for many years, and was the precursor of the "floater" fishery in Labrador during the nineteenth century.[164]

Salmon

Settlement brought opportunities to utilize fish resources other than cod. Atlantic salmon was one of the first new species to be exploited, at first mainly as food for the settlers. Export followed in short order, however, as demand for salmon in Europe and New England grew. Captain Cook reported early salmon fisheries on the south and west coasts, and inferred that the French fishery had heavily exploited some of the major runs for many years, especially on the Humber River.[165] Although salmon were nowhere as numerous as cod, they were easy to catch in nets stretched across spawning rivers. Curing salmon was also much easier than curing cod: salmon were simply split, gutted, and soaked in brine with no drying. By the 1620s, with several colonies established, the economy of the settlers had diversified to include a salmon fishery and other industries.

Salmon production may have been poor in Newfoundland and Labrador waters in the late 1600s and early 1700s, although there are no direct data to confirm this. Archaeological evidence did suggest a large increase in salmon numbers in New England, the southern limit of their range, during this period, perhaps as a result of the lower ocean temperatures that occurred during these decades.[166] Salmon landings in northern Russia declined rapidly during the same period, perhaps reflecting conditions in Newfoundland and Labrador.[167]

Temperatures and overall productivity increased rapidly after 1720, and the salmon fishery grew quickly in Newfoundland as English settlements advanced on the northeast coast to encompass the mouths of the Gander and Exploits rivers. In 1723, George Skeffington received the first exclusive rights grant to salmon fishing in an area that stretched from Gambo River to Gander and Dog Bay rivers.[168] Skeffington sold his concession six years later in 1729. In Bonavista and Gander bays, salted salmon production quadrupled from 1743 to 1757.[169] Over the next 100 years, salmon fisheries progressed north past Twillingate and into the salmon-rich rivers of Notre Dame Bay and the Bay of Exploits. Salmon fisheries spread to Labrador by the 1770s.[170] In 1763, Labrador and Anticosti Island were annexed to Newfoundland and opened to migrant English fisheries and liveyers. The merchants of Fogo and Twillingate were amongst the first to go north to Labrador, with other English and Newfoundland firms following in their wake. These initiatives, combined with the issue of ownership of Labrador, led Governor Palliser to construct Fort Pitt in Chateau Bay, the original landing spot of some early trans-Atlantic explorers.

The great runs of salmon in southern Labrador must have been well known long before this time. The earliest French cod fisheries were located adjacent to several major salmon rivers, including the Pinware River, one of the finest salmon rivers in the world, and Salmon Bay near Bonne Espérance and the ancient fishing station of Brest. New England fishermen took notice of the abundant salmon in Newfoundland's many rivers as early as the mid-1600s.[171] In one account, a New England vessel sailed to the well-known and "very much esteemed" fishing harbour of Renews in 1663 to fish for cod, but, finding cod scarce that year, loaded up with salmon taken from the Renews River at the head of the harbour.[172]

While most New England salmon boats fished close to home in the 1600s,[173] salmon were beginning to disappear from local rivers by the mid-1700s as settlement, river dams, land clearing, and exploitation increased.[174] New England fishermen turned north to Newfoundland and the lightly exploited rivers of Labrador. They described waters full of salmon, cod, and other fishes.[175] These early records also reported on the aboriginal settlements in coastal Labrador, and contributed to the establishment of the Moravian Missions in the 1760s[176] to serve the native people and, later, the Grenfell Mission to the Labrador fishery in the next century to serve the interests of liveyers who by then had settled at remote locations "on the Labrador."[177]

Walrus

Walruses were hunted for meat, hides, and ivory in small numbers from the mid-1500s onwards after Jacques Cartier reported their existence on the Magdalen Islands (then known as Ramea). The French took the early lead in the walrus hunt: in 1591, the French vessel *Bonaventure* reported taking 1500 walruses on the Magdalen Islands.[178] They would not get their catch back to France, however: it was wartime and the English captured the *Bonaventure*. The demand for animal oils and leather was strong, and the English regaled their spoils and the commercial potential of the walrus.[179] By 1593, English vessels were outfitted to hunt walruses at the Magdalen Islands, but their crews lacked experience and that year left England far too late to catch walruses on their breeding colonies. Few walruses remained at Ramea when they arrived. The more experienced French, on the other hand, had already taken their share.[180]

Seals

Survival in Newfoundland and Labrador was not possible without hunting the sea, and early liveyers, following hunting patterns set down since the first Archaic Indians, took seals along the coast for food whenever they could. The earliest records of liveyers hunting seals with heavy nets occurred in the late 1700s. Nets were stretched in winter from near shore to a maximum depth of forty fathoms (72 metres), and were used from Conception Bay to Labrador. The nets targeted migrating harp and hooded seals that followed the ice floes south. Seals that hauled out on land – the less numerous harbour and grey seals – were less important to the fishery, at least on the northeast coast. Seals figured prominently in early Newfoundland culture, language, and natural history, and early observations of the relationships between wind, ice, and animal movements were made in a letter from John Bland of Bonavista to Governor Gambier in 1802.[181] The different species of seals and their various life stages were also well recognized by this time.

Although they were originally used for subsistence living, seals had become an export fishery that fueled the oil lamps of Europe by the end of the 1700s.[182] Demand quickly outstripped the numbers of seals that could be supplied by the traditional shore fishery, which typically took less than 50,000 seals each year.[183] By 1793, a large-ship seal fishery had begun to hunt breeding seals on the ice. It was dangerous work, but employed several thousand men. It was thought at first that this fishery could not survive without a subsidy, but that would change: in the next century, the large-ship seal hunt would become a fixture and icon of Newfoundland life.

THE ECOSYSTEM

> The rivers also and Harbours are generally stored with delicate Fish, as Salmons, Peales [landlocked salmon], Eeles, Herring, Mackerell, Flounders, Launce, Capelin, cod and Troutes the fairest, fattest and sweetest, that I have seene in any part of the world. The like for Lobsters, Crabs, Muskles, Hens: and other varietie of Shell-fish great store…And also observe here, that in these places there is usually store of the spawn and frie of severall sorts of fishes: whereby the Sea-fowle live so fat, as they are there in the winter: And likewise the Bevers, Otters and such like, that seeke their food in the Ponds and fresh Rivers…The Seas likewise all along the Coast, doe plentifully abound in other sorts of fish, as Whales, Spanish Mackerel, Dorrel [uncertain species], Pales [plaice or founder?], Herring, Hogs [walrus?], Porposes, Seales, and such like royall fish [sturgeon]…But the chiefe commodity of New-found-land yet knowne…is the Cod-fishing upon the Coast.
>
> – RICHARD WHITBOURNE, *Discourse and Discovery of New-found-land*, 1620

Many early observers remarked on the abundance of fish and marine life on the Grand Banks and in coastal Newfoundland and Labrador waters. The 1600s were a period of high productivity in Newfoundland and Labrador waters until about 1660, when ocean temperatures declined. Whitbourne mentioned the common presence of warm-water species such as Atlantic mackerel and sand lance in the early 1600s. These species became less abundant in later cold periods – mackerel in particular were nearly absent from Newfoundland waters during the cold period of the 1800s, and likely in the less-well-documented 1700s. Cod were very abundant in the early 1600s; the exuberant descriptions of their availability may have reflected some euphoria on the part of early fishermen, but were not misplaced. Other species were also abundant. Richard Mather crossed the southern Grand Banks in 1635 and described seeing "mighty fishes [dolphins?] rolling and tumbling in the waters, twice as long and big as an ox…and mighty whales spewing up water in the air, like the smoke of a chimney, and making the sea about them white and hoary."[184] John Mason, the second governor at the Cuper's Cove colony, gave an early description of the seasonal biology of several marine fishes and confirmed Whitbourne's description of an abundance of mackerel. Mason wrote of the fishery in his *Briefe Discourse of the New-found-land* in 1619:[185]

> May hath herings one equall to 2 of ours, cants and cod in good quantity [the mention of cod in May and prior to the capelin migration suggests a coastal stock that overwintered and likely spawned in Conception Bay]. June hath Capline a fish much resembling smeltes in forme and eating and such aboundance dry on shoare as to lade cartes, in some partes pretty store of Salmond, and cods so thicke by the shoare that we nearlie have been able to rowe a boate through them, I have killed of them with a pike; of these three men to sea in a boat with some of shoare to dresse and dry them in 30 dayes will kill commonlie betwixt 25 and 30,000, worth, with the oyle arising from them, 100 and 120 pound. [This is clearly a reference to migratory shoals of cod pursuing capelin.] And the fish and traine in one harbour called Sainct Johns is yearly in the summer worth 17 or 18 thousand pounds. Julie and so till November hath Macrill in aboundance one thereof as great as two of ours. August hath great large Cods, but not in such aboundance as the smaller, which continueth, with some little decreasing until December…[and] Squides a rare kinde of fish, his mouth squirting mattere forth like Inke, Flownders, Crabbes, Cunners, Catfish…Of al which there are innumberable in the Summer season; Likewise of Lobsters plenty.

Early observers also remarked on the variability in the weather and ocean conditions – particularly ice and water temperature – and how these related to cod catches, which were interpreted to infer movements, migrations, and local variations in the abundance of fish. There were poor years and poor decades from the beginnings of the fishery – it was a rare year indeed when the fishery was good everywhere. Productivity declined drastically in the early 1700s and did not recover until around 1720; this decline matched a period of reported cold conditions in anecdotal records and a much-reduced cod fishery,[186] particularly in northern areas of Newfoundland.

The vast marine ecosystem from Labrador to the southern Grand Banks felt its first bump in the late 1700s as the rapidly expanding fisheries began to extract a toll on the marine environment. It was a mild bump, perhaps, but also a harbinger of things to come. English and French fisheries were expanding, and the New England fishery extended north to Labrador. These increased fisheries led to the first significant human impacts on the great cod stocks, but the effects were small and did not lead to stock declines. The first irreversible impacts of human activity were on the heavily exploited marine bird and mammal populations in the area. Seabirds had been routinely used as bait for cod in addition to being consumed as food, and egging and feather hunting destroyed the nests of hundreds of thousands of birds. Consequently, the flightless great auk was in serious decline by the end of the 1600s; one hundred years later, the great flocks that had been reported to be inexhaustible were nearly gone. The walruses in the Gulf of St. Lawrence and on the Magdalen Islands, the southern extent of their range, were also in major decline, and it would not be long before the local populations were extinct.[187]

These changes did not go unnoticed. By the late 1700s, concerns additional to those in Governor Guy's proclamation in 1611 were expressed for the fishery and its conservation. In 1775, a petition from merchants, boat keepers, and principal inhabitants of St. John's, Petty Harbour, and Torbay made several suggestions as to how the fishery might be improved, with specific condemnation of the destruction and wastage of small cod by seines.[188] There were also concerns regarding the slaughter of seabirds during the breeding season,[189] and a proclamation was issued against egg taking at the Funk Islands and enforced by the common punishment of public flogging.[190] There was also concern about the wanton destruction of large pines, which by 1775 were virtually gone.[191]

5.15
Newfoundland and Labrador ecosystem species abundance from AD 1578-1799 (warm and cold periods).

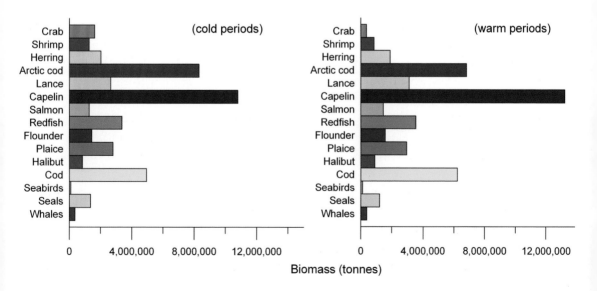

The ecosystem model suggests that most fish species other than cod and perhaps salmon were not impacted by human activities to any extent until 1800. The large cod stocks off Labrador and the northeast coast of the island of Newfoundland would have been least affected, as they retreated to then unknown offshore waters that were fringed or blanketed by ice during much of their overwintering and spawning period and were therefore inaccessible before the age of steel ships. Other cod stocks felt their first impacts, however, with small declines occurring on the Grand Banks and south coast where fishing began in ice-free waters in late winter and spring. Declines may also have been registered in the small coastal stocks. Overall, however, the cod stocks still measured between 6 and 7 million tonnes, with catches surpassing 200,000 tonnes in several years towards the end of the 1700s.[192] In addition, many cod still lived to a ripe old age and great fish of over a metre in length were a common component of the catches. At the end of the 1700s, for all intents and purposes, and as believed by most fishing parties, be they English or French, planter, migrant, bye-boat keeper, or settler, there were fish for all.[193]

NOTES

[1] The *Icelandic New Annal* for the period 1393-1430 indicated that, in 1412, there "came a ship for England east of Dyrhólmaey (Portland), people rowed out to them and they were fishermen from England" (Jónsson 1994, 10).

[2] See Karllson (2000, 118) and Jónsson (1994, 11).

[3] See Jónsson (1994).

[4] The Hanseatic League was a cartel of towns based in Germany whose goal was to promote and protect commerce. The League spread to Norway (Bergen was a centre) and then to Iceland in the 1400s, and entered the growing trade in salted cod.

[5] See Karlsson (2000, 126).

[6] See Karlsson (2000, 122).

[7] Cell (1969, 23) gave detailed accounts of the development of the English fishery in the late sixteenth century: "In Spain and Portugal the fishermen could take advantage of the generally flourishing state of trade during 1570s, and of the decline in the supply of fish from Iceland. English-made stockfish had long found a market in both countries, but a drift away from the Iceland fishery had begun when Denmark had regained that island."

[8] Kurlansky (2002, 180-181) stated that "Anglo Saxons called a saltworks a 'wich,' and any place in England where the name ends in 'wich' at one time produced salt." English place names indicate that there were many small saltworks.

[9] See Cell (1969) and Pope (2004).

[10] Cell (1969, 5) argued that salt limitations dictated the type of fishery developed by the English. In contrast, Pope (2004, 14-15) remarked that the "international market [in salt] made it easy for West Country supply ships to obtain suitable salt in southwest France or Portugal" and "dependence on imported salt is not a plausible explanation for West Country reliance on the dry cure, with its conveniently light requirement for salt. The technological choice was a consequence of consumer habits in England's markets." The French and Basque fisheries made much more heavily salted fish which found a ready market, but the lightly salted product was more valuable. Hence, a light-salted, air-dried shore cure fit several requirements of the English merchants, including cost and market.

[11] During the reign of Queen Elizabeth I (1558-1603), William Cecil, her secretary, wrote, "Remedyes must be sought to increase marrynors by fyshyng as a cause most naturall, easy and perpetuall to brede and mayntene marynors" (Cell 1969, 23).

[12] A letter from Anthony Parkhurst to Edward Dyer in 1577 stated: "For first and chefest I holde our trade of fysshynge, which might be made twyse, ye thryse, as good as yet yt ys, which thanked be god ys well amended within v yeres…Nowe for yf this cuntry wer inhabited, we might make salt ther mutch more cheper than in Inglond, for that owr wood and the caryage (that makes yt dere) would be saved. And possible not far thense to fynde some apte place to make salt, with the helpe of the sonne as in france and spayne, for the clyme will permit…Our salt beynge saved, which stands us more then the vittels and all that we cary, we might not only sell thynges better chepe, but mighte make grete store of dry fysshe with the bryne yt here is cast away. So might we save halfe the salt we spend, and make twice as mutch dry fysshe as we do" (Quinn 1979, 5-7).

[13] Cell (1969, 4) provided a description of early English fishery methods based largely on the observations of James Yonge, a ship's surgeon, in 1663: "Once at Newfoundland and the chosen harbour, the ship would be unrigged for the season and her crew would go ashore to cut timber and to build stages, flakes, cookrooms, and shacks. The stages were begun on shore and extended into the sea, so that the cod could be thrown up from the small boats in which the fishing was done. These fishing boats, usually of three

to five [tonnes], could hold between 1000 and 1200 fish and were handled by as many as five men. The skilled work of cleaning and preparing the fish was done on the stage, the waste being thrown into the sea and the livers saved to make train oil. An expert 'splitter' could bone some 480 fish in thirty minutes. It passed next to the salter who applied salt with a brush; this too was skilled work for too much salt 'burned' the fish, causing it to become wet and break, while too little make it turn red when dry. The elaborate drying process now began. The cod was first stacked in piles three feet high where it stayed for three to ten days depending on the weather. Next it was washed and laid in a second pile, skin side up, 'on a platt of stones, which they call a horse.' A day or so later the men placed it on flakes, erections of branches laid over a frame, where it dried in the air and sun. By night or in wet weather, the fish was made up into 'faggots,' four or five fish with the skin side up and a broad fish on top. When it was well dried it was put into a 'press pile,' where the salt sweated out leaving the fish looking white. After one more day of drying on the ground, it was finally stacked into a 'dry pile' and there it remained until the ship was ready to sail, when the cod was weighed and carried on board. The hold of the ship had to be completely waterproof and, once the fish was put down, the hatches were sealed and could not be opened again until the vessel reached its destination."

[14] See Whitbourne (1620, 42) and Cell (1969, 3-4).

[15] A quintal was the old measure of dried fish. In Newfoundland, a quintal was equivalent to 112 pounds or about 51 kilograms.

[16] Based on dry-to-wet conversion of about 3.5.

[17] See Cell (1969, 4).

[18] Cell (1969, 5-6) mentioned the first instance of direct trade to Europe in 1584.

[19] See Cell (1969, 5-6) and Pope (2004).

[20] Anthony Parkhurst gave an original account of the Newfoundland fishery in a letter dated November 13, 1578 (Prowse 1895, 60): "He had made four voyages to Newfoundland, and had searched the harbours, creeks and lands, more than any other Englishman. That there were generally more than 100 sail of Spaniards taking cod, and from 20 to 30 killing whales; 50 sail of Portuguese; 150 sail of French and Bretons, mostly very small; but of English only 50 sail. Nevertheless, the English are commonly lords of the harbours where they fish, and use all strangers help in fishing, if need require, according to an old custom of the country; which thing they do willingly, so that you take nothing from them more than a boat or two of salt, in respect of your protection of them against rovers or other violent intruders, who do often put them from good harbours."

[21] Sir Humphrey Gilbert was an Eton- and Oxford-educated Englishman who dreamed of a colony in Newfoundland. In 1578 he received a large patent from Queen Elizabeth I "to discover, occupy, and possess such remote heathen lands not actually possessed of a Christian prince of people as should seem good to him." His first expedition in 1578 failed, but he did succeed in reading his patent in St. John's in 1583.

[22] Sir Humphrey Gilbert sailed from England with five vessels and 250 men. Unfortunately, he did not have a good crew, and it included many tradesmen and dancers. According to Prowse (1895, 67), "Gilbert's crew consisted partly of the off scourings of the jails, and the result, as might be anticipated, was most disastrous; one ship took to piracy, and a great portion of his villainous followers deserted. On the 3rd of August, 1583, Gilbert arrived at St. John's with two ships and a pinnace – the *Delight*, Admiral's ship, 120 tons, the *Golden Hind*, 40 tons, and the *Squirrel*, 10 tons."

[23] Gilbert's vessel may have struck Chain Rock, which was used to anchor a large chain across the harbour entrance during early conflicts with the French and to secure an underwater fence to keep submarines at bay during World War II.

[24] It is evident that a southern route to Newfoundland was known and used by 1583. Sir Humphrey Gilbert's route would have taken him south across the Grand Banks to the Nova Scotia Banks, sailing southeast and perhaps using the strong currents of the North Atlantic Drift to cross to Ireland in good time.

25 Prowse (1895, 74) described the loss of the *Squirrel*, Gilbert's vessel, as follows: "Suddenly on Monday night we lost sight of the *Squirrel*'s light. Our watch cried out the General was cast away, which was too true; for in that moment the frigate was devoured and swallowed up by the sea."

26 The cover page and some text of Peckham's pamphlet on Newfoundland were reproduced in Prowse (1895, 76-77). Peckham wrote, "By establishing a safe harbour and head-quarters, and it is well known to all men of sound judgement that this Newfoundland voyage is of greater importance and will be found more beneficial to our country than all other voyages at this day in use and trade amongst us." Peckham wanted to make St. John's a fortified town – a real settlement, not just a seasonal fishing harbour.

27 The French fishery, which was at least equal to the English until the 1760s, still endures on the south coast from the islands of St. Pierre and Miquelon.

28 See Hakluyt (1598, 316).

29 See Hakluyt (1598, 332-333) and Markham (1889, 56).

30 See Markham (1889, 224).

31 Morison (1978, 345), citing earlier sources, stated of Davis' voyage in 1587: "Becalmed on the 25th, the weather marvellous extreme hot, master Bruton and several mariners went ashore to course dogs – the first hint that Davis had with him a few couple of hounds. They had not been let ashore earlier for fear of the Eskimo huskies, with the result that were so soft and fat from eating codfish, that they were scarce able to run. Here is the most comic episode of the voyage – musicians blowing horns, sailors trying to 'sic' the hounds onto a fox or hare, all hands shouting 'tally-ho!' and cracking whips, and the over-fed hounds waddling a short distance then lying down with their tongues lolling out. The Englishmen, furious at being robbed of their sport, had to call off the hunt, unsuccessful as their quest for the Northwest Passage."

32 See Markham (1989, 70).

33 The Spanish Armada was described as "such a mightie Navie…as never the like had before that time sailed upon the Ocean sea. It consisted of 64 Galeons [the largest warships of the era], nearly 100 other vessels, 8000 mariners, 20,000 soldiers and 2088 slaves" (Hakluyt 1598, 361-362).

34 Richard Whitbourne, an English West Country sailor, was present at Sir Humphrey Gilbert's proclamation in St. John's harbour, and would play a prominent role in early seventeenth-century Newfoundland. Prowse (1895, 82) stated that "Whitbourne was there in his own large ship, and some smaller vessels fitted out entirely at his own expense; Bristol, Biedford, Dartmouth, Plymouth, Topsham, Bridport, Exmouth, Poole, all had their share in the fight" in the battle with the Spanish Armada in 1588.

35 See Whitbourne (1620, 2).

36 Prowse (1895, 81) stated: "The result of the successful Spanish war was to give England complete control of the Transatlantic cod-fishery; the great Newfoundland Spanish fleet never came again after that terrible defeat, although stray Biscayans, protected by the French, continued to fish in Newfoundland up to a much later date. We shall see subsequently that they claimed the fishery as a right, which Pitt declared he would not acknowledge, even if the enemy were in possession of the Tower of London. From the Armada year until 1630 the English held the Newfoundland fishery practically without a rival; out of the whole number of fishing vessels England had more than double the fleet of her rivals, and all quietly submitted to her control." Well, not quite: the French would have their say for almost another 300 years, and it was they who finally excluded the Basques from the Newfoundland and Labrador fisheries in the late 1600s (Prowse 1895, 48).

37 See Whidborne (2005).

38 See Cell (1982, 35).

39 Guy's proclamations had little enforcement and no teeth. Whitbourne implemented the first true law.

40 See Seary (1971, 29).

41 See Whidborne (2005, 173).

42. See Hayman (1628, 2:14).
43. Sabine (1853, 36) claimed that "some forty or fifty houses for the accommodation of fishermen were built at Newfoundland as early as 1522." Sabine did not state his source for this information or the nationality of the fishermen.
44. This view was written into history by Prowse (1895), but is consistent with earlier work by Anspach (1827).
45. Sabine (1853) believed that piracy by Peter Eason and others was "not to be overlooked" as an obstacle to settlement.
46. See http://www.heritage.nf.ca/lawfoundation/articles/doc1_1634charter.html.
47. In their reply to the colonists, the merchants stated that "the Petitioners knowing better how to manage their fishing than the Planters [colonists] can direct, declare that they are altogether unwilling to be ordered by the Planters, or to join with them as they desire" (Prowse 1895, 100).
48. See Innis (1940, 99).
49. See Whitbourne (1620, 25).
50. See Rowe (1980, 112-117).
51. See Prowse (1895).
52. See Matthews (1988) and Pope (2004).
53. See Matthews (1988, 10).
54. Sabine (1853, 40), citing Sir William Monson, stated that "England may boast that the discovery, from year aforesaid to this very day, hath afforded the subject, annually, one hundred and twenty thousand pounds, and increased the number of many a good ship, and mariners, as our western ports can witness by their fishing in Newfoundland."
55. The Charter of Avalon to Lord Baltimore, 1623, stated: "Saving always and ever reserved unto all our subjects free liberty of fishing as well in the sea as in the ports of the province and the privilege of drying and salting their fish as heretofore they have reasonably enjoyed which they shall enjoy without doing any injury to Sir George Calvert or to the dwellers and inhabitants" (Prowse 1895, 132).
56. Innis (1940, 68) cited English reports from 1723-1737: "Nowe and for the tyme being for ever hereafter [they] shall and may from tyme to tyme and att all tymes for ever hereafter peaceably and quietlie have hould use and enjoye the freedome of fishinge…with full power and authoritie to goe on the shoare or land in or upon any place of the saide contynent of Newfoundlande aforesaide as well for dryeing, saltinge and husbandinge of their fishe on the shoares thereof, cuttinge off all manner of trees and woods for makeinge of stages shippes and boats, and making all manner of provisions for themselves."
57. See Pope (2004).
58. The term "planter" has had different meanings over time. In Newfoundland, a planter was denoted as "a certain class of settlers, those who owned boats and plantations (in a narrow sense) and employed other men" (Pope 2004, 1). Basically, planters were property owners, as distinguished from servants who owned little or nothing and fished from other men's boats and properties.
59. See Matthews (1988, 19).
60. Use of the word plantation conjures up visions of cotton, bananas, and southern climes, but in early usage planters were land or property owners, and their colonies were called plantations, no matter their location or economy. The name was somewhat misplaced in Newfoundland. They should have been called fisheries.
61. The first census of Newfoundland was in 1675, but as pointed out by Pope (2004, 1), servants were not listed "because they lacked a distinct economic personality."
62. See Prowse (1895, 59).
63. See Pope (2004, 436).

⁶⁴ Ignorance in England about Newfoundland continued until at least the early 1900s, if not to the present. J.G. Millais, a noted Victorian artist, hunter, naturalist, and writer, who developed an unabashed love for *Newfoundland and its Untrodden Ways*, stated that "the average Englishman images it to be a little bit of a place somewhere near the North Pole, which, with two or three other colonies, could be safely stowed away behind the village pump." Millais considered the true Newfoundlander was "the man of the outports – who throughout the year follows a variety of dangerous callings which build up characters of remarkable strength." He also made special mention of the Mi'kmaq who guided some of his trips: of all his companions, "the men I have employed were one and all the best of companions and good fellows." Millais wrote for a primarily English middle class readership and seemed determined to counter what he considered to be prevailing negative views of Newfoundland and its people.

⁶⁵ See Pope (2004, 435).

⁶⁶ See Prowse (1895, 97).

⁶⁷ Although most of Guy's colonists were not fishermen, Guy realized that they would have to have to fish to eat. In his first letter to backer Sir Percival Willoughby, dated October 6, 1610, he stated, "after the ship [the ship they arrived in] is departed we purpose to send to fishinge for ou[r] p[ro]vision" (Cell 1982, 60).

⁶⁸ Prowse (1895, 97) described the arrival of John Guy and his colonists at Cuper's Cove in August 1610: "They had a remarkably fine passage out, in twenty-three days they sighted their new home in the deep Bay de Grave (now Port de Grave), Conception Bay. In the bottom of this estuary lies the beautiful little land-locked harbour of Cupids. It was so far embayed that the resident fishermen, who were then sparsely scattered about Harbour Grace, Carbonear, and the bottom of the bay, had passed by this…[harbour]…as unsuitable for the fishery." Guy's initial houses and the centre of the colony were located at the top of Cuper's Cove. However, evidence of farms and mills has been found near the Southern River that drains into Bay de Grave.

⁶⁹ See Ryan (1994, 30).

⁷⁰ Presumably not all for himself.

⁷¹ See Pope (2004, 51).

⁷² Among the fanciful names are Heart's Desire, Little Heart's Ease, Heart's Content, Summerside, Comfort Cove, a fleet of Paradises – including Paradise, Little Paradise, Greater Paradise, and Paradise Cove – and the French-named Baie d'Espoir [Bay of Hope, ironically changed to Bay Despair in Newfoundland dialect].

⁷³ Hayman (1628, 2:9). See also Cell (1982, 18-19).

⁷⁴ See Cell (1982, 19).

⁷⁵ A detailed account of the life of Captain John Mason was given by Prowse (1895, 104-107).

⁷⁶ Pope (2004, 51) cited a letter, in which the colonists were said to scorn fishing, and unpublished records, which indicated that "archaeological work suggests that settlers occupied at least one of the structures erected for Guy's original colony until it was destroyed by fire about 1665."

⁷⁷ See Cell (1969, 69-96).

⁷⁸ See Hayman (1628, 2:12).

⁷⁹ See Cell (1982, 211).

⁸⁰ See Pope (2004, 437).

⁸¹ See Cochrane (1938, 52).

⁸² See Pope (2004, 437).

⁸³ The Kirkes seized Samuel de Champlain's *comptoir* in Quebec in 1627, made money in the fishery even in the worst seasons, invested in the Hudson's Bay Company in 1666, and rebuilt their premises after

the Dutch raid in 1673. Lady Sara died at Ferryland in the early 1680s after 40 years of residence. Her son George was the major planter on the southern Avalon Peninsula at the time (Pope 2004, 437). Fortunately for Lady Sara, she did not live to see the French raze Ferryland in 1696.

[84] See Pope (2004, 437).

[85] See Handcock (1977, 16). The West Country merchants retained considerable political power in England. A Poole merchant was blunt in giving evidence before a Parliamentary committee in 1793 (Handcock 1977, 18): "It is the wish of the generality of the merchants to keep no more men in the island of Newfoundland than the interest of their particular concerns and employment require; and so far from its being the desire of the merchants to incline at all to make the Fishery more a residence than can be avoided, their distresses arising from a great number of families already on the island…calls for the intervention and assistance of the Government of this country to prevent it and bring them home."

[86] Pope (2004, 435) argued that the view of farming as a lost cause in Newfoundland came largely from the outside, which apparently influenced the settlers. In Iceland, in contrast, outside opinions were either viewed with scepticism or disregarded completely.

[87] See Cell (1982, 62-63).

[88] Cell (1982, 58-59) stated that: "In Newfoundland it did not take long for the faulty assumption that underlay the first settlements to be exposed: settlement could not provide any overwhelming advantage in the exploitation of the fishery…[because] expenses always exceeded the income gained from fishing." However, Cell admitted that this was only a short term reality, that the prosettlement argument was correct, and that "Whitbourne has been proved right. Fishing could not support large-scale settlement, but a combination of fishing with subsistence farming could allow individuals to survive." Cell also failed to consider that trading became a large part of the economy that supported settlement, and that settled enterprises would play an important role in the expanding trade in the North Atlantic.

[89] Naval Officer John Berry, charged with evicting settlers after 1675, pointed out that many settlers would not be able to support themselves in England, but they could in Newfoundland – he was clearly referring to bye-boat men or servants who had settled there (Rowe 1980, 113).

[90] The term "families" is used here not to denote kinship, but a household that in the 1600s consisted of a planter and his or her servants, who were usually male and did the fishing. Some of the better-off households would also have had female servants who gardened, kept livestock, and performed other more traditional female tasks. As time progressed, more and more of the families would have become kin groups, consisting of a husband and wife, children, and servants. Planters could be men or women. There were several well-known and successful female planters, mostly widows, who apparently did not remarry – perhaps because men who matched their station were few and far between in Newfoundland at the time. See Pope (2004) for additional details.

[91] English Commodore John Graydon commented in 1701 that seal meat was something "which they and none but they could eat," adding "such people such stomachs" (Pope 2004, 427), based on Colonial Office (1701).

[92] Pope (2004, 56) gave an estimate of 200 for the total winter European population in the late 1620, including 100 at Ferryland. Other estimates are somewhat higher.

[93] See O'Flaherty (1999, 41-42) for details on these regulations. "The main intent of the regulations of 1671 was clear; new settlers were to be kept out. They were not directly forbidden to go to Newfoundland, but English fishing ships were forbidden to take them there and seamen on the ships were forbidden to stay there over the winter."

[94] Berry defended the settlers and their benefit to the fishery and to England. Pope (2004, 66) cited a letter from Berry stating that "some self ended persons have a mind to engage all into their own hands," that "those planters [settlers] are not soe bad as the merchants makes them," and that "in case those people should be removed out of this country, his Majesty's selfe in a few years would find the ill effects of itt."

[95] See Pope (2004, 63).

[96] See O'Neill (2003, 16).

[97] Matthews (1993, 29) argued that there were few or no impediments to settlement after 1677, implying that Newfoundland was just not worth settling and that "the more likely explanation for the lack of settlement is that there was no reason to live in Newfoundland year-round." However, this view ignores the attitudes of governments and merchants, which had considerable influence on life in seventeenth- and eighteenth-century Newfoundland, and that settlements did increase in number later on with no more reason to live there. The motivations of the settlers may remain obscure, but most were probably looking for a better life than they had in England and, later, Ireland. O'Flaherty (1999, 45) noted that "resistance to settlement would characterize official English thinking about Newfoundland until well into the 18th century. Indeed, traces of it lingered into the 19th, and combined with harassment by the same and by the French, must have impacted negatively on the charm of Newfoundland, and its attractiveness as a place of settlement."

[98] See O'Flaherty (1999, 115).

[99] See Matthews (1988, 8).

[100] Pope (2004, 56-57) related the lineage of some of the earliest Avalon settlers.

[101] Liveyers were residents as opposed to migrant fishermen. The term has been used in various forms and spellings, including liver, livyer, liveyer and liveyere, but all refer to people, both English and later mixed-race, who resided on the coasts of Newfoundland and Labrador. *The Dictionary of Newfoundland English* can be found online at http://www.heritage.nf.ca/dictionary.

[102] See Handcock (1977, 21).

[103] See O'Neill (2003, 46-51).

[104] Prowse (1895, 316) stated: "In 1762 the first custom house came into existence, and the first collector, Mr. Hamilton, was appointed; he was under the control of the department in Boston, Mass., then the capital of the British North American Colonies. The West Country merchants were so reluctant to pay him any fees that, after trying to live there for one season, he threw up his hands in disgust." Colonial records indicate that Hamilton also "found the people and climate so disagreeable" (Dunfield 1985, 54). Newfoundland is not for everyone. Hamilton was succeeded the next year by Alexander Dunn, a Scotsman, who appeared to have been "made of sterner stuff; either his ominous name, steady perseverance, or better defined legal rights enabled him to get in the fees."

[105] Prowse (1895, 296) related the history of Lieutenant Williams and quoted his report: "I found in Conception Bay 496 boats kept, and computed on an average each boat caught 500 quintals of fish (though many caught 750 and several at Trinity Harbour 990). These made 248,000 quintals; I allow for the shipping about 10,000, which I make 258,000 and allow for men, women and children employed in catching and curing the fish of each boat (as they all equally work) ten, which will make 4960 and for the shipping 300 making in all 5260 people…As the fishery of Conception Bay was reckoned equal to one quarter part of the whole fishery of Newfoundland from the year 1745 to 1752 i.e. Trinity, Bonavista, Catalina with the creeks thereunto belonging one quarter; Bay de Verd, Carboneire, Harbour Grace and the several creeks and coves thereunto belonging one quarter; Torbay, Kidivide, St. John's and Petty Harbour a quarter; Bay of Bulls, Firiland, Firmoses, Trepassey and Placentia another quarter part; so that the whole produce of fish and oil for one of the aforesaid years will be (exclusive of the whale and seal oil) of fish 1,032,000 quintals and oil 5,160 tuns." [1 tun = 953 litres and displaced 955 kilograms of water]. Approximately 200,000 tonnes of fresh fish would be required to make a million quintals of dry fish. Of further interest is the breakdown by region: many people assume that Conception Bay has never had a strong fishery, but the statistics of Lieutenant Williams argued against that opinion.

[106] See Appleton (1891). It is unlikely that this tax could be collected outside the English shore.

[107] A cartographic school developed in Amsterdam in the 1600s, with their information coming largely from Dutch and French exploration.

[108] Cushing (1988, 61), based on de la Morandière (1962), stated that "Governor de Meulees" ruled thus.

It is unclear if this ruling applied only to the French settlements at Cape Breton and in New Brunswick, but it likely applied in Newfoundland as well.

[109] See Handcock (1967, 135).

[110] Pope (2004, 311-313) related how Trepassey became an "interface with our friends the enemy" in the mid-1600s, how trade for French salt was practiced by English settlers, and that the winter allotting of French premises by English settlers was not openly tolerated by the English admirals, at least not for a time.

[111] See Pope (2004, 407).

[112] The English military expedition consisting of 1500 soldiers was sent out in 1697 after the French attack. According to their reports, they found "St. John's completely abandoned. The French had burnt, pillaged and destroyed everything movable and immovable in the once flourishing settlement. There was not a solitary building left standing, all the forts were razed to the ground; literally, there was not one stone left upon another" (Prowse (1895, 222). Additional detail of the sacking of St. John's is given by O'Neill (2003, 41-45).

[113] Captain Robert Land, a direct ancestor of the author, fought in these battles on the English side, alongside Captains Hugh Palliser and James Cook. Captain Land had a distinguished military career and, as was typical of the adventurous military officers of the day, eventually settled in New England. During the American War of Independence, Land's former military connections and his sympathy with Britain resulted in his persecution by the Americans. After their home was burned, Land and his family separated for safety and made their ways north – Land through the bush northwest to Niagara Falls, and his wife and children by way of Nova Scotia. Each assumed that the other was dead. Some years later – almost by accident – they were reunited. By 1800, descendents of the Lands had ventured widely over most remaining British colonies in North America.

[114] Under the Treaty of Paris, Spain "for ever relinquished all claims and pretensions to a right of fishing on the Island." Prowse (1895, 320) stated that "the Treaty of Paris, which Palliser (the Governor) was to put in force, extinguished the last hopes of the Basques to participate in the Newfoundland fishery."

[115] See Anglin (1970).

[116] From an unnamed French magazine, copied into *Scots Magazine*, in December 1762 (Prowse 1895, 321-322).

[117] See Prowse (1895, 320).

[118] See Prowse (1895, 597).

[119] See Hollett (1986).

[120] As late as 1893, an English naval vessel attempted to prohibit English settlers in St. George's Bay from selling their herring except to French vessels at French prices (Prowse 1895, 355).

[121] O'Neill (2003, 62-63) related details of this military engagement. By 1781, Fort Townshend was completed on a hill overlooking the town and harbour on the present site of The Rooms, the provincial museum and archives. Fort Townshend served as the summer home of the governor. The St. John's garrison at the time was comprised of 561 men from the Royal Artillery and the Newfoundland Regiment and 52 men of the Royal Newfoundland Volunteers. The French had 1500 troops. However, Bay Bulls resident John Morridge apparently deceived the French commander into believing there were 5000 troops in St. John's manning 200 cannons.

[122] See Bannister (1997).

[123] Prowse (1895, 228), citing a 1701 report from Mr. Larkin, a barrister, stated: "The then admiral of this Harbour, Caption Arthur Holdsworth, brought over from England this fishing season 236 passengers all or great part of which were bye boat keepers and they were brought, under a pretence of being freighters aboard his ship, though it was only for some few provisions for their necessary use. These persons he put and continued in the most convenient stages in the Harbour which all along since the year 1685 had belonged to fishing ships, insomuch that several masters of fishing vessels had been obliged to hire rooms of the Planters."

[124] See Prowse (1895, 226-227). It must be pointed out that Prowse used colourful language and may have embellished his claims.

[125] See Whitbourne (1620, 21-23).

[126] See Bannister (2003).

[127] Governor Palliser's submission to the Board of Trade in 1764 states that the admirals were "concerning themselves with nothing but what is for their own or employer's interest…[and]…who are the most part illiterate men, incapable of keeping the account required by [King William's] Act." To be fair, Palliser was a snob who wrote disparagingly about settlers as well, so his views may have been biased.

[128] See Bannister (2003, 57).

[129] See Bannister (2003, 59).

[130] See Matthews (1988, 98-99).

[131] See Bannister (2003, 58) and Matthews (1988, 136), respectively.

[132] Prowse (1895) painted a bleak picture of the fishing admirals, but this has been challenged by recent historical accounts (e.g., Bannister 2003).

[133] Matthews (1988, 8) stated that "the first struggles over the Island, as distinct from the fishery, occurred as men quarreled about who should take possession of the best fishing rooms."

[134] An argument could be made that John Guy or David Kirke were the first governors (Pope 2004).

[135] See Bannister (2003, 59).

[136] See O'Neill (2003, 391-394).

[137] See Prowse (1895, 301) for a list of the Justices and Constables in 1732.

[138] See Palliser (1765).

[139] See Innis (1940, 194).

[140] See Chadwick (1967, 8).

[141] See Hunt (1844, 72-73, 86). Palliser was ardently English and wanted to run the French out of North America.

[142] James Cook was born of very humble stock in Yorkshire, England, had little formal schooling, and ran away to sea at an early age. His brilliance as a navigator and geographer did not escape the attention of Captain Hugh Palliser, who became his patron. Palliser and Cook fought at the battles of Louisbourg and Québec in the 1750s, and Cook charted the St. Lawrence estuary. Later, after Palliser became governor of Newfoundland, Cook was brought to chart many of its coasts. Cook became the premier surveyor of his era, and many of his British Admiralty Charts are still in use. Cook's talents took him around the globe. He investigated Pacific waters from Alaska to New Zealand, and verified the existence of the Australian continent and many south sea islands, where he was killed. Palliser was so devoted to Cook that he erected a monument to him near his estate in Vache Park, Buckinghamshire, England, bearing the epitaph, "To the memory of James Cook, The ablest and most renouned navigator this or any country hath produced…He raised himself solely by his merit, From a very obscure birth to the rank of Post-captain in the Royal Navy…He possessed, in an eminent degree, all the Qualifications requisite for his profession And his great undertakings; together with the Amiable and worthy qualities of the best of men" (Hunt 1844, 121).

[143] The purpose of Palliser's Act was described by Prowse (1895, 344) as "to make the Newfoundland business a British fishery, carried on by fishermen from England and the King's dominions in Europe; American colonists were rigidly excluded from any participation in the bounties and other benefits granted by it." Even British Canadians were excluded.

[144] Some recent social and historical accounts have downplayed the role of authorities in discouraging settlement, suggesting that the damper on early Newfoundland settlements was the lack of interest in

145 living there year round (Matthews 1993, 29) and that Newfoundland was "nothing but a fishery" (Matthews 1968, 10-15). There may be some truth in these conjectures, but overall the attitudes and actions of the authorities in discouraging settlement are evident.

145 One estimate has a total resident population of about 6000 in 1729, and 3449 English and 3348 Irish in 1771 (Sabine 1853, 51).

146 In 1788, 19 fishing vessels from Bermuda made their way in St. John's harbour, loaded with fish from the Grand Banks. This was not legal at the time, but accommodations were made to these vessels so as not to raise any political hackles. The original letters describing these events were given by Prowse (1895, 416-418).

147 Hutchings and Myers (1995) reported that cod catches from 1500 to 1800 were in the range of tens of thousands of tonnes; these were considered to be incomplete and gross underestimates by Pope (1997, 2004). Pope (2004) stated that "European live catches might have been as much as 200,000 tonnes a season – approximately the total catch at Newfoundland implied by early administrative statistics collected by France and England in the 1660s, 1670s, and 1680s and already in the same order of magnitude as catches of northern cod in the late nineteenth and early twentieth century." The present reconstruction is more in accord with Pope than with Hutchings and Myers: average catches in the 1600s are estimated to be closer to 100,000 tonnes than 200,000 tonnes, although in some years the average catch would have been greatly exceeded, perhaps achieving 200,000 tonnes.

148 According to Cushing (1988, 67), "in 1603, George Waymouth reported that some cod from Newfoundland were 5 feet (1.5 [metres]) long and 1.5 [metres] round."

149 The earliest fishing methods used single baited hooks fished by hand. Fish could also be caught by actively jerking a weighted and unbaited hook, perhaps adorned with shiny metal or feathers, up and down near the ocean floor (jigging). Later, multiple baited hooks on lines weighted to the seafloor became a common method.

150 See Pope (2004, 424).

151 See Pope (2004, 425).

152 There are several references to a particularly severe cold period along the northeast coast from about 1713 to 1724. According to the ecosystem model, it was a short period of low productivity that saw a modest decline in the cod stocks. Catch declines were even more severe. Innis (1940), based on colonial reports, stated that: "A series of bad fishing years had begun with the severe and prolonged winter of 1713-14. It had chilled the water along the coast, and it had been followed by 'the worst season for many years.' Up to 1720 the average yearly catch was only 90,000 quintals…In 1714 and 1715, small ships were sent to the Banks." Head (1976) reports on the same events from an English perspective.

153 See Matthews (1988, 113).

154 Cod movements near shore are strongly influenced by water movements and temperatures. In some years cod will come in early, in others later, and in others still not at all, if conditions are not favourable (Templeman 1966, Rose and Leggett 1988).

155 Briere (1997) provided a French perspective and described the withdrawal of much of the French Shore fishery in Newfoundland from 1713 until 1724. Politics clearly played a supporting role in this movement. Briere (1997, 60) stated that "the effects of these phenomena was strengthened by the letters patent of 1717 that exempted from export duties supplies for ships fitting out for the French colonies in America: Newfoundland, a British territory [after the Treaty of Utrecht, 1712] was excluded." However, politics could not directly influence the fish stocks. By the mid-1720s, Newfoundland had once again become the hotspot (Briere 1997, 60): "From 1724-25 to 1733, there was an appreciable decline in the fishery on Isle Royale [Cape Breton], at Gaspé, and on the banks, while it picked up again on the coasts of Newfoundland and Labrador." France abandoned the export duties on Newfoundland in 1733 and French Shore fishermen moved back to Newfoundland where the fishery was better.

¹⁵⁶ A Colonial Office report from 1739 cited by Innis (1940, 150) stated that "to catch and load and then come in, put their fish ashoare to dry, and immediately out again leaving people ashore to cure it; and this manner of working has turned to prodigious account…ten hands employed in a barke this way where the whole amount of weare and teare of the vessel and wages of servants included has not been computed at above 70 pounds has catch'd 600 quintals and upwards whereas the service of seven boats employing in all 35 hands and wages to the amount of near 400 pounds has not caught an equall quantity…every fish brings its own bait with it to catch another with for by opening the maw you are always stock'd with fresh bait to proceed upon new purchase." Many of the poorer residents went to this fishery, especially if the coastal fishery had failed them, but there were apparently problems with quality despite having the fish split and salted the same day they were caught. Another Colonial Office report of the same year referenced complaints that the fish produced were of poor quality, "chiefly owing to the fishing ships who have left off keeping of shallops and fishing near the shore but send their ships and vessels on the banks for a month or five weeks, then bring the fish to land to cure; such fish as are caught at the beginning of the season are good, if rightly salted, but in height of summer and latter end of the year very bad." Quality issues with the fish could not be denied and would resurface over the coming centuries. By 1765, the poorer quality of the Newfoundland cure and the selling of it in Europe is an "as is" condition, without the strict grading of the Portuguese, led to loss of some of the high-value market to the French cure and the traditional stockfish from Norway and Russia (Innis 1940, 150-151).

¹⁵⁷ Colonial Office records cited in Innis (1940, 151) reported that "all rooms and conveniences now used for the fishery to the southward of Cape St. John are constantly possessed and kept by the same people for their own private benefit and are becoming private property…Yet all ships arriving from Britain directly call themselves fishing ships because they clear out as such; though they have no more men engaged to them than is necessary for their navigation nor more boats than one each employed in the fishery."

¹⁵⁸ Howley (1915, 45-49) reported notes on the "Red Indians" (Beothuks) from Captain George Cartwight's journal. Cartwright related much about the state of the Beothuk in the 1760s: "As far as I can learn there were many Indians on the island when first discovered by Europeans, and there are still fishermen living who remember them to have been in much greater numbers than at present, and even to have frequented most parts of the island. They are now much diminished, confining themselves chiefly to the parts between Cape Freels and Cape St. John. The reason I presume of their preferring that district to any other is because within it are several deep winding bays, with many islands in them, where they can more easily procure subsistence, and with greater security hide themselves from our fishermen. I am sorry to add that the latter are much greater savages than the Indians themselves, for they seldom fail to shoot the poor creatures whenever they can, and afterwards boast of it as a very meritorious action…I fear that the race will be totally extinct in a few years, for the fishing trade is continually increasing, almost every river and brook which receives salmon is already occupied by our people, and the bird islands are so continually robbed that the poor Indians must now find it much more difficult than before to procure provisions for the summer, and this difficulty will annually become greater. Nor do they succeed better in the winter, for our furriers are considerably increased in number, and much improved in skill, and venture further into the country than formerly, by which the breed of beavers is greatly diminished."

¹⁵⁹ Full accounts of the Beothuk people and the captures of at least three women, including Demasduit and Shanadithit, in the early nineteenth century are given by Howley (1915) and Marshall (1996).

¹⁶⁰ Cassini's narrative contained a detailed description of the French fishery in the late 1700s and some interesting comments on cod distributions (Prowse, 1895, 569-570): "The vessels destined for the bank fishery sail from France from the end of February to the end of April; happy those, however, who can get there by the middle of April. From that time until the 15th June the fishery is most plentiful, then the capelin draw away the cod, who forsake the Grand Bank[s] until the middle of September; the fishery again in September and October yields almost as much as it did in May and June. Many ships, consequently, go twice a year to the Great Bank[s], and employ the interval in returning to France to dispose of their cargo and recruit their provisions and salt; few ships, except those from Olonne, go twice

a year to Newfoundland; the rest are stationed there for six or seven months together, and never came home till they begin to be in want of provisions, unless they have made a speedy and plentiful capture, which is seldom the case…Before and after the war of 1744, prodigious shoals of cod flocked to the Banks of Newfoundland, and made the fortunes of fishermen and privateers; but since the last peace, the produce of the fishery is reduced to one-third of what it was before; doubtless, the bait of a small fortune has increased the number of vessels, and proportionally divided the profit."

[161] Innis (1940, 217), summarizing uncited earlier reports, wrote: "From Fleur de Lys to St. Anthony the fish were said to be smaller, but to the north and at Quirpon after the end of July they were as large, it was claimed, as those of the Banks. To the south, 50 or 60 quintals produced a barrel of oil; but to the north it took 90 or 120. The small size were taken with capelin to the end of July, and the large with herring during the remainder of the season." The season would have ended in late August or early September.

[162] Cushing (1988, 60-61), based on de la Morandière (1962).

[163] Innis (1940, 217) summarized earlier but unspecified reports of French Shore fisheries in the late 1700s: "In 1787 north of Hare Bay to Quirpon the fishery was tolerably successful, but elsewhere poor. St. Lunair was deserted. Some vessels were engaged at Port Saunders but 37 boats were drawn up on shore at Bay of Islands…From Bay of Islands around the north coast to Croc the fishery, in 1788, in contrast with the previous year, was a failure; but south of Croc, in Canada and Orange bays and East of White Bay to Cape St. John it was a success. Five vessels from Bayonne were at Port au Choix and Ferrole on the west coast, and took 8,400 quintals. In 1791 the fishery was described as very bad and, for the third year in succession, it was a failure in 1792, probably averaging less than 100 quintals to a boat (a saving voyage). At Quirpon the number of vessels declined from 10 to 3,500 tons in 1786 to 8 of 2,350 tons in 1792; at Fichot, from 7 of 1,900 tons to 3 of 240 tons; at Croc, from 7 of 1,120 to 3 of 690; at Cap Rouge, from 12 of 3,300 to 9 of 1,540; and, at Conche, from 10 of 2,500 to 3 of 470. The total number of men employed in boats dropped from 4,627 in 1786 to 2,007 in 1792. The number working on shore dropped from 3,232 to 1,390, and the grand total from 7,859 to 3,397."

[164] Whiteley (1977, 49) stated: "If fishing was poor at Bonne Espérance a schooner might be used as a fishing vessel [normally they were used as freighters] within the St. Paul Archipelago or be sent down the Labrador coast in search of fish. Usually such voyages did not extend beyond Sandwich Bay but there were occasions when the schooner would go almost as far north as Cape Chidley." This type of fishery contrasted greatly with the trap fishery, whose berths at Bonne Espérance were within earshot of the stages.

[165] Captain James Cook described the Humber River in 1767 as having "formerly been a very great salmon fishery" (Taylor 1985, 19).

[166] See Carlson (1996).

[167] See Lajus et al. (2005).

[168] Taylor (1985, 9), citing earlier reports, stated: "As to the right of property at Newfoundland…the salmon fishery, which had been carried on and improved by Mr. Skefflington between Cape Bonavista and Cape John in a part never frequented by any fishery ships; he had cleared the country up the rivers for forty miles and had built houses and stages. This person applied for an exclusive grant of this fishery for a term of years…The board accordingly recommended to his majesty, that a term of 21 years, in a sole fishery for salmon in Freshwater Bay, Ragged Harbour, Gander Bay and Dog Creek, might very well be granted…with liberty to cut wood and timber in parts adjacent, provided it were six miles distant from the shore." This right was granted in 1723. Thus, a form of fishery property rights came early to the salmon fishery in Newfoundland.

[169] Prowse (1895) described early developments in the salmon fishery in Bonavista and Notre Dame bays.

[170] One of the first English fisheries on the Labrador was established by George Cartwright. Cartwright was a keen observer of natural history, but apparently a poor businessman. He lost money and his business

171 The records of Captain John Smith indicated that the first New England vessel visited Bay Bulls to fish and trade in 1645 (Prowse 1895, 90).

172 James Young, a crew member of a New England vessel, wrote from Renews in 1663 that "the harbour we were in was very much esteemed for a good fishing place – the Barnstable men prefer it above any – yet we had poor fishing and made not above 130 quintals per boat, and 3 pounds 5 shillings a share. At the head of this river [Renews River] are many salmon; we caught abundance and our master saved several hogsheads and dried abundance in the smoke" (Neary and O'Flaherty 1974, 29).

173 Dunfield (1985, 31), citing the records of Charles Taylor in 1680.

174 The southern New England salmon fishery was in decline as early as the mid-1700s. A Swede named Peter Kalm remarked in 1753 that "many old people said that the difference in the quantity of fish in their youth in comparison with that of today is as great as between night and day" (Dunfield 1985, 43). Cod may also have been scarce in southern New England coastal waters by these times (Elliot 1887, 21).

175 The captain of a Massachusetts fishing vessel wrote of Labrador in 1758: "The coast is full of islands, many of them large, capable of great improvement as they have more or less good harbors, abounding in fish and seal, water and land fowls, good land covered with woods, in which are great numbers of fur beasts of the best kind. Along the coast are many excellent harbors, very safe in storms; in some are islands, with sufficient depth of water for the largest vessels to ride between, full of codfish, and rivers with plenty of salmon, trout and other fish" (McFarland 1911).

176 The Moravian church was Protestant and originated in Czechoslovakia in the 1400s. Moravian missionaries were sent to Labrador to work with aboriginal peoples and remain there.

177 Goode and Collins (1887, 137) reported that, north of Cape Harrison in northern Labrador, there were "about sixty resident settlers in the deep fjords, most of whom have been in the service of the Hudson's Bay Company or the fishing firms already named, and some of them are married to Eskimo women." No exact date is given for this, but it was likely in the early 1800s. The Grenfell Mission was established by Dr. Wilfred Grenfell with funding from the United States and Britain to provide medical care and some education for the liveyers in northern Newfoundland and Labrador, who lived in fairly primitive conditions. The headquarters of the Grenfell mission was located in St. Anthony on the Great Northern Peninsula.

178 A narrative describing the voyage of the French ship Bonaventure in 1591 stated that "the coast stretcheth three leagues to the West from Lisle Blanch, or the white Isle, unto the entrance of a river, where we slewe and killed to the number of fifteene hundred Morses or Sea oxen, accounting small and great" (Quinn 1979, 58-59).

179 A letter written by merchant Thomas James of Bristol in 1591, after describing the capture of the French ship, stated that: "The Island lyeth in 47 degrees, some fiftie leagues from the grand Bay, neere Newfoundland; and is about twentie leagues about, and some part of the Island is flat Sands and shoulde; and the fish commeth on banke (to do their kinde) in April May & June, by numbers of thousands, which fish is very big: and hath two great teeth: and the skinne of them is like Buffes leather: and they will not away from their yong ones. The yong ones are as good meat as Veale. And with the bellies of five of the saide fishes they make a hogshead of Traine [oil], which Traine is very sweet, which if it will make sope, the king of Spaine may burne some of his Olive trees" (Quinn 1979. 59-60).

180 The *Marigold* carried 20 men, including three coopers and two butchers, and sailed from Apsham, England, on June 1, 1593, to "flea the Morsses or sea Oxen (whereof divers have teeth above a cubit long & skinnes farre thicker then any buls hide)." It arrived at the Magdalen Islands too late and took "nothing such numbers as they might have had, if they had come in due season." The *Marigold* encountered a "shippe of St. Malo three parts freighted with these fishes." From a narrative from Richard Hakluyt reprinted by Quinn (1979, 61-63).

[181] A letter from John Bland, resident of Bonavista, to Governor Gambier in St. John's, dated September 26, 1802, provided a description of the early seal fishery (Prowse 1895, 419): "The seals upon this coast are of many species, they are classed and distinguished by names only to be found in the Newfoundland nomenclature, and only understood by the Newfoundland naturalists. Tars, Doaters, Gunswails, and many others, breed upon the rocks in the summer season, and may be called natives, but these make but little part of our fishery, our dependence rests wholly upon Harps and Bedlemers, which are driven by winds and ice from the north-east seas. The harp in its prime will yield from ten to sixteen gallons of oil, and the bedlemer, a seal of the same species, only younger, from three to seven. About the middle of March the female harp whelps upon ice, and, in the course of a few days, becomes reduced to less than half of her largest bulk. The male, also, from this period reduces, but not in the same proportion. There is, therefore, an evident advantage in catching the seals in winter…In the winter of 1712, a succession of hard gales from the north-east brought the seals in great numbers, before the middle of January, unaccompanied by any ice, a circumstance that rarely occurs. In Bonavista about two hundred men might have been employed in attending the nets, and the number of seals caught amounted to about seven thousand. The entire catch at Bonavista Bay may be taken at ten thousand, and two-thirds of the whole reckoned Harps. The harps yielded thirteen shillings each, and the bedlemers seven shillings and sixpence, which at that time afforded the merchant a large profit in the English market." Regarding the spring fishery on the ice which was just beginning: "The sealing-adventure by large boats, which sail about the middle of March, has not been general longer than nine years…From two to three thousand men have been employed in this perilous adventure, and it may excite surprise that so few fatal accidents have happened." This foreshadowed the tragic accidents in the seal hunt that would come in the early 1900s.

[182] The killing of seals in Newfoundland has always been termed a fishery – not because anyone thought that seals were fish, but because of the methods employed. The original method used nets, just as the fisheries did. Nets were replaced with "hakapiks" (wooden clubs with a hardwood or metal spike at the end) and firearms in the nineteenth century. The first sealing muskets were horrible things that were cheaply made, inaccurate, and likely to blow up in the user's face. The modern seal hunt uses high-powered rifles.

[183] In John Bland's letter, the number of seals taken at Bonavista was estimated to be between 5000 and 10,000 in all years from 1791-1792 to 1801-1802, with the exception of 1800-1801 when an incursion of young hooded seals increased the take to about 20,000 seals. Bland stated that the seal fishery in "the ports on the coast of Labrador, on the northern shore, now possessed by the French, and southward to Fogo, have been out of all comparison more successful in this fishery by nets." Hence, it is likely that the seal fishery took somewhat less than 50,000 seals annually during this time, although in some years the take may have been higher.

[184] See Sabine (1853, 42).

[185] This passage represented one of the first biological descriptions of the seasonal movements of fishes in Newfoundland waters. The comings and goings of cod, herring (hering), capelin (Capline), Atlantic mackerel (Macrill), and Atlantic salmon (Salmond) are all consistent with current knowledge of these species. See Prowse (1895, 106-107) and Cell (1982, 95).

[186] See Innis (1940).

[187] Prowse (1895, 336) wrote that "the New Englanders still gave trouble on the Labrador [in the 1770s]. A small 'Behring Sea difficulty' arose at the Madgalen Islands, owing to the reckless and barbarous way in which the New Englanders were killing the sea-cows [walruses], driving them away from their quiet haunts, and preventing them from breeding."

[188] A seine is a small meshed net that is dragged through the water and is intended to trap fish either against the shore or within a bag formed by the encircling net. There are various forms of seines. The first cod seines were fished near shore, and attempted to trap cod against the beaches and rocks.

[189] In a petition to the House of Commons in London by a group of merchants, boat-keepers, and principal inhabitants of St. John's, Petty Harbour, and Tor Bay in 1775, it was stated (Prowse 1895,

341-344): "Codd Seans [cod seine nets] we deem a great nuisance as by them we destroy a great quantity of small fish, which after being inclosed in the sean (and not worth the attention of the person who hauls them) are left to rot, by which means a multitude of fish that would grow to maturity, perish…Contiguous to the Northern Part of the Island are a great many Islands where Birds breed in vast abundance which were of great service to the inhabitants residing near them, for food in the winter, and also for bait in catching of fish during the summer, of which valuable resource they are now almost entirely deprived, as great part of the birds are destroyed within a few years by the crews of men who make it their business to kill them in their breeding season, for their feathers (of which they make a Traffic) and burning the carcasses, we have applied to get this with many other grievances redressed but have yet only retained a partial relief, therefore pray that an entire stop may be put to destroying the birds otherwise than for food or bait as before excepted."

[190] Laws were harsh in the eighteenth century. John Reeves, the first chief justice of Newfoundland, "tried some fishermen for taking eggs at the Funk Islands, which was forbidden by Proclamation; it was proved that one of the culprits, Clarke, lived at Greenspond; he was in want of food for his family, and the eggs were taken solely to obtain some for his wife and children. Whilst sentencing the other prisoners to be publically whipped, he solemnly ordered that, out of regard to these mitigating circumstances in Clarke's case, he was only to be privately flogged. We do not think this unfortunate victim of a cruel law appreciated the distinction" (Prowse 1895, 359).

[191] A petition of the merchants, boat-keepers, and principal inhabitants of St. John's, Petty Harbour, and Tor Bay in 1775 stated that "lumber at present is a scarce commodity…regard should be paid to the Timber Trees growing in this Island, which if not wantonly cut down would in a few years, become large spars, for Masts of Ships and other uses, as well as to saw into boards" (Prowse 1895, 341-342).

[192] Cushing (1988, 75) gave estimates of catches from this period: "An average catch of 2000 quintals per vessel, [with] up to 250,000 tonnes of fish fresh from the sea were taken from about 1580 to 1750."

[193] The estimates by Pope (1995) for the decades of the late 1600s are consistent with Cushing and Rose (2004). All of these estimates are considerably higher than those reported in other works (e.g., Hutchings and Myers (1995)).

[194] Earlier speculations that the overfishing of cod may have begun as early as the late 1600s (see references in Pope (2004, 420)) are thought to be overstated. Fishing was substantial, but so too were the standing stocks. The harvest rates indicated in the present analyses do not indicate overfishing. The sole exception to this conclusion might have occurred at very local scales – i.e., small groups of coastal fish that might have been excessively harvested by the growing settlements.

*[Settlers] are making roads in Newfoundland,
next thing they will be having carriages and driving about.*

PETER OUGIER, West Country merchant, ca. 1825

*The best policy for Newfoundland is to cherish and develop her
cod fishery, for in it her people have a reliable mainstay.
The danger lies in overfishing any one locality and taking
immature fish before they reach the reproductive stage.*

REVEREND DR. MOSES HARVEY, 1900

SETTLEMENT, SALT FISH, SEALS AND SCIENCE
AD 1801 TO 1949

There is not the slightest sign of overfishing in Newfoundland waters, which are probably capable of producing a higher average annual yield than is exacted from them.

HAROLD THOMPSON, 1943

The blunt truth is that Newfoundland did not... properly manage to develop its maritime economy.

DAVID ALEXANDER, 1977

Settlements in the Early 1800s . . .	269
Salt Fish Trade and War . . .	272
Home Rule . . .	275
International Politics and the Fishery in the 1800s . . .	277
The French Connection . . .	277
Canada and Confederation . . .	281
New England Fisheries . . .	282
Ecosystems in the 1800s . . .	284

Fisheries in the 1800s . . .	285
Cod . . .	285
Salmon . . .	294
Seals . . .	295
Whaling . . .	298
Lobsters . . .	301
Atlantic Halibut . . .	301
Pelagics . . .	303
Other Species . . .	308
The Invention of the Cod Trap . . .	308
Fishing Ships . . .	313
Cruel Years: 1880-1900 . . .	316
Early Science . . .	319
The First Newfoundland Fisheries Commissions . . .	322
William Coaker . . .	325
World War I . . .	327
Ecosystems: 1900-1949 . . .	328
Fisheries: 1900-1949 . . .	330
The Rise of Post-World War I European Fisheries on the Grand Banks . . .	336
Post-World War I Science in Newfoundland . . .	338
Commission of Government, Fisheries Policy, and Confederation . . .	346
Newfoundland Associated Fish Exporters Limited (NAFEL) . . .	350
The State of the Ocean Fisheries at Confederation . . .	351
Protecting the Continental Shelf Ecosystem . . .	353

SETTLEMENTS IN THE EARLY 1800s

The early 1800s brought a new prosperity – and burgeoning settlements in Newfoundland. During this era, Newfoundland ceased to be an oversized fishing station and became a true settled colony.[1] From 1790 to about 1816, immigration from England and Ireland led to a dramatic increase in the number of outports and a doubling of the resident population each decade.[2] Women and children, once a rare sight outside of the few major centres, became a larger proportion of the colonists, and with that, settlement became permanent. The period of intense immigration was brief: by the mid-1830s, few immigrants were coming to Newfoundland, and population growth had been largely internalized as families strengthened and grew.[3] After 1850, immigration made up only a small portion of the population increase. Over this entire period, from the late 1700s until the final decades of the 1800s, almost all of the population – old hands and newcomers, males and females, old and young – were engaged in the fishery.[4]

While the migratory fisheries of the 1600s and 1700s had relied primarily on transient and largely male crews working either for English merchants or bye-boat keepers, the fishery of the 1800s was based on the growing number of resident family enterprises. In resident families, everyone had a role in making salt fish: men and boys caught and split them, women and girls cured and dried them, and merchant families owned and sold them. Each part of the operation was crucial to the quality of the product, the market in which it could be sold, and the price it would fetch. Merchants traded fishing and household goods and foodstuffs to the settlers for their fish.

In the 1800s, St. John's and Harbour Grace vied for commercial predominance. Both became prosperous if rough-edged towns and housed the first urban culture in Newfoundland. These were first and foremost sea and sailors' towns, paid for with fish. Drinking establishments lined the waterfronts. There were dozens of licensed taverns in St. John's in 1807, among them the Ship Inn, the Flower Pot, the Red Cow, the Happy Fisherman, the Blue Ball, and the Bird-in-Hand. Rowdy taverns passed as community centres of a sort; to help keep order, licensed proprietors were made Constables of the District of St. John's.[5] Some of the earliest tavern names, including the Ship Inn and the Happy Fisherman, live on in St. John's establishments today.[6]

As settlement increased and expanded, a distinct outport way of life developed in hundreds of small coves around the island. Many outport people, later called baymen, lived an isolated existence and seldom, if ever, travelled far. Their fish may have gone to St. John's and Harbour Grace, but they did not. Those who travelled away seldom returned. Conversely, few urban folk, or townies, ever visited an outport. Two solitudes developed, held apart by geography but united by salt cod production, which was the common currency of the economy as well as the glue of life and the developing culture in both the outports and towns in Newfoundland. The salt fish product remained a major commodity in international markets throughout the 1800s.

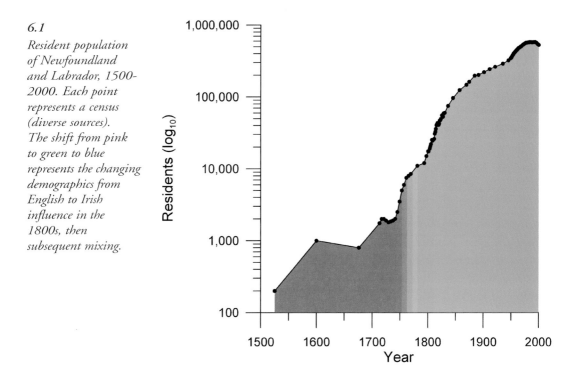

6.1 Resident population of Newfoundland and Labrador, 1500-2000. Each point represents a census (diverse sources). The shift from pink to green to blue represents the changing demographics from English to Irish influence in the 1800s, then subsequent mixing.

William Epps Cormack, the first European-Newfoundlander to traverse the island's interior on foot in 1822,[7] provided vivid descriptions of the outport of Bonaventure in Trinity Bay in the early 1800s. Cormack believed that Bonaventure was much like all of outport Newfoundland at the time.[8] About a dozen families lived in Bonaventure, all of whom were totally dependent on the cod fishery. Each family had an average annual production worth the not-inconsiderable sum of about 100 pounds sterling. They traded their fish and cod liver oil to Trinity Bay merchants for goods from Europe and America. They grew potatoes and other vegetables. Life was hard but sustainable. Bonaventure, like most fishing outports, was located near a headland where large shoals of migrating cod from the offshore banks were intercepted each summer. Bonaventure was also strategically sited near the seasonal tracks of coastal fish moving along the shore to feed. Few of the original outports were located deep in the bays or fjords, where the coming of the great schools of cod was far less certain.

Increases in population meant that new fishing grounds and rooms were needed. In western Trinity Bay, a new settlement was established south of Bonaventure at Riders Harbour on Random Island in the 1760s,[9] and many additional satellite outports – including British Harbour, Kerley's Harbour, Ireland's Eye, Traytown, and the Thoroughfare – were established before the mid-1800s in adjacent coves and islands. This pattern was repeated near most of the earliest settlements around the island, especially on the English shore.

It was inevitable that the growing population would expand their searches for new coves and fishing grounds beyond the established settlements of the Avalon Peninsula, and that conflict with others who claimed the cod grounds would increase. French authorities hotly contested English settlement north of Cape St. John throughout the 1800s. Some would-be English settlers were forcefully evicted by the French navy, assisted at times by the British. The Labrador coast, however, was open to settlement, and the first permanent residents came to the Strait of Belle Isle region in the late 1700s and early 1800s. But even the faraway Labrador was not free of trouble: in July 1832, William Buckle wrote from Blanc Sablon to the Governor of Newfoundland complaining that New England vessels and a (probably English) fishing admiral had harassed him and his family and wrecked much of his gear.[10]

The fish trade largely changed hands in the 1800s. At the beginning of the century, most outport merchants were agents of English companies. By 1850, however, Newfoundland-based merchants in Harbour Grace and St. John's began to make inroads into this commerce, and the formerly dominant English outport firms were in decline. By the end of the century, few of the original English West Country outport firms remained in business and, one by one, ceased their operations in Newfoundland. The Slade Company had operated at Battle Harbour on the Labrador coast since the 1770s, but sold their interests to local merchants (Baine Johnston) in 1871. William Cox and Company closed their premises in Fogo, Greenspond, and Twillingate by 1868, ending 250 years of commercial involvement of English firms from Poole in the Newfoundland fishery. Newman and Company, the last English holdout, sold their sole remaining holding at Harbour Breton in 1907, bringing to a close over 300 years in the salt fish trade in Newfoundland.[11] From then on, St. John's and Conception Bay firms controlled the trade, with St. John's becoming the undisputed commercial centre of Newfoundland and the Labrador coast.[12]

Newfoundland merchants maintained control of the fishery using the "truck" system inherited from the English merchants. The truck system was a form of indenture where merchants advanced supplies to fishing families in spring and were repaid later in the season in fish equivalents. Each outport or cluster of outports had a merchant, most of whom were agents of larger companies based in St. John's or Conception Bay.

The merchant set the prices of both commodities and fish, and so could scale the prices to his benefit – merchants kept families going, but also in constant debt.[13]

Much has been written condemning the merchant-run economy, with views at times reflecting political rather than economic agendas.[14] To be fair, merchants were necessary for the trading of fish and the provision of goods to the growing resident population, and while the truck system was clearly unfair and repressive, it did have benefits. Fishing families needed a means to sell their product in international markets, which they could not do on their own, and they required credit, which they could only get from a successful merchant. A more co-operative or competitive system of commerce might have been preferable, but – as Coaker and Smallwood would learn 100 years later – the independent and deeply conservative nature of most fishing families was antithetical to such organization. The fortunes made by a few merchants became a point of social discontent in Newfoundland, but their prosperity was far from universal and many merchants were as much in hock as fishermen. Prices received for salt cod were quite beyond the merchants' control, and while some undoubtedly became wealthy using questionable business practices, many others went bankrupt. Perhaps the best that can be said is that there were good and bad merchants – those with some sense of social responsibility and those with little. The good merchants made money, but demonstrated an enlightened self-interest – it was not unusual for them to carry credit for outport families who made little attempt to repay their debt.[15] There is little evidence that merchants used strong-arm tactics to extract payments from recalcitrant debtors. In bad times, many merchants would assist the poor, and some ended up as destitute as other outport families.

The truck system was supposedly nullified on many occasions during the 1800s and into the early 1900s. In truth, however, it outlasted these declarations, receiving support from both merchants and fishermen. Merchants and fishing families had their differences, but fishing and the truck system formed a common bond: as the fishery went, so went the economy, although quite arguably not in equal proportion to merchant and fisherman.[16]

SALT FISH TRADE AND WAR

The growing population of the 1800s conducted an international trade in salt cod, salted barrelled salmon, pickled herring, and, increasingly, sealskins and oil. Imports from England and the newly independent United States of America included goods of all kinds as there was little farming or manufacturing in Newfoundland. European clothing and Caribbean rum poured into St. John's, along with shiploads of bread, flour, pork, and beef.[17]

Trading patterns depended on political alliances. The defeat of the Spanish and French coalition fleet by the Royal Navy under Admiral Nelson at Trafalgar in 1805 helped establish British sea supremacy and ended Napoleon's plan to invade Britain.[18] Nevertheless, the war with Napoleon's France was not over, and conflict was rekindled in the New World by 1809. Labrador and Anticosti Island were repatriated as part of Newfoundland under an act of the English parliament in order to restrict American and French interests there. War broke out with the United States in 1812, and for the first time in its history, the Newfoundland cod fishery was well protected. Royal Navy

6.2 The Thoroughfare, Trinity Bay, in the early 1950s, prior to resettlement. This was a typical small Newfoundland outport (Tom Mills collection).

6.3 Merchants weighing in fish at the end of the season in the early 1900s (Gordon King collection).

warships occupied many berths in St. John's and, with sails full and cannons roaring, coursed the coasts regularly.[19] The outports were secure and proudly flew the Union Jack. American and allied French vessels were out-manned, out-gunned, and regularly seized by Royal Navy vessels. St. John's clerks spent Sunday afternoons firing rounds at looted French champagne bottles placed at the end of the wharf; the man who knocked the head off a bottle won the case.[20] Nevertheless, victories were not all one-sided, and skirmishes between English-Newfoundland and American and allied French vessels were common, and sometimes bloody, especially on the south coast.[21] Although the cod and seal fisheries did well during the war, American privateers plundered the Labrador fishery of George Cartwright and salmon catches declined.

Despite some losses, perhaps for the first time war brought an added level of prosperity to Newfoundland. Catches rose, and prices for both fish and cod liver oil were high.[22] Under the watchful protection of the Royal Navy, Conception Bay merchants established a migratory fishery for Labrador cod – with the seal fisheries also growing, work was provided for the increasing numbers of new immigrants.

The Napoleonic war with France and the War of 1812 with the United States were over by 1815. Although history might record that the English won both conflicts, the victories did not resolve the conflicts over the fisheries in Newfoundland. St. Pierre and Miquelon were governed alternately by England and by France in a dance of political niceties. Fishing rights on the west coast of Newfoundland and in the Petit Nord – the area ranging from Cape St. John to Cape Bauld – were returned to France. In 1818, a convention with the United States gave the New England fleet fishing rights in the same region, on the coast of Labrador, and on a part of the west coast of Newfoundland that came to be known as the American Shore.[23] England's continued concessions to France and the United States were not appreciated in Newfoundland.

6.4
Farming in Newfoundland in the early 1900s (John Wheeler collection).

The prosperity of the early 1800s was short-lived. The heavy dependence on a single product – salt cod – and the lack of local produce left the colony vulnerable to the whims of international markets and trade. Less than two decades into the 1800s, although landings of cod had increased, prices crashed to half of what they had been a few years earlier. To make matters worse, the seal hunt, which was playing an increasingly important role in the economy of the northeast coast, failed miserably during the long, cold winter of 1817. Then, as if to complete some hideous chorus, fires in St. John's in the winters of both 1816 and 1817 left many people homeless and desperate. Merchant enterprises failed, bankruptcies were commonplace, and imports of basic foodstuffs became inadequate. Starvation came to many communities, including St. John's.[24] By the 1820s, the lack of local economic development sufficient to sustain the growing population was becoming a political issue, and demands for local and representative government grew.[25]

HOME RULE

As early as 1802, Governor Gambier had urged Britain to grant Newfoundland home rule, as had been done years earlier with Nova Scotia and even tiny Bermuda, but the British would not hear of it. English merchants plotted to retain their influence over the Newfoundland fishery. Despite the Napoleonic Wars, the French proprietary notions in Newfoundland were seen by some to be less of a threat than were the growing numbers of resident English and Irish Newfoundlanders. After Gambier, Newfoundland endured a series of governors who, for two decades, resisted any form of self-government.

In the late 1820s, Newfoundland needed a saviour – and found one in Sir Thomas Cochrane, who governed from 1825 to 1834. Cochrane quickly realized that – home rule or no home rule – a permanent settled population needed both local transport and food supplies. He can be credited with major road building both within St. John's and radiating outwards to Topsail, Bay Bulls, Torbay, and Portugal Cove, accomplished, for the most part, with relief labour.[26] Farms and estates sprang up around St. John's, providing local produce. The remaining merchants who objected to local agricultural development for fear of losing control of the produce trade had little influence on Cochrane or his policies.

Greater self-reliance led to increasing demands for local government. An odd coalition formed by Dr. William Carson (a Scotsman), Patrick Morris (an Irishman), and a group of Newfoundland merchants demanded home rule. Any form of local government was still strongly opposed by some English merchants; in 1825, for example, merchant Peter Ougier stated that Governor Cochrane and the uppity settlers were "making roads in Newfoundland, next thing they will be having carriages and

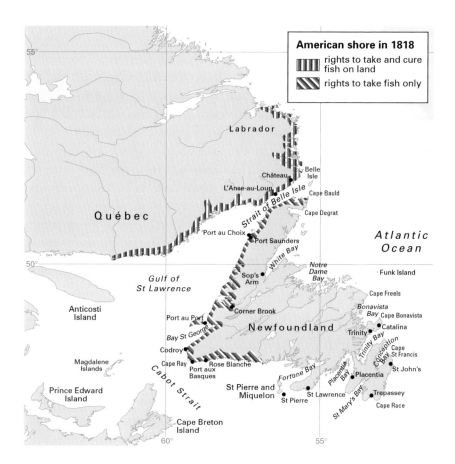

6.5 The American shore in 1818.

driving about."[27] Central to the political divisions on this issue was that home rule advocates wanted local control of the fishery and restrictions on foreign rights to it, whereas opponents wanted to see a return to the days of the migrant fishery. However, the migratory fishery became increasingly untenable for both political and economic reasons: the settlers were less and less content to see the profits of their labour carried back to Britain, and local merchants were simply out-competing their West Country counterparts.

Many local merchants supported home rule, particularly after Britain imposed new duties on imports to the colony in 1828, and rallied under the banner of "no taxation without representation." In addition, a developing free press spurred open debate on the future of Newfoundland, and this led to increasing support for local government.[28] In 1832, a bill was finally passed in the English House of Commons granting representative government to Newfoundland. A general election followed in August of that year. Surprisingly, Dr. Carson, who had spearheaded the home rule movement and had suffered at the hands of a succession of governors, was defeated – perhaps because he was a Scot – in favour of two relatively inexperienced and untested Irishmen who were supported by the Catholic bishop. It was the beginning of sectarian politics in

Newfoundland that would last for more than 100 years.[29] In 1855, a further step towards autonomy was taken with the institution of responsible government, making the executive branch of government responsible to the elected legislature (which they had not been under representative government).

INTERNATIONAL POLITICS AND THE FISHERY IN THE 1800s

The French Connection

The installation of local government led to renewed calls to restrict foreign fisheries, particularly the French.[30] The motivation was two-fold: first, to reduce competition in the primarily European marketplace for Newfoundland fish, and, a distant second, to conserve the fish stocks. These efforts were largely unsuccessful. Even after the implementation of responsible government, Britain kept a firm hand on the international aspects of the fishery. In their wider dealings with the generally hostile France of the 1800s, the fishery was used as a game piece to bolster British interests, often at the expense of the colony.[31] The lack of sovereignty over its lands and fisheries put the Newfoundland government in a compromised and, at times, humiliating position when foreign nations occupied up to one-third of its coast, actively excluding Newfoundland settlers and controlling the fishery.

Foreign fisheries influenced the prices and marketing of salt fish – both key problems in a fisheries economy that was heavily impacted by international events well removed from Newfoundland's modest influence. Newfoundland's fish products were marketed almost exclusively within the British international trade complex and were subjected to "quite prohibitory" duties in some European countries.[32] Cod caught off Newfoundland by the heavily subsidized French fishery competed with Newfoundland-cured fish in the same markets, keeping prices artificially low. Under British control, Newfoundland failed to develop a marketing system that was responsive to world demands for fish and directed towards the long-term prosperity of the fishery and the local economy. This would cost Newfoundland dearly in the 1900s.

The migratory French fishery remained a thorn in the side of Newfoundland development for 200 years. Four major treaties spanning more than a century – Utrecht (1713), Paris (1763), Versailles (1783), and Paris II (1815) – had failed to curtail the French fisheries or their subsidies. In each case, the military defeats suffered by the French did little to stop the irrepressible French fishermen and their backers from pursuing the Newfoundland fishery. French rights had been revived in 1802 with the minor treaty known as the Peace of Amiens, and most of the old French rights dating back to the Treaty of Utrecht in 1713 were re-established. The islands of St. Pierre and Miquelon were returned to France, traded by Britain for concessions in trade and colonial agreements in other parts of the world. Newfoundland protested, but Britain had gained several new acquisitions from France under the Peace of Amiens, including the Caribbean islands of Tobago and St. Lucia, which outweighed any concerns about Newfoundland.

The French fishery became a preoccupation of the growing Newfoundland population in the 1800s.[33] Eight Anglo-French commissions were appointed between 1846 and 1886 to deal with French fishing rights in Newfoundland. The French claimed exclusive rights to fishing and curing on their shore. The English consistently showed little inclination to rebuff the French, at best arguing that French rights were concurrent and not exclusive, and only weakly challenging the French salmon fisheries that had penetrated many of the best rivers.

The English-French Convention of 1857 is of particular interest, approved without the representative government of Newfoundland even knowing of its existence. This convention not only upheld the exclusivity claimed by the French in Newfoundland, but granted similar rights in Labrador.[34] Britain also signed on to excluding Newfoundland settler buildings and facilities on the French Shore. Even more offensive to Newfoundland, French officials were to be allowed some jurisdiction over the Shore, and the French Navy could expel any vessels violating their exclusive right to fish.

The English-French Convention was greeted with incredulity and rebellion in Newfoundland. British flags, so proudly flown only decades earlier, were lowered to half-mast and many American flags took their place. It was no Boston Tea Party, but it marked the beginning of the end for the old colonial ways. Perhaps for the first time, Newfoundlanders from all walks of life were united against British insensitivity. A select committee of Newfoundland's representative government met shortly after the Convention was signed, decrying the English as "either fools or something worse" who would "throw away the Colony and the interests of its people with the greatest indifference."[35] In a rare show of unanimity, 90 merchants from St. John's signed a petition against the English-French Convention.[36] Britain's representatives in Newfoundland had little choice but to recant, and recommended that no agreement could be made without consent of the local legislature.[37] The treaty was in effect dead.

Nonplussed, the French still claimed exclusive fishing rights to the Petit Nord in the final decades of the 1800s, and, at times, to all of the French Shore as established by the Treaty of Utrecht of 1713. They claimed rights not only to cod, but to lobsters and salmon as well. Their salmon weirs caused Newfoundland geologist Alexander Murray to write to the Governor of Newfoundland in 1875 complaining that French salmon fisheries in the Petit Nord were barring all access to the spawning rivers.[38] This practice continued until 1904, as bounties were paid by the French government to bring salmon back to France. French warships routinely contested any Newfoundland structures constructed on what they considered to be *their* shore.

In the mid-1880s, with the Newfoundland fisheries moving westward, the French again threatened to implement their claim of exclusivity. Britain once again wavered, and soon informed France that if the unruly Newfoundland colonials persisted in asserting themselves, they would be left to their own devices. This came to be known as the "stew in their own juice" policy.[39] After some introspection, the Colonial Office thought that such outright spinelessness on the part of Britain towards France would do much harm to her colonists in Newfoundland: if the French implemented exclusivity, what with the bounties already paid to them by their own government, the Newfoundland fishery would be put at severe disadvantage.[40] If the French fishery had an Achilles' heel, it was bait. The French line fishery required it, so Newfoundland sought to restrict their access to it. In 1887, the Newfoundland Assembly put forth a law, the *Bait Act*, prohibiting Newfoundland fishermen from supplying any French ship with bait. In response, another Anglo-French commission disallowed the Newfoundland law and framed a so-called "bait clause," which gave French fishermen the right to buy herring and capelin anywhere in Newfoundland, without any duty or restrictions, from April 5 onward each year.[41] There was some disagreement regarding the outcome of these actions – statistics were used to evaluate the effectiveness of the *Bait Act* in 1888, but they were not entirely convincing. Nevertheless, French salt fish production on St. Pierre and Miquelon peaked in the 1880s, but declined rapidly thereafter.[42]

The political mood turned ugly in the 1880s. New settlements had sprung up over much of the island and in Labrador after the 1830s. By the 1870s, most good fishing sites along the English portion of the Newfoundland coast were taken, and for all intents and purposes, the inshore fishery could expand no more.[43] Settlers were moving north, which increased conflicts with the mostly migratory French. For their part, the French openly referred to the 12,000 Newfoundland liveyers who had settled what they considered to be their shore as "foreigners."[44] France had steeled its will to hold its interests in Newfoundland and, far from backing off, claimed additional rights, including those to the increasingly lucrative lobster canning industry.

By the early 1880s, the developing Newfoundland lobster fishery had extended well into the French Shore. In 1882 and 1883, canneries were established at St. Barbe and Port Saunders near the major lobster grounds in the Gulf of St. Lawrence. France protested and a French warship destroyed parts of the plant at Port Saunders in 1886. In 1889, another factory was established midway between St. Barbe and Port Saunders, and was subsequently attacked by another French warship.[45] It was not only the lobsters that had reached the boiling point. According to the treaties, French rights extended only to "fish," and the Colonial Secretary stated that the said crustaceans were not fish.[46] Academics were consulted and thoroughly muddied the waters by expounding that the meaning of "fish" depended on the level of language being used, which in turn was disputed. After much hand wringing, Professor Flowers of the British Museum claimed that the term "fish" in ordinary use could mean any animal that lives in the water.[47] The spat was not resolved. In any case, back in Newfoundland, "fish" meant cod.

There was continual strife regarding lobsters, salmon, and cod on the French Shore for the next two decades. The French position had further hardened, as expressed by a French official in a letter to the English about lobsters. France claimed that "the exclusive right of fishing…the right of France to the coast of Newfoundland reserved to her fishermen is only a part of her ancient sovereignty over the island which she retained in ceding the soil to England, and which she has never weakened or alienated."[48]

An English map published in 1890 contained a great deal of information about the French fishery. Across the area called the "Bank of St. Pierre, Green Bank, and the Great Bank [Grand Banks] of Newfoundland" was written "the 'Nursery' for French Seamen is on those 'Banks,' where the French now carry on their fishing, and to which the Treaties do not relate." The French fishery was substantial, and the map recorded that "a fleet of over 200 sail of French fishing vessels from 100 to 400 tons [displacing approximately 95 to 381 tonnes] arrive here every spring from France, and make it their headquarters for the fishing season." The resentment of Newfoundland towards this fishery was expressed there as well: "France levies a heavy duty upon fish exported from abroad into her markets, thus preserving the home markets for her own fishermen. She also gives bounties equivalent to 10 [shillings] per quintal (112 lbs) [52 kilograms] upon French-caught dried codfish exported from French ports into foreign markets [Portugal, Spain, and Italy]. These bounties equal seventy-five percent of the actual value of the fish, and in consequence the price of Newfoundland fish, coming into competition with the French fish, was so depreciated as to threaten the destruction of the trade of the colony."[49]

The Anglo-French Convention was signed in 1904 and effectively removed the French claims in Newfoundland that had been disputed for nearly 200 years. France retained St. Pierre and Miquelon against the wishes of the Newfoundland government.[50] French fishermen retained their right to fish, but not to cure their catch on land, on the former French Shore. For this concession, French property owners were compensated 1.4 million francs by Britain. As the French Shore became part of history in 1904, so did the St. Pierre-Miquelon salt fishery. The French fishery remained irrepressible, however, and would re-emerge – this time with a fleet of trawlers – only a few years later. The retention of St. Pierre and Miquelon as the last bastion of a lost colonial French empire in North America also ensured that the issue of French fishing rights would resurface 70 years later in disputes over the fisheries and ownership of the continental shelf.

Canada and Confederation

The issue of union with Canada first arose in 1867, when Confederation brought together the former colonies of Upper Canada (Ontario), Lower Canada (Quebec), Nova Scotia, and New Brunswick, and again in the 1890s. In both instances, Canada made tepid advances towards the older but quirky island colony,[51] but it was not to be. Canada was viewed with some suspicion, an easy political sell in a place so isolated. After a decade of debate, anticonfederates, led by a coalition of the Newfoundland Irish, the Catholic Church, and many St. John's merchants, won a landslide victory in the Newfoundland election of 1869.

In the 1890s, Sir Robert Bond attempted to secure better markets for Newfoundland fish by developing a free-trade reciprocity agreement with the United States: basically, fish for American manufactured goods, with no duties. Canada was livid, perceiving a loss in markets for its manufacturing industries. With Britain maintaining control of Newfoundland's international dealings, there was little chance that a U.S. reciprocity agreement would be supported. Britain sided with Canada, perhaps concerned with American annexation of Newfoundland or other parts of North America. In any event, Newfoundland lost a deal that could have helped its struggling economy. A fishery trade war resulted between Newfoundland and Canada, with Newfoundland demanding licences and refusing bait to Canadian fishermen.[52] In return, Canada imposed duties against Newfoundland fish products. The trade war with Canada continued unabated for a decade until an issue arose that both countries could agree on: helping Britain in the First World War.

The relationship of Newfoundland with the United States and Canada remained complex. Canadian influence in Newfoundland increased greatly after 1895, when Canadian banks and money came to Newfoundland. In 1902, another attempt was made to negotiate trade reciprocity with the United States, but by this time the U.S. Senate had turned to protectionism and Newfoundland was rebuffed. Newfoundland responded by harassing American fishermen. Protests from the United States led to Britain siding against Newfoundland, adding further humiliation to the local government.

Canada–United States relations during this period with respect to the fisheries were just as unsettled, with Newfoundland always affected. In the 1880s, American tariffs were countered by Canadian restrictions on American vessels in Canadian waters, but in the end, "American fishermen [were given] just about all they wanted…[including entrance to] the bays and harbours of the Atlantic coasts of Canada and Newfoundland, upon payment of an annual license fee of $1.50 per ton (of fish)…and the Canadian fishermen got nothing in return."[53]

New England Fisheries

> Newfoundland is connected with some of the most interesting events to be found in our annals. Cabot saw it before Columbus set foot on the American continent. There came the first men of the Saxon race, under the first English Charter, to found an English colony. Visitors to, or residents upon, its shores, were the noble Gilbert, and Raleigh, the father of colonization in this hemisphere, Mason and Calvert, the founders of two of the United States. — LORENZO SABINE, 1853

Newfoundland and New England vied for prominence in the economy of the New World in the 1600s. By 1800, New England had far surpassed Newfoundland in all forms of developing society and economy, but also had depleted her once-abundant stocks of fish. In the 1800s, the fishing fleets of New England were focusing on Newfoundland waters, and by mid-century, American fleets fishing for cod, mackerel, Atlantic halibut, and herring were common in the Gulf of St. Lawrence and in Labrador. Young McGill doctor Pierre Fortin was hired by a newly formed Canada East (previously Lower Canada, then Quebec) and Canada West (previously Upper Canada, then Ontario) legislative committee to investigate the fisheries of the Gulf of St. Lawrence and the Labrador coast (Labrador then described both the north shore of the Gulf of St. Lawrence and the Atlantic region north of the Strait of Belle Isle). Fortin became the first Canadian "fisheries officer" in the Gulf of St. Lawrence, reporting that more than 100 American schooners were fishing for herring near the Magdalen Islands in 1852 and that "foreign fishermen [Americans] had…made themselves masters of everything, and frequently drove away our fishermen."[54] Fortin also saw 49 schooners, including 10 American vessels, at Bradore Bay near the Strait of Belle Isle, 29 American vessels at nearby Salmon Bay, and 50 American vessels fishing for mackerel at Sept Isles. New England bait vessels ventured to at least Conception and Bonavista bays on the northeast coast, and at times there was great competition among the vessels to buy squid and herring.[55] In 1867, Fortin noted 400 American schooners from Gloucester, Massachusetts, fishing cod and mackerel off the Magdalen Islands.[56]

Fortin was keenly aware that the American and French fleets – the main competitors to the fisheries in Canada and Newfoundland – were highly subsidized by their home countries, whereas local vessels were not. A subsequent committee report foretold a problem that would resurface again and again over the next 100 years: "The liberal encouragements granted by the governments of France and the United States of America to their Fisheries have the effect of sending, year after year, during the fishing season fleets of their bankers and fishing craft into the gulph [sic], where they carry on extensive, and no doubt, profitable Fisheries, constituting nurseries for their respective navies, and enabling them to compete with, and underselling us, in the article of fish, the products of our own waters, in foreign, and even in our own, Upper Canada, markets."⁵⁷ The same report concluded that the foreign fishery was severely impacting Canadian fisheries and communities of the Gaspé Peninsula and the north shore of the Gulf of St. Lawrence.

In 1871, England and the United States ratified a treaty whereby British subjects in Canada and Newfoundland were granted fishing rights off New England and access to American markets in exchange for access by the New England fleet to the Nova Scotia and Grand banks. Protests to the treaty ensued in Newfoundland, where it was argued that access to New England grounds would never be used by Newfoundland vessels and was therefore without value, while access to American markets benefited New England as much as Newfoundland. No simple agreement could be reached, and in 1877, a commission met in Halifax to judge the case. The result was a US$5.5 million settlement, of which Newfoundland received US$1 million. Solicitor-General William Whiteway of Newfoundland thought the payment to be of great significance, not only for the immediate cash, but, more importantly, because it established that access to the fishery had considerable value.⁵⁸ In the outports, however, this agreement may have gone unnoticed. In the fall of 1877, with the treaty in place, 26 American vessels voyaged to Long Harbour, Fortune Bay, to secure bait. The herring were scarce; when they did show up for a short period, a few American vessels attempted to fish with their own seines, apparently to bar herring from shore – an illegal act in Newfoundland. This led to a riot that involved about 200 Newfoundlanders and resulted in the destruction of a considerable amount of American gear.⁵⁹ The dispute was not settled until 1881. In the following years, the New England frozen herring fishery was mostly translocated to the Grand Manan Island region of southern New Brunswick, where fishermen apparently were more receptive to an American fishery.

In the 1800s, Newfoundland and the fisheries off its coast remained pawns in an international political and economic fish game played primarily to benefit European colonial powers. With increasing settlement and local representative government, however, a view emerged in Newfoundland that the ancient and outmoded treaties between England and France (and now with the United States) should no longer be allowed to dictate its future. Many Newfoundlanders would have liked to tear up the Treaty of Utrecht that had been implemented by a colonial and nonrepresentative government, and boot the French out. The American colonists had done so long ago and, for good measure, had also evicted the English in 1762. At issue in both the New England and Newfoundland colonies were sovereignty and the ability to direct and protect their own policies and destiny. But the will to take charge of her own affairs was far less strident in Newfoundland than it had been in the American colonies, and Newfoundland's so-called independence, and hence its abilities to regulate the fisheries, was in reality quite limited.

ECOSYSTEMS IN THE 1800s

The last gasps of the Little Ice Age brought atypically cold conditions and lower productivity in the northwest Atlantic region in the 1800s. From 1810 to 1840, temperatures and tree growth across northern North America declined rapidly, with rebounds in productivity not occurring until the final decades of the century. The effects of this cold period were widespread and could be felt all the way south to New England, where in the bitterly cold year of 1816, "it snowed in June, and then killing frosts continued until August." There was widespread famine in Newfoundland and elsewhere in North America. On the grave of New Hampshire farmer Reuben Whitten, who died in 1847, it states that in the "cold season of 1816 [he] raised 40 bushils of wheat on this land whitch [sic] kept his family and neighbours from starvation."[60]

The 1800s were not a productive time for cod in the northwest Atlantic. Far to the north, off western Greenland, cod had nearly disappeared by 1850, and "for the remainder of the nineteenth century we have no notices of large occurrences of cod."[61] In the oceans off Newfoundland, cod stocks declined to half of what they had been in the 1700s. Some years and decades were exceptionally poor for the fishery. It was written in the 1860s that "after an indifferent fishery in 1861, the Island was visited in the spring of 1862 by a blockade of ice of unprecedented severity. For 52 days the wind blew continuously from the north-east, driving the ice onto the land…it was not until 1869 that a remunerative catch [of cod] was obtained."[62]

Ecosystem-wide changes in climate are expected to affect many species, not only cod, but there are few records from the 1800s for most species. Those that exist confirm a severe period of climate. Fishery exports from northern Labrador showed a rapid switch from a predominance of Arctic char in the late 1800s to salmon and cod in the warmer 1900s. Although market conditions may have contributed to this shift, it was consistent with major abundance and distribution changes in these species during the late 1800s, with the cold-water char declining in the waning years of the Little Ice Age and the warmer-water salmon and cod increasing.[63] Warm-water migrants that visited the Grand Banks and Newfoundland in summer – in particular Atlantic mackerel, but also saury and tuna – were scarce or nonexistent in Newfoundland waters in the 1800s. Mackerel catches were near nil for nearly 100 years despite their value as bait and food.

Although there is little direct evidence, some species – such as the cold-loving Arctic cod, snow crab, and pandalid shrimp – may have increased in abundance and expanded their ranges southward during this period. With their chief predator, the Atlantic cod, in decline, crab and shrimp may have increased in abundance. Harp seal numbers likely rose as favourable ice conditions on the pupping grounds enhanced survival, with adult seals spending more time farther south near the growing settlements. A greater abundance and availability of harp seals to the settlements may have played a role in the quickly expanding seal fishery of the early to mid-1800s. There is also evidence of a changing climate from the distributions of seabirds: until near the end of the 1800s, gannets were absent from their present colonies at Funk Island, Baccalieu Island, and Cape St. Mary's, and their numbers were low on the Bird Rocks in the Gulf of St. Lawrence. As conditions warmed, pelagic fish communities would have become richer; gannet colonies followed the fish and advanced from south to north, with Cape St. Mary's colonized first in 1879, then Baccalieu in 1901, and Funk Island in 1936.[64]

EXTINCTIONS

By 1822, the great auk – the northern penguins of earlier literature – had been slaughtered to extinction.[65] Seabird species that could fly fared better in the face of human settlement, but did not escape entirely. Puffins were under intense pressure from egg collectors who destroyed tens of thousands of incubating eggs so that fresh ones would be laid, which were then taken. William Cormack decried this destruction and predicted the end of seafowl if this "cruelly-begotten temporary subsistence" continued.[66]

The Atlantic walrus disappeared from Newfoundland waters during the mid-1700s. The walrus did not become extinct, as these waters were only a southern extension of their Arctic range, but unique populations in the Gulf of St. Lawrence and at Sable Island – first observed by Jacques Cartier in 1534 – were lost.[67] It took 200 years for the multitudes of walrus to be extirpated by commercial export harvesting. It cannot be stated for certain when the last one was killed, or if a few last survivors retreated to the north.

FISHERIES IN THE 1800s

Cod

> The heart of the matter in this dominion was fish…almost everybody's livelihood was tied to the fishery, which meant the cod fishery, for to most people the word "fish" meant cod. When for some reason or other the fishery slumped, the whole society felt the blow. Had fish struck in down north? How was the voyage in Conception Bay? What price was cod fetching in St. John's? In Spain? Everyone asked such questions, for so much depended on the answers. – PATRICK O'FLAHERTY, 2005

The first problems with sustainability of the cod fisheries arose in the 1820s as a consequence of increasing catches and a decline in ocean productivity. The fishery increased roughly in proportion with the rising population of the 1800s and had

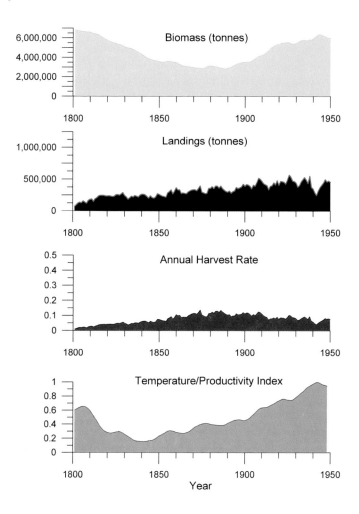

6.6
Cod biomass, landings, annual harvest rate (proportion of biomass landed each year), and the temperature-productivity index based on tree-ring analyses, from 1801 to 1949.

expanded to almost every cove around the island by 1880. Meanwhile, the ability of the cod stocks to withstand the fishery was compromised by lower productivity. These factors, as well as increasing competition with the fisheries of France, New England, Canada, and other nations,[68] made it harder for Newfoundland fishermen to catch fish than in earlier times. To compound the problem, the growing numbers of fishermen resulted in constant or slightly increasing total catches as stocks weakened, thus causing further declines and initiating a cycle of overfishing wherein fishermen worked harder to maintain catches of declining stocks. By 1815, cod landings had surpassed 200,000 tonnes per year, and total harvest rates continued to march upwards, if somewhat erratically, until about 1880. On the other hand, the catch rates per man and per boat, which might be expected to better reflect stock abundance, were plummeting: by 1880, the average annual catch per fisherman was about one-third of what it had been in 1800.

Despite the decline in cod in the 1800s and the routine occurrence of disastrous fishing years,[69] the stocks did not crash. The power of the fishery to kill fish was limited until the late 1800s because fishing gear was restricted to jiggers,[70] baited hooks, bultows (lines of many baited hooks), and cod seines.[71] Most importantly, the fishery was limited to four months of the year in most areas,[72] and the major overwintering and spawning grounds of the largest stock, the northern cod, had no fishery at all. For most of the year, and over most of the grounds, the fishery was in effect closed. As a result, even under the most adverse natural conditions, the cod stocks had large numbers of fish of many different ages to sustain them until conditions improved. This imparted resilience to the stocks and to their breeding potential that precluded total crashes, even in times of low productivity.

Until the end of the 1800s, inshore fishing methods were similar in most areas of Newfoundland and Labrador. Single or multiple hooks baited with mussels, capelin, herring, or squid were the most common gear (the use of seabirds as bait had become increasingly frowned upon). The Newfoundland-based bank fisheries that developed during the late 1870s initially mimicked the early French and English fisheries and used baited single lines, but shortly began to use longlines to counter declining catch rates.

The problems of lower catches per person in the cod fishery in the 1800s were compounded by market prices that crashed in the early 1820s and remained low for the next 80 years. As a result, the total export value of the Newfoundland fishery increased hardly at all during the entire 1800s, despite the increasing population and total catches. The lack of any significant price increases in salt fish product robbed Newfoundland of its chance of self-sufficiency.

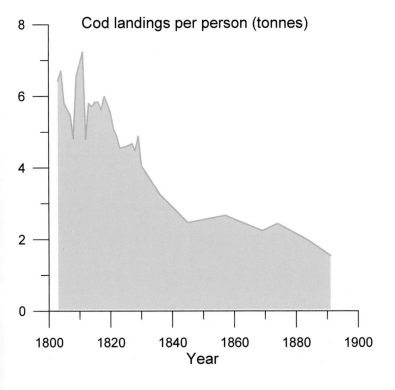

6.7
Cod landings per resident in the 1800s. Until the 1880s, virtually everyone in Newfoundland worked in the fishery.

SALT FISH

Fish curing methods differed greatly from region to region – there were only *ad hoc* standards for handling and preparing the fish. On the northeast coast, the fishing season lasted four to five months at most, even in the most southerly areas, and the Labrador season was much shorter. The best salt fish was generally produced on the northeast coast, where summer weather was typically windy and moderately sunny – ideal for curing fish. Most families fished close to home. Each cove housed a cottage industry where the fish were lightly salted and air-dried on wooden flakes near the fishing stages. In Labrador, where the fishing season was shorter, the weather more severe, and the shores barren, fish were dried on rocky beaches or salted on board and brought to Newfoundland for drying. The Labrador salt fish product – known as the notorious "Labrador cure" – was of poor quality: this was a fishery of quantity, not quality.

On the south coast – the home of the graceful banking schooners that fished the Grand and St. Pierre banks – the fish were dried on vast beaches. The bank fishery was organized differently from the shore fishery in that merchants owned both the schooners and the fish, while the skipper and crew worked for a share of the catch. To avoid spoilage, bank fish had to be heavily salted for the voyage home. On shore, after the fish were washed, women took over: a "boss woman" employed by the local merchant supervised a crew of "beachwomen" who cured the fish. There was both co-operation and competition among crews, and a great deal of pride in curing the best fish.[73] Despite the best of care, however, the heavy salting and transportation of the bank fish, combined with the foggier weather on the south coast, meant this product could never attain the quality of the light-salted product of the northeast coast.

With the decline of the English firms and the withdrawal of the New England fleet, the Labrador salt fish trade in the late 1800s was controlled largely from the Conception Bay towns of Harbour Grace and Carbonear under two dominant firms, Ridleys and Munns. Conception Bay firms accounted for nearly 65% of exports of Labrador fish in 1865. St. John's controlled the higher quality northeast coast salt fish exports, with 72% of exports. Walter Grieves and Company of St. John's was the largest exporter of shore fish, but Ridleys of Harbour Grace exported the most salt fish of all kinds. Ridleys went bankrupt in 1873, with liabilities of 250,000 pounds sterling (C$1.2 million). It was a major blow to the economy, in particular to the Labrador fishery in which Ridleys was the major player.[74]

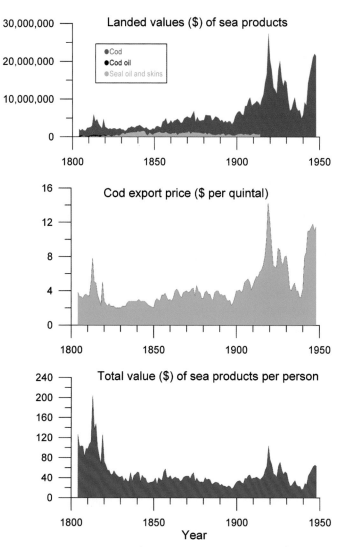

6.8
Landed value of sea products, the export price of cod, and the total value of sea products per resident in Newfoundland and Labrador from 1800-1950 (data from various sources).

In response to declining catch rates and increasing human population, the Newfoundland fishery expanded increasingly northward. There are few records of fisheries north of Cape Harrison before 1860,[75] but a major salt fish industry developed in northern Labrador after that time. As cod stocks rebounded in the late 1800s, both Newfoundland and New England vessels ventured north and there were reports of an "amazing fish wealth."[76] Unlike the southern stocks, the northern Labrador cod were a virtually untapped resource in the early to mid-1800s.

The New England fishery had expanded well beyond what could be sustained by local stocks as early as the mid-1700s, and their vessels ventured north to fish cod and herring in the southern Gulf of St. Lawrence. In the early years of this fishery, baited handlines were commonly employed, fished from a drifting ship. The bait was nearly always herring or mackerel and mostly caught by the fishing vessels. By the mid-1800s, line trawls were used almost exclusively in the Gulf fishery and much larger fish were taken, many of them weighing up to 100 pounds (45 kilograms). The large fish did not last, however, as they "were nearly all caught up in time."[77] American vessels pushed farther and farther north to the Labrador coast. The first American vessel to pursue the Labrador fishery sailed from Newburyport, Massachusetts, in 1794. Over the next 100 years, vessels from several states attempted to fish the Labrador coast.[78] In 1815, a fishing company was established in Gloucester, Massachusetts, to pursue the Grand Banks and Labrador fisheries. About ten schooners were built for the Labrador fishery, each carrying four or five 23-foot (10.8-metre), twin-masted skiffs called "Labrador boats" used for fishing. Fishermen on Labrador boats used either capelin-baited handlines or jiggers, if bait was scarce. The Labrador coast was attractive to the New England fishery: it was wild and beautiful in summer and had a well-protected shoreline in many places. It was generally agreed that "a voyage to Labrador, unlike a trip to the Banks of Newfoundland, is not without pleasant incidents even to a landsman…It is preferred to any other on account of its security, and a general certainty of affording a supply of fish."[79]

Many New England fishermen were literate men. In 1820, N.E. Atwood was a crew member on a schooner out of Provincetown, Massachusetts, on a voyage to Labrador. He wrote:[80]

> We sailed from Provincetown on June 6, 1820. We went to the coast of Labrador but…we were unfortunate in getting codfish. Our men were not the best of fishermen…I don't know how far north we went, but it was to the locality familiarly known to us as Grosswater Bay. Our mode of fishing then was to let the vessel lie in the harbor and send the boats out. At the time Provincetown had not a single vessel on the Grand Bank…all the fishing vessels were on the coast of Labrador. We carried four boats. We used one to get capelan for bait when capelan were plenty during the capelan school. The bait boat would seldom go fishing. The fishing boats were baited out of her. We had one of the crew to throat, one to head, and one to split, and a salter in the hold of the vessel salting them as they came down. On our arrival on the coast of Labrador very few could be caught until the capelan came in, and then the capelan schools of cod came in also. The capelan school lasted about three weeks. After they went away we picked up fish very slowly. After the capelan had finished spawning the fish slacked off, and we used to say that the cod were "capelan sick."
>
> In 1823 I again shipped for the Labrador fishery in the schooner Favorite. I think we carried 160 hogshead, or 1280 bushels of salt. We sailed from Provincetown about the middle of May, and proceeded first to the northern coast of Newfoundland [actually the west coast], making a stop at the Bay

of Islands where we commenced fishing with clam bait, which we carried with us. We were too early for the capelan school. After fishing her eight or ten days we proceeded and arrived at Indian Harbor, on the north side of Grosswater Bay. Soon after we arrived the capelan came on the coast, and while they remained we wet nearly all our salt. The school lasted about three weeks. Having some salt left, we proceeded homeward, stopping at the Straits of Belle Isle at a place called Henley's Isle [previously called Chateaux by the French]. The capelan were gone, and we were compelled to fish with launce, or sand-eels. We used up all the rest of the salt, excepting a few bushels; left the coast and proceeded on our voyage homeward, arriving about the 20th of September.

Charles Hallock visited the Labrador coast on a New England schooner in 1861, an average fishing year that preceded a very cold period and concomitant poor fisheries.[81] By then the Labrador fishery was "equally divided between the Provinces [mostly the Newfoundland migrant fishery] and the States [the New England fishery], though the number of men and vessels employed by the former is much the largest in proportion to population." Hallock was amazed by the activity on the Labrador coast: "Little idea has the world of the populous community to be found on the Labrador coast from the 1st of June to the end of September. Every little harbor as far up as latitude 56° is filled with vessels, and fleets are constantly moving from place to place, following the vagaries of the fish…many parties have salting rooms and dressing stages on shore, but the majority of vessels cure their fish on board."[82] The capture of the bait was done in the evening, when:[83]

> after the labors of the day, the seine boats go in quest of capelan [bait], carefully searching the little coves and inlets and creeping along the shores; three men pulling in the usual way, an oarsman in the stern standing up and pushing, while he scans the surface of the water for the ripple of passing schools, and a lookout in the bows, motionless as a figure-head, resting upon his elbows and peering into the depths before him. Now one gives warning, and over goes the seine smoothly and noiselessly, and with a rapid circuit the bait is impounded and quickly hauled aboard. One cast is generally sufficient, for the capelan swarm in millions, swimming so densely that often a dip-net can be filled from a passing shoal. They keep near the shore to avoid their finny pursuers, and are left floundering upon the rocks by every reflex wave. The cod often leap clear of the water in their pursuit, and at such times may be taken by the hook almost the instant it touches water. The capelan is a delicate fish, about 6 inches [15 centimetres] in length and not unlike a smelt; his back a dark olive green, sides of changing rainbow hues, and belly silvery white.

The Labrador fishery had two components: "stationers," who made a temporary camp on shore and fished for the duration of the season, and "floaters," who made no shore station, but fished from a schooner. Most returned home to New England and Newfoundland settlements in the early fall, although a few stayed and settled in Labrador. Most Newfoundland floaters and stationers came from the outports of Conception, Trinity, Bonavista, and Notre Dame bays, and the contribution of the Labrador migrant fishery to economies of the burgeoning settlements in Newfoundland was vital.[84] The heyday of the Newfoundland-based Labrador fishery lasted only a few decades, but vestiges of it would persist until the 1940s.

Women may have been a rare sight on the earliest schooners, but were involved in the Labrador fishery by the 1860s. As related by Hallett, women "are wont to stand in tubs while at work, protected from the filth and offal by long gowns – cod-liveries – of oil-cloth extending to the floor; and when their task is done they emerge from these like butterflies from their chrysalids, clean and intact…An expert will split 8,000 fish per day, or head twice that number. The lodging shanties [stationers] are constructed of spruce poles or sheds, generally boasting but a single compartment, and here both sexes occupy in common, the only partition being that mathematical one which excludes all objects not within the line of vision."[85]

The New England interest in the Labrador fishery declined greatly in the 1870s, largely due to changing markets and increasing competition from Newfoundland. In the same decade, with the ice-ridden 1860s now past, Newfoundland vessels ventured even farther north. In 1876, about 400 vessels passed Cape Harrison, carrying on average eight men and three fishing boats. This northern fishery used jiggers and caught about 18 million fish averaging three pounds (1.4 kilograms) apiece.[86] Cod were small in northern Labrador; however, larger cod were believed to be on the offshore banks: "It is well known that only the smaller sized cod come into shallow water, the larger fish remaining to feed on the banks outside and in deeper water. Very few attempts have been made to fish on the Labrador Banks (nowhere do they rise much less than 200 [metres] – too deep for handline fishing), but when tried, I have been assured by trustworthy persons that large fish have always been taken with bait."[87] This was only partially true, as the Labrador fish grew more slowly and were of much smaller size than those in the stocks farther south.

The cod fishery on the Grand Banks was largely prosecuted by French and American fleets, with a Newfoundland fleet – largely from the south coast – entering the Banks fishery in the 1870s. Henry L. Osborn described a trip to the Grand Banks in July 1879 made by a Gloucester schooner and indicated that average-sized fish were 45 inches (114 centimetres) in length, although some measured up to 63 inches (160 centimetres) and had girths measuring 35 inches (89 centimetres). These were huge fish, and it was well understood that the shore fishery "never caught any as large as the average Grand Bank[s] fish."[88] The fishery followed a seasonal pattern related to the migrations and aggregation patterns of the fish:[89]

6.9 Beach women making fish on the south coast in the early 1900s (Gordon King collection).

6.10 Making fish on the stages of Newfoundland in the early 1900s (Gordon King collection).

In December and January none are taken. Toward March those who go thus early to the Banks begin to take a few, and as the year advances the fishing steadily improves. Those who fish early in the year anchor their vessels on the most southern and eastern edge of the Banks. Later, as the year advances, the fishing fleet move further north and west, till finally in July most of the vessels are anchored in the neighborhood of the Virgin Rocks, latitude 46° 27', longitude 51° 6'. After this, as the year progresses, the vessels begin a movement back again towards the south and east, until at last those who have remained till November are again fishing on the very outer edge of the Bank[s].

Osborn described two types of fish on the Grand Banks: the large and well-fed migratory "school fish" that were fussy about bait, and the smaller and less well-fed "gurry fish" that would take any bait, including guts of their own kind. The gurry fish were thought not to migrate, but to spend the year on the same grounds of the Banks. Fish in spawning condition were not observed during this late summer trip, and cod eggs, or pea, were never observed this late in the season. Cod stomachs contained large quantities of sand lance, spider crabs, and toad crabs, as well as sea urchins, anemones, and unspecified invertebrates.[90] Of note, there was no mention of capelin or snow crab.[91]

Salmon

> The salmon here is excellent and in great abundance from June to August, it is taken in nets, placed along the shores in bays and large harbours.
> — LOUIS AMADEUS ANSPACH, 1827

From its earliest beginnings with aboriginal fisheries, the highly prized salmon was landed in relatively small quantities that varied from year to year and decade to decade. In the final decades of the 1700s, the growing number of settlements and interest in salmon led to an increase in the average annual landings.[92] The increases were not sustained, however, and catches remained relatively low and variable throughout most of the 1800s, despite continuing increases in population and fishing effort. Average catches increased somewhat in the late 1800s, but there were some very poor years during that time. Although the low catches of the early to mid-1800s may have been partly due to local overfishing,[93] poor salmon fisheries were consistent with the notion that salmon did not fare well in Newfoundland waters during the Little Ice Age. After the 1860s, as temperatures and productivity improved, salmon landings increased. Salmon exports from northern Labrador grew substantially in the early 1900s.[94]

Seals

> The fishery...became entirely residential and, furthermore, [the settled] population actually increased – all because of the discovery of the seal resources. – SHANNON RYAN, *The Ice Hunters*, 1994

Although cod remained king, the growing settlements of the early 1800s needed more to survive. The most important new enterprise was the seal fishery.[95] Seals had been taken in small numbers by native peoples and liveyers for centuries, but a rapid expansion of the hunt paralleled the settlement of the island. The sealing industry began as an outport industry, prosecuted primarily by outport firms and outport men in outport-built, wooden sailing ships. By the mid-1800s, over half a million seals were landed annually. In 1844, Newfoundland exported over 685,000 sealskins and about 10,000 tuns (9,530,000 litres) of oil valued at more than 300,000 pounds sterling (equivalent to about C$1.5 million at the time). Nearly 9000 men, mostly from Conception Bay and St. John's, sailed in 300 vessels having a value of over 200,000 pounds sterling (approximately C$1 million).

In the 1830s and 1840s, the seal hunt was worth 50%-60% of the value of the cod fishery. The seal hunt provided cash to fishing families – for many the first they had ever had – but they did not acquire it easily. Merchants wanted to impose the truck sytem on the seal hunt, as with the cod fishery, but the sealers refused to accept this and won their first battle to gain a fair return from their labour.[96] The prosperity that this new income brought to Newfoundland was seminal to the increased settlement and the political agendas of local government.[97] Newfoundland sealers – the ice hunters – became world renowned for their skill on the ice. They were uniquely at home "copying" from ice pan to ice pan. It was treacherous work, and a true *esprit de corps* and black humour developed among sealers. Prowse related a story of competition between Newfoundland and Canadian sealers: "There was a Canadian on board one of our sealers [vessels] who was very smart at going on the ice; there were also some boys [from Newfoundland]... a great rivalry existed between them. One day the master called out that there were seals on the starboard bow; over the side went the Canadian, followed by the boys; there was a channel just caught over, which the Canadian did not see, and down he went; as he rose the boys, seven in number, came to the edge, and successfully copied over the Canadian's head and shoulders."[98] Whether or not this story is literally true is irrelevant: the important point is the pride in which Newfoundland held its sealers and their skills on the ice. If the story is true, the "boys" undoubtedly retrieved the wet Canadian, their point well made.

The seal industry was dominated by St. John's firms by 1865. Sealing and the Labrador fishery became joint enterprises, conducted by the same people and ships in different seasons. Vessels could be used both in the seal fishery in late winter and to transport stationers to Labrador in summer. The cod fishery remained the dominant

6.11

Landings of salmon in Newfoundland and Labrador. Data are from several sources, including Dunfield (1985), Taylor (1985), May and Lear (1971), and DFO statistics.

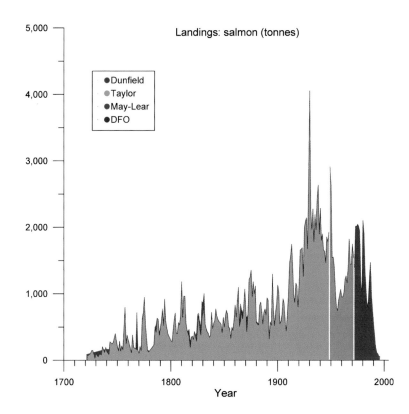

6.12

Landings of harp and hooded seals (where data exist 95% harps, 5% hoods). Data from several sources, including Sergeant (1991) and DFO statistics.

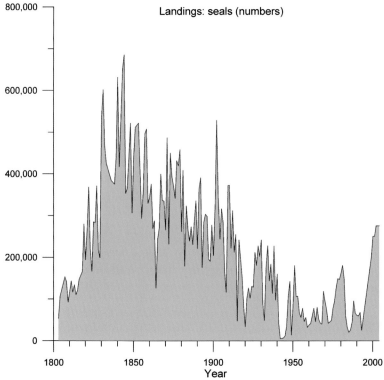

6.13 On the ice with the steamer seal hunt in the early 1900s (Gordon King collection).

industry, but the seal fishery provided cash. To outport families, it gave new hope in spring, the means to gear up for another fishing season, and perhaps even a luxury or two for wives and children, who generally saw few. The hunt became an icon and the captains became national heroes. A good captain meant a good catch and wages. Newfoundland developed the most prominent sealing fleet in the world, as well as the first fleet of icebreakers. As Cassie Brown stated of Arctic exploration, "men like Robert Peary might reap the international renown, but it was the Newfoundland sealing captains who guided their ships through apparently solid fields of ice to within five hundred miles of the North Pole itself."[99]

Whaling

Whaling in Newfoundland and Labrador waters was an ancient endeavour: it had been practiced by aboriginal peoples for several thousand years. Aboriginals hunted whales for meat and fat, concentrating on smaller species, such as the pilot whale, with their small craft and harpoons.[100] The Basques introduced commercial whaling to the Strait of Belle Isle region in the mid-1500s. The Basques wanted oil and baleen, and hence hunted larger whales; they concentrated on the Atlantic right whale, a large and docile creature that fought little and floated when dead.[101] The Basque whaling industry did not last long – it was in decline by the late 1580s, largely as a result of local overexploitation.[102]

Dutch whalers revitalized the Arctic whaling industry in the 1700s and made excursions off Labrador. In the same century, New England whalers came to Newfoundland in search of right and sperm whales, and continued this practice until the early 1800s. Several attempts were made to develop a local whaling industry in Newfoundland with the assistance of government bounties. In the 1840s, the firms of C.F. Bennett and Job Brothers of St. John's had whaling vessels on the south coast, and a small factory worked at Gaultois in Hermitage Bay from about 1840 until the end of the century. At its peak, Gaultois processed 40-50 whales each year.[103] During this era there was little whaling on the northeast coast of Newfoundland or in Labrador until the late 1890s. It is not known if whales were less common in northern Newfoundland waters during this period or if their range had shifted south. It is possible that the location of the whaling stations on the south coast in the mid-1800s reflected a more southerly distribution of whales and their prey during the Little Ice Age.

Norwegians brought modern whaling to Newfoundland and Labrador in the late 1890s, after having depleted their own stocks in less than 15 years with their new hunting methods. In the warming waters of the era, whales that had not been previously heavily hunted were numerous on the northeast coast. With the populations of right whales much reduced, whaling turned to these species. However, the traditional method of hand-throwing harpoons from small craft (a dangerous pursuit at best) was only suitable for the more docile and slower whales. The more aggressive and powerful cetaceans, such as the humpback, fin, and blue whales, might well drag attackers for miles and tow them under.[104] Even the most daring whalers avoided these species until the invention of rocket guns with exploding harpoons that were fired from the bow of a large whaling ship. With such modern whaling techniques, even the largest and feistiest whales could be taken.

Adolphus Nielsen, a Norwegian who had been the first superintendent of the Newfoundland Fisheries Commission and had instigated the first research on cod and lobster, was behind much of the Norwegian whaling initiative after the termination of his research and until his death in St. John's in 1903. Nielsen became the whaling manager, fourth largest shareholder, and board member of the newly formed Cabot Steam Whaling Company in 1896, with the Harvey family and several Norwegians holding the bulk of the shares. Their first factory was operational at Snook's Arm in Notre Dame Bay in 1898, and a second was built shortly thereafter on the south coast at Balaena near McCallum in Hermitage Bay. The whalers were mostly Norwegian, while Newfoundlanders did the "lower-paying and often repugnant processing tasks onshore."[105] These factories prospered at first, a fact that did not go unnoticed in St. John's. The Newfoundland Steam Whaling Company was formed in 1900 by a group of Newfoundland merchants and bankers, aided by a Scottish merchant, with a station established at Reuben's Cove on the south coast west of Balaena. The total numbers of blue, fin, humpback, and minke whales killed and processed increased quickly from 91 in 1898 to 258 in 1901, with the total value of whale products increasing from about $1500 to nearly C$70,000.[106]

By 1902, only four years after modern whaling had begun, the first signs of overexploitation became evident. The *Act to Regulate the Whaling Industry* was passed by the legislature in the same year, at least in part to protect merchant interests (merchants and lawmakers were often the same people). Consequently, the act read more like industry protection than conservation, with no protection for breeding females and no limit to the number of factories, licences, or catches.[107]

The Cape Broyle Whaling and Trading Company was also established in 1902; initial shareholders included several St. John's merchants, such as John Munn, Michael Cashin, and the Cabot Steam Whaling Company, who were joined two years later by Bowring Brothers, the estate of Adolphus Nielsen, and lawyer-politician William V. Whiteway. In addition, in 1902-03, there were 45 new applications for whaling stations from Cape Chidley south and around the island of Newfoundland. Not all of these were granted or activated, but whaling grew with few controls in the early 1900s. The quick expansion of whaling did not go unchallenged: citing the Norwegian example of stock depletion in a very short time period, the Newfoundland Department of Fisheries advised against any further expansion of the industry just five years after it had begun. This advice was not heeded, however, as the government was focused on employment and licence fees and whaling companies were interested in profit.[108]

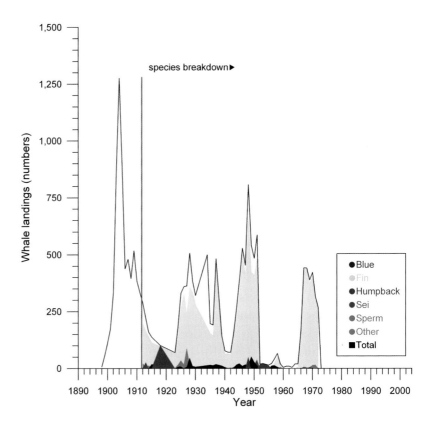

6.14 Whale landings in Newfoundland and Labrador waters. Data from International Whaling Commission and Sanger and Dickinson (2005). No data on small whales; species breakdown of the large whale catches after 1912 (red line).

For a short time, both government and industry reaped the benefits of whaling – the value of whale products increased to nearly C$300,000 in 1902-03 – but such profits would not last for long. Newfoundland shore stations caught 1275 whales in 1904, but despite the many new stations and expansion of the industry to Labrador, the total catch declined to 409 whales in 1906.[109] Many stations were closed and their whaling vessels sold off to countries as far away as Japan. Modern whaling did not last any longer in Newfoundland and Labrador than it had in Norway: by the beginning of World War I, both the whale stocks and the industry were in free fall. Most of the Newfoundlanders employed by the industry went back to fishing cod and, by 1916, it was all over. Daniel Ryan, the president of the Labrador Whaling and Manufacturing Company, summed up what had happened to the whaling industry: "Take the whole business. We had some parties [that] introduced this, and they got along very well. Then we had two or three factories make a big success and then all wanted to get into it. I remember at the time being a member of the Government when at one time we had twenty-three applications for those factories in one batch. Fortunately for most of these people the applications were not granted. The few who got them lost all."[110] The pattern of authorities advocating caution in expanding fisheries but being overruled by short-term political and business interests would be repeated many times in the Newfoundland and Labrador fisheries.

Lobsters

In the mid-1800s, Newfoundland's need to diversify its economic base led to the development of a lobster fishery. Lobster canning had begun in New England around 1840, but the stocks were overfished within a few decades and the potential of Newfoundland was explored. At the time, most Newfoundlanders and Canadians considered lobsters unfit to eat, and they were "often…used in manuring the fields…people were so ashamed of eating lobsters that old fishermen in Canada [Nova Scotia, New Brunswick, and Prince Edward Island] have told me that if they wanted to eat a lobster they would bring it home at night so that the neighbours would not know about it."[111] When the fishery in Canada and Newfoundland began in the late 1800s, lobsters were plentiful and fishermen received about 50 cents for one hundred lobsters.[112] The European and American markets for lobster were strong. From 16 to 18 million pounds (7-8 million kilograms) were caught annually in the peak years of the late 1880s and until the First World War. Catches began to decline after the turn of the century as a result of fishing down what had previously been essentially unfished stocks.[113]

Atlantic Halibut

The fishery for Atlantic halibut was developed primarily by New England fishermen. The halibut fishery did not develop in Newfoundland or in Canada until the 1930s. Halibut became "one of the most valuable of our food fishes."[114] But it was not always so – in the early 1800s, halibut were regarded in New England as a "decided nuisance by cod fishermen,"[115] and they were usually killed and discarded. By 1825, a market had developed for fresh halibut in the growing city of Boston. At first, there was no shortage of fish: in 1837, "they were so plentiful that off Marblehead, Maine, four men…caught 400 halibut in two day's fishing, for which they obtained…nearly [U.S.]$600."[116] Another boat took 15,000 pounds (about 7 tonnes) in two days. These coastal fisheries expired quickly, and New England fishermen expanded their range to nearby Georges Bank, where they found an abundance of both halibut and cod. Halibut were said to be so abundant that, in the words of one fisherman, "they were like sand on the beach, the more you catched [sic] the more there was."[117] The Georges Bank halibut fishery was exhausted by 1850 and the ever-growing fleet searched farther north for fish. By 1879, 40 vessels from Gloucester, Massachusetts, were employed exclusively in the fresh halibut fishery. Their range included waters as far north as the Grand Banks, the Gulf of St. Lawrence, and Bonne Bay, Newfoundland.[118]

The 1870s were banner years for the New England halibut fishery. The Gloucester fleet landed about 12 million pounds (5500 tonnes) in 1878, valued at about U.S. $313,000. The origin of most of this catch is uncertain, but it is known that the channels around St. Pierre and Green banks, where halibut were abundant, were important

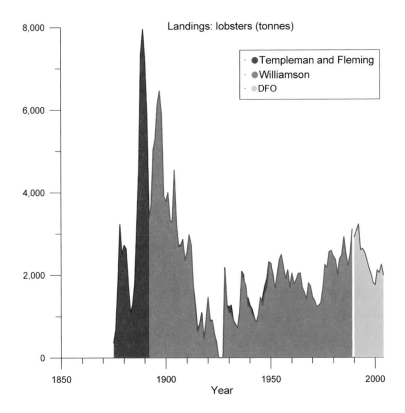

6.15
Lobster landings in Newfoundland and Labrador waters. Data from Templeman and Fleming (1954), Williamson (1992), and Department of Fisheries and Oceans sources.

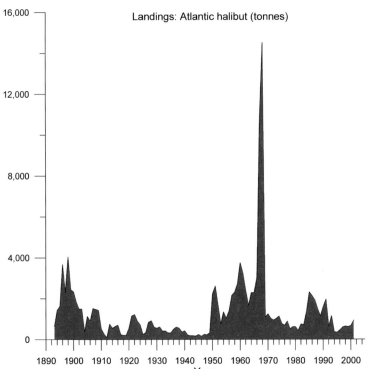

6.16
Landings of Atlantic halibut. Data from International Commission for Northwest Atlantic Fisheries and Northwest Atlantic Fisheries Organization reports.

grounds, as were the southern Grand Banks. The halibut fishery, like other fisheries, was inconsistent from month to month and from year to year, primarily due to weather changes. The journal of Master J.W. Collins of the halibut schooner *Marion* from Gloucester, Massachusetts, described in detail the day-to-day travails of fishing on the western edge of the Grand Banks and Green Bank in the cold and windy winter of 1878. The *Marion* did not do well on several trips from February to March of that year despite the fact that "last year at this time halibut were very plenty in this vicinity."[119] In an account of the largest halibut catch ever made, landed in Gloucester on March 22, 1882, although the exact fishing location is not known, it is mentioned that the vessel was frozen up for eight days at Canso, Cape Breton Island, close to the Gulf of St. Lawrence and south coast of Newfoundland fishing grounds. The catch included 98,825 pounds (about 45 tonnes) of halibut and 3000 pounds (about 1.3 tonnes) of cod, and sold for over US$6,000 U.S.[120] Such enormous early catches were short-lived and within a decade had stabilized at a much lower level.[121]

Pelagics

The cod fishery of the 1800s required bait – mostly herring, capelin, and squid. These species for the most part were "not considered sufficiently remunerative to warrant the fishermen in devoting any considerable portion of their time to them."[122] About 19,000 tonnes of herring were landed in New England from shore fisheries in 1879; about 25% were used as bait, while the rest was pickled, smoked, or eaten fresh. A similar amount of herring was taken in Newfoundland waters and used mostly for bait by American fishermen. While fishing the Grand Banks for halibut and cod, they "depend almost wholly upon fresh herring, with the exception of the summer months, when capelin (*Mallotus villosus*) and squid (*Ommastrephes illecebrosa*) are used. The entire supply, with few exceptions, is obtained along the coast of the British Provinces, the greater part being secured at Newfoundland and Nova Scotia."[123]

A New England fleet also developed a fall herring fishery in the 1820s. Early records indicated that the first American vessel, sailing from Isle au Haut, Maine, fished the Magdalen Islands in the Gulf of St. Lawrence in 1822.[124] Its catch was pickled, and the fishery expanded rapidly: by the 1860s, there were 100-500 vessels fishing the area each season. Within a few years this fishing was considered to have "rapidly declined."[125] In an early example that demonstrates the difficulty of separating overfishing from natural fluctuations, there was considerable disagreement on the cause of the failure. The changes in herring catches did not enable any resolution of the debate: if anything, they suggested fluctuations rather than overall declines in catches, but as no information existed on how many men or vessels were fishing, how this might have related to abundance was uncertain.

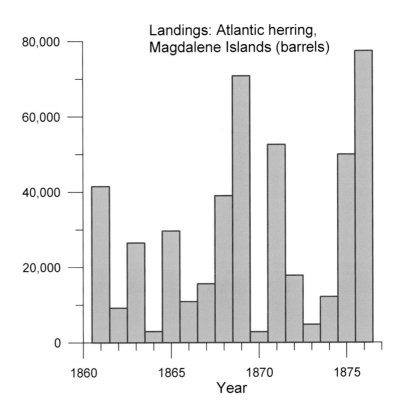

6.17

Herring landings at Magdalen Island from 1860-1876. Data from Earll (1887).

At the same time as the fisheries off the Magdalen Islands were in decline, there were increases in the New England herring fisheries at Fortune Bay on the south coast of Newfoundland, at Bonne Bay and Bay of Islands on the west coast, and on the Labrador coast as far north as Cape Harrison. Catches of herring in Newfoundland waters rose to an average of about 25,000 tonnes in the later decades of the 1800s, with much of this amount traded to New England bait boats for various goods. On the Labrador coast, the New England boats caught their own herring, primarily with gillnets. The herring bait fishery in Newfoundland was always contentious and subject to several legislative initiatives during the 1800s.

In the mid-1850s, the New England fleet began a new practice with frozen bait. In 1854, Captain Henry O. Smith, a Gloucester, Massachusetts skipper, ventured to the south coast of Newfoundland in winter to catch and freeze halibut. Smith did not find much halibut, but he did find a plethora of cod and herring. He switched his voyage to cod, but conceived the idea of taking frozen herring back to Gloucester to sell to schooners as bait. The following season, vessels using the frozen Newfoundland bait made record catches of cod and a new industry was born.[126] The New England frozen herring fishery in Newfoundland became centred in Fortune Bay on the south coast, which was "frequented by immense schools of herring during the winter months and spring months."[127] For the most part, the American fleet bought fish from settlers so as not to deprive them of "the chance of obtaining money from the capture and sale of the fish which these [U.S.] vessels require."[128] Many Placentia and Fortune Bay residents depended on the winter herring fishery for the American bait market, and "the ones who have been most successful are those who have catered most largely to the American trade, spending the summer in the cod fisheries and the winter and spring in supplying the American and French vessels with herring and capelin for bait."[129] Unfortunately, the supply of bait from Newfoundland was not always sufficient to satisfy the American schooners, and some began to seine their own herring.

In the 1860s, large overwintering schools of herring were found in the waters off Grand Manan Island in New Brunswick, near the Maine border. The Newfoundland fish was preferred by cod and halibut fishermen, because it was firmer and held a hook better, hence commanding a better price. Nevertheless, by 1880 the Newfoundland fishery was nearly abandoned because it was cheaper to exploit the nearer New Brunswick stocks and because of the growing hostility to any foreign fishery on the south coast of Newfoundland.[130]

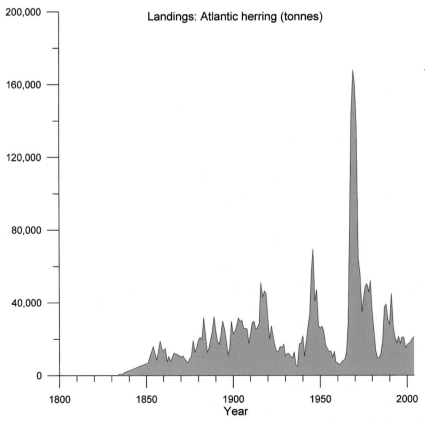

6.18
Herring landings in Newfoundland. Data from International Commission for Northwest Atlantic Fisheries and Northwest Atlantic Fisheries Organization reports.

6.19
Hauling capelin, mostly for the garden, in the mid-1900s (Gordon King collection).

Capelin were also used as bait in both the Newfoundland-based and foreign cod fisheries.[131] Capelin began to be used around the middle of June when the spring herring bait supplies were running low, and the capelin were running to shore to spawn and easily caught. Capelin were also used in Newfoundland and Labrador for food and for garden fertilizer. Capelin catches were not well documented, but perhaps reached 20,000-25,000 tonnes each year and likely had little effect on the stocks.[132]

Mackerel landings were very low in the 1800s. A few catches were recorded around 1875, but this warm-water migrant was scarce in most Newfoundland and Labrador waters until the mid-1900s.

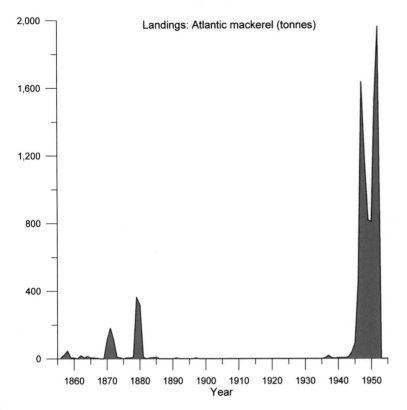

6.20
Landings of Atlantic mackerel, 1860-1955. Data from Templeman and Fleming (1954).

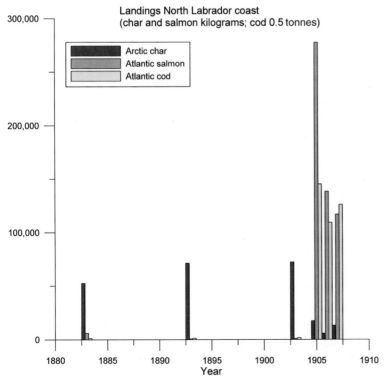

6.21
Arctic char, Atlantic salmon, and cod landings in northern Labrador around the turn of the twentieth century. Data from diverse sources.

Other Species

Greenland halibut – known in Newfoundland as turbot – fisheries began sometime in the late 1800s, likely in Trinity Bay where large fish were common in the deep trenches. This was a minor fishery and its beginnings were not well documented.[133]

Arctic char had been fished in Labrador as a rich food source since the arrival of the Dorset Eskimo people. Moravian missionaries to Labrador[134] introduced nets to this fishery in the 1700s, but it was not until 1860 that the first exports were recorded from Labrador, about the same time as the cod fishery was expanding there. During the late 1800s and early 1900s, records indicated that significant quantities (50-75 tonnes) of Arctic char were exported from Labrador. In the early 1900s, interest in the char fishery declined as the salmon and cod fisheries expanded rapidly.

THE INVENTION OF THE COD TRAP

There were few technological changes in the methods of the Newfoundland and Labrador cod fisheries over nearly 400 years. In the mid-1800s, unbaited jiggers, single handlines with bait, and longlines were the dominant fishing gear, as they had been since the 1500s. With the decline in catch per fisherman in the late 1800s, however, entirely new fish catching technologies were investigated. Chief among these was the cod trap.

There have been several accounts of the development of the cod trap,[135] the most plausible of which attributes its invention to Captain William "Bossy" Whiteley.[136] Whiteley was an American-turned-Newfoundlander who operated a fishery at Bonne Espérance Island on the Labrador coast, west of the Strait of Belle Isle, in what is now the Lower North Shore region of Quebec, and near the site of the original French fishing port of Brest. The potential to "box up" cod within a seine near the shore and take them out almost at will led Whiteley to test a fixed box or "pound" net with a leader at his Bonne Espérance fishery in the early 1870s. The first cod traps may have been attachments to salmon weirs, but they soon became stand-alone gear. On August 15, 1876, Whiteley made an application to the Department of Fisheries to use his invention at Bonne Espérance,[137] although he had been using the trap for at least five years prior to this. Nearby fisheries were quick to seize upon this invention, and by 1877, American-made traps were being used at Indian Harbour.[138] The trap proved very effective. The *History of the New England Fisheries* mentioned that these early cod traps could catch up to 100 quintals (5 tonnes) of fish in one haul.[139] One hundred quintals would have been considered a good catch for a small station for a full season prior to this time. The trap was a passive gear that relied on fish coming right to shore to the site, or berth, of the trap, exploiting the repeated migration tracks of cod. Some berths were known to be the best producers. After 100 years of use, the original trap berths at Bonne Espérance were still being fished in the mid-1980s.[140]

6.22 Was this the fishing crew that first applied the cod trap? The crew at Bonne Espérance at the Whiteley fishery in 1885 (Albert Whiteley collection, courtesy of Cheriton Graphics, Ottawa).

6.23 Women and children at Bonne Espérance at the Whiteley fishery in 1885 (Albert Whiteley collection, courtesy of Cheriton Graphics, Ottawa).

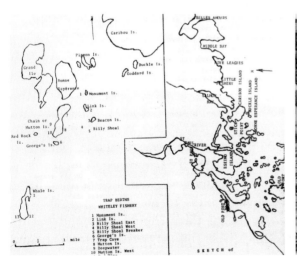

6.24 The original trap berths at Bonne Espérance. These same berths were in use in the mid-1980s when the author was conducting research in the area (Albert Whiteley collection, courtesy of Cheriton Graphics, Ottawa).

6.25 Flakes and houses at Bonne Espérance around 1885 (Albert Whiteley collection, courtesy of Cheriton Graphics, Ottawa).

6.26
Hauling the trap near St. John's in the 1950s (Gordon King collection).

The cod trap created quite a stir in the 1880s, and the House of Assembly appointed a committee to investigate the use of traps and their effect on the failing fishery. The committee reported that the trap was not the cause of stock declines, but that mesh size limits (4 inches or 10 centimetres) should be mandatory and traps should be placed at least 100 fathoms (183 metres) apart. Only the mesh regulation was implemented, but a similar placement regulation – which required traps to be placed 80 fathoms (146 metres) apart – was established in 1884. The issue did not die there: in 1888 the Newfoundland Fisheries Commission remarked that "the general conviction now among those best acquainted with the fisheries, seems to be that they [traps] are highly injurious, and that the use of them should be prohibited by law…when they are planted thickly the fish are prevented from coming in on shore, young fish are destroyed when unfit to be taken…the effect on the fishermen who use them is demoralizing. They require to develop [sic] little skill or energy in working them and there is the danger that they may lose those qualifications for which, as fishermen, they have long been honorably distinguished."[141] These views led to passage of the *Cod Trap Act* in 1888, which placed a ban on their use as of May 1890. Some interpreted this act as a concession to the French, who did all they could to prohibit the use of traps on the French Shore.[142]

In 1889, the newly elected government of Newfoundland, under William Whiteway, immediately repealed the *Cod Trap Act*. In its place, a series of regulations was implemented that limited the setting of traps prior to June 15 in Newfoundland and June 25 in Labrador, and banned traps altogether in some harbours in Placentia Bay.

Were traps really harmful to the stocks? There is little doubt that they could catch many small fish, but that was dependent on the mesh size used. Fisheries superintendent and scientist Adolphus Nielsen thought that the small and illegal meshes used in many traps were the main problem. Fisheries department reports in the 1890s were riddled with complaints about illegal traps, mostly based on small mesh.[143] The increase in fish abundance from 1900 to 1950, during the main period of trap usage, suggests that the benefits of the trap, which caught fish live and for the most part in proportion to their abundance, far outweighed their failings.[144]

6.27
Drying up the trap near St. John's in the 1950s (Gordon King collection).

Despite the controversies, the cod trap quickly became the dominant gear in the Newfoundland and Labrador fisheries. There were 2588 traps in use by 1891, 4182 by 1901, 6530 by 1911, and 7365 by 1921. About 5000 traps were in use each year throughout the 1930s and 1940s. The cod trap would seal the fate of the aged handline fishery. In Labrador, stationers soon converted to traps, which made their fishery more competitive with the handlining floaters. The number of schooners, mostly from Conception and Trinity bays, decreased from 825 in 1894 to 470 by 1898, but increases in the fish stocks, better prices in the early 1900s, and a switch to traps led to a temporary resurgence in the Labrador floater fishery with 1432 vessels participating in 1908.[145]

Additional changes came to the fishery by 1914 with the advent of steamship service to Labrador and the availability of small boat engines. Stationers could now travel easily from their homes in Conception and Trinity bays to the Labrador fishing grounds. Small boat engines assisted the trap fishery; in a few years, over 4000 trap skiffs were equipped with motors that reduced the backbreaking labour of this fishery considerably, and also increased mobility. Within a few decades of its introduction, the cod trap could be found along the northeast coast, the Avalon Peninsula, and even into Placentia Bay on the south coast of the island of Newfoundland.

6.28 The Labrador coast near Blanc Sablon with fishing shack and killicks (anchors used to set cod traps). Photograph by the author in 1985.

In the Labrador fishery, use of the trap exacerbated quality problems. The success of trap fishing relied on the feeding patterns of cod – in particular, the middle-age groups (fish of about three to eight years of age), which were the heaviest capelin feeders. Older and larger cod seldom chased capelin right to shore; therefore, unless large meshes were used in the trap, there was a tendency to catch the smaller fish of the run. On the Labrador grounds, cod were slow growing and small in size compared to those in more southerly stocks – the Labrador trap fishery therefore caught the smallest of small fish, and small fish brought lower prices. To compound this situation, the short Labrador summer limited the drying season and the relatively large catches of the traps were difficult to deal with quickly.[146] These factors conspired to make fine curing in Labrador nearly impossible – salt cod from that region was of poor quality and fetched low prices.[147] Profits depended on the availability of vast amounts of fish – which the trap delivered. As reported by A.B. Perlin in 1930, "two million quintals at five [Canadian] dollars a quintal is better than one million at seven dollars."[148] With a goal of quantity, not quality, this fishery persisted into the mid-1900s, and then in modified form with the addition of gillnets until the stocks collapsed in the late 1900s.[149]

FISHING SHIPS

By 1850, the Newfoundland shipbuilding industry was in decline. The ageing fleet of Newfoundland-built sealing schooners (a vessel might last for six to seven years) was being replaced with vessels built in Prince Edward Island.[150] Landings of seals declined steadily over the next 50 years, as the initial seal populations could not sustain the high harvest rates of the mid-1800s. The industry became more and more consolidated into a few St. John's firms that used larger and larger vessels.[151] This trend was not entirely welcomed. In 1855, a St. John's reporter wrote: "We think the old mode of prosecuting the seal-fishery, with superior class vessels, is the best and only mode of doing it successfully. In this way fortunes have been made, and may be made again, with the same kind of perseverance and industry which marked the days of our fathers. Business carried on in a fast manner is very likely to end in the same way as the career of fast young men – in ruin."[152]

By the mid-1860s, Scottish-built, wooden-walled steamers, proven in the seal and whale fisheries off Greenland and Iceland, were brought to the Newfoundland seal hunt by Walter Grieve and Company and the closely related Baine Johnston Company.[153] They were equipped with both sail and steam, and were more maneuverable than any sailed vessel – a steamer could force its way through ice that stopped sail. The first two steamers – the *Wolf* (Grieve) and the *Bloodhound* (Baine Johnston) – were sent to the ice in 1863. Neither did well, which no doubt bolstered the critics. Nevertheless, the

6.29
Island outport in Placentia Bay in the early 1900s with schooner and punts (Gordon King collection).

writing was on the wall, and the other large St. John's firms, including Bowring Brothers, Job Brothers, and Harvey & Company, quickly followed suit. Over the next two decades, many more wooden steamers were added to the Newfoundland fleet, and the age of sail on the front was over. The schooner seal fishery held on longer on the south and west coasts, hunting the Gulf of St. Lawrence herd. In 1898, in an attempt to stay its passing, the Newfoundland government enacted a bounty to encourage expansion of the schooner seal fleet, but it had little effect.

The use of steamers would destroy the sealing industry, but the technology seemed both inevitable and irresistible. Prowse wrote that "when Mr. Walter Grieve sent the first sealing steamer to the ice it was a poor day for Newfoundland. The only consolation we can lay to our hearts is that steam was inevitable; it was sure to come, sooner or later, the pity of it is that it did not come later. Politics and steam have done more than any other cause to ruin the middle class, the well-to-do dealers that once abounded in the out-ports. It was as bad for the merchants as for the men."[154] Prowse did not mention that the advent of steamers would also further decimate the already depleted seal herds. The development of long-distance steamers enabled foreign enterprises to try their hand at seals. In the 1870s, Scottish, American, and Nova Scotia firms sent steamers to the ice, mostly with Newfoundland crews. They did not stay long, as profits were low.[155]

In the end, the profits that accrued to the few merchants with their larger and more powerful vessels were short-lived. Wooden-walled ships gave way to steel steamers in the early 1900s. By 1915, not only was World War I underway, but the overcapitalized seal fleet – now expanded to include about 10 steel-hulled steamers with icebreaking capacities – was proving uneconomical to operate: the price of seals could not justify the expense of the vessels. Those that hung on in the seal fishery used far fewer boats. With tough times, tempers grew short, and there were many accusations that pelts "flagged" and "panned" by one vessel were stolen by another.[156] Such tawdry tactics were unheard of a generation earlier. As World War I began, owners took the opportunity to sell the steamers off or convert them to passenger service. Some of the best-known vessels, including the *Beothic*, the *Adventure*, the *Bellaventure*, and the *Bonaventure*, went to Murmansk, Russia. The *Stephano* was torpedoed by a U-boat in 1916, and the converted *Florizel* struck a shoal near Cape Race in 1918 on her route to New York with a loss of 99 people.[157] The seal hunt was much reduced thereafter.

6.30 Fishing schooner on the south coast in the early 1900s (Gordon King collection).

*6.31
Schooner on drydock for repairs in early 1900s (John Wheeler collection).*

Between the World Wars, the seal hunt reverted to smaller, more economical, wooden ships. The early era of the seal hunt was drawing to a close. The beginnings of the hunt were rooted in the demand for seal oil to light the lamps of Europe, but Europeans now had access to much cheaper supplies of fossil fuels. A revived seal hunt decades later would see no more of the large ships, and seal oil would be processed for human consumption rather than for lamps.

The conversion from sail to steam occurred more slowly in the cod fishery than in the seal fishery. The advantages of the steamers to the seal hunt were not as readily apparent to the cod fishery, but it was soon realized that the steamers had sufficient power to pull large nets through the water. This led to the advent of trawlers, also known as draggers, in the early 1900s. Trawlers were clumsy at first, but became increasingly efficient at catching fish. From 1908 until the outbreak of World War I, France sent about 236 sailing ships and 25 steam trawlers to Newfoundland each year. In the years between the two great wars, the numbers of trawlers increased only modestly, but France, Portugal, Spain, and the United States all entered this new type of fishery.[158] Italy bought the largest trawler that Britain had in 1938 and brought her to the Grand Banks. Until the mid-1950s, most deep-ocean trawlers were either side trawlers – vessels that deployed and retrieved their net over the side of the vessel – or pair trawlers – two vessels that worked together and towed the net between them.

CRUEL YEARS: 1890-1900

By the 1880s the declines in the cod and seal fisheries had taken their toll on Newfoundland. To make matters worse, the year of 1892 brought several disasters. On February 28, dozens of sealers were caught in a storm in Trinity Bay and froze to death on the ice. The growing city of St. John's burnt nearly to the ground on July 8, leaving almost 10,000 people homeless (a similar but less extensive fire nearly levelled St. John's on June 9, 1846).[159] Conditions in the seal hunt grew worse for the sealers, who were paid a pittance while daily risking their lives. After a series of modest protests in the mid-1800s, the Great Sealers' Strike took place in St. John's in March 1902. Several thousand sealers gathered at the corner of McBride's Hill and Water Street. Alfred Morine, the sealers' advocate, was initially rebuffed by the merchants, but after four days they agreed to meet with him. After tough negotiations, the sealers' pay was increased by almost 50%.

6.32 St. John's from the inner harbour in the early 1900s (Gordon King collection).

The Newfoundland economy languished in the 1880s. Bankruptcies were commonplace and the major banks crashed in 1894. John Munn and Company of Harbour Grace, the Conception Bay fishing giant, folded days after the crash. Munn had been in trouble for a number of years: subsidized French-cured fish from the Banks had been underselling their Labrador-cured fish in Europe, and with the decline in the seal hunt, steamers were not even paying their expenses. In the end, however, it was competition with St. John's firms that finished Munn. The bankruptcy of the Munn enterprise coincided with a general decline in prosperity on the northeast coast in the 1880s and 1890s. Even well-established and affluent outports such as Trinity were hard hit. In 1886, *The St. John's Evening Telegram* published the following:[160]

When you look back, some 20 or 30 years, to the days of the sailing vessels, what a contrast! Then the harbour of Trinity was alive with industry. Just about this time of year there would be a couple of fine brigs launched, and the fitting out of these would afford plenty of employment to the people... Now, however, things are changed and all this is merely a memory of our departed greatness. Well may we say the steamers are the curse of the country! It seems to me such a glaring piece of injustice that a few merchants should reap the entire profits of this fishery and make fortunes in the same, while the great bulk of the people are in a semi-starving condition...

Only Bowring Brothers of St. John's appeared to have escaped the bank crash of 1894. Other firms, even some of the most prosperous, were not so lucky. Baine Johnston and Job Brothers were broke, but managed to survive.

Much has been written on the poverty of the outports in the 1800s. It would be wrong, however, to conclude that Newfoundland was an economic wasteland at this time. There was extreme hardship when the cod and seal fisheries failed: the increasing numbers of people dependent upon the fishery in a period of declining productivity and

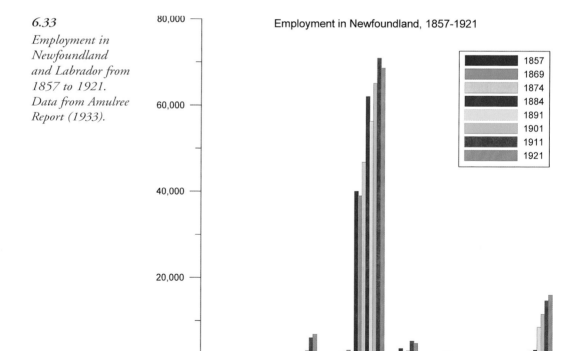

6.33 Employment in Newfoundland and Labrador from 1857 to 1921. Data from Amulree Report (1933).

the low market prices during most of the century made economic gains nearly impossible for most people. In Newfoundland and Labrador, only the merchants and town-based professionals had much money; however, the cod and seal fisheries employed more than 80% of the workforce at a level comparable to most other types of work in Newfoundland and the rest of North America.[161] J.G. Millais, the famous English hunter-naturalist, wrote in 1907 that, "On the whole Newfoundlanders, except the poor of St. John's and the islanders of the east coast, are exceedingly well off in the literal sense of the word, and would be in clover were it not for the over-powering taxes, for which they get absolutely nothing in return."[162]

J.P. Howley, a Newfoundland-born geologist who traversed much of the island in the late 1800s, gave similar comments of the lives of people in the isolated outports. Many families and communities were described as "well off" and "well to do," and possessed "an air of comfort and abundance." Howley visited the fine cod grounds of inner Placentia Bay in 1868. At North Harbour, "lived a man named Emberly. His was the only family here. He had a nice clearing, good house, and quite a lot of cattle. We got some beautiful fresh butter from him, and plenty of milk to drink." Howley then proceeded to the local merchant's home on nearby Sound Island, where "Old man Brown is well off and has a fine house…has a large shop and store and has a good deal of fish collected…owns two fine schooners, one of which he built himself."[163] There is no doubt that failures in the fishery produced hard times, and not all outports were as fine as those described by Howley, but circumstances in Newfoundland were arguably no more difficult than those faced by colonists of the same era in other parts of the world, and there were good years too.[164]

EARLY SCIENCE

Natural science typically begins with observation. By the mid-1800s, the collective observations of seafarers and fishermen over 300 years had described the general features of the Grand Banks and the waters around Newfoundland, as well as the migrations of the cod.[165] This knowledge helped sustain high catches, but inevitably the increasing numbers of fishermen would search out new grounds to the north off Labrador. The coastal fishing grounds of Labrador north of Cape Harrison were estimated to be nearly as extensive as all the other fishing grounds of Newfoundland combined.[166] Descriptions of the Labrador coast, the Labrador cod fishery, and the ecology of the cod were summarized in an American account published in 1911:[167]

> A succession of deep narrow fjords along the coast has for an outer fringe a vast multitude of islands about twenty-five miles in stretch from the mouths of the fjords seaward. About fifteen miles outside these fringing islands are numerous banks and shoals, the feeding ground of cod from the middle of June to October, while outside the shoals there appears to be a second range of banks, where, probably, the cod feed in winter. This island-studded area forms an immense codfishing ground, which has been estimated at 5200 square miles [13,470 square kilometres] in area.
>
> The mean length of the fishing season for cod…on the northeastern coast of Newfoundland has been estimated to be 142 days, for southern Labrador… it is 87 days, and for northern Labrador…it is 52 days. The cod does not travel far in its annual migration on this [the Labrador] coast. After the spawning season the fish retires to deeper water offshore. Each year the cod returns to its birthplace with the school, and haunts the same neighborhood the short season of its inshore life. The school of cod arrives on the coast about a week later for every degree of latitude farther north. For a period of about forty days the codfishing goes on simultaneously during August and September throughout the length of a coast line, extending from latitude 47° to latitude 58° 30', or more than 700 miles [1126 kilometres].

Such detailed descriptions of Newfoundland fisheries were rare, but the troubles of the 1800s led to thoughts that a more systematic examination of the fish stocks might help the fishery become more prosperous. Few such efforts had been made since Lieutenant Griffith Williams attempted to collect better data on landings and effort in 1765. During the final two decades of the 1800s, a revolution in science took place in Europe and North America, and the field that would become fisheries science developed in Norway, Britain, the United States,[168] and Canada. Newfoundland lagged well behind its competitors in this respect. Many thought that science was a short-lived fad and was certainly not worth paying for.

In the early 1880s, with its economy in crisis and the cod stocks likely at their lowest point since the beginning of the fisheries, the Newfoundland government took its first tentative steps towards a science program by appointing a committee to compare the Newfoundland fisheries to those of other countries such as Norway, Canada, and Britain. In 1883, the Great International Fisheries Exhibition was held in London. The Honourable Ambrose Shea represented Newfoundland at the Exhibition, while W.H. Whiteley, the inventor of the cod trap, was sent ahead to prepare the Newfoundland exhibit. Whiteley's exhibit won first prize.[169]

The London Exhibition featured the "Great Debate" about the fluctuations in the sea fisheries. As early as 1854, overfishing was an unpopular explanation for why fish catches could not be held constant.[170] In truth, there were other explanations for year-to-year variations in landings, such as changes in fish migration routes and in the environment. There were also yearly variations in markets and fishing effort. At the end of the debate, English scientist Thomas Huxley made his infamous conclusion "that the cod fishery, the herring fishery, the pilchard fishery, the mackerel fishery, and probably all the great sea-fisheries, are inexhaustible; that is to say that nothing we do seriously affects the number of fish. Any attempt to regulate these fisheries seems…to be useless."[171]

To be fair, Huxley provided several qualifications to his conclusion that were quickly forgotten. These included reservations about the generality of his conclusion and, more importantly, that it was based on the modes of fishing of the 1800s. Given these qualifications, Huxley's conclusion did not appear so far off the mark. However, Sir Ray Lankester, an English biologist with mostly freshwater experience, countered Huxley's statement: "It is a mistake to suppose that the whole ocean is practically one vast store-house, and that the place of the fish removed on a particular fishing-ground is immediately taken by some of the grand total of fish, which are so numerous in comparison with man's depredations as to make his operations in this respect insignificant."[172]

The debate was important not so much for who was right or wrong, but because both sides agreed that fisheries science was needed to answer basic questions about changing fish stocks. As a result of the London Exhibition, systematic studies of fisheries fluctuations were begun in Norway and Scotland. The Fishery Board of Scotland conducted a series of closed-area experiments in the North Sea to test the effects of trawling – one of the first relatively long-term studies of a marine fishery. The Scottish Research Board concluded in 1893 that "it has now been made clear by statistical and scientific investigations that the seas around our coasts are not the inexhaustible storehouses…that they were thought to be less than a generation ago. The doctrine that the operations of man cannot disturb the balance of life in the sea, and diminish or exhaust the supply of valuable food fishes, is now abandoned by fishery authorities everywhere."[173]

Their conclusion did not go unchallenged. W.C. McIntosh, a noted Scottish scientist regarded as "one of the leading fishery scientists of the world,"[174] refuted the methods and findings of the study, and argued for Huxley's case that the fisheries had little effect on the stocks.[175] The matter would not be fully settled for another 100 years, and it would take the total collapse of major herring and cod fisheries on both sides of the North Atlantic to do it.

Marine science was also developing rapidly across the North Sea in Norway. Scientists such as G.O. Sars and Johan Hjort asked difficult questions about why fish stocks fluctuated and urged the authorities to develop a scientific understanding of the fisheries.[176] Hjort put forward the idea that observed fluctuations were caused by variations in the numbers of young fish surviving to a fishable age each year, a concept which came to be known as fish recruitment.[177] He also dared to ask the grand question of how many fish were in the sea. Under the leadership of Sars and Hjort, Norwegian fisheries science developed as a service to the fishing industry, assisting in its prosperity, but also collecting knowledge for its own sake. These twin motivations led directly to the formation of the International Council for the Exploration of the Sea (ICES) in 1900. The first ICES research programs were mandated to investigate migration and overfishing, and drew on the experience of developing marine biological stations in Norway, Sweden, Denmark, Germany, Russia, Holland, Scotland, and England.

By the 1880s, Canada had hired E.E. Prince,[178] a protégé of W.C. McIntosh, as Commissioner of Fisheries. At first, the Canadian Department of Fisheries focussed almost exclusively on the freshwater fisheries of central Canada. It took a young professor at the Ontario Agricultural College in Guelph, Ontario, to point out the incongruity of that approach. J.P. McMurrich, who later would lead the Canadian Fisheries Research Board, had studied fisheries at the University of Toronto and in New England. His article "Science in Canada," published in 1884, is worth citing in some detail:[179]

> Another source of revenue to the country is in great need of encouragement and protection by scientific investigation. The Department of Fisheries has become of great importance to Canada and something has already been accomplished by the establishment of fish hatcheries, etc., but this affects only our inland waters, the sea fisheries receiving little or no benefit. The life histories of our various food fishes, their manner and times of migration, their spawning localities, their food, their personal enemies, the destroyers of their food, all these should be properly investigated. True the Americans have done much for us in this line, but there is much yet to be done; in fact, the entire fisheries of the western coast are yet to be studied. Stations established on Vancouver Island, on the Gulf of St. Lawrence and on the Nova Scotia coast with facilities for investigations…would in a very short time repay the expenditure…by enabling us to adopt measures for the increase of our fisheries by informing us of their real extent, of which we are yet comparatively ignorant, and by preventing their wanton destruction.

THE FIRST NEWFOUNDLAND FISHERIES COMMISSIONS

The notion that systematic study of the fisheries might pay off was slower to develop in Newfoundland than in other countries. In 1888, the first Newfoundland Fisheries Commission appointed Adolphus Nielsen as its superintendent. In 1892, Dr. Moses Harvey, secretary of the Commission, made a key presentation to the Royal Society of Canada. Thinking that Newfoundland could not on its own support a research station to study the fisheries, he argued for a joint effort by Canada and Newfoundland. Of note, he concluded that "all our regulations of our fisheries must have their basis in a scientific study of fish-life. Failing such accurate knowledge, our legislation regarding the fisheries will be largely groping in the dark."[180]

In the mid-1880s, Nielsen established cod and lobster hatcheries at Dildo with the intent of restocking Trinity Bay, and launched investigations into various aspects of the biology of other species. From 1890-1893, nearly 424 million cod fry were released into Trinity Bay, perhaps with some success as "fishermen report having seen large shoals of fry and small codfish during last summer in various localities where they were never seen before, in such number and in such conditions as to satisfy them that they were the product of the Hatchery."[181] Nielsen made many other contributions to fisheries science in Newfoundland, including experiments on cold tolerance in cod, and some of the first descriptions of cod movements in coastal waters and of spawning distributions in Trinity and Placentia bays.[182] Nielsen's far-sighted ideas about the role of science in fisheries stand out beyond the technical excellence of his early research. Ten years after the London Exhibition and the Huxley debate, Nielsen wrote this wordy but important note:[183]

> The old theory advanced by Professor T.H. Huxley, and a few more authorities on fish,…for a generation it has had its effect on the minds of a good many people on both sides of the Atlantic, and which has been so readily adopted and cultivated, especially in England, not alone among fishermen and people directly interested in the fisheries, but also among statesmen and legislators, I presume on account of its cheering character, and thereby worked injury to that industry, this theory, namely: That the resources of the sea are inexhaustible, and that it did not matter where, when and how fishing was carried on, or what engines [fishing gear] were at work, man was unable to desturb [sic] the equilibrium of nature, and that all that man could do in the way of destruction, let him do his best, would only be equal to a drop in the sea, compared with the natural reproduction and fecundity [number of eggs] of fish…

> This old theory, now that knowledge has accumulated during the last decades, is entirely abandoned by the majority of administrators, scientists and practical men. It has been proved beyond all doubt in all countries, where fishery has been carried on to a large extent, that man plays a most important part in the diminution of fish life at sea, and that…the natural reproduction of the various species of fish, is not so very great after all, and can not keep steps with the forced destruction of man.

> The progress that science is making, and has made, during the last two decades, and the vast amount of knowledge gained in a comparatively short

period respecting the life, habits and food of various species of fish, as well as of the physical condition of the element in which they live, is *remarkable*, considering the difficulties under which such work often has to be carried on.

Nielsen made many recommendations on how the Newfoundland fishery might be improved. Among these, he advocated the protection of spawning cod – especially larger females who produced the most eggs – and the protection of juvenile cod by the enforcement of a law banning small meshes in cod traps. Nielsen believed that "the destruction annually [of small cod in the Newfoundland fishery] *is simply enormous* [emphasis added]."[184] Nielsen also made extensive recommendations to protect the lobster fishery, including the imposition of gear and seasonal restrictions, and the involvement of the lobster factories in growing and releasing of young lobsters.

Despite Nielsen's considerable research and undeniable enthusiasm, the Newfoundland government lost interest in fisheries research within a decade and Nielsen moved on to the whaling industry. In 1898, the fisheries laboratory was closed and the buildings sold. It was an inauspicious start to fisheries research in Newfoundland.

The rejection of fisheries science in Newfoundland in the late 1890s requires explanation. At least two factors appear to have been at play. The idea that the fishery did not hold the promise to better the economy of Newfoundland was significant. It was rightly thought that the increasing population could not depend solely on one industry, the beleaguered cod fishery. This belief had been expressed repeatedly since at least the 1840s when Governor LeMarchant advocated diversification of the economy. LeMarchant correctly predicted that "the fishery could not be thought of as the sole support of the rapidly increasing population, for despite the increase, cod production was stationary from year to year. The numbers of labourers grew but the profits of the individual are decreasing."[185] The fact that both the cod and seal fisheries were in a depressed state in the late 1800s did not help. Salt fish prices were low and increased foreign competition was making markets difficult. A second factor may have been the generally poor education levels in Newfoundland, the high rates of illiteracy, and the resultant lack of appreciation of the scientific method and its sometimes abstract concepts.

But for all that, the fishery still employed most of the people of Newfoundland and produced 90% of its exports in the final decade of the 1800s. Nevertheless, the government chose, for all intents and purposes, to ignore it. Many politicians believed that the only way forward was industrialization – the first step being a railway across the island. A decade of debate ensued in which many were firmly convinced that "no material increase of means is to be looked for from our fisheries, and that we must direct our attention to other sources to meet the growing requirements of the country."[186] Critics of that view, such as Sir Robert Bond, argued that the terms of railway construction by the R.G. Reid Company "represented an unwarranted sell-out of the country's resources and assets to an outside contractor."[187] In the end, there was massive spending on the railway project and the election of 1889 confirmed the industrial platform of William Whiteway. The fisheries took a back seat.

In hindsight, despite the difficulties with the fisheries and the rightness of developing other industries, it made little sense to cut loose and set adrift the most important part of what had been the mainstay of the economy of Newfoundland and Labrador for hundreds of years. The social historian David Alexander was perplexed by this, and compared Newfoundland with Iceland, which took a very different tack with very different results:[188] Iceland's fisheries-based economy flourished, while Newfoundland's withered.

WILLIAM COAKER

In the early 1900s, the Newfoundland fishery enjoyed a short period of renewed prosperity.[189] William Coaker organized the first local of the Fishermen's Protective Union at Herring Neck, Notre Dame Bay, in 1908. By 1913, 26 permanent and 7 temporary union stores were scattered around Newfoundland, selling salt, flour, coal, butter, tobacco, salt beef, and salt pork. Cod catches and prices were on the rise. There was an attempt to wrestle the truck system to the ground and bring a cash economy to fishermen, but this initiative failed. By the election of 1913, the union had spawned a political party; with William Coaker at the helm, 11 members were elected to the Newfoundland legislature and became the official opposition. Coaker was guided by the grandly titled Bonavista Platform, which recommended a complete reorganization of the fishery and its outmoded mercantile system, and sought to increase markets and prices for Newfoundland cod products.

DEATH ON THE ICE

In March 1914, the death on the ice of 78 sealers would forever change Newfoundland.[190] Fathers, sons, and brothers from many outports became icons for the harsh treatment of hard-working people by an uncaring and aloof mercantile and political system. It was not so much that the men died – death was no stranger to the outports – as it was the way that they had died: it seemed as if their lives did not matter, not even to the ships' captains who were sworn by ancient laws of the sea to look after their crews. Captain Abram Kean – perhaps most responsible for the disaster – under direction from the large firm of Bowring Brothers, refused to interrupt the hunt to bring the dead men home until he was ordered to do so by the government.

> This callous action changed many attitudes towards the mercantile elite. In St. John's, it was traditional for the returning sealing fleet to be greeted by half the town, with the "high liner," the vessel with the largest take of seals, playing the starring role. In 1914, the high liner entered St. John's harbour with flags flying and whistles blowing. She was greeted by no one.
>
> William Coaker was aboard one of the sealing vessels in the disastrous 1914 hunt observing the conditions under which the men lived. They were horrid: the vessels were filthy and there was little food. Coaker based much of his platform for reform on these experiences. Thus it was that the 1914 sealing disaster steeled many Newfoundlanders against the merchants and colonialism – who were now seen as tyrants.[191]

Coaker's initiatives bore some early fruit. In 1914, another commission was appointed to study the fisheries. The lack of scientific investigation caused the commissioners to write in 1915 that "some attempt should long ago have been made to investigate in an intelligent, comprehensive, and scientific way, the waters and fishing grounds contiguous to Newfoundland."[192] In the same year, the famed Norwegian scientist Johan Hjort traveled to St. John's and gave a public lecture about how investments in fisheries research had benefited Norway. In 1916, the Newfoundland Board of Trade called for "a much larger share of the public monies…[to be] expended in scientific investigation and practical experiment"[193] and indicated that a fishery school with a vessel properly equipped for research around the coasts was needed. Unfortunately, this scientific momentum was cut short by World War I.

Coaker joined a national war coalition government in 1917 as Minister of Fisheries, but wartime expenditures restricted the implementation of the Bonavista Platform. After the end of World War I, Coaker was Minister of Marine and Fisheries under the newly elected government of Richard Squires, but by then his platform had by and large been forgotten.

Coaker tried to change hundreds of years of history and concomitant attitude towards the fisheries in a decade. Coaker wanted democracy in Newfoundland and in the fishery, but the fishermen demanded dictatorship.[194] The outports wanted deliverables, not deliverance. Coaker became increasingly cynical and corrupt, turning his union into a personal cash cow based on a return to the old truck system, the very institution he initially set out to reform. In the end, Coaker turned against the outports, the fishermen, and, some may argue, against Newfoundland.[195] He bought an estate in Jamaica and mocked Smallwood in the 1930s for trying to create co-operatives. In a revealing and bitter comment about Smallwood's efforts, Coaker jeered, "He'll talk and he'll talk, and he'll work and he'll work – but he'll never get a cent of dues out of them."[196] On this issue, Coaker was proven correct. He understood well the deep conservatism of the Newfoundland fishery, born of centuries of isolation, servitude, and lack of trust. It would prove again and again to be very hard to change.

WORLD WAR I

World War I brought tragedy to Newfoundland. On July 1, 1916, in one bloody morning, over 700 of its young men lay dead or wounded at Beaumont Hamel, France. Over the course of the war, Newfoundland's per capita casualties were among the highest of the Allied countries. The dead came from all walks of life, but whether they were merchants' or liveyers' sons, they were mostly of the fishery. In addition to the loss of life, the financial cost of the war to Newfoundland would plague its future and play a major role in loss of sovereignty in the 1930s. The total cost of the Newfoundland war effort was over C$35,000,000, equivalent to billions in today's dollars.[197]

Wartime landings of cod and prices for salt fish were high, and despite massive borrowing to support its war effort, Newfoundland experienced a period of unprecedented, if short-lived, prosperity. The first cold storage facilities were introduced in St. John's and Rose Blanche, and merchant groups attempted to enhance the markets in Spain and Italy. In 1919, Fisheries and Marine Minister Coaker attempted "a remarkable experiment":[198] legislation was put forth to control the prices and export of all fish. Prices were drawn up for various markets and grades of salt fish, and no fish could be sold at any other price. Trade commissioners were put in place in all the major market countries, as was a bureaucracy of licences, permits, and export affidavits, all under the control of Minister Coaker.[199]

The experiment lasted for only a year. Price controls might work in a national market, but they proved to be unworkable internationally.[200] Difficulties with exchange of currencies (Newfoundland had been using the Canadian dollar since the bank crash in 1894), export taxes imposed by England, and international resistance led to a recall of the trade commissioners and an abandonment of the price controls system. It was a hard lesson. Newfoundland and Labrador, despite possessing enormous fish stocks and high levels of cod production, could not control world market forces or even the price of its traditional cod product.

Meanwhile, some of Newfoundland's main competitors were flexing their muscles in the world cod markets. World War I did not devastate Norway and Iceland, either in terms of dead young men or war debt, as it did Newfoundland. Moreover, their fisheries had the advantages of proximity to European markets, longer seasons, a quality product, and more unified and co-operative mercantile systems. Perhaps most importantly, far from rejecting their fisheries as a means to economic growth, Norway and Iceland embraced them and invested heavily in them.[201] They saw their fisheries as their future, the foundation of a pyramid of marine-based industries that included the production of fishhooks, electronics, and shipbuilding, and any and all products related to seafaring and fishing. Their products came to be used in fisheries all over the world. Norway and Iceland also realized the importance of protecting their fish stocks. Iceland, with a population of only 250,000 people, challenged the fishing fleet and great maritime power of the United Kingdom to protect its cod stock, and won, while Norway has persisted in not joining the European Union, turning down enormous trade advantages in order to protect its fishery.

ECOSYSTEMS: 1900-1949

Climate and marine conditions improved markedly in the early 1900s after the recession of the Little Ice Age. As the century advanced, the warming conditions resulted in increasing fish stocks: cod were many million tonnes strong and increasing, with both large and small fish abundant in every stock, and there were perhaps 10 million tonnes of capelin, 4 million tonnes of redfish, and nearly a million tonnes of haddock on the Grand Banks.

The warming conditions in the early to mid-1900s also led to distribution changes in several pelagic species. Capelin may have shifted to the north: although capelin were plentiful on the south and southwest coasts until 1929, few spawned there after that until at least 1941.[202] Mackerel suddenly appeared in great abundance along the northeast coast in the 1940s. At fishing communities outside Smith Sound, Trinity Bay, "mackerel were almost unknown…prior to the mid-1940s but since then have made their appearance in considerable quantities and a significant mackerel fishery has developed especially during the months of September and October each year."[203] A bar seine fishery for mackerel developed in the region around Random Sound in Trinity Bay in the 1940s.[204] Dr. Wilfred Templeman, who had grown up in the fishing community of Bonavista in the 1920s and 1930s and was very familiar with its trap fishery, witnessed the arrival of the mackerel and verified that they had not been present there until 1940.[205] Saury were unknown in many areas of Newfoundland in the 1800s and early 1900s; many fishermen saw them for the first time in the mid-1940s. On the west coast, a large school of saury was encountered in 1945, with a report that "this fish is a newcomer," heard of only recently. In the fall of 1947, saury were present in large schools in Conception Bay and at Harbour Breton on the south coast, where they had not been seen before.[206]

Some invasions were less wanted. Dogfish moved north in large schools in the late 1930s and early 1940s, and were generally regarded as a pest.[207] Based on local knowledge gathered in coastal southern Labrador, C.L. Summers reported in 1942 that "dogfish were first reported in this area about five or six summers ago having been unknown until then. In warm summers they are more plentiful and come earlier than September."[208] In 1942, "the dogfish were so thick at West St. Modeste [Labrador Straits] last August that the fishermen could not get their trawls to bottom before every hook was full of them. The fishermen had to give up fishing until the dogfish slacked."[209] Farther north, fishermen at Smokey Harbour, Indian Harbour, and Cape Chidley had never seen a dogfish in the previous decades, but in 1942 they were commonly caught. The abundance of dogfish and the many complaints from fishermen resulted in research into both the life history of this little shark, about which little was known, and its potential as a commercial species.[210]

6.34 *Newfoundland and Labrador ecosystem species abundance from AD 1800-1949 (warm and cold periods).*

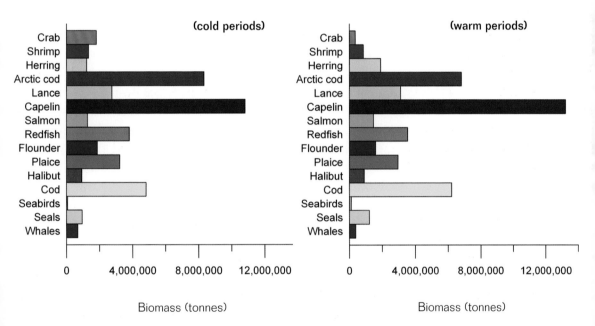

Haddock stocks may also have increased on the southern Grand and St. Pierre banks at this time. There is little reference to these species in the early banks fisheries, but they were a dominant species there by the mid-1900s. The generally warming conditions and resurgence of southern migrants – such as mackerel – on the northeast coast may also have resulted in increasing seabird populations and the colonization of Funk Island by gannets after 1936.[211]

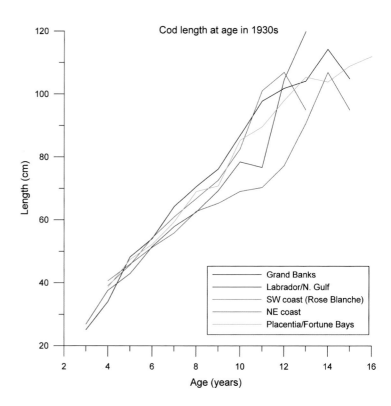

6.35 Cod length versus age in the 1930s. Note that fish from Labrador likely came from both northern cod and northern Gulf of St. Lawrence cod stocks, as fish were measured at St. Anthony, Battle Harbour, and Bonne Espérance. Data from Frost (1938b).

Four hundred years after the first export fisheries, and after 50 million tonnes of cod, 25 million seals, and numerous millions of other species had been taken from the marine ecosystems of Newfoundland and Labrador, the ecosystems were still functioning much as they had prior to 1450. Cod, haddock, capelin, herring, sand lance, and American plaice still dominated the continental shelves, and redfish and Greenland halibut inhabited the deeper waters of the shelf edge.

FISHERIES: 1900-1949

Although cod did not migrate to each bay and cove along the coast each year, there was no shortage of fish in the first half of the 1900s: stocks were rebounding, with catches improving to average over 400,000 tonnes per year. Landings surpassed half a million tonnes in 1907 for the first time in the long history of the fishery. Prices were good. Stock growth more than compensated for the increased catches and annual harvest rates declined, fostering further growth. By the mid-1900s, annual harvest rates were back down to less than 10% of the stocks, as they had been for centuries prior to the late 1800s. Several other fisheries were also doing well. Herring landings were steady, providing a good source of bait, and showed no signs of overfishing. Capelin were present for the taking. The warming conditions led to new fisheries for mackerel and a developing interest in the virtually untouched haddock stocks on the Grand Banks.

Some fisheries, however, were less fortunate. The lobster fishery was closed in the mid-1920s in response to declining catches. Salmon landings increased dramatically throughout the first half of the century and were never higher for a time, but these harvest levels proved to be unsustainable. Whale landings were high for the first decade of the 1900s, but then collapsed. Seal harvests decreased from about 500,000 animals per year in the mid-1800s to 100,000 animals in the first decades of the 1900s. All of these declines were almost certainly related to overharvesting. Interactions within marine ecosystems are difficult to specify, but the decline in the numbers of the predatory harp and hooded seals may have been a factor in the strong rebuilding of the cod stocks over this time period.

Management of the fisheries was minimal from 1900-1949. The French Shore issue was permanently resolved in 1904, for the first time granting all of Newfoundland to Newfoundlanders. Some local management restrictions were put in place: gear and season restrictions were used to limit some fisheries, particularly lobster;[212] cod traps were prohibited on the northeast and south coasts in response to lack of fish and a decline in quality; and longlines were prohibited in parts of Placentia Bay.[213] Other than those minimal actions, which could be viewed as efforts to protect existing fisheries from competition rather than as conservation, it was mostly a free-for-all. Any problems associated with the fishery were thought to be of purely economic and social origin.

The total export value of cod increased to record levels by 1918,[214] then declined due to lowered prices until the early 1940s. Many outport fishing families were plagued with poverty as prices for their only products fell. World markets were changing, and there were thoughts that Newfoundland's fishery was outdated and that, to avoid destitution, fishermen had to produce more and different products.[215] A report on trade between Great Britain and the colonies issued in 1927 by the Imperial Economic Committee made it clear that Britain wanted fresh, not salt, fish. To drive the point home, the committee president addressed the Board of Trade in St. John's shortly afterwards, delivering the message that there was no future in salt fish, except in poor countries, and that those that could pay wanted their fish fresh.

Despite these sentiments, the demand for Newfoundland fish, even the poorest quality, was still very strong.[216] At the top end, Newfoundland's dry-cured cod had no peers in the best markets in Europe. Nevertheless, by the early 1900s, the skill and workmanship needed to produce the finest, lightly salted, dried cod was being lost,[217] to be replaced by the cheap and poor quality Labrador cure, which was more vulnerable to foreign competition. The Labrador trap fishery and the so-called *tal qual,* or ungraded, buying by American and Nova Scotia firms did nothing to encourage product quality, but despite the poor quality and price, the sloppy salt bulk fish was easily produced and sold.[218] Even if cheap and unprofitable, there was a lot of it – the fishery emphasizing quantity over quality. Producers watched as the old European markets for the finest dry-cured Newfoundland salt fish were forfeited to the competition in Norway and Iceland.

To replace traditional markets, the cheapest possible fish was produced for the Caribbean and Brazil.[219] From 1907 to 1932, the proportion of salt fish exported from Newfoundland to the West Indies, the poorest market, increased from about 6% to 25% of production.[220] Icelandic cod competed directly with Labrador cure in Europe and elsewhere, and Iceland was moving towards large-scale production methods.[221] Iceland's production of dried fish increased from about 15,000 tonnes in 1907 to 75,000 tonnes in 1930, with the new trawler fisheries increasing production by 50% from 1925 to 1930 at low cost.[222] In short order, Icelandic trawler fish began to replace Labrador trap fish in the markets of Italy and Spain. According to Macpherson's report in 1935, a European market still existed for the best quality fish at the best prices,[223] but Newfoundland forfeited that market to Iceland, which was by then producing better quality fish at lower prices. Newfoundland had missed an opportunity to develop better markets for its main product, while Iceland was preparing itself for independence from Denmark and laying its groundwork well.[224] The first calls for a change to trawlers in the Newfoundland fishery were heard.

TRAWLERS

In the last year of the 1800s, the first trawler operated by a Newfoundland company, Bowring Brothers, fished the Grand Banks, but failed to be profitable. Much later, Harvey & Company brought the trawler *Cape Agulhas* to Newfoundland where it fished from 1925 to 1930, again without great success.[225] The *Cape Agulhas* later became the first research vessel of the Newfoundland Fisheries Research Institute at Bay Bulls. Crosbie and Company purchased the larger trawler *Imperialist* in 1935, but it too fished without great success. Trawlers would not become profitable until the market for frozen fish overtook that for salt fish in the mid-1900s.

The advent of steam side trawlers introduced "anytime, anywhere" fishing, and required a reassessment of Huxley's view that the great sea fisheries were inexhaustible. Canada and Newfoundland called for a ban on the use of steam trawlers in 1911,[226] arguing that trawling threatened young fish and their habitat, and wrecked fixed gear. However, Britain was still in charge of the affairs of Newfoundland, and had great influence in Canada. British officials argued that the issue of trawling was complex and, in any event, banning any type of fishing outside the limited territorial waters would not be feasible. Nova Scotia proceeded to limit the number of trawlers in its waters nonetheless. This policy was reviewed in the 1930s, at which time it was concluded that it was naïve for Canada to limit its own trawling efforts when European nations were fishing without any restrictions. The Bates report of the 1940s recommended a reversal of the limited-trawling policy and strongly favoured trawling over traditional capture methods.

6.36 *Making fish in the early 1900s (Gordon King collection).*

6.37 *The Bay Bulls laboratory of the Newfoundland Fishery Research Institute in the 1930s (formerly fishing premises of Harvey and Company).*

The decline in salt fish quality hurt the Newfoundland and Labrador fishery. As a consequence, regulations on fish grading and quality were put in place in 1932. This was perhaps the first example of needlessly complex and confusing rules to be implemented in the management of the Newfoundland and Labrador fisheries.

> Other than the *tal qual* system, salt fish was traditionally sold by fisherman in three categories based on a visual examination of the dried fish: merchantable, Madeira, and cullage. The worst quality fish were cullage – these were fish that were "broken, slimy, or appeared to have too much salt; and all fish which, because they would not sell in other markets, were shipped to West Indian markets." Merchantable fish were "well-cured sound fish with an even surface, free of blood spots at nape and sound bone…perfectly split, with no salt showing on the face." Madeira fish were in between.[227]
>
> In 1932, there were at least 23 types of dried codfish exported from Newfoundland, and some standardization was needed. Classification was based on both the source of the fish and the curing method. There were four main categories and many subcategories[228]:
>
> 1. **Light-salted Shore and Medium-salted Bank Cod.**
>   The traditional shore-cured, air-dried, and light- to medium-salted dried fish (average salt content 19.9%).
>
>   - Choice or Number 1 Spanish: Sound quality cod, extra thick, light amber colour, even surface thoroughly clean on both back and face, not showing blood stains, clots, liver or gut; well split and not showing excessive salt on the face.
>   - Prime or Number 2 Spanish: With the same qualities as Number 1, but of fair thickness.
>   - Merchantable: As Number 1 and 2, but not thick.
>   - Madeira: Any cod not passing as choice, prime, or merchantable; rough in appearance.
>   - West Indian: Cod that is broken, sunburnt, slimy, dun, oversalted, or otherwise defective.
>   - Damp fish: All shore or bank cod must be thoroughly hard dried. Otherwise, they are classified as damp (e.g., damp merchantable or damp West Indian).
>   - Fish caught in Labrador or in the Straits, but light- or medium-salt cured as shore or bank codfish shall be described as Labrador Shore Cure or Straits Shore Cure and come under the same classifications.

- Size classifications for all shore and bank cod for sale are: Tomcods (8-11 inches, 20-28 centimetres); Small (11-18 inches, 28-46 centimetres); Medium (18-20 inches, 46-51 centimetres); and Large (over 20 inches, 51 centimetres).

2. **Genuine Labrador Soft Cure Codfish.**
   The heavier-salted fish from Labrador (average salt content 33.4%). The term "Genuine Labrador Soft Cure Codfish" shall be applied only to fish caught at and north of Blanc Sablon, off the coast of Labrador, and cured according to the following standards. All Genuine Labrador Soft Cure Codfish shall be dried sufficiently to stand export shipment without loss of weight.

- Choice or Number 1: Sound quality, well split, thoroughly clean; must show no excessive blood stains, clots, liver or gut; must be thoroughly salted and firm; if not quite white on the face, must have a clean, clear, even surface.
- Prime or Number 2: Badly split fish, extremely thin fish and fish showing excessive blood or liver clots; uneven surface, but in other respects similar to Number 1.
- Cullage: All fish not up to the standard of Number 2.

3. **Heavy-salted Soft Cure Newfoundland Codfish.**
   Cod caught off the shore of the island of Newfoundland or off the shores of Labrador south of Blanc Sablon, cured in Labrador style, shall be known as "Heavy-salted Soft Cure Newfoundland Codfish" and shall be purchased and sold as such, and culled under the same standards as applied to Genuine Labrador Soft Cure Codfish.

4. **Salt Bulk or Wet Salted Fish (from any location).**

- Number 1: Sound quality, well split, must show no excessive blood stains, clots, liver or gut: must be thoroughly salted and firm; must have a clear, even surface.
- Number 2: Similar to Number 1, but badly split.
- Cullage: Not up to the standard of Number 2.

The first Newfoundland venture into frozen fish was with salmon. In 1927, the Hudson's Bay Company acquired an interest in the St. John's firm of Job Brothers, and the large-scale export of frozen salmon began. It was the end of the pickled salmon fishery. Other St. John's firms quickly followed suit: over 1136 tonnes of frozen salmon were exported in 1929 and over 2000 tonnes in 1930.[229]

Whaling reappeared for a brief time in the late 1920s and early 1930s, largely in Labrador. The British-Norwegian Whaling Company established a factory at Grady in 1926, but it was closed by 1934. Most of the animals processed at Grady were fin whales. The Labrador Whaling and Manufacturing Company also operated during this period at Hawke Harbour, Labrador, and at Rose-au-Rue in Placentia Bay. Operations were reduced after 1930, although a few larger whales were still taken as late as 1950, and pilot whales were taken after this time to supply meat for mink ranches.[230]

By 1940, it was known that Greenland halibut were plentiful in the bays that had deep trenches, but absent in shallower and rockier bays. Little more was known of the biology of this species. Longline fisheries for Greenland halibut developed in Trinity, Green, and Fortune bays, with about 500 tonnes of barrelled, salted halibut shipped to the lumber camps of Newfoundland, Maine, New Brunswick, and Quebec.[231] Statistics on Greenland halibut landings are incomplete, but it is known that large fish weighing 20-25 pounds (9-11 kilograms) and average fish weighing 5-10 pounds (2-4.5 kilograms) were eaten fresh, smoked, or salted.[232] American plaice were not fished, despite being available in "great numbers," being of fine quality as a food fish, and increasing American demand. American plaice were often caught as bycatch in cod fisheries – as Frost related in 1940, "at present it is largely scorned as an article of food and probably millions of pounds are caught yearly by cod fishermen only to be thrown overboard again."[233]

THE RISE OF POST-WORLD WAR I EUROPEAN FISHERIES ON THE GRAND BANKS

The French fishery changed greatly in the early 1900s with the loss of the French Shore, continuing problems securing bait, and changing demands for fish products in Europe. The number of schooners at St. Pierre declined from 151 in 1904 to a single boat in 1915. The first French trawlers appeared on the Grand Banks in 1904, the same year that the Anglo-French Convention removed their shore fishery. By 1909, 32 French trawlers were on the Banks. Some of these were new vessels, while others had previously fished at Iceland. France was determined to increase their place in this fishery and offered a bounty of 10 francs per quintal for Banks fish. The war years interrupted the French Banks fishery, but it grew aggressively after that. In very short order, the French catch increased from 26,000 tonnes in 1918 to 157,000 tonnes in 1925.[234] The growth in the French trawler fisheries in both Newfoundland and Icelandic waters created additional competition for Newfoundland soft-cured fish in European markets.[235]

Spain and Portugal eyed the French initiatives with great interest, and following World War I, adopted a new strategy to supply their markets with fish. After centuries of being content for the most part to buy Newfoundland product, first from French and

English merchants and then from Newfoundland (as well as Iceland and Norway), Spain and Portugal decided to increase their own small fleets of trans-Atlantic longliners and side trawlers to exploit the Grand Banks. Portuguese cod catches from the Grand Banks increased from about 3000 tonnes in 1919 to nearly 20,000 tonnes by 1924.[236]

The depression of the 1930s brought political changes worldwide, many of which impacted the Newfoundland fisheries. Of particular note, the dictator António de Oliveira Salazar took control of Portugal in the late 1920s and his policies dominated until the revolution in 1974.[237] In the 1930s, Portugal embarked on a national policy to further expand its fleet.[238] The result was the famous White Fleet that reached its zenith in the 1940s and 1950s. Most were whitewashed three- or four-masted combination steam and sail vessels that fished with dories, with upwards of 50 dories to a ship. In later years, the White Fleet was accompanied by a hospital ship. Conditions on board the *bacalhoeiros* – the fishing boats – were poor, with little regard for the well being of their fishermen.[239] From the late 1940s until the early 1970s, the vessels of the White Fleet made St. John's their port of call and could be seen tied up three or four deep in the harbour.[240] The Portuguese used a one-man-to-one-dory system, fishing with traditional handlines, and many were undoubtedly lost in the fogs of the Grand Banks. It is worth pointing out that in contrast to other foreign fleets, especially trawlers, this fleet was much beloved in Newfoundland. Despite the different languages and cultures, there was a universal language of fish and fishermen, especially poor fishermen, and a respect for the sea and hard way of life catching cod by hand. Portuguese soccer games were cheered on the St. John's waterfront and are still remembered by many Newfoundlanders.

British migratory fishing vessels came to Newfoundland for the last time in 1845. Deprived of the Newfoundland fishery for the first time in 300 years, English merchants reverted to their old Icelandic trade with a new fleet of trawlers. By 1901, British trawlers had become "a serious nuisance in Icelandic fishing grounds."[241] Denmark controlled Iceland in the early 1900s and negotiated a three-mile limit with Britain in 1901.[242] European-caught fish was in direct competition with Newfoundland salt cod in the European markets, and at first the impact of these expanding European fisheries was mostly economic. Neither Iceland nor Newfoundland raised very strong objections to the foreign fisheries until much later in the 1900s. The reason for this was likely the belief that the offshore fishing grounds of Iceland and Newfoundland were essentially inexhaustible, the lessons of early science having no firm hold.[243] As these new fisheries began, harvest rates were low and for a time there was little biological impact on the stocks. Nevertheless, the rebirth of a trans-Atlantic fishery foreshadowed what was to come after the Second World War, when vast fleets of European trawlers would invade almost every part of the North Atlantic.

For Newfoundland, the increases in foreign fleets exacerbated the loss of the more lucrative European markets and the selling of its fish in Brazil and the Caribbean.[244] More serious damage would follow shortly, and be done to the fish themselves and the ecosystems in which they lived. Iceland stopped the destruction in time to save its stocks, but Newfoundland could not. The European trawler fisheries would change the fish and marine life of the Grand Banks and continental shelf almost beyond recognition, and bring a virtual end to cod stocks that had existed for 10,000 years and a fishery that had been sustained for over 400 years. But I am getting ahead of the story.

POST-WORLD WAR I SCIENCE IN NEWFOUNDLAND

After World War I, fishery management was largely restricted to market controls and attempts to manipulate the economics. There was little investigation of the fish stocks or the ocean. In 1921, as a follow-up to Canadian-U.S. fisheries initiatives, James Davies was sent from St. John's to the first meeting of the International Committee on Deep Sea Fisheries that included representatives from the United States, Canada, and Newfoundland. The annual report of the newly established Department of Marine and Fisheries under Minister Coaker stated that it was "humiliating that Mr. Davies representing the oldest fisheries in the New World was not possessed of any information of a scientific or hydrographic nature."[245] The next meeting occurred in 1926, but Newfoundland did not send a representative.

Memorial University College was established in St. John's in 1925. J.L. Paton, its first president, was bullish on the need for fisheries research and highly critical of Newfoundland's lacklustre approach to fisheries. In a moment of exasperation about the lack of science and knowledge, he fumed: "No pisciculture – no stations for marine biology. No understanding of the life-story of the cod. Troubled with bait – no study of the question. No guidance. All rule of thumb. As it was in the beginning so it is now. Same as to drying and curing of fish – they tell me the Norwegians are far ahead of us, but we go on in the same old rut."[246]

In 1926, Paton established the Department of Biology at Memorial with fisheries as its focus. He hired George Sleggs, an English fisheries scientist who had been doing a doctoral degree at the Scripps Oceanographic Institute in California, as its first professor. Sleggs developed the first fisheries lab at the College and was also seconded to the Newfoundland Department of Marine and Fisheries. His first research was on the distribution and migration patterns of capelin and cod, and he conducted the first drift bottle experiments in Newfoundland waters to test surface current patterns and their influence on fish distribution. His monograph on Newfoundland capelin was the first of its kind and remains a valuable reference.[247]

Paton furthered his attempts to establish fisheries science in Newfoundland on two fronts: he attempted but failed to get funding to endow a chair in Memorial's

Department of Biology, then pushed for a government fisheries laboratory in Newfoundland, but by 1929 had failed to raise the necessary funds. Paton was above all a pragmatist, and was on good terms with A.G. Huntsman, the noted Canadian marine scientist. Together they sought funds for a joint Canada-Newfoundland laboratory, but the political support in Newfoundland for this initiative was weak. In a letter to Huntsman in 1929, Paton was livid about comments made in the Newfoundland legislature about Sleggs' early research. With respect to the oceanographic research practice of using drifters to track surface currents, a method commonly employed in Europe and the United States, a member who refused to agree to funding asked, "What's the good of paying a man to be chucking bottles into the sea?"[248]

The Imperial Economic Committee suggested that a Canadian laboratory might take on Newfoundland projects as co-operative efforts.[249] Paton realized that support for an independent laboratory in Newfoundland was weak, so he agreed and proposed establishing a laboratory under the direction of the Canadian Fisheries Research Station in Halifax. But even this was not to be: Newfoundland would put forth no funds and Canada did not want to support Newfoundland research. Paton despaired of the conservative attitudes of the Newfoundland fishery, the lack of willingness to change, and the resistance to science, but he never shed the opinion that the fishery could be the basis of a prosperous economy.[250]

In 1930, with financial assistance from the British Empire Marketing Board until 1935, another Newfoundland Fishery Commission was appointed in which Paton was a member. The Commission hired Scottish scientist Harold Thompson to undertake a survey of the fisheries. Thompson insisted he needed a laboratory and a research vessel; with Paton's backing, he got both in 1931 in an arrangement with Harvey & Company at their premises in Bay Bulls. The vessel leased from Harvey & Company was the former side trawler RV *Cape Agulhas*. Thompson's research program had three parts: a survey of Newfoundland's fisheries resources, including life histories and stock assessment statistics; fish processing; and utilization of fish products. This work began in earnest in the early 1930s at the Bay Bulls laboratory.

The 1930s finally brought fisheries science to Newfoundland. The Bay Bulls laboratory was purchased from Harvey & Company by the Newfoundland government and put under the Department of Natural Resources. The first Newfoundland fisheries scientists were trained by a Memorial University College firmly committed to the fisheries. Nancy Frost became one of the first Memorial University College students and Newfoundland fisheries scientists in the late 1920s. Bonavista-born Wilfred Templeman was offered a Canadian Fisheries Research Board scholarship after completing studies at Memorial University College and Dalhousie University.[251]

WILFRED TEMPLEMAN, 1908-1990

Wilfred Templeman was born in 1908 in Bonavista, Newfoundland, and attended primary school there before completing his senior matriculation at the Methodist College in St. John's. He became one of the first students in biology at Memorial University College in St. John's which he attended for two years, as was the custom of the times. He completed a BSc degree (with distinction) at Dalhousie University in Halifax, Nova Scotia, in 1930. Upon graduation, he was awarded a Canada Fisheries Research Board graduate scholarship which he took up at the University of Toronto under Dr. A.G. Huntsman. Templeman graduated with an MSc in 1931 and PhD in 1933 at the age of 25.

Dr. Templeman had a distinguished career as a fisheries scientist. His first job was as a lecturer at McGill University, a position he held from 1933-1936. At the urging of J.L. Paton, the first president of Memorial University College, Templeman returned to Newfoundland, first to head up the Department of Biology at Memorial, which focused on fisheries biology. He was made a full professor in 1943, and then became the director of the Newfoundland Government Fisheries Laboratory, which evolved into the Biological Station of the Fisheries Research Board of Canada after Confederation with Canada in 1949. Dr. Templeman was director of both laboratories until his official retirement in 1972; as such, he oversaw the transfer of authority over the fisheries from Newfoundland to Canada. His achievements were many, and he maintained his attachment and commitment to fisheries science at Memorial throughout this period, holding a visiting professorship from 1957 until his retirement and remaining as J.L Paton Professor of Marine Biology and Fisheries from 1972-1982. Perhaps his greatest disappointment was not being able to prevent the decline in the great cod stocks in the 1960s and 1970s. From 1983 until his death in 1990, he continued his work at the newly formed Northwest Atlantic Fisheries Research Centre in St. John's, under the Department of Fisheries and Oceans, which succeeded the Canadian Fisheries Research Board.

During his long and distinguished career, there were few subjects in marine fisheries to which Dr. Templeman did not contribute. His work and publications were legion, and included more than 250 scientific papers. His work was marked by a combination of scientifically derived knowledge and hard-won wisdom about the marine ecosystems and fisheries of Newfoundland and Labrador, and was rooted in boyhood observations from the wharfs of Bonavista. Templeman's publications included seminal works on species as diverse as lobsters, salmon,

> capelin, and haddock, as well as papers on rare fish species, seabirds, and oceanographic influences on fish. Although cited as an authority on many fisheries topics, Templeman is best known for his work on groundfish, in particular Atlantic cod. His bulletin titled "Marine Resources of Newfoundland" will remain an all-time classic scientific work on the Newfoundland fisheries. In recognition of his many contributions to marine and fisheries science, Dr. Templeman was awarded the Order of the British Empire in 1948 and became a Fellow of the Royal Society of Canada in 1950. He received an Honorary DSc degree from Memorial University of Newfoundland in 1982 and a research vessel took his name.
>
> I first encountered Dr. Templeman's work at McGill, where my PhD dissertation largely consisted of exploring and testing several theories that he had put forward decades earlier on cod migrations and movements in coastal waters. Like him, I returned to my birthplace in Newfoundland to take a job in the fisheries. Little did I know that my first office would be just down the hall from the man people called "Old Temp." Although in his 80s and long retired, "Temp" worked every day, even on weekends, more often than not clad comfortably in slippers.
>
> Dr. Wilfred Templeman was one of the most accomplished of a generation of natural scientists. He was the father of fisheries science in Newfoundland and Labrador. I shall remember him best as "Old Temp," a brilliant and humble man, who possessed an unparalleled knowledge and interest in the fisheries of Newfoundland and Labrador.

Paton hired Templeman to succeed George Sleggs as head of Memorial's Department of Biology in 1936, further strengthening fisheries science.[252] The future looked bright. Regrettably, in the same year, Harold Thompson's contract at the Bay Bulls laboratory expired and he was offered only a year-to-year extension. Thompson subsequently went to Australia to head the fisheries section of the Council for Scientific and Industrial Research in New South Wales. It was almost an omen: on April 19, 1937, the Bay Bulls laboratory burned to the ground, with great loss of material and information. Fortunately, Thompson had taken some of his research results, including monumental works on cod and haddock, to Australia, which he later published with help from both Newfoundland and Australia. These publications remain essential references to the Newfoundland fisheries.[253]

HAROLD THOMPSON'S 1943 PUBLICATION OF "A BIOLOGICAL AND ECONOMIC STUDY OF COD (*GADUS CALLARIAS*, L.) IN THE NEWFOUNDLAND AREA INCLUDING LABRADOR."

Harold Thompson wrote his epic work on Newfoundland cod while serving as Director of Fisheries in Australia. It summarized his five years of research using the RV *Cape Agulhas* from 1931-1935 and included the work of scientists under his charge in Newfoundland, especially Nancy Frost. It was published by the Newfoundland Department of Natural Resources in 1943. He advanced the knowledge of cod biology and ecology markedly. Among his conclusions were the following:

1. Based on migrations of tagged cod, racial characteristics, size and growth, and age class composition, "there is substantial basis for recognizing local sub-divisions of the total cod stocks."

2. Based on tagging, young cod "are comparatively stationary, there is no evidence of mass migration out of the region in which the period of adolescence is passed," but "after maturity is reached, there appears to be increased migration and intermingling."

3. Based on tagging, "no great degree of interchange of cod between the Banks and the coastal area was demonstrated." There were exceptions to this, including "from Grand Bank[s] and St. Pierre Bank to the coast." Somewhat in contradiction to this, "interchange appears to be possible between all parts of the cold Labrador and eastern Newfoundland regions," but not between the Gulf of St. Lawrence, the northeast coast, and Labrador.

4. St. Pierre Bank is a mixing area of various stock components.

5. "The majority of cod in the Newfoundland area attain first maturity between the sixth and the ninth years of life, at a size of from 60 to 80 [centimetres]. In this respect they somewhat resemble Iceland and Norwegian cod, and differ from cod of warmer seas – e.g. the North Sea – where maturity occurs at an earlier age than the above. Most spawning occurs in spring, but the spawning season can extend from March to October."

6. "The distribution of the eggs and fry of cod varies according to the type of season – i.e., whether or not the Arctic current is of normal, or of greater or less than normal, strength. The bulk of the eggs and fry occur off the east coast of Labrador and Newfoundland and on the [Grand Banks], the center of distribution being rather more northerly in warmer years, and more southerly in colder years."

> 7. The temperatures where the fishery is best vary by region. Catches can be good in sub-zero temperatures in northern regions, but not to the south. Trap catches are better in cold years. Water temperatures where cod have been found in Newfoundland range from nearly -2°C to 12°C. Spawning may take place in waters with temperatures of 1.5°C-7°C, but mostly occurs in waters with temperatures between 3°C-5°C.
> 8. Food supply is better on the Grand Banks than off Labrador or in the Gulf of St. Lawrence.
> 9. Trawling on the Grand Banks can be more successful in a four-month period than in any other known region of cod fishing.

Prior to the fire, the Bay Bulls laboratory and the RV *Cape Agulhas* were seldom at rest. By the mid-1930s, the work of key scientists, including Harold Thompson, N.L. Macpherson, Nancy Frost, Sheila Taylor Lindsay, Anna Wilson, W.F. Hampton, and Wilfred Templeman, was being published in many forms, including Service Bulletins for the fishing industry, and Economic and Biological Research Bulletins of the Newfoundland government.[254] The prestigious British journal *Nature* published its first Newfoundland-based fisheries science paper in 1933 titled "Vitamin A Concentration of Cod Liver Oil Correlated with Age of Cod," authored by N.L. Macpherson.[255] According to the Amulree Royal Commission of 1933, the station was "recognized both in Canada and the United States as a leading authority on the deep sea fishery on the Western North Atlantic. Its potential importance to the industry can hardly be exaggerated."[256]

NANCY FROST

Nancy Frost entered Memorial University College in 1926. She was one of the first students to be awarded a scholarship at Memorial, followed by a full scholarship to finish her degree at Acadia University in Nova Scotia. After completing a Master of Arts, she returned to Newfoundland to take a position at the Bay Bulls laboratory and was a pioneer in fisheries science in Newfoundland. Her early papers covered a wide range of topics from plankton to salmon to shrimp to cod. She provided the first descriptions of the early life history of several species and wrote the first Service Reports that attempted to describe the biology of many commercial fishes to industry and fishermen. Frost was also a remarkable artist and illustrator. She was the

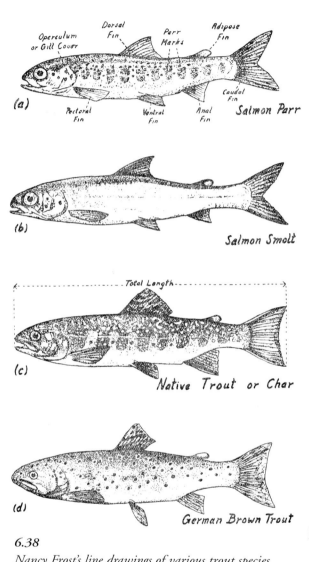

6.38
Nancy Frost's line drawings of various trout species (Frost 1938a).

first to use plankton species to index oceanographic changes in the marine ecosystems of Newfoundland – the very earliest use of what would much later be called an ecosystem approach. In 1938, she wrote:[257] "The first requisite towards the intelligent exploitation of any fishery is a knowledge of the life histories of the fishes concerned. The greater the intensity of the fishing in any area the greater the necessity for adequate fishing Regulations which will protect immature and breeding fish…the quantities of cod on the Newfoundland banks are such that there is probably little fear of depleting the stocks at the present rate of fishing. The time may come, however, when it is found that even our stock of cod, if misused, will not prove inexhaustible."

Frost's instincts about conservation did not stop at the tidewater. She also studied Newfoundland's freshwater systems, expressing concern about the decline in trout populations and the need for involvement of fishermen in conservation:[258] "There is the necessity of pointing out to the angler how urgently his co-operation is needed if the trout fishery in this country is to be preserved. It seems to be generally agreed that a large number of ponds, particularly near St. John's, are becoming rapidly 'fished out'…it can be prevented to a large extent through the immediate co-operation of the anglers."

A committee was struck in 1937 after the Bay Bulls fire to provide recommendations on the future of fisheries research in Newfoundland. In consultation with the public, the committee was tasked with deciding whether or not fisheries research should be continued, and if so, how could it best work with other scientific or educational organizations and what type of facilities should be provided.[259] The committee unanimously recommended the continuation of fisheries research in Newfoundland, reasoning that "investigational studies firstly of the natural history of marine species, and of the effects of human depredations thereupon; secondly of the physical conditions in the oceans and of the effects of variations in such conditions upon the occurrence and abundance of sea life; and thirdly of the preparation and preservation of marine products, need no justification in our modern world…[because] the importance of the sea fisheries in the economic framework of Newfoundland is obvious."[260] The committee recommended that a fisheries institute be established in St. John's, along the lines initially conceived by Thompson in 1930, to provide "intensive experimental, consultative and educational aspects of fisheries work."[261]

In 1940, space was secured for the remaining fisheries researchers from the Bay Bulls laboratory in the new Newfoundland government laboratory on Water Street in St. John's. William F. Hampton was acting director until 1943, when Wilfred Templeman became its first permanent director. New scientists were recruited during the early years, including Arthur Blair, who worked on salmon, and Noel Tibbo, who worked on herring. A new research vessel, the 82-foot (25-metre) *Investigator II*,[262] was commissioned in 1946. Much of the work at the laboratory in the late 1940s was directed at mapping the distributions of haddock, cod, redfish, and American plaice, the most valuable commercial species. Ironically, this pioneering work would shortly be misused by foreign nations to fish out those very stocks.

6.39
The Water Street laboratory of the Newfoundland Biological Station in the 1940s.

COMMISSION OF GOVERNMENT, FISHERIES POLICY, AND CONFEDERATION

In the late 1920s and early 1930s, Newfoundland's economy was still reeling from the debts incurred in World War I. By 1930, a full half of annual government revenues went to pay interest on the debt, 40% of which was incurred to send the Newfoundland Regiment to fight in France.[263] To make matters worse, the 1930s brought the Great Depression that gripped the western world. The price of salt fish fell to 1800s levels, and the value of the fisheries plummeted from more than C$10 million in 1929 to C$3 million in 1931. Although the cod stocks and catches were increasing and environmental conditions continued to improve, the fishery was in a state of crisis.

Salt cod and seals, the two main economic pillars that had enabled permanent settlement and at least semi-independence for nearly 100 years, were no longer of sufficient value to sustain the struggling nation. With the failure of industrialization to positively impact the economy,[264] the lack of attention to the development of the fisheries had come home to roost,[265] despite the best efforts of Paton and the Newfoundland Fisheries Commission. Poverty reigned, as it did in much of the world, and by 1932, with a debt of nearly C$100 million, the possibility of national bankruptcy loomed large. As Richard Gwyn pointed out, "a government of geniuses could not have prevented it."[266]

The fisheries featured prominently in a pre-election manifesto written by Frederick Alderdice, a politician who defeated the government of Richard Squires in the election of 1932. Alderdice promised strong action to revive the fisheries, "still the most important industry to the country." He further stated that "there are many who profess to regard the fishery as a thing of the past, who assume it is no longer possible to make it a paying proposition, or that it holds out any attraction to our industrious young people. I am not one of those."[267] But strong action of any sort was difficult for a nearly bankrupt nation.

With bankruptcy looming, Britain appointed the Amulree Royal Commission to recommend on the future of Newfoundland.[268] The Commission was comprised of Lord Amulree, assisted by two Canadians, C.A.Magrath and Sir W. Stewart, and an English civil servant, P.A. Clutterbuck.[269] In 1933, after consultations in Newfoundland, Canada, and Britain, the Amulree Commission produced what Alexander termed "a devastating document":[270] it determined that Newfoundland could not pay its debts, blamed corrupt and incompetent politicians for much of the mess,[271] and recommended that responsible government be suspended and the country governed by a new commission, without an election on the matter. In a dramatic move, Newfoundland complied, relinquishing its representative legislative and independent status, and opting to be ruled by a Commission of Government under Great Britain in

1934. In reality, it was a return to colonial rule. In theory, this arrangement was to last only until Newfoundland got back on its feet financially – assisted by Britain – and only until the people requested a return to responsible government.[272] In reality, it lasted for 15 years and resulted in Confederation with Canada.

The Commission of Government brought about many administrative changes with respect to the fisheries. Alderdice's manifesto may not have been forgotten. Under the *Salt Codfish Act* of 1933, the industry and its marketing system were overhauled. The Ministry of Marine and Fisheries (absorbed by the Department of Natural Resources in 1935) would oversee a Saltfish Board to control exports. The Board had inspectors and cullers, trained at the Bay Bulls Research laboratory, working in 11 fishery districts around the island in an attempt to standardize and improve the curing of the fish. Bait depots became better organized. These were difficult years for the market, as low prices for Newfoundland fish products meant that little cash was passed down to fisheries families.[273]

QUOTES FROM THE AMULREE REPORT, 1933

Newfoundland "has always been, first and foremost, a fishing country." (Chapter V, Item 198)

"The fisheries of the island…measured by the total annual catch on all parts of the coast over a long period of years, is unsurpassed in any country of the world." (Chapter VI, Item 247)

"Had the fisheries been placed in the past under the control of an independent Board or Commission, composed of disinterested persons and free from political interference, it is conceivable that their record would have been one of continuous progress and development instead of stagnation. In practice, however, the control of the fisheries has always been political." (Chapter VI)

"Those responsible for the conduct of the industry have dissipated their energies in jealousy and intrigue instead of concentrating on the development of the fisheries on rational and scientific lines…while successive Governments, embarking on ambitious schemes of industrial expansion and neglecting the fisheries except in so far as they impinged on the immediate political horizon, allowed the industry to drift. In the years since the War, loans were raised amounting to [C]$50,000,000…of this formidable total, less than [C]$1,000,000 was spent on the stimulation of the fisheries." (Chapter VI, Item 303)

The Commission of Government that governed Newfoundland after 1933 lasted longer than originally intended, prolonged by World War II. At the end of the war, it was clear that Newfoundland would either have to regain independence or join another country (Canada or the United States).[274] There were camps in favour of all three options, much campaigning and counter-campaigning, and allegations of subterfuge and secretive scheming by Britain and Canada that persist to the present day.

During the Second World War years, air and naval bases holding both Canadian and American military personnel were established at several sites in Newfoundland and Labrador. Britain made all the arrangements. For the first time, many Newfoundlanders got to know people from both countries, and many Newfoundland women married American servicemen. As Richard Gwyn wrote, "In contrast to the generous, gregarious Americans, Canadian servicemen tended to be aloof and patronizing; they poked fun at the island accent and coined the hated nickname, Newfies."[275] Union with the United States was unlikely, however: there was too much political baggage with the British and Canadians. Despite the obvious and historical trade relationship with New England, the United States, if interested at all, was cautious. On the other hand, Newfoundland and Canada shared a British heritage, something the United States had long since rejected.

Newfoundland had been using Canadian banks and currency since shortly after the bank crash of 1894. The Bank of Nova Scotia, the Bank of Montreal, and the Merchants Bank of Halifax (which later became the Royal Bank of Canada) were all in business in Newfoundland by early 1895. With these economic ties, it is not surprising that the strongest opinions favoured either independence or confederation with Canada. However, opinions on the matter were split both geographically and socially: St. John's and the Avalon Peninsula wanted independence, while the outports, strongly influenced by Joey Smallwood's promises of baby bonuses and other payments in cash from Canada, favoured confederation with Canada. The support of the outports for Smallwood is ironic, as he had little interest in the fishery, their only industry. In a famous refrain, he advocated that outport people "burn your boats."[276] Smallwood was almost perversely disinterested in the fisheries and firmly convinced that the future of Newfoundland and Labrador was with land-based industries.[277] During the National Convention debates about Newfoundland's future in 1946, he was accused of deliberately berating Newfoundland's past and its poverty, and by implication the fishery, for political ends.[278] With Britain and Canada maneuvering on the sidelines, a second referendum went in favour of confederation with Canada by less than two percentage points.

In April 1949, Newfoundland became the tenth province of Canada. The draft terms of union contained "no special fisheries clause,"[279] and the final terms contained only a short passage on the 450-year-old fishery. There was as much attention to the sale of margarine.[280] Newfoundland had forfeited the fishery, its main industry: Newfoundland's Fisheries Board would retain control of the fishery for five years, after which both the fish and the responsibility to manage the fishery would be given exclusively to Canada. The wisdom of a jurisdiction giving up any say in the conduct and management of its principal industry has puzzled many since.

The influence of Canadian fisheries policy in the 1940s was to be felt in the Newfoundland fishery well after Confederation. A seminal report in 1944 by then Nova Scotian Deputy Minister of Fisheries Stewart Bates, a Scottish economist who went on to influential posts in the federal government, recommended the economic rationalization of the fishery.[281] Bates was one of a cadre of new and highly educated bureaucrats who came to run Canada in the post-war years. So-called "experts" controlled the brave new Canadian bureaucracy, and the views of fishermen and industry were seldom considered relevant. The fishing culture and traditions of Atlantic Canada could not withstand their critical economic gaze.

Bates' recommendations stressed modernization of the fishery. According to his thinking, markets were a problem, but so was the low productivity of the fishermen.[282] Bates decried the use of schooners and dories as an outmoded and stagnant relic of backward peoples selling product to an even more backward salt fish market.[283] To Bates, the future was in trawlers and frozen product sold to American markets. He suggested that the salt fish industry was a hindrance to economic development, and that without its demise, the Atlantic fishery could never join the mainstream North American economy. Bates had little time for romantic notions of the fishery and seafaring as a way of life. He wrote sardonically that for "fishermen…seafaring is necessary, living is not necessary."[284] He also stressed the need for education of the generally unschooled and often illiterate fishing families.

Bates' policies might have worked – his views were not so different from those in Iceland. In his monumental publication on Newfoundland cod in 1943, Harold Thompson stated that "it is through the moderate development of trawling [a more modern fishing method] that progress in Newfoundland fisheries is likely to be made."[285] Few could quarrel, and fewer in hindsight would quarrel, with the notion of modernization. However, modernization depended on labour being easily transferable from one industry to another, and this proved to be difficult in Newfoundland.[286] In addition, a mainstay of modernization policy is a program to develop and protect the traditional industry [and in this case the fish] as it transitions to a more modernized condition. Iceland implemented full measures to do this and modernization worked.[287] Newfoundland and Canada, on the other hand, settled for half measures, and the modernization policy would soon fail.

Newfoundland Associated Fish Exporters Limited (NAFEL)

Several organizations, including with the Salt Fish Exportation Board and various co-operative initiatives, worked to implement the Amulree Royal Commision recommendations regarding the marketing of salt cod to several countries, especially Portugal and Spain.[288] In 1947, the Newfoundland Fisheries Board created the Newfoundland Associated Fish Exporters Limited (NAFEL). Under the elected chairmanship of Dr. Raymond Gushue, NAFEL was granted exclusive rights to export fish from Newfoundland. Membership in NAFEL cost C$10,000 and entitled members to the complete services of NAFEL, which included overseas market negotiations, import-export dealings, currency transactions, and payment to the exporter for the salt fish product.

NAFEL kept track of salt fish production and inventories in Newfoundland and Labrador, and attempted to maximize markets and sales of the fish to the benefit of the members. NAFEL wasn't entirely without its problems. Some members attempted to circumvent NAFEL's regulations and deal secretly with overseas clients themselves, but overall this was an uncommon occurrence.[289] There were criticisms that NAFEL perpetuated the old *tal qual* buying system, with its attendant quality problems, and that it legitimized the middleman power of the merchants with their questionable tactics. Some of these criticisms were undoubtedly valid. In bankrupt 1940s Newfoundland, however, with another World War to fight and a highly conservative colonial political regime "administered by civil servants from a withering British Empire,"[290] NAFEL might have evolved into a progressive organization. Alexander concluded that, all in all, "if they had been left to their own devices, the Newfoundland Fisheries Board and the trade might have evolved still further the capitalist reorganization that they had begun in 1935. Yet even as it was, the system that existed in Newfoundland in 1947 was far superior to anything [else] available to Canada's east coast fishery."[291]

THE STATE OF THE OCEAN FISHERIES AT CONFEDERATION

The warming ocean conditions of the 1940s fostered increases in the stocks of cod and other species. The reduction of fishing during World War II lowered removal rates and added further growth to the stocks. By the late 1940s, strong production in cod and haddock led to expanding numbers of large fish on the Grand Banks and adjacent areas of the continental shelf. Cod stocks were strong in all regions, especially to the north off Labrador where the long-protected and largest spawning aggregations occurred. Haddock on the Grand Banks were abundant, perhaps near a million tonnes. Redfish and American plaice stocks were virtually untouched, and the vital capelin – the most important food for all the other fishes, seabirds, and whales – were likely at all-time highs. The warming conditions led to changing migration patterns of some southern species, which in turn led to new fisheries. The ecosystem was highly productive, with an abundance of many species.

Notions of modernization of the fisheries were based on the assumption that the fish would continue to be abundant. Lobsters, whales, seals, and salmon were already overharvested by the 1930s and 1940s, but there was no opinion that cod were being overfished. Harold Thompson, the leading authority of the day, stated that "there is not the slightest sign of overfishing in Newfoundland waters, which are probably capable of producing a higher average annual yield than is exacted from them."[292] Hindsight suggests that, in the 1940s, he was right.

Catch declines combined with difficulties experienced in Norway and Iceland, the other major supply regions, created a shortage of salt fish during the years of World War II. Prices went through the roof. The total value of Newfoundland cod tripled in a decade and reached record highs by the late 1940s. High stock levels and high prices should have meant prosperity for Newfoundland, but the fishery failed to provide a reasonable standard of living for fishermen and their families.[293] This failure reinforced the suggestion that the fishery was a dead end, an idea that first surfaced in the 1880s, giving it traction and recharged influence in political discussions of Newfoundland's future.

In the years after World War II, while both Newfoundland and Canada were downplaying the potential of the fisheries, distant countries – among them some of Newfoundland's traditional competitors and markets – were doing just the opposite. In the 1940s, a million tonnes of haddock schooled on the Grand Banks, and Spanish pair trawlers began fishing for them soon after they found their spawning and feeding grounds. At first the fishery was small, but the growing fleet of Newfoundland trawlers joined the Spanish after World War II, and by 1949, total annual landings of haddock measured about 179,000 tonnes.[294]

At the same time, almost 7 million tonnes of cod coursed the banks and bays of Newfoundland and Labrador, sustained by at least 10 million tonnes of capelin. There were more than 4 million tonnes of redfish – a Russian favourite which had never been fished before – and an abundance of American plaice, Greenland halibut, flounder, mackerel, and squid.

THE SECOND EAST COAST FISHERIES CONFERENCE, Québec City, February 1948

Under the auspices of the Minister of Fisheries of the Province of Quebec, in co-operation with the Fisheries Council of Canada, a conference was held to discuss problems in the fishing industry in 1948. Newfoundland was not yet a province of Canada, but attended the meeting. Below are several quotes from participants at the conference:[295]

> "The French Government…is building trawlers…to produce this European self-sufficiency in fish…If the European countries do establish a fishery of these dimensions, the repercussions so far as Canada is concerned will be quite serious, also in Newfoundland."
> (C.J. Morrow, Chair)

> "All of the European producing countries are building ships." (F.A.J. Laws, NAFEL, Newfoundland)

> "Let us not forget, gentlemen, that our territory is actually the fishing field of many European countries which, after having depleted the resources of the North Sea and some areas of the North West Atlantic, are nearing our coasts with highly mechanized means of mass production…if we do not immediately remedy the problem, there will come a day when that immense industry, the Atlantic Fisheries, will not be sufficient to nourish the population…Is it not the proper time to state and proclaim that only North American countries will have the right to fish within a certain distance of the coast?" (C.E. Pouliot, Minister of Fisheries, Quebec)

> "If there were a failure of fisheries in…Canada or the United States, it would hit very hard…in Newfoundland, it would be a national disaster. The majority of the people in Newfoundland make their living from fisheries – they are steeped in fisheries." (Dr. R. Gushue, Chair of Newfoundland Fisheries Board)

PROTECTING THE CONTINENTAL SHELF ECOSYSTEM

The Icelandic fish stocks had been given a respite during World War II, but fishing effort greatly increased after it was over. From 1945 to 1949, Iceland doubled its tonnage of fishing vessels and foreign vessels returned to its cod grounds. Catches quickly declined under the pressure of the increased fishery.[296] The outdated notion that the major cod stocks were inexhaustible and could withstand nearly any level of fishing was losing credibility, while the idea that fishing grounds belonged to the coastal state was gaining credibility. In 1945, the United States declared that its continental shelf was to be put under American jurisdiction and might be limited to American fishing. In 1948, Iceland enacted the *Law on the Scientific Protection of Fishing Grounds on the Shelf of Iceland* whereby regulations on fishing on its continental shelf would be made unilaterally by Iceland. In the next year, the 1901 treaty that allowed Britain to fish within three miles of the Icelandic coast was withdrawn. The new Icelandic law was used as the legal authority for all subsequent extensions of jurisdiction and the coming cod wars with Britain in the 1950s. The stage was set. A proverb in Iceland stated, "When all is said and done, life is first and foremost salt fish."[297] This sentiment, that the fish were all important, was put into action. Icelandic policy recognized that the fishery was its future, and it meant business: the new law announced to the world that Iceland belonged to Icelanders, as did its continental shelf and its fisheries.[298] The law also enabled unilateral protection of the full continental shelf ecosystem.

In the waters of Newfoundland, a different approach emerged. In 1948, the same year that Iceland passed its law on protection of its fishing grounds, planning was underway for the first meeting of the multilateral International Commission of the Northwest Atlantic Fisheries (ICNAF), to be held in St. John's. ICNAF would regulate the Newfoundland and Atlantic Canadian fisheries, and the Newfoundland-Canadian plan for the fishery was to share it with almost any comers.

The incursion of new fleets of European vessels after World War II was bad news for all northwest Atlantic fish and fisheries, but much worse was to come. In the late 1940s and early 1950s, hundreds of long-distance, factory-freezer trawlers were on the drawing boards of shipyards first in England, then France, Germany, Spain, and Portugal. Some clever industrial espionage led to Russia pirating the English design, and German shipyards became busy building a fleet hundreds strong for Russia and its allies. Their main target would be the Grand Banks of Newfoundland.

As the ink dried on the Confederation documents in St. John's and Ottawa, with Newfoundland becoming Canada's tenth and most easterly province,[299] the burgeoning stocks of cod, haddock, redfish, and capelin,[300] thought by some to be of limited importance to the future of the region, were without any protection. They were about to be assaulted by an onslaught the likes of which could hardly be imagined.

6.40 *An abundance of cod in the early 1950s. The traditional inshore fishery would soon be nearly eradicated by the onslaught of foreign trawlers from then Eastern Block countries and Western Europe (Gordon King collection).*

6.41 Large cod were common in the late 1940s and early 1950s. This one appears to be at least 1.5 m in length and likely weighed over 25 kg. It was caught near Random Island in Trinity Bay. Such large and old fish were an essential component of the reproductive potential and social behaviour of cod stocks. The lack of such fish in contemporary stocks limits their potential and is a hallmark of their current depressed state (Tom Clenche collection).

NOTES

[1] Many historians and politicians have referred to Newfoundland as Britain's oldest colony, and considered it to be so from the early 1600s, but true settlement, which the term colony implies, did not take place until the mid- to late 1700s.

[2] Prowse (1895, 404) stated that "eleven thousand Irish came to St. John's between 1814 and 1815. Their sufferings in crossing must have been terrible, only exceeded by the horrors of the middle passage on board an African slaver." This is an incredibly high immigration rate. The resident population of the entire colony in the early 1800s was only about 20,000 people.

[3] Mannion (1977, 7) wrote: "In terms of the growth of a permanent population, and the expansion of sedentary settlement, the actual volume of the nineteenth-century migrations was not as important in the long run as the modified sex ratio of the migrants. In contrast to the predominance of single adult males and boys in the eighteenth-century seasonal and temporary migrations, the subsequent influx was marked by an increase in the number of women and even young children, although their percentage was still exceedingly low. However, the resulting increase in the availability of female spouses had a profound impact on natural demographic growth in Newfoundland through the nineteenth century."

[4] Mannion (1977, 12) stated of the population increases of the nineteenth century: "Labour in the fishery rose almost in proportion to general population growth until the 1880s when incipient industrialization and out-migration helped reverse this trend…One of the underlying problems of outport society was the fact that by the 1880s the annual average production of fish per fisherman had declined to roughly one-quarter the early nineteenth century catch."

[5] See Prowse (1895, 378-381).

[6] Today, the Ship Inn and the Happy Fisherman are both relatively genteel establishments that bear little resemblance to their rowdy namesakes that catered mainly to fishermen and sailors in the early nineteenth century.

[7] Cormack's jumping-off point to the interior was near present-day Shoal Harbour. Cormack described the local waterfowl and evidence that vessels had been built near the sand bar at the head of Smith Sound, but did not mention cod or cod fisheries. This was a place of timber, not fish. Smith Sound would play a very different role in the cod ecosystem of the northeast coast in the late twentieth and early twenty-first centuries. The oral history of the Smith Sound–Random Island area is consistent with Cormack's older writings and confirms that the earliest settlements in Smith Sound were for timber. Some of the earliest clearings were farms of fishing families from Bonaventure and British Harbour (Tom Clenche, Petley, Random Island, personal communication).

[8] Cormack (undated), reprinted in Howley (1915, 133) wrote: "The inhabitants of Bonaventure, about a dozen families, gain their livelihood by the cod fishery. They cultivate only a few potatoes, and some other vegetables. Which were of excellent quality, amongst the scanty patches of soil around their doors; obtaining all their other provisions, clothing, and outfit for the fishery, from merchants in other parts of Trinity Bay, or elsewhere on the coast, not too far distant, giving in return the produce of the fishery, viz., cod fish and oil. They collectively catch about 1500 quintals, or 300 tons of cod fish, valued at 12 [shillings] per quintal, 900 pounds [sterling]; and manufacture from the livers of the cod fish, about twenty-one tuns of oil, valued at 16 pounds [sterling] per tun, 336 pounds [sterling]; which is the annual amount of their trade. The merchants import articles for the use of the fisheries from Europe and elsewhere to supply such people as these, who are actually engaged in the operations of the fishery. The whole population of Newfoundland may be viewed as similarly circumstanced with those of Bonaventure."

[9] See Martin (1990).

10. On July 27, 1832, liveyer William Buckle wrote to the Governor of Newfoundland from Blanc Sablon (Whiteley 1977, 15-16): "Dear Sir; I beg you will excuse my taking this liberty of addressing you on the following subject. Having been a planter on this coast for these Forty Years and being at the head of a large family who almost all depend on my fishing post in Middle Bay (Labrador) I laid out my seal frames this spring in the hopes of doing the fishery when on the ninth of June last an American Schooner commanded by Capt. Sampia entered my frame altho' I entreated him not to and at the same time weighed anchor where they hooked and broke our nets and hawser whereby I lost my best. Two days afterwards the American schooner the James Morrow Capt. Brodie and the Admiral Captain Templeton [presumably a fishing admiral, not navy] entered my frame although I told them not to as they kept me from taking my seals. Still they persisted on many days which caused me great loss. I have therefore humbly to request you would take this in consideration and be kind enough to let me know whether these vessels had a right to do so and if not what I can do to recover the loss which I have incurred. I remain, Your most humble and dutiful servant, William Buckle." Here was one liveyer who does not fit the stereotype of an illiterate and lawless lout. The fact that Buckle appealed to the Governor of Newfoundland for help is interesting, as the region west of Blanc Sablon was given to Canada East (Quebec) in 1825, with Newfoundland having no jurisdiction.

11. Newman wine cellars still exist near the waterfront in St. John's. The vaults last held the famous Newman's port in 1966, but the port was aged at another Newfoundland location until 1996. In 1998 the Newfoundland Historic Trust hosted a "Farewell to Newman's" port tasting inside the vaults in honour of the last bottling of the product in Newfoundland.

12. Ryan (1986, 62-65) told the story of the decline of the United Kingdom–based outport firms and the rise of the St. John's merchants. He concluded that "the close individual control that was possible in the capital and the willingness of the St. John's merchants to adapt combined to make St. John's the sole commercial center of the colony by 1900."

13. Many protests were made over the truck system. A typical petition of this era from Fogo read (Prowse 1895, 379): "For a number of years we have been struggling with the world, as we suppose, through the impositions of the merchants and their agents by their exorbitant prices on shop goods and provisions, by which means we are from year to year held in debt so as not daring to find fault, fearing we may starve at the approach of every winter. We being at the distance of seventy leagues from the capital, where we suppose they arrogate to themselves a power not warranted by any law, in selling to us every article of theirs at any price they think fit, and taking from us the produce of a whole year at whatever price they think fit to give. They take it on themselves to price their own goods and ours also as they think most convenient to them." In 1933, the Amulree Royal Commission stated that the truck system "fosters inefficiency and laxity, raises prices of essential commodities, lowers the standard of living and keeps the fisherman in a condition bordering on servitude" (Amulree Royal Commision, 1933).

14. See O'Flaherty (2005, 5).

15. Sweeny (1997) gave an analysis of the trading practices of the Ryan and Templeman premises in Bonavista from 1889 to 1891. The analysis indicates that the informal economy in outport Newfoundland was of great importance to the local standard of living, as people grew food and built houses, boats, and other things without any recognition by the formal economy (this is still the practice in much of rural Newfoundland). The merchants knew this and based their businesses on capturing as much of the trade that underpinned the informal economy as possible. Credit was part of this. Many debtors of these firms did not repay much in fish and there were many uncollected debts.

16. See O'Flaherty (2005, 5-6).

17. Somewhat surprisingly, Newfoundlanders caught and salted fish, but preferred to eat imported salt beef and pork, a tradition that continues to this day. This could be an extension of the negative view conferred on the fishery and local products, first by outsiders, then internalized.

18   Worrall (2005, 55) related the story of the Battle of Trafalgar and how "nearly 5,000 men died in one of the great battles of sailing ships, among them Britain's legendary naval hero [Admiral Lord Nelson]." It is very likely that many British sailors, and perhaps some of the French as well, had cut their teeth in the Newfoundland fishery as it was one of the main training grounds for navy sailors.

19   Prowse (1895, 387) stated that, in 1812, "in St. John's alone there were three sail of the line and twenty-one frigates, with thirty-seven sloops, brigs and schooners of war."

20   Prowse (1895, 387) reported on the booty taken by English vessels during the war of 1812: "On board the captured vessels were all sorts of valuable freights – Lyons silks and whole cargoes of champagne. The clerks at Hunt, Stabb, Preston and Co....spent their Sunday afternoons firing at champagne bottles on a gumphead at the end of the wharf; the man who knocked the head off the bottle won a case, the one who missed had to pay for one."

21   See Prowse (1895, 389).

22   Prices for cod in St. John's from October 1813 to October 1814 were 20-26 shillings per quintal. The first exports to Brazil were made during this period (Prowse 1895, 403).

23   See O'Flaherty (2005) for details of French and American involvement in Newfoundland in the 1800s. The Convention of 1818 stated "that the inhabitants of the United States should have for ever, in common with the subjects of His Britannic Majesty, the liberty to take fish of every kind on that part of the southern coast of Newfoundland extending from Cape Ray to the Rameau Islands; on the western and northern coast, from Cape Ray to the Quirpon Islands; on the shores of the Magdalen Islands, and also on the coasts, bays, harbours, and creeks, from Mont Joli on the southern coast of Labrador to and through the straits of Belleisle; and thence northwardly, indefinitely along the coast, without prejudice, however, to any of the exclusive rights of the Hudson's Bay Company," and "that the American fishermen should also have liberty for ever to dry and cure fish in any of the unsettled bays, harbours, and creeks of the southern part of the coast of Newfoundland above described, and of the coast of Labrador" (Prowse 1895, 409-410). American fisheries involvement in Canadian waters caused similar problems around the same time, as "ever since the American Revolution, the question of port, inshore, and onshore privileges to United States fishing vessels in Canadian waters and territory, has been a continuous subject, and at times it threatened the peaceful relations of the two countries" (Desbarats 1918, 7).

24   The winter of 1817-1818 was referred to as the "Winter of the Rals," or rowdies. People from all walks of life were affected. A grand jury wrote that "many families, once in affluence, are now in absolute want. Within these two days, two men have been found perished of cold, and many hundreds must inevitably experience a similar fate if humanity does not promptly and effectually step forward to their relief" (Prowse 1895, 405).

25   Prowse (1895, 398) cited a pamphlet of Dr. William Carson, arguably the founder of constitutional government in Newfoundland: "The only remedy against the evils flowing from the present system, will be found in giving to the people, what they most ardently wish, and what is unquestionably their right, a civil Government, consisting of a *resident* Governor, a Senate House and House of Assembly" [emphasis added]. Governor Sir Richard Keats called Carson's pamphlet "poisonous…and vicious." Governor Duckworth, reflecting anti-home rule sentiments, considered Carson's work to be of "very libelous character concerning the authorities and the system of government in the Colony."

26   Governor Cochrane distributed relief to the poor, but demanded work in return. That labour built the first road system around St. John's, including Portugal Cove Road which linked to a ferry service across Conception Bay to Harbour Grace. Cochrane also had Government House built, which still houses the Lieutenant Governor of Newfoundland.

27   See Prowse (1895, 428).

28   See Bannister (1997).

29   See O'Flaherty (2005, 3).

30 See Prowse (1895, 467-468). The first administration in 1855 performed quite well despite some friction between Protestant and Catholic representatives. In one instance, a member challenged Receiver General Glen, but eventually conceded that he agreed with Glen and would support him. Glen replied, "I don't want your support when I am right; it is when I am wrong you must back me up." Thus began a long history of colourful Newfoundland politicians.

31 O'Flaherty (2005) described many examples of British concessions to foreign nations at the expense of Newfoundland. For example, the convention of 1857 over French rights in Newfoundland was a give-away to France, and Britain initially threatened Newfoundland that if it did not accept these concessions, Britain would no longer enforce any rights on the French Shore. Nevertheless, Britain "clearly expected little resistance in Newfoundland" (O'Flaherty 2005, 77-78). Britain was wrong.

32 See Mitchell (1864, 323-326). In the 1850s, France had stiff import duties on English fish of several species, including cod, herring, and mackerel. These duties were "greatly reduced" in 1860, but notably, cod were "excluded from the reduction of duty." English cod would have been mostly Newfoundland product.

33 O'Flaherty (2005, 82) stated that "how to compete with the French fishery off their coasts, control it, and if possible curtail it, was to be a preoccupation and defining struggle in Newfoundland for the rest of the nineteenth century."

34 See O'Flaherty (2005, 78).

35 See O'Flaherty (2005, 80).

36 See O'Flaherty (2005, 80).

37 Lord Derby's dispatch in 1857 stated (Chadwick 1967, 34): "We deem it our duty, most respectfully, to protest in the most solemn manner against any attempt to alienate any portion of our fisheries or our soil to any foreign power, without the consent of the local legislature. As our fishery and territorial rights constitute the basis of our commerce and of our social and political existence, as they are our birthright and the legal inheritance of our children, we cannot under any circumstances, assent to the terms of the convention: we therefore earnestly entreat that the Imperial Government will take no steps to bring this Treaty into operation, but will permit the trifling privileges that remain to us to continue unimpaired."

38 Alexander Murray wrote: "In 1864, while I was in Croque Harbour, I saw the crews of two French Men-of-War engaged in sweeping the mouths of the brooks which empty into that harbour, with long seines, for salmon, the same brooks moreover being closely barred from shore to shore. In that same year, at Canada Bay, I witnessed a bar set, said to have been set by a settler, which contained something like eighty salmon, the larger proportion of which were in a high state of decomposition" (Taylor 1985, 17).

39 Colonial Office (1886), cited in Chadwick (1967, 40).

40 A Colonial Office report stated that the so-called arrangement "would leave Newfoundland under such grave commercial disadvantage that H.M. Government could not press the Newfoundland Government and Legislature to adopt it on its present lines…the French Government will not fail to perceive that it would be altogether unreasonable to expect the Colony to sanction the sale of bait to French fishermen who by the use of it, and subsidised by liberal bounties, are enabled to carry Newfoundland fish to distant parts of the world and sell it at prices as the Newfoundland fishermen cannot compete with" (Colonial Office 1886, cited in Chadwick 1967).

41 This bait deal was signed at Paris on November 14, 1885 (Chadwick 1967, 39).

42 Ryan (1986, 95) cited the British consul in St. Pierre: "Looking back…I find that 1886 was a record year with an export of 45,556 English tons [911,120 cwts (a cwt is one hundred pounds) or 41,415 tonnes]…but since 1900 cod has gradually but surely become scarcer. In 1902…an average of 1945 cwts [88 tonnes] per schooner. With 1903…983 cwts. [45 tonnes] per vessel, and in 1904…825 cwts [38 tonnes]." The comment that cod may have been scarcer around St. Pierre after 1900 is interesting,

as it conflicts with the overall increases of the early 1900s and with the increases in the French fishery some years later. It is possible that the Little Ice Age resulted in a concentration of fish in the southern waters around St. Pierre in the late 1800s, producing a relative decline as overall conditions improved and cod shifted north.

43 See Alexander (1977).

44 See Chadwick (1967, 41).

45 See Innis (1940, 444).

46 Correspondence from the Colonial Office to the Foreign Office in 1889 indicated that "Lord Knutsford is of the opinion that Her Majesty's Government could not admit that crustacea are fish." The French believed that "'Fish' s'applique à tous les produits de la mer et le verbe 'to fish,' qui est employé dans la rédaction du Traité de Paris de 1763…possède encore dans la langue courante de notre époque une valeur générale qui exclut toute restriction…La France a non seulement le droit de pêcher du homard mais encore celui de le préparer industriellement sur place" (Colonial Office 1889, cited in Chadwick 1967, 43). This translates to "Fish applies to all sea products and the verb to fish, which is used in the Treaty of Paris of 1763…possesses in the current language of our epoch a value without restriction…France has not only the right to fish for lobster but to industrial preparation on site." "Industrial preparation" referred to lobster canning. Translated by the author.

47 Is a lobster a fish? This was the answer from Professor Flowers of British Museum of Natural History in 1891: "Scientifically lobsters are certainly not fish in the sense in which term is now used by all zoologists, but belong to a totally different division of the animal kingdom. Fish, or animals belonging to the class 'Pisces,' a division adopted by naturalists of all nations, are capable of distinct definition. As a primary distinction from lobsters, they belong to the great division of vertebrates or backboned animals, whereas the class 'Crustacea,' to which lobsters belong, are members of a totally distinct main division, the Arthropoda, and they are really more akin to beetles and spiders than they are to fishes. But as to ordinary language, before the knowledge of the classification of animals according to their real structure and affinities had been attained, popular usage had taken the true fishes, the best known of aquatic animals, as the type of all or almost all others which resemble them in one of their most obvious attributes, i.e., living in the water, and consequently as falling under the Johnsonian definition of fish as 'an animal that inhabits the water'" (Flowers 1891).

48 Prowse (1895, 541), citing a letter from an unidentified French official to Lord Salisbury.

49 Quotes from a map of Newfoundland prepared from an Admiralty chart by: Sir J.S. Winter, K.C.M.G. and Q.C.; P.J. Scott, Q.C.; and A.B. Morine, M.L.A., June 1890 (map courtesy of the Memorial University of Newfoundland).

50 See Anglin (1970).

51 Sir John A. MacDonald, whose vision of Canada was from sea to sea and who was of the view that the joining of Newfoundland to Canada was inevitable, still held the view in 1869 that "the acquisition of the island itself is of no importance to Canada" (Gwyn 1972, 80).

52 See O'Flaherty (2005, 168-170).

53 See Johnstone (1977, 21).

54 Fortin (1852), cited in Johnstone (1977, 11).

55 Osborn (1887, 162) related the voyage of a Gloucester, Massachusetts, schooner that searched for bait to buy for its Grand Banks fishery as far north as Bonavista Bay, where squid was available. Earll (1887, 454) cited Captain D.E. Collins of New England who visited Carbonear in Conception Bay in 1879, and described the competition for squid bait. Apparently in this case there were 25 bankers competing for the available squid, and many inducements were made to the fishermen to sell, including free meals and much attention. Earll had a poor opinion of Newfoundland fishermen, whom he described as being backward, illiterate, and living in extreme poverty. He also mentioned thievery from American vessels.

The illiteracy aside, which was unquestionably true, Earll's account did not match those of the English hunter-naturalist J.G. Millais who visited Fortune Bay some 25 years later, or Newfoundland geologist J.P. Howley who described life in Placentia Bay around the same time. These authors wrote much more favourably about the life and character of Newfoundland fishermen. Millais (1907) described the south coast outport of St. Lawrence: "Above high-water mark stands the village of wooden houses, many of them built on trestles after the Norwegian fashion. Some of these small crofts have a little hayfield surrounded by wooden palings, in a corner of which stands the cow-byre, whilst all possess on the sea front large staging and store-houses for the drying and curing of cod. The houses are roofed with wooden slates; they are of two stories and possess a loft. The best ones have little gardens, in which grow potatoes and cabbages, or, if the owner is sufficiently well to do, flowers." He found the people "amiable and polite. It is a rare thing to pass a man or woman who does not wish you good day, and the children too, are equally well-mannered. They are kind, sociable, and by no means reserved. The people of the outports make friends at once." Millais did note poverty, illiteracy, and superstition (a trait shared by all fishing cultures), but in terms of character, he states of his guides that "better men to go anywhere, and turn their hands to anything, I have not found."

[56] Fortin (1867), cited in Johnstone (1977, 15).

[57] Fortin (1867), cited in Johnstone (1977, 12).

[58] Whiteway wrote, "It has been decided after a most rigid investigation that the right to fish along a portion of our coast for 12 years…is worth one million dollars (U.S.). We have now an established basis, and I look upon this as of great importance, and that upon which we cannot set too high a value." Whiteway must have had in mind the French claims to fishing rights in Newfoundland waters, which had never been subject to such a fee. The full text of Whiteway's statement is given by Prowse (1895, 506).

[59] An account of this event was given by Earll (1887, 446-450). His account was politically charged and no doubt biased towards the American perspective, but he did give interesting details of the south coast herring fishery of the 1870s, including the use of gillnets and seines, and the dependence of south coast fishermen on the bait trade with the Americans and the French at St. Pierre Miquelon.

[60] See Stommel and Stommel (1981, 176, 183).

[61] See Hansen (1949, 6).

[62] See Item 57 in Amulree Royal Commission (1933).

[63] LeDrew (1984, 542) described the market changes in the char fishery in northern Labrador.

[64] See details in Fisher and Vevers (1954, 57). The extended decline of gannets at the Bird Islands colony likely resulted at least in part from too many birds being killed for bait, eggs, and feathers.

[65] Cormack (1822), reprinted in Howley (1915, 132) stated that "penguins, once numerous on this coast, may be considered as now extirpated, for none have been seen for many years past."

[66] Cormack (1822), reprinted in Howley (1915, 132) stated: "Baccalao Island, formed of a horizontally stratified rock, apparently gritty slate, is famous for the numbers of sea fowl that frequent it in the breeding season, principally the puffin, called on this coast the Baccalao or Baeulieu bird. The Island has one landing-place only, on its east side, and no resident inhabitants; but is visited by men in boats and small schooners called Eggers, who carry off cargoes of new laid eggs. The end of the profession of these men will be the extermination of the sea fowl of these parts for the sake of a cruelly-begotten temporary subsistence. The destruction by mechanical force of tens of thousands of eggs, after the commencement of incubation, precedes the gathering of a small cargo of fresh-laid eggs." This was one of the first clear expressions of how short-term subsistence demands could reduce longer-term benefits from marine natural resources.

[67] Loughrey (1959, 12), summarizing earlier reports from the 1800s, wrote: "One of the earliest authentic records for North America was that made by Jacques Cartier in 1534 at Sable Island. Other early records indicate that the walrus bred (or at least hauled out) on that island and on the

Magdalen Islands, Cape Breton Island, and Miscou Island in the Gulf of St. Lawrence. It was hunted extensively at those locations during the sixteenth and seventeenth centuries, but was exterminated in the gulf by the middle of the eighteenth century, except for occasional passengers on the polar pack ice drifting in through the Strait of Belle Isle."

68 In the 1820s, the rights of all British subjects, including those in Nova Scotia, Upper and Lower Canada, and Bermuda, were affirmed in Newfoundland waters, as were the rights of the French and Americans. These fisheries were entirely unregulated, and the modern principle of adjacency, under which those closest to the resource have priority claim on it, was unheard of. The "rules" of the time favoured the nonadjacent.

69 A remarkably bad fishery occurred in 1831 and was followed by an unusually cold winter, the combined effects of which were "great misery, sickness and deprivation" (Colonial Office 194/83; O'Flaherty 1999, 146).

70 The jigger was jerked up and down near bottom to attract cod. This gear was created for use where bait was scarce and may have been a French invention (Leslie Dean, personal communcation, 2005).

71 Cod seines were long and deep nets fished from small boats near shore (or attached to shore) and designed to encircle fish. The seine would be closed and either lifted in place or dragged to land, where the entrapped fish would be removed.

72 Fishing in most areas commenced in May or June as the ice receded; in some areas of the northeast coast, the first fisheries would have been based on the coastal fish which overwintered deep in the bays. The large fisheries would begin in mid- to late June when the migratory fish would arrive at the coast in search of capelin. This fishery would last through the summer, but, as the waters warmed and the capelin spawning cycle was completed, the cod would descend to deeper waters offshore. Where conditions permitted, handlining would continue in the deeper waters, but this is a difficult method where waters are more than 50 metres deep. By September, most cod would have retreated to waters outside the range of this type of fishing. This pattern continued year after year in most parts of Newfoundland. A similar pattern existed in Labrador, but with the fishing season generally extending over a shorter period and a later start date as a result of the colder waters and later capelin arrival. On the south coast, the fishery extended over a much longer period, as this coast is ice free year round and large numbers of fish overwinter in the inshore waters of Placentia and Fortune bays.

73 Boyd (1997) described the working lives of beachwomen that cured the fish at Grand Bank in the early 1900s. It was serious work: the drying and "haypooks" of fish had to be constantly monitored because they could be ruined if they got too wet or received too much sun. In the words of Dulcie Grandy, "What they would do if the fog would come in and the rain, they'd ring the bell, the fish bell, you see…and when you hear the bell ring, you'd have to drop everything and go, no matter what you were doing. Everybody be going…you'd see them going down, I mean they wouldn't walk, they'd fairly run. They figured they'd get there to the beach to take up their fish because that was just as precious to them as if they owned it, you know. That was their bread and butter. You wouldn't take your time and say, well now, I've got to finish…I got bread in the oven…you'd just go on. Drop everything and go" (Boyd 1997, 176). It was back-breaking work, but many families and women who had lost their husbands at sea (not an uncommon occurrence) had few alternatives. As Becky Rose put it, "There was nobody on the beach that had plenty because they wouldn't be there if they did…when they went to the beach they had to go" (Boyd 1997, 178).

74 Ryan (1994, 156-158) described the financial troubles of the 1860s and 1870s, and refers to the time as being a "depression."

75 Goode (1887, 134), citing an earlier report, stated of the Labrador fishing grounds: "Those extending from Sandwich Bay to Cape Harrison or Webeck have also been visited by fishing craft for a generation or more; but north of Aillik, about 40 miles from Cape Harrison, the coast has only been frequented by Newfoundland codfishing craft during the last fifteen years. A Quebec and London house have possessed detached salmon-fishing stations as far north as Ukkasiksalik or Freestone Point (latitude 55°

53', longitude 60° 50') for about thirty years, but these have all passed into the hands of the Hudson Bay Company."

76 Goode (1887, 134) stated that Labrador "from its amazing fish wealth, promises to become a very important commercial adjunct to Newfoundland."

77 See Goode and Collins (1887, 138-139).

78 The most bizarre case was the schooner *Charlotte*, of Rochester, New York, that in 1858 sailed from Lake Ontario down the St. Lawrence River, through the canals, into the Gulf of St. Lawrence and then on to Labrador. The success of this voyage remains in doubt, but in any event this fishery did not persist.

79 Sabine (1852), cited by Goode and Collins (1887, 143).

80 See Goode and Collins (1887, 138-142).

81 See Item 57 in Amulree Royal Commission (1933).

82 See Goode and Collins (1887, 139).

83 See Goode and Collins (1887, 139-140).

84 Martin (1990) described the history of Random Island and its settlements adjacent to Smith Sound in Trinity Bay. From the 1890s to the 1940s, "the Labrador fishery made an important contribution to the economic base of most settlements on Random Island."

85 See Goode and Collins (1887, 140).

86 Based on data in Goode and Collins (1887, 142).

87 The comments of Professor Hinds in 1876 regarding the northern Labrador fishery were cited in Goode and Collins (1887, 143).

88 See Osborn (1887, 166).

89 See Osborn (1887, 167).

90 See Osborn (1887, 167-168).

91 See Osborn (1887, 168).

92 See Taylor (1985, 43).

93 See Taylor (1985, 43).

94 LeDrew (1984, 543) summarized early commercial records from northern Labrador.

95 Ryan (1994, 25) wrote: "Successful European settlement in North America was always heavily dependent on the resources that prospective settlers discovered. The waters around Newfoundland offered unlimited supplies of cod fish (but little else) to European fishermen who visited annually during the summer months. However, while the migratory fishery encouraged and supported limited and scattered settlement in Newfoundland from the end of the sixteenth century, the short fishing season, combined with an inadequate agricultural base, prevented settlement from growing and acquiring an existence independent of the migratory fishery until the nineteenth century. However, the population of Newfoundland expanded during the early years of the nineteenth century because of the unusual circumstances of this period; and accompanying this expansion, there was the development of a new industry – the seal fishery."

96 See O'Flaherty (1999, 146).

97 Ryan (1994, 414) emphasized the importance of the seal hunt to the development of Newfoundland: "The fact remains, the seal fishery created and maintained a society, a culture and a colony and left a legacy that has enriched Canada's tenth province. While the work of the men and women in the Newfoundland cod fishery must never be underestimated, cod alone did not lead to the growth of permanent settlement on a viable scale. It was only after they became ice hunters that the migrant fishermen settled and literally sculpted a colony from the Arctic ice floes."

98 See Prowse (1895, 452-453).

99 See Brown (1972, 13).

100 The small pilot whale occurs in schools or pods near shore and can be driven by boats onto the beach where they can be easily killed. This practice is ancient and continued into the mid-1900s in Newfoundland and Labrador waters.

101 The Atlantic right whale was given its name because it was the right one to kill. It posed little danger to harpooners in small craft, unlike the more aggressive humpback and sperm whales and the much larger blue whales (which also were much more likely to sink after being stuck with a harpoon). Right whales yielded large quantities of oil that was used for various purposes and baleen for the corsets of ladies in Europe and America.

102 Dickinson and Sanger (2005, 8) stated that Basque whaling on the Labrador declined as a consequence of "overexploitation, the development of more lucrative whaling at Spitsbergen, growing competition with Dutch and English whalers, domestic strife in Spain, and European wars."

103 See Dickinson and Sanger (2005, 11).

104 See Dickinson and Sanger (2005, 13).

105 See Dickinson and Sanger (2005, 31).

106 Newfoundland Annual Fisheries Reports (1903, 1904), cited in Dickinson and Sanger (2005, 39).

107 See Dickinson and Sanger (2005, 44).

108 Dickinson and Sanger (2005, 63) citing an article in the *St. John's Evening Herald*, March 2, 1904, quoted a government minister: "There is every indication and every assurance at present that the coming year will see the whale fishery fleet largely increased, which will not only have the effect of giving additional employment to our people, but must necessarily swell the exchequer of our colony."

109 Summary of data in Dickinson and Sanger (2005, 67) based on Newfoundland Annual Fisheries Reports from 1904-08.

110 *St. John's Evening Herald*, St. John's, July 28, 1917, cited in Dickinson and Sanger (2005, 83).

111 See Templeman (1940, 5).

112 See Templeman (1940).

113 Data from Templeman and Fleming (1954).

114 See Frost (1940, 11).

115 See Goode and Collins (1887, 5).

116 See Goode and Collins (1887, 29).

117 Comment attributed to Captain W.H. Oakes of Gloucester, Massachusetts, and cited in Goode and Collins (1887, 35).

118 See Goode and Collins (1887, 3-4).

119 The text of the Marion's journal was reported in Goode and Collins (1887, 65-89). The quote is from the entry dated February 27, 1878.

120 See Goode and Collins (1887, 25).

121 Goode and Collins (1887, 3) summarized the depletion of the halibut stocks: "Since 1879, at which time the fresh-halibut fishery was at its greatest activity, there has been a very marked decline in the quantity of fish taken if not in the number of vessels employed. The product of this fishery has seldom been larger than it was in the above-named year, but since that time halibut have gradually become scarcer on all of the old and well known grounds, until now a 'big trip' – of 70,000 or 80,000 pounds of fish – is seldom made, and 40,000 to 45,000 pounds of halibut constitute a 'good fare.' In 1885, at the time this

paper is being printed, the fresh-halibut fleet does not probably include more than 40 vessels, and the total catch is estimated not to exceed 5,000,000 pounds; for the first three months of the year the Bank[s] fleet landed only 612,000 pounds [15,300 pounds per vessel]."

[122] See Earll (1887, 426).

[123] See Earll (1887, 436).

[124] See Earll (1887, 466).

[125] See Earll (1887, 467).

[126] *Cape Ann Advertiser*, February 23, 1877, cited in Earll (1887, 440).

[127] See Earll (1887, 442).

[128] See Earll (1887, 445).

[129] See Earll (1887, 445). This account may have been biased towards the American point of view, but nevertheless did provide a view of the need for an economy beyond the seasonal cod fishery, even on the south coast. Earll cited the *Cape Ann Advertiser*, January 24, 1862, as reporting "a sad case of destitution among the inhabitants, especially in the vicinity of Placentia Bay."

[130] See Earll (1887, 442).

[131] Osborn (1887, 184) stated that capelin "are taken in immense numbers by the 'liveyers' and furnished to the bankers."

[132] See Sleggs (1933) and Templeman (1948).

[133] See Frost (1940).

[134] Moravians are a Protestant group originating in the former Czechoslovakia. German Moravians came to the Labrador coast in the early 1800s and set up missions to assist the aboriginals and liveyers with schools, some medical services and religion.

[135] See Newfoundland Inshore Fisheries Association (1991, 28-29).

[136] In truth, credit for the invention of the cod trap should probably go to fishermen who first deployed the trap at Bonne Espérance and who worked for Bossy Whiteley in the 1860s and 1870s.

[137] Whiteley's application in 1876 read in part as follows, and made specific reference that his trap did not catch salmon (pound nets for salmon were by this time illegal): "I have been for the last six years in the habit of fishing for codfish at this place with an enclosed pound set at the end of my Salmon weir in the manner shown in the Diagram. You will please see by this sketch that salmon are not able to pass into the Pound, for the nets prevent them, and also we do not catch any salmon in the salmon weir to speak of...I have improved on the mode from time to time, giving my whole attention to it, giving up my seal fishing, that I might give my whole capital to this, whereas even years ago I employed only four men and took about 100 quintals fish, now we employ 20 men and could take 3000 [quintals], if fish were plenty. As I read the Fishery Act it does not prohibit fishing by Pounds except for salmon, we do not take any salmon" (Whiteley 1977, 38).

[138] In a letter to Prowse in 1899, Whiteley gave the following account of the invention and use of the cod trap: "Looking over my records I find that 1871 was the year that I first fished 'the cod trap.' For three years previous I had used Salmon Nets (6-inch mesh) reaching the bottom for the purpose of meshing codfish when schooling – finding this worked well the bottom net or 'trap' was first tried in 1871 with great success – six men and myself caught, dressed and dried 1080 [quintals] taking the first fish 20th June and the last 28th July, a record we have never beat yet. About four years after Mr. S. used the first trap in Newfoundland made by American Net & Twine Co. Boston for Job Bros & Co. from plan furnished by Capt. James Joy – dimensions taken from my trap at Bonne Espérance – the following year the same firm had traps made for Indian Harbour these were the first traps used on the Labrador say 1876 or 1877" (Whiteley 1977, 40).

[139] McFarland (1911, 5) stated: "On the Newfoundland coast as well as on the Labrador fishing grounds the fishermen employ the cod trap for catching the fish. The contrivance is a large room with floor and walls of twine arranged with an opening on the landward side through which schools of cod may enter but can not pass out again. As many as one hundred quintals of cod have been caught at one haul on many occasions; on the other hand, a whole fishing season may pass without a school of cod entering the trap."

[140] Rose and Leggett (1989) showed that, in the original trap grounds just inside the Strait of Belle Isle, trap catches related more strongly to fish abundance than did the catches of the mobile gillnet fishery. In addition, Rose (1992) showed that, at larger scales, trap catches could mimic cod stock abundance. These relationships are thought to relate to the passive nature of the trap and the consistency of the limited good berths, which to some extent stabilizes fishing effort.

[141] *Newfoundland Inshore Fisheries Association* (1991, 4-5).

[142] In 1886, a French naval officer seized Newfoundland cod traps set on the French Shore, claiming they were interfering with legitimate French fishery interests (NIFA 1991, 7).

[143] A detailed discussion of these events was given in *Newfoundland Inshore Fisheries Association* (1991).

[144] See Rose (1992).

[145] Statistics from Innis (1940, 457-458) and in *Newfoundland Inshore Fisheries Association* (1991, 20).

[146] See Macpherson (1935, 5-6). The best salt cod is that which can be restored most closely to the original wet condition. "By far the greater part of the codfish catch of Newfoundland is 'light-salted.' The idea behind the production of these lightly salted fish is this: the fish have not to lie long in salt bulk and it is intended, and very necessary, that they should subsequently be dried to a low water content. This type of preservation gives a valuable product and at the same time one which requires great care in the process of making because it is so perishable. Of all the salt fish types the light salted shore cure reverts back nearest to the original" (Macpherson 1935, 6). Macpherson also pointed out that "fish caught on the Island of Newfoundland littoral proper are thicker fish than the thin Labrador fish, and that heavy salting of these fish does not produce the same product as Labrador cure."

[147] Gosling (1910, 424) stated: "The use of traps is now universal on the Labrador and the fish taken generally small and owing to the shortness of season cannot be made into hard dry salt fish. It does not keep well and is all rushed off to market together, with the result that the markets are always glutted and the returns small…The fishery has now become a trapping voyage only."

[148] *St. John's Evening Telegram*, June 5, 1930, cited in Dickinson and Sanger (2005, 63).

[149] Innis (1940, 459) gave a description of the Labrador fishery in the early twentieth century: "In 1930 the steamers carried planters, crews, and supplies from Conception Bay down to stations on the Labrador coast. The crews hired generally on a share basis. From northern bays, such as Notre Dame, schooners carried down their salt and supplies, thus saving freight, and caught fish along the northern part of the Labrador coast. Fish were exported directly from the coast in chartered steamers, particularly by planters, crews and stationers, which, it was alleged, involved dumping on the European market; or, particularly in the case of 'floaters,' the catch was taken to their home ports or to St. John's and sold as 'Labrador,' or cured in Newfoundland (12 hogsheads of salt to 100 quintals) and sold as 'Labrador shore.' The powerboat enabled families to come down on the steamships on payment of a nominal fare and carry on the fishery in their own interest. This trend continued as a result of dissatisfaction with the share or planter system, in which one merchant employed several crews. With the race down in the spring, berths for traps were the objective, and were obtained in part by drawing lots. The cash system enabled the fishermen partially to avoid being dependent on the planter or middlemen who received supplies from the large merchants. The sale of fish to the supplier or planter, and in turn to the merchant, necessitated classification for various markets; and the absence of an efficient grading system weakened the selling position of the product."

150 Prowse (1895, 451-452) wrote: "The effect of the seal fishery was to add materially to the wealth of the various settlements where it was carried on. First of all there was the building of the ships, as prior to about the Forties [1840s], when the Conception Bay merchants began buying the slop-build vessels from the Provinces [especially Prince Edward Island], nearly every vessel in the seal fishery was native built; the crews belonged to the place; in many cases the seals also were manufactured into oil in the same harbour – Twillingate, Fogo, Greenspond, Trinity, besides the Conception Bay ports, had vats. All this…brought strength and wealth to the out-harbours, and nourished the growth of a great middle class – the traders and sealing skippers. Enormous amounts of money were made in those days." Judge Daniel Woodley Prowse knew well the importance of the seal hunt to the Conception Bay middle class. Prowse was born into it in 1834 in Port de Grave, near the original settlement of John Guy. He went on to become a lawyer, judge, and noted character, and wrote the *History of Newfoundland* in the late nineteenth century.

151 Ryan (1994, 144) related that there were about 30 St. John's firms engaged in sealing and about 15 establishments in Conception Bay in 1853. By 1914, eight firms operating 16 steamers controlled the entire industry.

152 Report in the St. John's *Patriot*, December 24, 1855 (cited in Ryan 1994, 148). In the end, the reporter was right. The use of larger and more-expensive-to-operate vessels did keep the seal hunt going for a few decades, but harvests were reduced, wages were lower, and in the end the steamers were sold or sent to other duties in World War I. This pattern of overcapitalization of the seal fishery, and the destruction of stocks, would be repeated a century later with cod and other groundfish.

153 Ryan (1994, 147-164) gave a detailed account of the change from sail to steam in the seal fishery.

154 See Prowse (1895, 453).

155 See Ryan (1994, 175).

156 See Ryan (1994, 181).

157 See Ryan (1994, 200).

158 See Blake (1997, 209).

159 Prowse (1895, 521) wrote of the fire: "Thrice our capital has been destroyed by the devouring flame; the conflagration of the 8th and 9th of July 1892 far exceeded all former calamities in suddenness and in the immense value of the property destroyed. The great fire of 1846 began with the upsetting of a glue-pot…the still greater of 1892 commenced in a stable, and was, in all probability, caused by the spark from a careless labourer's pipe."

160 *St. John's Evening Telegram*, March 19, 1886, cited in Dickinson and Sanger (2005, 63).

161 Alexander (1976, 32) wrote: "Until the last decade of the nineteenth century, the codfishery employed more than 80% of the labour force and, contrary to popular assumption, provided a standard of living in conjunction with other market and non-market sources of income, which was not particularly inferior or less stable than that enjoyed among working people elsewhere in the Western World."

162 See Millais, J.G. (1907, 151).

163 See Howley (1997, 25).

164 J.P. Howley travelled much of the coast of Newfoundland in the late 1800s. Among his geological records are many descriptions of outport life. They are often humorous, but seldom depressing, and paint a favourable impression of life in most years.

165 McFarland (1911) provided a summary of the Grand Banks: "Shoals are found at Virgin Rocks and Eastern Shoals, located in the northern part of the bank. The Virgin Rocks consists of a group of small rocky shoals, the most important being Main Ledge, with three to nineteen fathoms depth, Briar Shoal, Southwest Rock, and Bucksport Shoal with a depth of $4^{3/4}$ to 11 fathoms. The average depth of the bank as a whole is between 35 and 45 fathoms…The southern part of the bank has a bottom of fine sand of

varying colour; a middle section consists of sand, gravel and pebbles over certain areas; the northern part is of gravel, pebbles and rocks. The eastern edge of the bank descends rapidly into deeper water. The southeast edge of the bank is the best region for halibut. Cod are found principally on the southern half of the bank and on the central and northern part. The fishing season lasts from April to October. Early in the season the fishing is carried on at the southern part of the bank. As the season advances the fleet moves northward to the vicinity of the Virgin Rocks. In June capelin make their appearance, and the shoal of cod found following them has been termed the 'capelin school' by fishermen. In July the body of cod on the ground is called the 'squid school,' since the fish no longer take the herring and capelin offered them for bait, but are attracted to the quantities of squid that arrive at this time. Hand-liners report seeing cod in large numbers lazily swimming along in the clear shoal water of Virgin Rocks and refusing to touch hooks baited with herring."

[166] Goode (1887, 135) compared the areas of different Newfoundland fishing grounds. The area of the northern Labrador boat fishery – Cape Harrison to Cape Mugford – was estimated to be 5200 square miles (13468 square km), with southern Labrador at 1900 square miles and the total of Newfoundland at 6204 square miles (16068 square km).

[167] See McFarland (1911).

[168] The American Fisheries Society held its first meeting in 1870 and became the main professional organization for fisheries scientists and managers in North America.

[169] See Whiteley (1977, 62).

[170] Cleghorn (1854), cited in Smith (1994), first used the term overfishing with reference to the North Sea herring fishery.

[171] See Smith (1994, 53).

[172] See Smith (1994).

[173] Esselmont (1893), cited in Smith (1994, 89).

[174] See Johnstone (1977, 24).

[175] McIntosh (1899) argued in his book, *The Resources of the Sea*, that the Scottish study was badly flawed and showed nothing other than trawling had no adverse effect on the fish.

[176] In 1874, G.O. Sars stated that "we now come to the difficult and so far very obscure question, what causes the irregularities in the…fisheries…which have been observed from time immemorial?…This phenomena, like everything else in nature, must have its natural causes, which can be found… only…from scientific point of view" (cited in Smith 1994, 70).

[177] See Hjort (1914).

[178] In the 1920s, after 40 years in Canada, Prince remained a staunch Victorian colonialist and would end lectures at the St. Andrews Biological Station with a slide of King George V, saying, "Our King, and a good King he is, too" (Johnstone 1977, 109).

[179] McMurrich (1884), cited in Johnstone (1977, 24-25).

[180] A more extensive citation is given by Johnstone (1977).

[181] See Harvey (1894, 5).

[182] Nielsen (1894, 26-45) described in detail the progress in growing cod and lobster at Dildo Island, Newfoundland. Nielsen got his spawning fish each spring in the bottom of Trinity Bay, and at times from Bull Arm, where "from this deep water we obtained a fine lot of large spawners." He also described "quite a number of spawning codfish with their sexual organs fully matured were caught in the bottom of Trinity Bay this summer [1893], from the latter part of August and till about the middle of September (a thing which is most unusual)." Nielsen stated that "at the head of Placentia Bay…the increasing use of cod-nets set in deep and shoal waters…coupled with the use of bultows…are catching a large quantity of spawning fish, and will be the means of ruining the cod-fisheries in these Bays." Apparently there were

very strong feelings among local fishermen that such fishing was very destructive, as it was practiced from "the latter part of April and until the month of December…as far in as…Kelly's Point in Piper's Hole."

[183] See Nielsen (1894, 29-30).

[184] See Nielsen (1894), 48.

[185] Colonial Office (1886), cited in O'Flaherty (2005, 23).

[186] Joint Commission (1880), cited in Prowse (1895, 506).

[187] See Hiller (2002).

[188] Alexander (1974, 12-13) stated: "The major conundrum of the Newfoundland economy in the early twentieth century is why the re-orientation towards export-led growth did not concentrate more effectively on the fishing industry. The accepted view of the time, and since, is that there was no alternative other than a development of land-based primary industries to absorb surplus labour from the rural economy. But to demonstrate that the fishing industry did not provide sufficient employment at satisfactory income levels is far from proving that it could not, either alone or in conjunction with other economic activities. A significant increment in fishing income might have floated the country over its railway, war and other development-debt burdens, and introduced a dynamic into the well-being, self-confidence and initiative of the country. The fact that Iceland, a staggeringly impoverished and exploited country in the nineteenth century, with no significant commercial fishery before 1890, and with fewer alternative resources than Newfoundland, was nonetheless able to establish itself during this century as an independent and prosperous country on the basis of the North Atlantic fishery, invites a fresh examination of the opportunities that existed in Newfoundland."

[189] Chadwick (1967, 112) stated: "Between 1905 and the end of the First World War, Newfoundland enjoyed a longer period of uninterrupted prosperity than at any previous time. The world price of fish increased: catches on the whole were good and, with the introduction of a large pulp and paper mill and fresh ventures in the mining field, the economy seemed at last to be on the road to a healthy diversification."

[190] For a complete account of the 1914 disaster, see Brown (1972).

[191] Brown (1972, 11-12) described Newfoundland in the early twentieth century: "In the early 1900's, the first steel ship joined the sealing fleet. She was the Adventurer built to the specifications of a St. John's merchant, Alick J. Harvey, so that she was the first ice-breaker in the world. The other Newfoundland merchants, impressed by the new steel ship, began adding their own steel ships to the fleet of 'wooden-walls.' By 1914, Newfoundland was the only country in the world with a fleet of ice-breakers." Brown also commented on the mercantile system: "Men like Alick Harvey were the real rulers in Newfoundland, not the King in far-off England, not the English governor, not the Upper and Lower Houses in the Legislature in St. John's. The aristocracy of the island, the merchant families of Water Street, had first made their money from trade in fish and later from trade in seal hides and fat. These profits had enabled them to branch out into other fields – buying steamship lines and so on – until all contact with the outside world seemed to be made through them. Many of these merchants had import and export links with as many as fifteen foreign countries. On the island they ran everything. They ran the fisheries, the shipping lines, the seal hunt; it seemed that every time money changed hands on the island it rolled inevitably into their tills. Not that money did change hands very often; the St. John's merchants controlled all of the communities outside the town through the outport merchants who bought on credit and sold on credit, perpetuating the system that kept the fishermen in bondage. It was as if the entire island was a 'company town' and every store a company store – the men of Water Street quite content to keep things that way."

[192] Newfoundland Fisheries Commission Report (1915). Additional details are provided in Baker and Ryan (1997, 162).

[193] Newfoundland Board of Trade (1915, 4).

[194] See Gwyn (1999, 28).

195 See Webb (2001).
196 See Gwyn (1999, 28-29).
197 See Hiller (2004).
198 See O'Flaherty (2005, 299).
199 See O'Flaherty (2005, 300).
200 See O'Flaherty (2005, 301).
201 See Arnason (1995, 15), who stated: "Within a period of a few decades the country [Iceland] was transformed from a backward farming community to an industrialized market-oriented society. The driving force of this transformation was the fisheries."
202 See Templeman and Fleming (1954, 84).
203 Attributed to L.G. Hodder of Ireland's Eye, Trinity Bay, by Templeman and Fleming (1954, 81).
204 Leslie Dean, personal communication.
205 Templeman and Fleming (1954, 81) reported that "while they were doubtless present on the west coast of Newfoundland, mackerel were apparently absent or extremely scarce from the east coast of the island for many years prior to their recent appearance in numbers. In the senior author's [Templeman's] early years at Bonavista in the middle of the east coast, although numerous cod traps were used with mesh small enough to catch mackerel, he can remember seeing only one mackerel."
206 See Templeman and Fleming (1954, 84).
207 Hampton (1938, 3) stated: "During recent years and probably at various periods in the past, dog-fish have been found to occur in the waters around Newfoundland in what seem to be very large numbers. Their appearance inshore during the summer codfishery is familiar to all our fishermen and their habit of attacking and driving off codfish is too well known to require further comment here." Hampton went on to investigate some possible uses that might be made of dogfish, but concludes that oil is the only likely market.
208 See Templeman (1944, 58).
209 See Templeman (1944, 7).
210 See Hampton (1938) for the utilization and Templeman (1944) for the biology of dogfish.
211 Tuck (1960) related changes in seabird numbers to environmental changes and uses the exports of mackerel as an indicator of warm periods. Gannet success, and even re-establishment of colonies, appears to follow environmental changes. Tuck shows earlier data that breeding gannets in the Gulf of St. Lawrence declined from over 100,000 pairs in 1829 to about 5000 in the late 1880s, after which there was a slow increase. Gannet colonies on Baccalieu Island may have come and gone over hundreds of years.
212 Lobsters were the focus of some of the earliest conservation strategies in Newfoundland. According to Wilfred Templeman (1940, 31), "the most important lobster laws are those regulating the close [sic] season, the liberation of berried lobsters and the size at which lobsters may be caught." These early measures paid off, as despite the early decline, the lobster fishery remains an important component of the Newfoundland fishery in many regions.
213 Innis (1940, 456) stated that "the depression of the eighties and nineties was marked not only by Newfoundland's assertiveness in excluding Canada, the United States, and France from her fishery and trade but also by her efforts to increase her own production."
214 Poor weather and the Spanish Influenza caused the 1918 cod fishery to decline in parts of the Newfoundland and Atlantic Canadian region, but on the Lower North Shore of Quebec (Labrador) there "the fish were met in 'phenomenal' quantities, in schools at the surface of the water" (Desbarats 1920, 22, 38).

[215] "Our fishermen have been too long accustomed to carry on their voyages in large schooners with costly traps and skiffs and with no trawling gear…It is very questionable whether the trap fishery has paid its way in recent years and it is undoubtedly true that trap fishing is far from being as remunerative as it was in years gone by" (Newfoundland Board of Trade 1925, cited in Innis 1940, 469).

[216] Innis (1940, 471) stated that "the demand for Newfoundland fish manifested itself plainly in a marked increase in the number of foreign fish merchants buying in the local market and doing direct exporting." No source for this information was given.

[217] "Bank fish cannot be compared to shore fish (the small boat shore cured traditional salt cod) at any time. There is no fish in the country better than that produced at Fogo. Bonavista comes second…I consider the whole fishery of Newfoundland is badly handled and the fish is cured worse than it was 40 years ago. People are getting away from the good standard of curing fish. The old-fashioned people worked on the fish themselves from sunrise to sunset. These people made good money because they made good fish" (Templeman 1920, cited in Innis 1940, 474).

[218] Macpherson (1935, 9) described the *tal qual* system: "The fisherman brought his fish to the buyer. He was generally limited to the man who supplied him on credit with his necessary outfit for the fishing season. This buyer had given that credit conditionally, not upon the fisherman paying his debt off the proceeds of this catch, but upon the fisherman selling his catch to the buyer. The buyer made a cursory inspection of the fish and declared a flat rate of so much per quintal for the lot."

[219] For the years 1915, 1916, and 1917, the prices per quintal for cod cured on the northeast coast were C$7.50, C$8.50, and C$10.50. For Labrador cure, prices were considerably less: C$6.30, C$6.70, and C$9.00 (Department of Marine Fisheries 1917, cited in Innis 1940, 464). In 1927, about 200,000 quintals of cod were shipped directly to Brazil in 1925; this amount had almost tripled by 1929. The percentage of Newfoundland fish exported to the Caribbean increased from 6% in 1907 to 26% in 1931 (Innis 1940, 472).

[220] See Macpherson (1935, 13).

[221] The Hawes Report on the Salt Fish Trade for 1927-1928, reported that "in the case of Newfoundland, markets are steadily being lost to Iceland, its chief competitor in the soft-cure trade. It is true that the shore-fish or hard-cure markets appear to be impregnable against all competition; this is due to the undoubtedly better eating qualities of Newfoundland fish. There is no reason why these qualities should not be of the same advantage to Newfoundland in its soft-cure trade, except that the consumers of this cure are influenced to a much greater extent by the appearance of the fish. It is no exaggeration to say that Newfoundland is behind the times in every branch of its salt-fish industry" (cited in Innis 1940, 471).

[222] See Macpherson (1935, 14-15).

[223] See Macpherson (1935).

[224] See Arnason (1995).

[225] Thompson (1943) was convinced that trawling would be necessary to increase catches, and believed it was only inexperience that limited the success of earlier trawling ventures. He pointed to regular voyages of French, Spanish, and Portuguese trawlers after World War I, and approved of voyages by British trawlers in the mid-1930s. In 1933, non-Newfoundland vessels were taking about 700,000 quintals (140,000 tonnes round or wet weight) of cod on the Grand Banks while Newfoundland vessels took 100,000 (20,000 tonnes round). The Newfoundland shore fishery took about 700,000 quintals (140,000 tonnes round) and Labrador fishery 320,000 quintals (64,000 tonnes round). Newfoundland catches were not declining, and foreign catches were increasing without restriction. Apparently, few could see where this was heading.

[226] See Blake (1997, 213).

[227] See Macpherson (1935, 9).

[228] Data and details summarized from Macpherson (1935).

[229] See Innis (1940, 475). Pickled salmon exports peaked at about 764 tonnes in 1922. Frozen salmon exports were near 2050 tonnes by 1930.

[230] Shannon Ryan, personal communication.

[231] Leslie Dean, personal communication.

[232] See Frost (1940, 12).

[233] See Frost (1940, 12).

[234] See Innis (1940, 447).

[235] See Innis (1940, 449).

[236] See Innis (1940, 469).

[237] Salazar died in 1968, but his policies lived on until 1974 (Blake 1997, 212).

[238] Doel (1992) repeated several contradictory inferences that Portuguese fishing was hundreds of years old on the Grand Banks. There is little evidence for this. The Portuguese fishery in the sixteenth century petered out quickly, and there is little evidence of any substantial Portuguese fishery in Newfoundland waters again until the 1930s. During the intervening centuries, Portugal, like most of continental Europe (except France), bought most of its fish from Newfoundland fisheries. This only changed in the mid-twentieth century.

[239] Jose da Silva Cruz published his memoirs as a doryman of the White Fleet from 1933 to 1941. "Life aboard the *bacalhoeiros* was a living hell. There, even the dogs were treated better…[we] were real slaves from every point of view. We, the fishermen, didn't have any rights at all. We were not permitted to quit the ship, for any reason whatsoever. Meanwhile the Captain…under orders of the Gremio [the State organization that ran the Portuguese fishery], could fire whoever they wanted to, even without reason" (cited in Doel 1992, 26-27). Apparently, da Silva Cruz was one of Portugal's most accomplished dory fishermen of the era, so he was a credible witness. Under Salazar, the Gremio apparently also did some good, such as providing fishermen with housing and social benefits, and building fishing schools (Blake, 1997, 212).

[240] See Doel (1992, 28).

[241] See Karlsson (2000, 342).

[242] The imposition of the Danish-British treaty in 1901 forbidding foreign fishing within three miles of the Iceland coast was actually a withdrawal from the protection of 16 miles from the coast that had existed previously. This limit was not for conservation, but to protect Danish merchants, as earlier laws in Newfoundland were put in place largely to protect English merchants (Karlsson 2000, 342).

[243] Karlsson (2000, 342) offered an opinion for why Iceland did not resist the foreign fleets in the early twentieth century: "The reason may be that competition from foreign fishing did not seem very damaging in the first half of the twentieth century, and the growth of the Icelandic fishing industry was mainly in deep-sea fishing, where the grounds seemed inexhaustible."

[244] Innis (1940, 472) reported that exports to Brazil increased from about 5700 tonnes in 1925 to 16,000 tonnes in 1929, and exports to the Caribbean increased from about 6% of production in 1907 to 26% in 1931.

[245] See Department of Marine and Fisheries (1921, 18).

[246] Memorandum from J.L. Paton to George Russell, Memorial University of Newfoundland, President's Office, Box PO-9. Additional detail in Baker and Ryan (1997, 163).

[247] See Sleggs (1933).

[248] Letter from Paton to Huntsman, March 8, 1929, Memorial University of Newfoundland President's Office, Box PO-9. Additional detail in Baker and Ryan (1997, 165).

[249] See Baker and Ryan (1997, 164).

250 In response to the Empire Marketing Board in 1929, Paton wrote: "The plain facts of the case are as follows: Those engaged in the fish business of Newfoundland – merchant, planter and fisherman alike – are, in their conservatism, persisting in putting new wine into old bottles. With little (or indeed frequently no) modification they are following the methods of drying, curing and marketing the fish that have been inherited as it were from past centuries…science throws her light on many things…Greater knowledge of her fisheries is necessary, new methods, new markets are imperative. There is no doubt that from the harvest that lies at her feet, Newfoundland could be prosperous; the potential value of her fisheries is surely immense" (Letters from Paton to Barnes, March 1, 1929, and to Huntsman, March 8, 1929, Memorial University of Newfoundland President's Office, Box PO-9). Additional detail may be found in Baker and Ryan (1997, 165).

251 Wilfred Templeman followed the academic route of the era for young Newfoundlanders, which was 2 years at Memorial College (non-degree granting), followed by 1-2 years at a foreign university, typically Canadian or British, from which a degree would be obtained.

252 Templeman (1940). On some of Templeman's earliest publications, he gave his address as "Fishery Research Institute and Memorial University College."

253 See Thompson (1943).

254 Key works included: Lindsay and Thompson (1932), Sleggs (1932), Frost (1938a), Thompson (1939), Templeman (1938), Thompson (1943), Templeman (1947), Macpherson (1937), and Frost (1938b).

255 See Macpherson (1933).

256 See Amulree Royal Commission (1933).

257 See Frost (1938a, 5).

258 See Frost (1938b, 3).

259 See Department of Natural Resource (1937).

260 See Department of Natural Resources (1937, 13-16).

261 See Department of Natural Resources (1937, 19).

262 The *Investigator I* was commissioned in the same year at the Pacific Biological Station in Nanaimo, Bristish Columbia. A photo of this vessel is included in Johnstone (1977, 177).

263 See Gwyn (1999, 51).

264 O'Flaherty (2005, 294), based on government statistics, stated that after World War I, "the realization of the country's dependence on its fishery hit home with great force. Published figures, ones reflecting the distortions of wartime trade, showed fish products in 1917-8 accounting for 85% of the country's exports, dwarfing the forest industry and mines."

265 McDonald (1987, 96) related that of about C$50 million borrowed by Newfoundland for economic development from 1918-1933, only about C$1 million was spent on fisheries.

266 See Gwyn (1999, 51).

267 A more complete text of Alderdice's manifesto can be found in Fitzgerald (2002, 2-9).

268 See Amulree Royal Commission (1933).

269 There were no Newfoundlanders on the Amulree Royal Commission.

270 See Alexander (1977, 2).

271 O'Flaherty (2005, 404) offered a much more charitable view of Newfoundland politicians.

272 Hiller (1998) gave a detailed account of the events of the 1930s and 1940s that led to Confederation with Canada.

273 "The codfishery…is almost unique as an industry in that the class which owns the capital in it has managed…to throw the whole risk, or very nearly the whole risk…on to the shoulders of the

[274] R.B. Job, chair of the Fisheries Committee, favoured some form of union with the United States. Other prominent Newfoundlanders were also in this camp, including Chesley and Bill Crosbie, the Perlin brothers, Don Jamieson, Lewis Ayre, and Andrew Carnell.

[275] See Gwyn (1999, 83).

[276] It is difficult to get a firm historical record of Smallwood having used these exact words, although it would not have been inconsistent with other pronouncements on economic development in Newfoundland. There are still people living that remember hearing Smallwood say these words, and there is no reason to doubt their memory (VOCM *Open Line* call-in radio show, 2005).

[277] Smallwood's disinterest in the fishery was puzzling, especially given his interest and support in rural Newfoundland. Perhaps his early failures to organize fisheries left scars. Gywn (1999, 205) put it this way: "The lack of affection for things maritime is reflected in what is perhaps the largest puzzle of Smallwood's administration: its lack of attention to the fishery. Although the fishery supports one Newfoundlander in three, right down to the late 1960s it has been the most neglected industry in the province. For twenty years, Smallwood has sought to turn his people away from the sea."

[278] See Jamieson (1989, 59) and Hiller (1998, 21), citing remarks about Smallwood's rhetoric during the National Convention in 1946.

[279] See Hiller (1998, 37).

[280] See FitzGerald (2002) for the Terms of Union. The issue of margarine was included because Newfoundland allowed margarine to be coloured the same shade as butter, but Canada did not. The Terms of Union stipulated that Newfoundland could continue to manufacture coloured margarine, but could not export it to the rest of Canada.

[281] Wright (1997) provided details of Stewart Bates and his influence on Canadian fisheries policies in the 1940s and 1950s. Bates described the salt fish industry as a predatory competition that "does little for consumers or producers, or for the march of economic progress."

[282] Alexander (1977, 5) stated: "Central to Bates' analysis and recommendations was his firm contention that Maritimers were wrong in believing that depressed and inadequate markets represented the core fishery problem…[that] the depressed incomes of east-coast fishermen…reflected their low productivity…outmoded…technology, organization, and products."

[283] See Bates (1944).

[284] See Bates (1944, 107).

[285] See Thompson (1943, 6). Although there were trawlers in Newfoundland in the 1940s, they did not usually fish for cod (Templeman and Fleming 1954). A trawler cod fishery had not proven itself to be economical. The onslaught of foreign trawlers that would fish for cod was just on the horizon.

[286] Henry Mayo was a Newfoundland-born Rhodes Scholar who completed a PhD at Oxford in 1948. His analyses suggested that the Newfoundland fishery had distinct properties and Newfoundland would not easily make the Canadian-style industrialization transition, either in the fishery or to other industries. His views were largely ignored, but have proven to be substantially correct. Alexander (1977, 13) noted that "Mayo doubted there were significant investment opportunities in Newfoundland outside the fisheries." Mayo would be proven wrong in this view, but can be forgiven for having no knowledge of the oil potential of the Grand Banks, or the ample hydroelectric power and mineral wealth.

[287] Iceland's strategy was based on hard work, efficiency, government intervention in the economy, and driving all foreign fisheries from Icelandic waters (Arnason 1995, 4).

[288] See Newfoundland Fish Marketing Agencies, Maritime History Archive, http://www.mun.ca/mha/holdings/findingaids/nafel_10.php.

[289] See Alexander (1977, 35).

[290] See Alexander (1977, 37).

[291] See Alexander (1977, 37).

[292] Thompson (1943, 110).

[293] Alexander (1977) gave a detailed account of the ups and downs of marketing Newfoundland and Labrador salt cod in the late 1940s. Smallwood's revolution, which aimed to "take the boys out of the boats" (Gwyn 1999, 205) and give them year-round jobs on land, more or less threw the baby out with the bath water.

[294] See Templeman (1964, 6).

[295] Second East Coast Conference (1948).

[296] See Karlsson (2000, 343).

[297] Laxness (1936), translated and cited by Karlsson (2000, 357).

[298] Former Iceland President Vigdis Finnbogadottir, CBC radio, June 2005.

[299] Fitzgerald (2002, 149-150) provided many documents on the Newfoundland Act, which confirmed and gave effect to Terms of Union agreed between Canada and Newfoundland. Among the Terms of Union is item 22(1) regarding the fisheries: this brief section stated that the existing Newfoundland Fisheries Board and the previous relevant acts governing the fisheries would stay in place for five years, funded by the federal government, thereafter reverting to full Canadian jurisdiction. The Terms of Union may also be found at http://www.solon.org/Constitutions/Canada/English/nfa.html.

[300] All evidence points to a high abundance of cod, haddock, redfish, and flatfish from the 1930s to the 1960s, including production indices and subsequent catches. The short-term decline in catch and exports of salt cod in the 1930s and 1940s had little to do with a lack of abundance or availability of cod, but with a collapse of the fisheries and their markets. Both the Labrador and Grand Banks fisheries were greatly reduced as fishermen gave up fishing to work in construction or other occupations, such as at military bases, in the lead up to and during the Second World War.

THE LOSS OF THE FISHERY
AD 1950 TO 1992

Look at Mother Nature on the run in the 1970s...

NEIL YOUNG, "After the Gold Rush," 1972

The oddity of a province, located on a major world fishing resource, that is unable to support decently some 15,000 fishermen and that is encouraged by the national government to reduce its fishing commitment requires some explanation.

DAVID ALEXANDER, 1977

Post Confederation Crisis . . .	381
The Russians (Among Others) Are Coming . . .	383
Early Extensions of Coastal Jurisdiction . . .	388
International Convention for the Northwest Atlantic Fisheries (ICNAF) . . .	391
The View from Newfoundland . . .	395
Farming and Fishing . . .	401
Joey Smallwood and the Walsh Commission . . .	402
The Dawe Comission . . .	403
Declining Fisheries . . .	407
Haddock . . .	407

Redfish . . .	410
Cod . . .	412
Flatfishes . . .	417
Grenadier . . .	417
Pelagics . . .	419
Salmon . . .	421
Seals . . .	421
Underutilized Species . . .	422
The Ecosystems in the Mid-1970s. . .	422
The Political Seascape . . .	423
Resettlement . . .	423
The Law of the Sea and 200-mile Limit . . .	426
Northwest Atlantic Fisheries Organization (NAFO) . . .	429
Who Owned the Fish? . . .	431
Irrational Rationalization after 1977 . . .	434
The Nova Scotia Connection . . .	436
Licensing . . .	438
The Fisheries, Food, and Allied Workers Union (FFAW) . . .	438
Watershed – The Kirby Report . . .	441
Post-Kirby . . .	446
Storm Clouds Gathering . . .	447
Science . . .	450
Developments in Fisheries Science in the 1950s . . .	450
The Rise of Fisheries Economics . . .	450
Stock Assessments of Northern Cod . . .	455
The Harris Report . . .	462
Newfoundland Science . . .	464
Memorial University of Newfoundland . . .	465
The "Perfect" Demise of the Northern Cod . . .	469
Overfishing or Environmental Effects? . . .	475
The Ecosystems in the Early 1990s . . .	475

FROM 1950 TO 1992, after sustaining an intensive fishery for over 350 years, many of the commercial stocks on the Grand Banks and continental shelf of Newfoundland and Labrador were fished to near annihilation. Haddock and redfish were first affected, then grenadier and American plaice, and finally the most numerous and productive species, the Atlantic cod. Cold and dry scientific facts can be summoned to document this destruction, but give little insight into why it occurred. Perhaps predictably, it began during a period of great fish abundance. In the late 1940s and 1950s, cod stocks were ebullient, growing to levels not seen since the late 1700s, and many other species, including redfish, haddock, and American plaice, fared equally well. Such abundance was a magnet for the huge international fleets built after World War II. They swarmed to the Grand Banks in the 1950s and moved north to the Labrador banks in the 1960s, mustering a fishing effort the likes of which the world had never experienced, and will likely never be seen again.

The depletion of so many fish stocks within 40 years cannot be understood as a rational process. The fishery fell to a cruel mix of unrestrained industrial development, particularly in post–World War II Europe and the Soviet Union, and later in Canada, which overwhelmed the limited marine conservation and fisheries management of that era in very short order. In addition, as a hangover of the 1884 London debate, there remained unrealistic scientific and political views about fisheries ecosystem productivity and the robustness of fish stocks to withstand intense fishing pressure.

Misinformation about the fishery was framed not only by scientific inadequacies, but also by political prejudice. A patronizing attitude towards the fishery persisted in Newfoundland, a stepchild of the raised noses and silk hankies waved long ago at a puzzled Patrick Morris among cracks of "Can you smell the fish, boys?" In Canada, indifference developed towards the fishery, as well as a prejudice in some quarters towards Newfoundland and Labrador and its people.[1] The conviction that the fish stocks and fishing industry were a basis – perhaps the best one – for economic growth and prosperity was resisted, if not discounted, after 1880 as an irrational corollary of the perfectly sensible notion that not everyone could make a living from it. Over many decades, the fishery was neither protected nor developed.

The productive state of the Grand Banks and continental shelf around Newfoundland and Labrador was unsurpassed in the 1950s, and nothing short of a goldmine in an increasingly food- and protein-hungry world. Although many nations, including Canada, were turning to agriculture and a chemically based "green" revolution, other countries, as diverse as France, the Soviet Union, Iceland, and Japan, were turning to the sea. With an annual production potential of perhaps half a million tonnes of healthy and "blue-ocean" protein, the Grand Banks and adjacent ecosystems would not escape attention. In Newfoundland, however, the fishery was thought to be – and remained – the basis for development of little more than a subsistence economy. As David Alexander wrote in 1977, downplaying the fishery made little sense:[2]

> Newfoundland's best chance for economic stability, full employment, entrepreneurial development, and the restoration of self-confidence lay in the fullest development of fishing and related marine industries. The industry badly needed a diversified product base, wider markets, and greater output to utilize the resource and discourage foreign fishing. By carelessly discarding the saltfish sector and mishandling the frozen trade, a major world ocean resource was forfeited to distant-water fishing fleets and a skilful marine labour force was dissipated, for whom alternative employment opportunities were poor. In the 1950s the ground work was laid for the massive economic and social problems of today…It has been said, and it will be said again, that the saltfish trade was hopelessly antiquated and had no place in a "modern industrial state." Our competitors, who were no less modern and in many instances more prosperous, thought otherwise.

David Alexander's views about the importance of the salt fish industry echoed those of the Amulree Royal Commission of 1933, but were not shared by others involved in the fishery in the early 1950s. The accepted economic logic of Stewart Bates gave assurance that the Newfoundland and Labrador fisheries must turn away from the past (salt fish) and towards a new economy and society (trawlers and frozen fish). It is noteworthy that both Alexander and Bates, despite having opposing opinions on the most suitable type of fishery, shared the view that the lack of prosperity in the industry was to a large extent a result of poor marketing, which no doubt was correct as far as it went. Unfortunately, there was more at stake than markets in terms of what limited the prosperity of fisheries – the ultimate limit would be the supply of fish.[3]

POST-CONFEDERATION CRISIS

Confederation with Canada was less than a year old when troubles in the fishery re-emerged. The heart of the problem was marketing, as Bates had indicated, but the narrow Canadian trade policy worked against the traditional role of Newfoundland as a near-worldwide exporter of fish.[4] The necessity of international trading of fish products appears to have been poorly understood or appreciated in Canada.[5] After Confederation, Newfoundland salt fish was sold in Canadian dollars, and NAFEL[6] was no longer as free to wheel and deal in foreign currencies, making Newfoundland fish more difficult to sell in the traditional European markets that lacked Canadian dollars after World War II. In several instances, sales were lost or compromised. In 1949, Italy disallowed purchases of cod with Canadian or American dollars. Even the cheap markets for Labrador cure in the Caribbean and Brazil were made difficult, as Norwegian stockfish were making inroads in Brazil and Norway did not demand dollar payments. Brazil soon informed NAFEL that it would no longer pay for cod in dollars. At the end of 1949, Newfoundland was holding nearly one-third of a year's production of salt cod – about 10,000 tonnes of light-salted and 5000 tonnes of heavy-salted fish – with no market prospects. Near the end of the season in 1950, NAFEL still was overstocked with the 1949 fish, and prices had to be slashed to sell it.[7] The salt fish industry was in deep trouble.

In 1950, with the crisis deepening, various delegations scurried back and forth between St. John's and Ottawa. NAFEL argued that Newfoundland must continue to produce about 50,000 tonnes of salt cod per year to keep the fishery afloat, and that although the expansion of the fresh-frozen industry was necessary, it should not be looked upon as a substitute for the salt fish industry for which there was a good market. NAFEL stated that "it must not be regarded…that salt fish is of diminishing demand. It has a background of centuries of trade and the per capita consumption in certain markets is increasing. Any policy that would work towards the elimination of salted cod in Newfoundland trade would only mean acceleration of European production."[8] But few in Ottawa were listening and the days of salt fish appeared to be numbered.

A decade later, at the end of the 1950s, the Newfoundland salt fish industry was "dying."[9] By the mid-1960s, with the demise of the uneconomical Labrador shore fishery, trade in heavy-salted cod was near nil and shipments of the light-salted product had been reduced to a remnant of historic volumes. Ironically, the demand for quality salt fish had increased significantly, but NAFEL could no longer supply the traditional upscale European market.[10] By 1967, almost 70% of production was shipped to lower-value markets in Jamaica, Puerto Rico, and the United States.[11] Most of the old Newfoundland companies withdrew from the salt fish industry and from NAFEL. Some – like Job Brothers – moved to the frozen fish industry, but most left the fish business altogether: Baine Johnston sold insurance, Bowring Brothers sold household

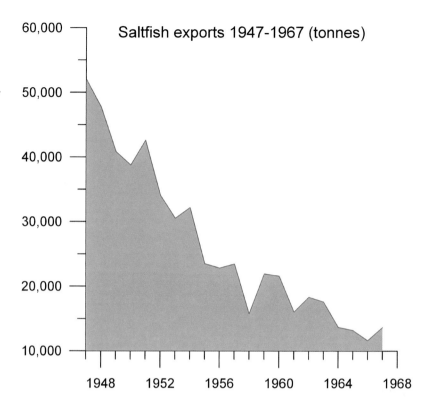

7.1 Newfoundland and Labrador salt fish exports, 1947-1967 (data from Alexander 1977).

wares, Harvey & Company manufactured soft drinks, and the Crosbie Group sold many products.[12] NAFEL, now under the gun both in Newfoundland and Canada, lasted only until 1957. Premier Smallwood promised that a new and "far superior" federal marketing system would immediately replace the Newfoundland-based NAFEL,[13] to invigorate the industry, but it didn't happen. The salt fish industry in Newfoundland went out with a whimper, provoking little sympathy in Canada. Modernization theories were firmly in command, and the salt fish industry was "regarded by federal authorities as an ancient relative that should not be encouraged."[14]

In post-war Soviet Union and Europe, the demand for fish was increasing, and the currency problems of the international salt fish trade stoked the fires of a movement to build national fishing fleets. A despondent Raymond Gushue, then chair of NAFEL, foresaw the implications of this expansion clearly and predicted the subsidizing of these fleets that would result: "Under normal conditions…such competition would…[make these fleets]…uneconomic, but once established, it can operate only if protected against the fisheries of countries such as ours."[15] Gushue resigned from NAFEL in 1952 to become the president of the Memorial University of Newfoundland. In a very few years, Gushue's ominous prediction would be proven correct: heavily subsidized foreign fleets, some from countries with little or no fishing history but that were willing to bankroll new shipbuilding and fishing industries, would invade the Grand Banks.

Several Newfoundland companies followed the path to modernization, and although there was no pell-mell rush to trawlers, some companies ventured into the much-heralded market for frozen fish in the United States. In 1938, the first frozen cod was shipped to the United Kingdom by Russells of Bonavista. In the early 1950s, Fishery Products, Job Brothers, and the Lake group tried to edge into the U.S. market, which previously had been supplied almost entirely by Nova Scotia, and by 1956, Newfoundland had captured about 50% of the U.S. frozen fish market. A side trawler fleet developed in Newfoundland to fish haddock on the Grand Banks in the 1950s, largely for the U.S. frozen market, but by 1963, with the haddock virtually gone, there was no trawler fishery based in Newfoundland with the exception of a small (about 11 tonnes per year) fishery for hake on the southern Grand Banks.[16] The frozen cod still came from the inshore trap and line fisheries, and for a time, it seemed that the Batesian ideas of modernization would bear fruit and that frozen fish was the way to go.

Then, however, came fish sticks. By the mid-1950s, the battered and fried concoctions of ground-up fish that could be oven-cooked in short order became the second-largest selling frozen item in the U.S. market. This might seem to have been a good thing, but it resulted in market substitution of a cheap product for high-value fillets or salt fish without any increase in fish consumption. Moreover, fish of both lower quality and lower price could be used to make fish sticks. Americans, like Canadians, were essentially meat eaters, and the market for quality fish in those countries was, in reality, quite limited.[17] The "bottomless pit" strategy of marketing to the United States, barely a decade old, had come up short.

THE RUSSIANS (AMONG OTHERS) ARE COMING

After World War II, the Soviet Union and many other newcomers came to fish off Newfoundland and Labrador,[18] joining the rapidly expanding fleets of France, Portugal, and Spain. The attraction was obvious: fish stocks were abundant and dollars were not needed to get them. Moreover, Newfoundland and Canada could – or would – do little to keep anyone out, as these were high seas. As a consequence, entire fleets of new entrants came from around the world to compete for Grand Banks fish.

As the foreign invasion gathered steam in the 1950s, the Newfoundland and Labrador fisheries declined in nearly direct proportion – a fish can only be caught once. Art May, a Newfoundland fisheries scientist, reviewed fishing catches from 1954 to 1963, prior to the highest foreign catches, and found a nearly 50% decline in catch per man in the inshore fisheries on the northeast and Labrador coasts. He concluded that "increased competition from the offshore fisheries is having a definite influence on the availability of fish to the inshore fishery."[19] By 1966, a more extensive report indicated that inshore cod landings in Labrador had declined to about one-third of the size of 1930s landings, while (largely foreign) offshore landings had become nearly an order of magnitude higher than inshore landings. May concluded that "it is most likely that the increase in offshore fishing since 1959 has resulted in decreased stock abundance," and argued that previous estimates of "maximum sustained yield" made by respected British scientist Ray Beverton in 1965 were overly optimistic[20] and that the northern cod was being overfished by the early 1960s.[21] May's concerns were not acted upon, and over a dozen competing fleets drove annual harvest rates above 40% of the stock population by the late 1960s. By comparison, harvest rates had seldom if ever exceeded 10% in the previous 400 years.[22] In addition to wreaking havoc on the Newfoundland and Labrador fisheries, the expansion of the long-distance fleets had become a dire threat to the fish stocks themselves, as they were on many major fishing grounds worldwide.[23] It was not long before there were far fewer fish available to be caught by anyone.

The expansion of the Soviet fleet is of particular interest because it had severe impacts on northwest Atlantic fish stocks. After World War II, food shortages in the Soviet Union led to a national policy to harvest the world's oceans, which inevitably brought them to the northwest Atlantic. By 1951, the Soviets had purchased several British side trawlers and sent them on "unofficial" exploratory fishing expeditions to the Grand Banks.[24] The Soviet allies of Poland and East Germany developed similar policies, but it would take a British invention to realize them.

In 1954, a pivotal year for the fisheries, Britain introduced stern trawling and on-board freezing to fisheries in the North Sea. The first stern trawler was the *Fairtry I*. By the standards of fishing ships of the day, *Fairtry I* was both huge (85 metres and 2600 gross tons [2642 tonnes]) and luxurious.[25] It had a ramp cut into its stern within which a large net could be deployed and retrieved. Stern trawling was fast and effective in all types of weather, and with a few modifications could be used even in ice fields. On her maiden voyage, with Newfoundlander Jim Cheater as first mate, *Fairtry I* voyaged to the Grand Banks. She could stay at sea for 70 days (although she seldom needed that length of time to fill up her holds) and brought back up to 660 tonnes of cod fillets (derived from 2032 tonnes of fresh fish) per trip.[26] In the same year, the first "official" Soviet expeditions to the northwest Atlantic were made aboard the older side trawlers *Sevastopol* and *Odessa*. Their main objective was the exploration of the southern Grand Banks. The trawlers spent 60 days at sea, 23 of which were devoted to trawling and exploration. The Soviets regarded the Grand Banks as "one of the most important and productive fishing regions of the world" and as fair game for them. Furthermore, they took little notice of the historical fishery, believing that the entire Newfoundland region was underharvested and that "intensive exploitation of the fishery resources of this region began only recently, 25-30 years ago."[27] They were about to redefine intensive exploitation. With Canada on the sidelines, the Soviets saw the Grand Banks as a grand opportunity: free fish and lots of it. A plan was put in place to begin exploitation in 1956 with a new fleet of stern trawlers, copied from the British design. Over the first five years of their plan, the Soviets took a relatively modest 200,000 tonnes of groundfish (mostly redfish) from Newfoundland waters. This, however, was only the beginning.

In 1956, the Soviet Union sent two brand new factory freezer stern trawlers, the *Pushkin* and the *Sverdlovsk*, to the Grand Banks, both nearly exact replicas of the British *Fairtry I*. The *Pushkin* went fishing for fish and the *Sverdlovsk* for political influence. The *Sverdlovsk* visit was part of a Canada-Soviet bilateral engagement that included invitations for Canadian Minister of Fisheries James Sinclair to visit the Soviet Union and for the return visit of the *Sverdlovsk*, carrying Soviet Minister of Fisheries Ishkov. The *Sverdlovsk* docked in St. John's, tours were offered, and people marvelled at the huge and modern factory vessel. It was great public relations. The Soviets, who wanted access to Newfoundland and Labrador fish and inclusion in the International Commission for Northwest Atlantic Fisheries (ICNAF), convinced the Canadian government that their intentions were harmless, even helpful.[28] Ishkov soothed conservation concerns by stressing that Soviet interests were restricted to research and experimental fishing, from which "both countries can learn much from each other: [through] an exchange of scientists, fishermen, and periodicals." When asked if the Soviet Union was not intent on increasing its fishing effort in the main cod and haddock fisheries on the Grand Banks, Ishkov responded, "No, we catch all the cod and haddock we need in many other fishing grounds." To keep the door open, he added, "We are, however, interested in the ocean perch [redfish] that can be caught here." Ishkov also baited Canada with inferences of the huge markets in the Soviet Union.[29]

Many people in Newfoundland were concerned about this potential new menace to the already troubled fisheries. The Soviet Union was an enormous country, much larger than Spain, Portugal, or France. They had no historic interest in the northwest Atlantic and appeared to be developing a huge distant-water fishing fleet, though just how big no one at the time could guess. Minister Sinclair assured Newfoundlanders that there was no need to be concerned: the Soviets just wanted to help with local research, and if they caught a few fish in the process, what harm would it do? In response to questions about the potential for Soviet fisheries expansion, Sinclair rationalized that the Soviets were no threat to Newfoundland fish and repeated almost verbatim Soviet Minister Ishkov's response that "it was more economical for Russia to fish in waters nearer home, the Grand Banks is a long way from Russia."[30] As if to validate Sinclair's statement, the *Pushkin* and *Sverdlovsk* returned home with a mere 3000 tonnes of cod and 13,000 tonnes of redfish, taken mostly from the Flemish Cap – a minor catch that was less than 8% of that of the Portuguese, Spanish, and French fleets.[31]

The next summer, in 1957, both the *Pushkin* and *Sverdlovsk* returned to the Grand Banks – but not alone. Research scouting trawler *Odessa* "found heavy concentrations of redfish *Sebastes mentella* on the northeastern slope of the Newfoundland Bank. During the period October–December a group of big refrigerating trawlers operated in this region."[32] The West German shipyards at Kiel had been busy with Soviet shipbuilding. In 1958, about 35 Soviet trawlers fished the Newfoundland banks[33] guided by "three searching trawlers, *Odessa*, *Kreml*, and *Novorossiysk*."[34] In the same year, the Soviets were welcomed into ICNAF. By 1965, less than a decade after the *Sverdlovsk* basked in attention in St. John's, the Soviet fleet had reached armada proportions and included 106 factory trawlers, 30 mother ships, and 425 side trawlers. The fleet and catches were still growing, bearing little resemblance to the initial modest redfish fishery off the Flemish Cap. By the mid-1960s, the Soviet fleet fished from Greenland to as far south as Georges Bank off New England and even beyond. Few vessels went home empty, and the Soviet catch approached a breathtaking 900,000 tonnes per year, well above that of the Spanish, Portuguese, and French fleets combined. That total included approximately 280,000 tonnes of cod and haddock. Canada and ICNAF had been duped. And the worst was still to come.

Soviet scientific documents belie the political manoeuvrings of fisheries ministers Ishkov and Sinclair in 1956 in St. John's. The Soviets regarded the northwest Atlantic region as being underharvested and underinvestigated, and they meant to change that.[35] They were undoubtedly aware of the dangers of overfishing, as had already been established in the waters around Europe.[36] Their fisheries science was advanced, if politically motivated, and well beyond the simplistic notions of the London debate of 1884.[37] Perhaps they sensed soft targets. Their megafleet damaged not only Newfoundland and Labrador stocks and fisheries, but also fisheries off Greenland, Nova Scotia, and New England, and in the open ocean east of Iceland, where they contributed to the decimation of the Norwegian spring spawning herring, once one of the largest of all marine fish stocks.[38] It is notable that the Soviets avoided Iceland itself, perhaps because the feisty Icelanders – who knew a valuable resource when they saw one – were too much bother to deal with.

The Soviet blitz included a full marine research program with the goal of enabling their fisheries. In the late 1950s, at the height of the cold war, the Soviets conducted research nearly year round in Newfoundland waters and became a dominant presence on the Grand Banks and the adjacent shelf.[39] They mapped the bottom, studied currents, temperatures, plankton, bottom animals, and, of course, the distribution and movements of the main commercial fish species. They claimed interest in "rational exploitation of commercial fish resources" and gave lip service to conservation, stating that the "intensive fishing of small rosefish [redfish] in the Newfoundland banks requires immediate control measures in order to prevent the harmful effect which overfishing would have on the rosefish reserves in the Newfoundland area."[40] But they failed to act on their own advice.

By the 1960s, the Soviets had the largest distant-water fishing fleet in the world. It included scientific survey vessels which, under their Northern Fisheries Survey

Administration, were used to search for new sources of fish. Soviet surveying was relentless. From 1965 to 1968, about 1500 miles (2780 kilometres) of the continental slope from the northeast slope of the Grand Banks to Baffin Island were surveyed. By then the stocks of redfish and cod were nearly exhausted, so the Soviets went deep and found large stocks of grenadier. Their vessels mastered new trawling techniques for very deep waters – down to 1500 metres – within a few years.[41] The grenadier stocks did not last long; by the mid-1970s, they were virtually gone.

The West German shipyards that mass-produced the British-designed stern trawlers for the Soviets were also building ships for themselves. The *Heinrich Meins*, the first West German stern trawler, sailed to the Grand Banks in 1957 and was joined by 30 similar vessels within a few years. The Germans applied their engineering excellence and were innovators in fisheries technology, outfitting their vessels with new electronic fishing and navigation equipment. Fish-finding echo sounders located schools of redfish, cod, and herring with unerring precision. Giant midwater trawls equivalent to the length of three football fields, combined with forward-looking searchlight sonar to steer them, made it easy to find and catch the big schools. The ancient troughs and channels of the Grand Banks and continental shelf that had provided sanctuary for juvenile and adult fish since the ice ages,[42] and from fishing over the previous 400 years, were now easily exploited.

Spanish shipyards were also busy after World War II. The Spanish high-seas fleet became the third largest in the world, after the Soviet Union and Japan. The Spanish perfected the pair trawling method, with the first pair arriving on the Grand Banks in 1948. It was dangerous work in the often-rough and foggy waters of the southern Grand Banks, but it was productive. By 1959, about 50 Spanish pairs were working the Grand Banks. Both side and pair trawlers typically turned broadside to the sea swell to retrieve their trawl, a laborious task that was dangerous in rough weather and nearly impossible in ice. These vessels therefore specialized on the aggregations of fish on the shallow and ice-free southern Grand Banks. To gain access to the more northerly grounds, the Spanish government provided easy loans to build larger factory-freezer stern trawlers that could fish under more difficult sea conditions and in ice. Spain's frozen fish production skyrocketed from 4000 tonnes in 1961 to 500,000 tonnes in 1972. Spanish vessels took an average of 200,000 tonnes of fish from the northwest Atlantic each year between 1961 and 1972.

The Portuguese White Fleet was the last of the ancient dory fleets. The one-man, 16-foot (3.5-metre) dories could still be seen on the Grand Banks in the late 1960s, even though Portugal also had about 25 stern trawlers by the 1950s.[43] The days of the White Fleet were numbered, however: even the Portuguese, with their devotion to tradition, gave up dory fishing in favour of trawlers in the early 1970s. There would be no more dories and no more boats under sail working silently in the fogs of the Grand Banks. It would be all trawlers now, and lots of them.

The development of factory-freezer trawlers meant that distant-water fishing fleets could exploit unprotected fishing grounds around the world. Japan built the world's second largest distant-water fishing fleet, regularly sending vessels halfway around the world to fish on and around the Grand Banks. Countries as diverse as France, East Germany, Poland, Bulgaria, Greece, the Netherlands, Belgium, Romania, and Israel built or bought trawlers and sent them to the Grand Banks in the 1960s. Still more would come later, heavily subsidized by their home governments.

EARLY EXTENSIONS OF COASTAL JURISDICTION

In response to the incursions of distant-water fleets after World War II, several nations began proceedings to extend coastal jurisdiction further offshore to protect their fish stocks and fisheries. In 1945, President Truman made a unilateral proclamation claiming exclusive national control of the subsoil and seabed of the continental shelf adjacent to all United States coastlines, effectively stating that they were to be regarded as a part of the landmass of the coastal state. Truman also declared that the United States considered that it was appropriate to establish fishing conservation zones within its waters.[44] Newfoundland publicly supported the American action, but could take no action on its own as it was still ruled by the British Commission of Government. Within a year, Mexico and several Latin American countries followed the American lead with similar unilateral actions. The next year, Argentina claimed control of its entire continental shelf and the sea above it, and by 1947, Chile and Peru had extended their jurisdiction to 200 nautical miles to protect their fisheries. In 1950, Ecuador did the same. These actions by South American countries cited Truman's proclamations, but by the early 1950s the United States was having misgivings, based on a wish to preserve worldwide freedom of the high seas. Unrepentant, Chile, Ecuador, and Peru united to assert their authority over the fisheries and seafloor to 200 nautical miles from their shores.[45]

In the North Atlantic, extensions were slower to be implemented. In 1950, Iceland unilaterally extended its limit to four miles in the northern parts of the island and drew the limit straight across the mouths of all bays, as had been the practice in Newfoundland. In 1952, the Icelandic four-mile limit was applied to the entire island; the limit was extended to 12 miles in 1958. These actions were taken without the support of international law on the basis of conservation of natural resources, both beneath the seabed and in the waters above.[46]

In 1948, just prior to confederation with Canada, Newfoundland's Commissioner of Natural Resources recommended that Newfoundland unilaterally extend management of the fisheries beyond the existing three-mile zone.[47] However, this recommendation found little favour either in London or Ottawa. Instead, Britain and, with some reservations, Canada[48] wanted the fisheries to be put under a multilateral commission. Over the following three decades, the government of Newfoundland repeatedly urged the federal government to extend jurisdiction first to 12 miles, then to 200 miles. In 1963, as the foreign trawler fleets were building, the provincial government went to Ottawa with these words: "The great majority of fishermen in Newfoundland are dependent on the small-boat inshore fishery. Their livelihood is seriously jeopardized by the encroachment of foreign fishing vessels on their established fishing grounds, especially during the very short season when these inshore fishermen must make their catch. The encroachments not only reduce the fish resources available to Newfoundland fishermen, but they also cause serious damage to gear and equipment. We ask that immediate steps be taken to conserve, protect, and develop these coastal fish resources by action to establish the twelve-mile limit for Canada's national waters."[49]

The first United Nations Conference on the Law of the Sea (UNCLOS I) in 1958 saw heated debates on the merits and legality of extending coastal state jurisdiction. The countries that had already undertaken extension held their ground, but there was much opposition. The initial Canadian position rejected the views of the "extensionist" states. Canada defended a meagre three-mile limit, with a somewhat ambiguous nine-mile "contiguous zone" added on.[50] Neither Britain nor the United States would agree, and proposed a six-mile territorial zone with six-mile fishing zone beyond. In the end, despite much political rhetoric, nothing could be agreed upon for the fisheries, but UNCLOS I laid the groundwork for coastal state jurisdiction over the full continental shelf itself,[51] including sedentary living resources (e.g., shellfish), even if it was not ratified by most nations involved in high-seas fisheries.

In 1960, the second United Nations Law of the Sea Conference was convened to resolve the territorial issues. Canada suggested that the three-mile territorial limit be increased to six miles, and after some dispute, joined with the United States in a joint proposal to extend the limit to six miles with an additional six-mile fishing zone, with a ten-year phase out period for countries that had utilized the fishing zone in the previous five years. Even this modest proposal found insufficient support and was not adopted.[52]

Canada was in a bind with regard to its territorial waters in the postwar years of the 1950s and 1960s. Philosophically, Canada advocated multilateral thinking and had used its military presence to invent peacekeeping as a legitimate role for the United Nations and international forces. But in the early 1960s, the failure of the multilateral approach to resolve the issues of the fisheries was all too obvious. In 1963, Canada abruptly changed course, stating its intent to declare unilaterally a 12-mile exclusive fisheries zone, as others had done, and to use as the baseline the headland-to-headland rule traditional to Newfoundland. In 1964, the *Territorial Sea and Fishing Zone Act* was passed, but fell back to a three-mile territorial zone and a nine-mile fishing zone. The act was further weakened by an Order in Council that authorized continued fishing rights for foreign nations until bilateral treaties could be signed.

Why Canada was reluctant to act more boldly is uncertain. One view held that "the Canadian government was not willing to sacrifice good trading relations with these countries for fish."[53] Initiating military action, which may have been the only option open to oust the foreigners, was resisted, as it was in the United States. Whatever the cause of the lack of action, early optimism about the 1964 act and speedy implementation of bilateral agreements was badly misplaced.[54] At this critical time in the history of the Newfoundland and Labrador fisheries, Canada blinked. David Alexander wrote of this era:[55]

> Newfoundland was located on a major ocean resource and unless it established economic control over that resource, other countries would move in to fill the vacuum. To a small extent this process began in the 1920s when the Spanish and Portuguese began to build up distant-water fishing after a lapse of several centuries; but the vacuum was mainly filled by western and eastern Europe after World War II. A national government that effectively surveyed the economic opportunities in all regions under its jurisdiction would have focused on this issue and taken vigorous steps to preserve the resource and the resulting employment and income of its citizens…the process would not have been easy…but Canada was not even a paper tiger on this issue.

Alexander believed that Canada, and no less Newfoundland under Smallwood, thought the fishery was not worth a fight: "The fishery resource was abandoned to other countries…[and] a modest little trade with the United States was substituted for the restoration of Newfoundland's historic role as a major fishing nation."[56]

Despite Alexander's strong views, the problems facing Canada and other nations in dealing with the burgeoning Euro-Soviet fleets on the high seas after World War II should not be underestimated. The Grand Banks was not the only area ravaged. It is telling that the powerful United States, where protection of the marine fisheries was proclaimed as a national goal in 1945, did no better than Canada at resisting the first wave of Euro-Soviet fishing that came to New England waters. One thing is clear: by the time Canada did act in 1970, most of the fish were gone.

INTERNATIONAL CONVENTION FOR THE NORTHWEST ATLANTIC FISHERIES (ICNAF)

In 1943, with World War II raging, the United Kingdom convened a meeting in London to discuss conservation of all North Atlantic fisheries. Canada participated and initially supported the U.K. initiative to form a pan-Atlantic fisheries organization, but American opposition and the war stemmed any progress. In 1947, when the war was over, the United States developed an alternative proposal for joint management of the northwest Atlantic only. Initially, the agreed-to arrangement involved only the United States and Canada, with the focus on Georges Bank and southern Nova Scotia-New Brunswick waters.

Two years later, in 1949 – the same year that Newfoundland joined Canada – the United States put forth a much grander proposal, which came to be the International Commission for the Northwest Atlantic Fisheries (ICNAF). ICNAF would preside largely over fisheries off Canada, Newfoundland and Labrador (now part of Canada), and Greenland.[57] Five countries signed on: the United States, the United Kingdom, Canada, Denmark (representing Greenland), and Iceland (despite the fact that its waters would not be included in the arrangement). The ICNAF mandate came into effect on July 3, 1950, with responsibilities including "the investigation, protection, and conservation of the fisheries of the northwest Atlantic in order to make possible the maintenance of a maximum sustained catch."[58] ICNAF's membership[59] grew to 18 countries by 1975 with the additions of France, Spain, Portugal, Norway, the Federal Republic of Germany (West Germany), the German Democratic Republic (East Germany), Cuba, the Polish People's Republic, Italy, Japan, Bulgaria, the Soviet Union, and Romania.

ICNAF was a groundbreaking organization in many ways. The importance of the northwest Atlantic fisheries attracted a great deal of scientific enthusiasm and energy, and many of the world's more thoughtful fisheries scientists. ICNAF implemented the "world's foremost system for the collection and reporting of fisheries catch and biological data." ICNAF scientists were "at the cutting edge of fishery analysis" and were responsible for providing advice on up to 70 commercial fish stocks. They implemented some of the first scientifically determined quotas on fishing, and undertook the first brave attempts at what would much later be known as multispecies management.[60] ICNAF's scientific record, many publications, and extensive storehouse of information on fisheries were impressive by any standard.

In the beginning, ICNAF was thought by Canadian leaders – such as noted scientist Alfred Needler and the influential Stewart Bates – to be far-sighted and of great value. From a scientific standpoint, they were undoubtedly correct. However, the ability of ICNAF to regulate sustainable fisheries was severely compromised by the virtually uncontrolled growth of fishing power and vessels in the 1950s and 1960s.

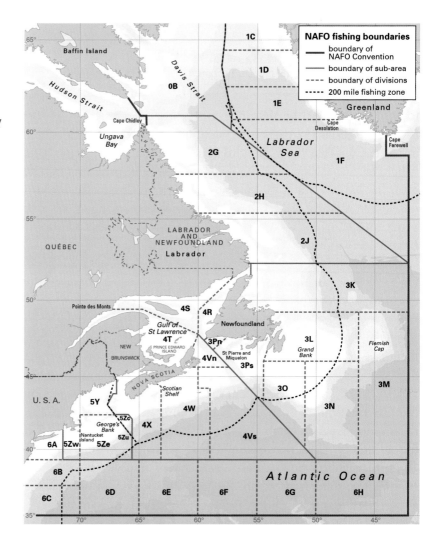

7.2
NAFO geographical fisheries divisions in the northwest Atlantic. These designations were originally designated by ICNAF.

This new fishing power was a leviathan that would not be tamed by scientific intention. ICNAF countries with heavily subsidized fleets vied for their share of the fisheries (which was more or less what they could catch), their claim to do so based on freedom of access to the seas and the fisheries. Scientists sat and watched – in hindsight with what appears at times to be questionable aloofness[61] – unsure how much the stocks could take. The adjacent coastal fisheries of Newfoundland and Labrador were given little consideration and no priority.

Until quotas were established for a few stocks in the mid-1970s, the only regulations put in place to control the fishery were limits to trawl mesh size. In theory, the meshes in the net would allow fish below a certain size to escape, ensuring a constant supply of young fish to the stock. Some scientists thought that mesh regulations were all that was needed to ensure survival of the fish stocks[62] and dismissed concerns that the massive trawler fisheries would negatively impact the shore fisheries. Despite evidence as early as the 1950s that the buildup of foreign fishing power directly correlated with the failure of the Bonavista and Burin Peninsula trap and longline fisheries,[63] as well as broader studies that linked the decline in inshore fisheries to the trawlers,[64] noted Canadian scientist William Ricker, based in British Columbia, was confident that Newfoundland concerns were overstated and that good catches should be expected for many years.[65]

Even the modest conservation regulations suggested by ICNAF were seldom agreed upon and were never effective due to inadequate enforcement.[66] The Soviets, for example, consistently used smaller mesh gear than what was recommended.[67] Under the ICNAF convention, any country could ignore any recommendation that displeased them by filing a formal objection. ICNAF scientists could recommend whatever they liked, but member states could not be made to comply.

The eleven ICNAF countries – most of them new participants in what had been only decades earlier a primarily Newfoundland fishery – did not only compete with the traditional fisheries, but developed many new ones. While the traditional fisheries had mildly assaulted the cod, harp seals, and perhaps a few other species, the new fisheries assaulted the entire ecosystem. Redfish, American plaice, Greenland halibut, yellowtail flounder, capelin, herring, hake, mackerel, and squid were fished. The strategy – often referred to as "pulse fishing" – was to fish an abundant stock as hard as possible until catch rates declined, then move on to another area or another species. It was a return to a migrant fishery, but this time a much more lethal one. Nothing was spared, not even the life-giving capelin that fed many of the other species. The fisheries records of the 1950s and 1960s read like a casualty list in a losing war.

Canada and the United States, the two main North Atlantic coastal states,[68] tried to inject some concern for conservation into the ICNAF proceedings in the 1950s and 1960s, but few other ICNAF members were concerned with conservation, and nonmembers, who were also free to fish, showed no concern whatsoever. Catches by nonmember states increased from 2000 tonnes in 1963 to nearly 100,000 tonnes in 1966, which represented a 5000% increase in just three years. Despite the best efforts of ICNAF data collectors, it was hard to tell who was fishing and how much was being caught: European fishing companies would flag some of their vessels in tax-haven countries where they essentially could not be controlled.

By 1970, most of the fish stocks in the Grand Banks and Newfoundland and Labrador regions were severely depleted, as they were in all of the northwest Atlantic. Haddock were commercially extinct.[69] ICNAF imposed the first voluntary quotas on catches shortly thereafter on the badly depleted Georges Bank (United States) and Browns Bank (Canada) haddock stocks.[70] Canada wanted quotas for all stocks in the northwest Atlantic, as suggested five years earlier by Wilfred Templeman and John Gulland,[71] and won some ground for a few species. However, even the most restrictive ICNAF quotas were set far too high and in most cases did little to reduce fishing effort or catch.[72]

ICNAF scientific reports are dry reading. During the height of the overfishing and stock decimation, there are few hints of the tragedies to come. On June 1, 1970, Alfred Needler, Deputy Minister of Fisheries and Forestry for Canada, welcomed the twentieth annual meeting of ICNAF to the Memorial University of Newfoundland in St. John's after five years – the years of the major decimation of the cod – of holding the meeting outside of Canada. Setting a pattern that would typify Canadian responses to foreign incursions for decades to come, Needler was far too polite to an organization that had just reduced some of the greatest fish stocks in the world, and Newfoundland and Labrador fisheries, to a subsistence level. Yet Needler expounded how "the Commission can look back on the accomplishment of a great deal of useful cooperative research…[although he acknowledged that] application to rational exploitation of the resource has been slower." Needler also stated that "we are extremely proud of the part which Canada has played [in ICNAF]."[73] At the same St. John's meeting, it was reported that off Labrador (ICNAF Subarea 2) "the Canadian [Newfoundland and Labrador cod] fishery again decreased dramatically from 18,000 to 5000 tons [5080 tonnes], the poorest catch record since 1939," while the foreign catch exceeded 400,000 tonnes.[74]

The following year, Needler greeted the ICNAF Annual Meeting in Halifax with the following words: "ICNAF has done a first class job…not only for Canada and Canadian fishermen, but also for the fishermen of all the countries which are represented here today." After this initial diplomacy, however, Needler changed his tack. His 1971 address focussed much more on the overfishing of the stocks than it had a year earlier, stating that "nature, it seems, cannot keep up with us," and expressing concern for inshore fishermen whose catches had "been cut roughly in half since the early 1950s…a trend which, if it continues, means real hardship for tens of thousands of Canadians living in Newfoundland and the Maritime Provinces."[75] Wilfred Templeman reported that "the Canadian [Newfoundland and Labrador] inshore fishery off Labrador was a failure due to lack of cod and decreased to only 2038 tons [2071 tonnes]."[76] By now, foreign catches were also in free fall. Needler, speaking for Canada, stated that "in the northwest Atlantic we…believe in an international fishery…[but it] need not be chaotic…[and] can be organized in such a way as to maintain stocks and increase the productivity of the individual fisherman."[77] This belief was not borne out.

In the end, from the standpoint of its goals of "protection and conservation," ICNAF accomplished little other than to provide a superb record of the decimation of the Grand Banks and northwest Atlantic fish stocks. Too late, Templeman, in his typically measured way, put it into perspective:[78]

> With ICNAF…you have 14 or 15 nations trying to agree and to share catches…this is not easy. Until recently, catches of cod have been rising, and while this is occurring, it is not easy to persuade nations that they have

reached their peak. The fact is that biologists didn't and couldn't know it. You must have fishing in order to prove your population. But this year, at least, they have got down to quotas. A quota, of course, is not the end…the trouble…might be the policing of them…one would like to cut fishing power. This would be the simplest thing. The nations are not very willing to cut fishing power, because the great nations with 3000-ton factory ships want to be able to take them somewhere else after a quota [or a fishery] in one place is used up.

THE VIEW FROM NEWFOUNDLAND

Beginning in the 1940's, there were concerns expressed in Newfoundland that the buildup of foreign fishing capacity would affect, perhaps ruin, the Newfoundland and Labrador fisheries. One of the earliest impacts occurred at Bonavista, one of the largest inshore fishing communities in Newfoundland for 400 years. Until the 1940s, Bonavista was a major salt cod producer; local fishermen fished mostly with traps and handlines, although a Labrador schooner fleet was also based there. Bonavista was a progressive community: it had the first freezing plant on the northeast coast in 1939 and attempts were made to increase production of the old top quality northeast coast salt fish using drying machines in 1950.[79] The trap and handline season was restricted to May to September in most years; in order to supply these new initiatives with fish over an extended season, the Newfoundland Fisheries Board laboratory undertook an experiment using larger longliners (37-55 feet, 11-17 metres) in the early 1950s. The longliners located large schools of cod in the deepwater glaciated trenches that ran from the Grand Banks north to southern Labrador. Off Bonavista, the trench was located within 20-50 nautical miles (about 35-90 kilometres) from shore, forming the inner extension of the Bonavista Corridor. This fishery proved highly successful at first, and detailed records were kept of catches, locations, depths, and water temperatures.[80]

Perhaps the research was too successful: before long, fleets of foreign trawlers, and even a few Norwegian and Faroese longliners, descended on the great storehouse of cod in the trenches off Bonavista. Within a decade, both the Newfoundland shore and longline fishery was in serious decline: fishermen noted that their catches were much lower than previously and that the fish were smaller. In the late 1950s, Templeman reported that "Canada's inshore fishery was less successful than usual with cod less available to shallow water traps." Although one year of poor cod catches was not unusual and could have resulted from unfavourable water conditions,[81] Templeman added that "the Canadian deepwater longline fishery off Bonavista on a stock also fished by European otter trawlers and longliners continued its steady decline from over 100 pounds [45 kilograms] per line

7.3
An early Newfoundland longliner near Bonavista in the 1950s (Gordon King collection).

from 1952-54 and 85 pounds [40 kilograms] in 1957 to 35 pounds [16 kilograms] in 1961."[82] In addition, few large cod were being caught by 1958.[83] These changes were not easily dismissed. The provincial Department of Fisheries was also collecting information: Colin Story presented a report to Deputy Minister Eric Gosse in 1958 with evidence that the inshore fishery was being depleted. Based on statistics similar to those used by Templeman, Story argued that foreign fishing was depleting the stocks and concluded that "the evidence is sufficient…or at least would be sufficient in other fishing countries who have experienced how rapidly depletion can take place."[84]

Gosse took these concerns to Ottawa in 1959,[85] but was told by federal officials that his apprehension was "not yet justified."[86] Several years later, as the foreign fishery was ramping up even further, Gosse was told again that there was no evidence yet of depletion.[87] A high level conference entitled "Resources for Tomorrow" was held in Montréal in October, 1961, opened by the Prime Minister of Canada and Premier of Quebec. The conference included two key sessions on the fisheries, the first on maintaining adequate stocks, and the second on attaining more efficient operations. The fisheries sessions began with a pessimistic view by J.A. Crutchfield, an economist from the University of Washington. Crutchfield believed that the fisheries were too important to ignore completely, but that they "would no longer play a leading and active role in Canadian economic development."[88] A comprehensive study was presented to the session on maintaining stocks by William Ricker, the influential "dean" of Canada's fish stock theorists.[89] Ricker's analyses indicated that the average yield of cod from 1951 to 1955 of approximately 340,000 tonnes could be increased to approximately 386,000 tonnes by 1980. Haddock was thought to have potential to increase from about 49,000 tonnes to 55,000 tonnes.[90] It was difficult to challenge Ricker on fish stock dynamics,[91] but Gosse remained unimpressed with Ricker's claims, responding that he "was so pleased to note from the article by Dr. W.E. Ricker that the productive capacity of groundfish off Canada's east coast will increase considerably during the next 20 years. However, I believe that the tremendous foreign fishing fleet now operating in the waters off the coast of Newfoundland will have an adverse effect on production."[92] Gosse continued: "Apparently the once prolific fisheries off the coasts of northern Europe have been so depleted that European countries have in recent years been forced to build larger ships and more elaborate equipment to fish on the Grand Banks and the coast of Labrador. If European fish resources could become depleted, surely it is inevitable, if the intensity is great enough, that the grounds on the this side of the Atlantic will eventually become overfished…remember that last year Canadian vessel tonnage operating off our Atlantic coast amounted to 26,742 tons, while the tonnage of other countries amounted to 581,138 tons. There is good reason to think that European countries will intensify their efforts in future and from a long-range viewpoint the signs are ominous …"[93] Indeed they were, and history would show that Gosse was right. But in the summary statement from the 1961 conference, there was not a mention of Gosse's concerns.

In the mid-1960s, it was clear in Newfoundland and Labrador that overfishing on the Grand Banks was responsible for declines in the traditional inshore fishery.[94] A decade later, in May 1975, the report of the Select Committee on the Inshore Fishery to Newfoundland's House of Assembly spelled it out:[95] "a serious and sometimes catastrophic decline in catches is evident in all areas of coastal Newfoundland and Labrador…the root of the problem is usually associated with too much offshore fishing effort and not enough management of a renewable but not limitless resource." The Committee supported the efforts of the Canadian delegation to ICNAF to implement quotas and to reduce fishing effort. However, they also sought much more rigorous management controls: "Members of this Committee…join the voices of fishermen and legislators alike who demand unilateral action by Canada in extending her fishery jurisdiction for a full 200 miles beyond her shores and to the very margins of the continental shelf."[96]

The Committee pulled no punches in assigning responsibility for the state of the fishery in Newfoundland:[97]

> By virtue of this existing legislative monopoly of the Federal authorities in fishery matters during the critical years when our resources were subjected to the most intensive effort at exploitation by the fishing fleets of many nations, this Committee holds that these same federal authorities through omission of timely measures to counteract the effects of this onslaught on *our* fish stocks, are the main cause of this present perilous state in which our industry finds itself…This Committee holds that the present division of powers in the realm of fisheries is a contributing factor to many of the existing hardships and concerns of the fishing industry. [They rued]…a policy which recognizes only "national objectives," and whose application has brought this fishing industry of ours to the brink of catastrophe. The people of this Province must obtain a greater voice in the management of *our* fishery…[emphasis added]

The Committee did not spare the provincial government. Commenting on the lack of support for the fishery within Newfoundland and the historical lack of resources put into developing the fishery, the Committee concluded that:[98]

> Notwithstanding the apparent absence of provincial legislative authority to deal with fishery matters…this Committee is convinced that the fishery is not yet receiving the provincial priority it deserves. The long history of a Provincial Government policy based on heavy industry and a "develop or perish" theory provided neither an incentive nor a real place for the fishery. Complete federal indifference to the problems of the inshore fishery, coupled with what can only be considered as a provincial lack of faith and effort in this selfsame industry have combined to stagnate and frustrate the efforts of inshore fishermen…Everyone agrees that the major fault lies directly at the doorstep of the Federal Government *but we as a Province are compelled to do more than complain,* and a good Provincial Government policy should recognize the fisheries' value to the economy by allocating a greater percentage of our budget to fisheries. *We can hardly shout of our concern for the inshore fishery if we limit our efforts to 2% of our gross budget*…Provincial and Federal expenditures on fisheries…for the past 20 years…demonstrate just how appalling the neglect of the industry has been and continues to be [emphasis added].

In the mid-1980s, the primary objective of the Newfoundland government was once again "to acquire a greater degree of Provincial control in the area of fisheries management." They regretted the fact that the federal government "currently holds almost exclusive control over the resource, the licensing of fishermen and fish allocation."[99] The province also continued its objections to foreign overfishing on the Nose and Tail of the Grand Banks and on the Flemish Cap, and the demands of other Canadian provinces for access to northern cod: "There are ever-increasing pressures, especially from fleets based in the Maritime Provinces and Quebec, which have never depended on this stock, to gain greater access to northern cod…this can only be achieved at Newfoundland's expense. The present Newfoundland catch is still far below historic catch levels and the needs of our industry."[100]

The federal government was unmoved by these arguments, and incursions of non-Newfoundland interests into the northern cod fishery persisted. As late as 1985, the European Economic Community – France, West Germany, the United Kingdom, and Portugal – were allocated a total of over 16,000 tonnes of northern cod, but they and other foreign nations caught over 44,000 tonnes. In 1986, 9500 tonnes were allocated, but nearly 68,000 tonnes were caught.[101] Although these catch levels were much lower than what they had been in the 1960s, so were the stocks. These catches undoubtedly contributed to the continuing decline of the stocks.

7.4 *Trawler catch of northern cod in the 1960s (Gordon King collection).*

CLOSURE OF THE HAMILTON BANK

As early as 1974, there was concern about overfishing of key overwintering and spawning concentrations of cod on the Hamilton Bank, and there were recommendations to close it to all fishing. A letter dated February 13, 1974, from Newfoundland Minister of Fisheries Harold Collins to federal Minister of Fisheries Jack Davis, stated:[102]

> **Re: Closure of the Hamilton Inlet Bank**
>
> As you are well aware, the Government of Newfoundland is very concerned about intensive foreign fishing fleet activity in Northwest Atlantic waters, sharp declines in catches from this area and the current unhealthy economic state of the Province's inshore fishery. There is every indication to believe that offshore fishing effort has been a significant factor affecting availability of the cod resource…it is proposed that strong steps be taken to reduce foreign fishing pressure which is directed towards the Hamilton Inlet Bank cod stock [which is the] major spawning ground of this stock…during the February-April spawning period…the resource is exposed to very intense foreign fishing effort with greater than 80% of the total catch…taken during this period of heavy resources concentration.
>
> I believe that closure of the Hamilton Inlet Bank to **ALL** fishing operations during the identified spawning period will help reverse downward trends in inshore cod landings.

The Newfoundland minister, and presumably the department who advised him, did not believe that the stock might be in trouble – their concern was that the winter fishery was interrupting the migrations of the fish: "total reduction of offshore fishing effort during the spawning period would result in larger numbers of cod moving shoreward during the summer months. At other times of year, the stock is widely dispersed and is not as susceptible to concentrated fishing activity." What follows was striking given that the main annihilation of the northern cod had already taken place: "the Province of Newfoundland's position advocating closure of the Hamilton Inlet Bank during the cod spawning season is based on social rather than biological factors. I realize that, biologically, there is no sound justification for closure of the Bank since the overall stock *per se* may not have been exposed to overfishing."

The province's position was somewhat illogical from a scientific perspective, as an offshore fishery that was not reducing the stock would likely have little impact on migration patterns. There was ample evidence of stock collapse by 1974, but perhaps this was the only argument they thought would fly in Ottawa. The following sentence appeared near the end of the letter – a good example of being right for the wrong reason: "It is our contention that Ottawa should have a strong position placed on the

> agenda of the upcoming June 1974 ICNAF meeting advocating closure of the Hamilton Inlet Bank during the identified cod spawning period on these grounds as a necessary first step to ensure protection of an important sector [the inshore fishery and longliners] of the Newfoundland fishing industry."
>
> No action was taken on this issue until 30 years later, when the Fisheries Resource Conservation Council recommended closing an area in the Hawke Channel on the southern edge of Hamilton Bank to all fishing except crab pots (cages set on the ocean floor and designed to catch crab; they are harmless to cod). This recommendation was implemented in 2003 within a 10-square-nautical-mile (18 kilometres by 18 kilometres) area, which was expanded to a 50-square-nautical-mile (90 kilometres by 90 kilometres) area in 2004 to protect spawning and juvenile cod and crab.

FARMING AND FISHING

Newfoundland grew by fishing, but Canada's central heartland grew by farming. The two occupations differed substantially not only in practice, but in the mindset of the practitioners. Farming required controlling natural processes and ecosystems, something that is occasionally wished for, but unachievable, in fisheries ecosystems. While farmers attempted to tame and control ecosystems, fishermen knew they could not. Hence, fishing societies must live within the bounds of natural production, whereas agricultural societies must stretch them. As a rule, fishing societies must have a much smaller impact on ecosystems to survive, but they are inherently less productive in economic terms than farming, and are subject to boom and bust episodes. Such differences in how ecosystems were used no doubt impacted cultural development and attitudes in human societies.

Agrarian societies tended to "share symptoms of homocentricity, illusions of omnipotence, hatred of predatory animals...lack of interest in non-economic plants or animals, and the willingness to drudge, with its deep, latent resentments...and absence of humor."[103] In contrast to farming, fishing was less certain, less of a drudge, and fishing cultures worldwide embraced an often black sense of humour. Newfoundland and Labrador was no exception. As Spanish philosopher Jose Ortega y Gasset related, "The hunter knows that he does not know what is going to happen, and this is one of the greatest attractions."[104] To fishermen, every cast of the net was an adventure – this is true even today with the advent of technology that takes much of the guesswork out of fishing, and it explains the nearly universal appeal of fishing for food and recreation.

In the 1950s and 1960s, there was an element of a farming attitude in the fisheries: production had to be maximized and unrealistic levels of stability were demanded of natural ecosystems. The present-day focus on aquaculture is a similar attempt to control nature for food production and to impose the agricultural ethic on the oceans.[105] Canada spent 11-14 times as much on agriculture as on fisheries in the early 1960s;[106] it was the only important fish-producing country in the world without a national fishery development program, although it had one of the finest national agriculture development programs.[107] It was inarguable that Canada was a farming country – fishing would never figure as high on the Canadian agenda as it had historically in Newfoundland and Labrador. Newfoundlanders and Labradorians, always the fishermen and seldom the farmer, had a hard time accepting this after Confederation.

JOEY SMALLWOOD AND THE WALSH COMMISSION

The role of Joey Smallwood in the decline of the fisheries in the 1950s and 1960s is at once critical and curious.[108] His farmland preferences and lack of interest in the fisheries were well known, but he apparently realized, if belatedly, that the fisheries were Newfoundland and Labrador's strongest natural asset. His begrudging acceptance of the fisheries may have been influenced by the failures of a series of expatriate economic "experts" brought in to develop new industries in Newfoundland. Most of their initiatives came to nothing.

Smallwood did make some early attempts to better the fisheries. In 1950, shortly after Confederation, Smallwood hired Clive Planta from British Columbia, installing him as Deputy Minister of Fisheries. Planta's job was to revive the fisheries. With Smallwood's blessing, Planta organized a commission to recommend on improving the state of the declining fisheries.[109] The Walsh Commission was appointed in January 1951 as a joint federal-provincial initiative and chaired by Sir Albert J. Walsh. The commissioners produced a comprehensive report and made many recommendations, but like so many before (and after), underestimated the deeply conservative attitudes and nonconformity of the Newfoundland and Labrador fishery. The Walsh Commission report was submitted to federal and provincial ministers in May 1953. It defined the problems of the Newfoundland fishery in these words: "The traditional equilibrium of the resource, the industry…and the market is breaking down…the objective of a development programme should be to establish a new equilibrium on the basis of a fully modernized fishing industry – one that, utilizing the resources to the best advantage and meeting market requirements to the fullest extent, provides the fishing population with a living conforming to the national standard."[110]

The Walsh Commission report summarized the state of knowledge of the fish stocks in Newfoundland and Labrador waters up to that time, based largely on the work of the Newfoundland Fisheries Research Board. They discussed the fisheries potential of some twenty-five species, and concluded with three classifications: species that were "incapable of much greater exploitation" (yellowtail flounder, winter flounder, salmon, pollock, lobster, large whales, and seals); "species on which information is inadequate to determine whether increased exploitation is possible" (mackerel, hake, shrimp, scallops, and mussels); and "species capable of expanded exploitation" (cod, redfish, haddock, American plaice, witch flounder, halibut, Greenland halibut, wolffish, herring, capelin, and squid). In very short order, increased exploitation would come at a scale hardly imaginable by the Walsh Commission, but it would not be the Newfoundland fisheries that would benefit. Many of the species listed as capable of expanded exploitation, including cod, redfish, haddock, most of the flatfish, wolffish, and herring, would be fished to near annihilation within two decades.

The clear message from the Walsh Commission report was that many species of fish were present in abundance in the early 1950s, and indeed they were. The problem, as they saw it, was the archaic Newfoundland fishery – not enough fish could be caught to support the numbers of people involved[111] – and a balanced strategy for modernization was recommended to fix the situation. The Walsh Commission admonished industry for failing to foster necessary export markets: "A programme of market research should be instituted as soon as possible…First priority should be given to markets for salted codfish and cured herring."[112]

In the 1950s, $C13 million were invested in private fish plants through provincial government loans. Nevertheless, the fishery languished, and a melancholy mood prevailed that was fed by weak markets, an uncaring and arrogant federal government, and the burgeoning foreign fleets. As foreign vessels caught more and more fish, the markets for Newfoundland and Labrador fish collapsed. In addition, a core of corruption and incompetence, as noted by the Amulree Royal Commission of the 1930s, was still alive and well in Newfoundland. Few if any of the fish plant loans would be repaid. Many firms went bankrupt, and at least one firm formed a U.S. subsidiary to receive all of its profits and hence claim inability to repay the Newfoundland government loan.[113] Planta left Newfoundland in 1954 amid unconfirmed allegations that he had accepted money from fish plant owners seeking government loans. David Alexander noted of this era that "a premium was placed on efficiency and imagination if any small country hoped to survive at acceptable standards of life. Newfoundland failed this test, and the responsibility for it is hers alone."[114]

The Dawe Commission

In 1962, Smallwood appointed another fisheries commission with a mandate similar to that of the Walsh Commission. Twenty-six people representing most aspects of the fishery and several Canadian economists with considerable experience in agricultural re-organizations were appointed to a commission under the chairmanship of Harold Dawe. The Dawe Commission reported to the government in April 1963, concluding that "the opportunities for improvement and development appear to us to be greatest among fishermen in…inshore operations."[115] Few alternatives existed for the majority of rural Newfoundland's still-growing labour force and there had been a 25% increase in the number of fishermen from 1956 to 1962. To deal with this situation, the Dawe Commission advocated full revitalization of the inshore salt cod fishery. The report contained a measured degree of optimism based on increasing demand for both fresh and salt cod, with increasing demand and prices for the latter. Unfortunately, the art of producing high quality northeast coast salt fish had, by and large, been lost – the typical product had declined well below the level acceptable in most European markets. Even the south coast bank salt fish, by the 1950s Newfoundland's best, was no longer readily available. A Greek buyer in 1961 rejected 90% of the fish cured without his supervision and eventually withdrew from buying any Newfoundland fish.[116]

Smallwood pushed federal assistance as the key to Newfoundland's future throughout his political career, but his view was short-sighted, self-serving, and far from universal. H.R.V. Earle, then chairman of NAFEL and a member of the Dawe Commission, skewered Smallwood's approach and what it implied for the fisheries. Earle's view was not one of assistance and subsidized markets but one of quality products:[117]

> We need to turn out a quality product of which we can be proud, there is no need of surpluses and bonuses…[They are] there…if we give the customers the right product. For goodness sake let's forget this trash [low-grade salt fish]…and give them [the customers] a good product and we will earn the bonus without the help of the Federal Government or anybody else… Fishermen…are getting too much for the old trash and not enough for the top quality fish. That trash should not be exported from this country… stamped "Canada Approved." They could just as well stamp "Canada Approved" on garbage.

On the subject of the growing fresh-frozen cod fisheries, the Dawe Commission was of two minds. There was a realization that "the inshore fishermen's needs might be obscured in a programme geared to the requirements of a frozen fish plant…made possible by a trawler operation," but the committee also recognized that "trawlers are essential in providing a continuity of supply to the shore plants and providing labour for a great many workers throughout the year."[118] Despite evidence that inshore, not trawler-caught fish, were in demand in United Kingdom markets, there was little movement in that direction.[119] There was an almost grim acceptance of a coming "arms race" in the fishery: "Our fishing banks are now being exploited by increasing numbers of Russians, French, Portuguese, Spaniards, and East Germans, and in order for Canada to obtain its fair share of fish, our offshore equipment must be equally as good as that supplied by other nations. It is our understanding that a Japanese exploratory fishing vessel is presently working on the Grand Banks. Should this vessel find sufficient quantities of fish, we may expect a deluge of Japanese crafts in addition to those from other countries."[120]

The Dawe Commission sounded one of earliest alarms, based on the work of the Newfoundland–Canada Fisheries Research Board, that fish on the banks and inshore were the same fish,[121] and expressed dire concern regarding what was to come for the Newfoundland fisheries:[122]

> Those in the fishing industry have viewed with increasing alarm the number of foreign boats which, in most instances, are far superior to those operated locally. In 1960 the Russians reputedly took 125,000,000 pounds [57,000 tonnes] of haddock on the Grand Banks, and in 1961, a total of 100,000,000 pounds [45,000 tonnes]; truly a significant quantity of fish. Last winter foreign boats fishing near the 100 fathom [183 metre] contour line along the northeast coast of Newfoundland took large quantities of codfish – as much as 40,000 pounds [18 tonnes] in a 20-minute tow. The Spaniards, fishing the very rough bottom of the northern part of St. Pierre Bank, also took tremendous catches of codfish and are fishing these grounds at the present time. In both these instances, the fish taken are from stocks which undoubtedly contribute to Newfoundland's inshore summer codfishery. Foreign boats fishing off the Cape Bonavista area have been correctly blamed for a decline in the inshore fishery of that area [a reference to the data and views of Templeman, Fleming, Story, and Gosse]. Our offshore resources are therefore of utmost importance to us, not only from a point of view of our trawling fleet, but to our inshore fisheries as well.

The Dawe Commission expressed concerns for the Newfoundland fishery, whose share of the catch had declined to 15%-20% of total landings by the mid-1960s, and for the fish stocks on which it depended. The Dawe Commission was far more aware of overfishing than the Walsh Commission had been a decade earlier, but some outdated theory persisted in the Dawe Commission report. On the Grand Banks, the haddock fishery was in free fall. Nevertheless, the Dawe Commission, no doubt based on contemporary science, held a misplaced faith in the reproductive potential of haddock by supposing that "given ideal survival conditions, there will always be enough haddock to provide a good year-class for, in effect, one can never deplete a fishery."[123] Such views carried over into government policy. In a presentation by Newfoundland to the federal government in 1963, following the Dawe Commission report, there was little mention of biological problems or the potential for overfishing the cod stocks on the banks. The problems with the fisheries were seen to be purely economic in nature, and it was stated that "the obstacles are not in the fish resources."[124]

With redfish, the Dawe Commission was more on the mark. The long-lived and slow-growing redfish were being fished out right before their eyes. Whereas the Walsh Commission had stated that redfish "can be fished very much more heavily than at present,"[125] the Dawe Commission stated that:[126]

> Due to the fishing out of accumulated stocks of fish, it is difficult to say just what the future holds for this species…In 1959, when we [Canada] took 14,000,000 pounds [6400 tonnes], other nations fishing off Newfoundland and Labrador landed a total of 645,000,000 pounds [293,000 tonnes]. For the future it is unlikely that the present rate of yield of redfish can be maintained as the accumulated older stocks are removed. On some of the longer-fished portions of the eastern and southwestern Grand Banks, the redfish has already become much scarcer and smaller in size…the same is certain to occur on the Flemish Cap and in the northern Labrador areas where the Europeans are reputedly fishing very heavily.

Overall, the Dawe Commission painted a cautiously optimistic view of the Newfoundland fishery, referring to the "very rich fishery potential, literally on our door-step." There were recommendations for a broad suite of research and education initiatives,[127] but also concerns about who would benefit from developing new fisheries: "We do not wish to see a repeat of our experience with the Hamilton Inlet Bank where the Canadian Government located a new red [sic] fishing area, and foreign interest realized at least 90% of its value."[128] The report contained some of the first stern warnings of the overfishing that would ensue within the next decade and a plea to Canada to take action to protect the fisheries of Newfoundland and Labrador:[129]

> The Commission is convinced that trawlers are seriously interfering with the operations of inshore fishermen…[and that this situation]…must be corrected as quickly as possible…[and]…must be excluded effectively from an area 9 miles from the shore line and base line across the mouth of the bays…[and that]…a 12-mile limit is necessary. It is generally recognized that there is little chance of any international agreement [as most but not all of the trawlers are foreign] in the near future and for that reason the Commission recommends that Canada take, unilaterally, such steps as are necessary to protect its shore fishing population…[with] penalties approximately on the same scale as are imposed in Iceland and Norway. The Commission wishes to draw attention to the fact that in the autumn there are increasing numbers of trawlers operating off the Labrador coast.

In 1963, a recommendation from the Dawe Commission to establish a Canadian Fish Marketing Board, especially for salt fish, was put forward to the federal government by the Newfoundland government, but the initiative went nowhere. A federal economic commission rejected the recommendation in 1964, and the emphasis on frozen fish and trawlers was reinforced.[130] Smallwood, for his part, remained equivocal in his support for the fisheries until the end of his term as premier. Richard Gwyn believed that Smallwood was "convinced the fishery...was gone beyond recall," a view perhaps established early in his career. In 1967, as Smallwood opened the new Marystown shipyard, he despaired of the inshore fishery workers, asking, "What are we going to do with them?"[131] He failed to mention that the unprotected fish stocks were being ravaged by hundreds of foreign freezer trawlers as he spoke, making profitable Newfoundland operations impossible. Gwyn's assertion that Newfoundland industry in the 1960s was "inefficient, uneconomic, and uncompetitive"[132] held a grain of truth, but in a few short years the supply of fish would be so depleted as to make these arguments superfluous: no industry can prosper with their raw material being literally stripped out from beneath them. As to the debate about salt fish versus frozen fish, the federal and provincial governments would be forced to bail out the frozen fish companies by the early 1970s, indicating that the problem was much more deeply rooted whether the product was "modern" or not.

DECLINING FISHERIES

Haddock

Haddock did not salt as well as cod and were not heavily fished until the advent of freezer trawlers. Haddock froze very well and a bottom trawl fishery commenced on the southern Grand Banks in the early 1940s.[133] Haddock shared the southern Grand Banks with cod, but haddock were more plentiful over most of these grounds, with an estimated abundance of about 600,000-1,000,000 tonnes.[134] The haddock fishery increased so rapidly that total annual catches approached 120,000 tonnes by 1955, about three-quarters taken by Spanish pair trawlers and the rest by Newfoundland side trawlers. There were two stocks: one centred on the St. Pierre Bank and a larger one on the southern Grand Banks. Haddock formed dense aggregations that were easily trawled and thus became the first casualty of the onslaught of fishing after World War II. The last strong year-class was born on the St. Pierre Bank in 1949 and on the Grand Banks in 1955. The haddock fishery ended in 1957 on the St. Pierre Bank.[135] By 1960, all the large haddock were gone from the Grand Banks; although large numbers of fish born in 1955 were present, they were barely large enough to catch. The arrival of the Soviet Union's fleet in 1960[136] sealed the fate of this stock. Templeman recalled an event in that year on the southern Grand Banks: "The *Cameron* [a Newfoundland research vessel] was fishing 30,000 [pounds] [13.5 tonnes] of haddock in 30 minutes, at the height of the capelin spawning. They were mainly small fish from the 1955 and 1956 [year-]classes, the last strong year-classes of Grand Banks haddock. The *Cameron's* skipper, the late Baxter Blackwood, was much given to radio chatter, like most Newfoundland captains. Excited, he started shouting the news to all his friends. The next morning the *Cameron* was surrounded by eight Soviet factory trawlers and four Spanish pairs."[137]

The Spanish fleet withdrew from the haddock fishery after 1960, but the small size of the fish did not deter the Soviet Union. To make matters worse, the haddock fishery was notoriously wasteful, with 50% or even 60% of fish discarded because they were too small.[138] The Soviets caught 68,000 tonnes in 1960 and 55,000 tonnes in 1961.[139] Newfoundland trawlers took 30,000 tonnes of the remaining mostly juvenile fish. By 1963, commercial trawlers could not find many haddock of any size anywhere on the Grand Banks. The haddock fishery was finished – the stocks had collapsed.

Haddock science got an early start on the Grand Banks. Harold Thompson wrote a major report titled *The Occurrence and Biological Features of Haddock in the Newfoundland Area* in Australia in 1939. The work summarized the results of the Newfoundland haddock studies from 1931 to 1936. At this time the haddock were only lightly fished, so the information in that report on growth, size, and distribution essentially described an untouched or virgin stock. Thompson was well aware of the vulnerability of haddock to overfishing, which had already occurred in northern Europe and on the major haddock grounds on Georges Bank off New England and on the Nova Scotian shelf.[140] A decade later, research surveys of haddock on the Southern Grand Banks were implemented by Wilfred Templeman and B.G.H. Johnson. These were the first scientific fish surveys on the Grand Banks. Unfortunately, they described the quick decline of the stock.

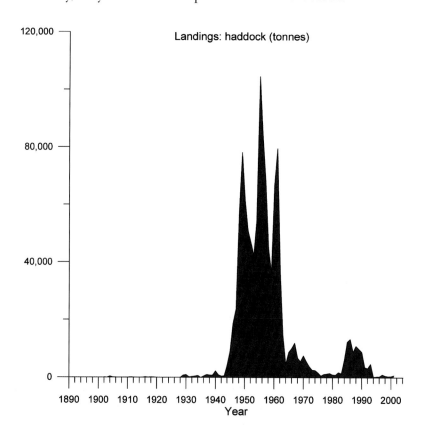

7.5 Landings of haddock (data from ICNAF, NAFO, and Department of Fisheries and Oceans sources).

A June 1957.

C June 1961.

B May 1959.

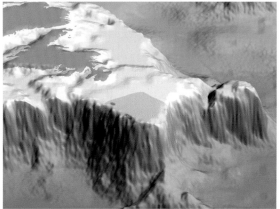

D May 1962.

Haddock catches (kg)
- >2000
- 1000-2000
- 500-1000
- 100-500
- 50-100
- 0-50

In the annual ICNAF report summarizing Canadian research in 1960, Wilfred Templeman concluded that "a crisis in the haddock fishery is evident with a rapidly declining population of haddock of commercial size in view at least for the period 1962-1964, and no significantly surviving year-classes of haddock later than those of 1955 and 1956 to provide a future commercial fishery."[141] In the same year, a research report from the Soviet Union dryly stated that "haddock caught in [ICNAF Division] 3N were mainly small, with the peak of the size curve at 34-39 [centimetres]."[142] Even as they fished out this stock, Soviet science recorded every detail: "The haddock stayed on a small bank with a sandy bottom covered with numerous mollusk shells. They fed on capelin eggs which were spawned there in July, and the great quantities of capelin eggs may have accounted for the stable concentrations and consistent catches of haddock during July."[143]

7.6
The first scientific survey of a Grand Banks fish stock (haddock) in the late 1950s and early 1960s, showing the scientific trawl catches and the abrupt decline in distribution and abundance (data from Newfoundland Biological Station reports).

In his 1961 report to ICNAF, Templeman flatly stated that "the trend in the haddock fishery over the next four or five years is expected to be strongly downward."[144] It was. A 1962 survey located few haddock, and by the early 1970s the stock was virtually gone.

At the time, science believed that the basic cause of the decline in haddock was the lack of survival of young fish,[145] but there was no evidence that young fish had ever been born in sufficient numbers to replenish the stocks. With the removal of all the large haddock by the mid- to late 1950s, it was likely that there were too few adults to produce a large number of young fish. There was evidence that relatively modest numbers of haddock can produce many young fish, more likely when there are many large breeding fish and nearly perfect environmental conditions. On the southern Grand Banks there were neither; large fish were scarce, and after 1960, production of young fish was dampened by deteriorating ocean conditions. Indeed, the abundance of haddock in the 1950s may have been atypically high, a consequence of the warm temperatures and strong productivity from 1940 to 1960 – the Grand Banks comprises the most northerly distribution of the species and the limits of its range. Thus, haddock on the Grand Banks were likely more vulnerable than those in their heartland off New England. One thing is entirely clear, however, that the Grand Banks haddock were fished out, the first Newfoundland stock to be fished to commercial extinction.

In 1980, conditions for survival of young haddock were unusually favourable, and the few remaining adult fish produced many surviving young. In the mid- to late 1980s, Canadian fishing interests learned of these fish and caught most of them before they reached maturity. A decade later, the few survivors were very large – nearly one metre in length – and could still be found in their old haunts at the edges of the Haddock Channel south of the island of Newfoundland, like pale ghosts of past glories.[146] Perhaps with some good fortune, they will reproduce successfully before they die out. But they must do so soon, as they are reaching the end of their life expectancy.

Redfish

The Soviets primarily targeted redfish in the 1950s.[147] In 1958, Icelandic vessels scoured the Grand Banks for redfish concentrations – despite their highly protective attitude towards their own fishing grounds, the Icelanders had no problem with sending their fleets to Newfoundland waters.[148] By 1958, the Soviet, East German, and Icelandic fleets[149] took at least 71,000 tonnes of redfish annually and in some years much more.[150] The redfish stocks declined rapidly, and landings decreased to 25,000 tonnes per year by 1961 and only 8000 tonnes by 1962. Having depleted the redfish, these fleets then turned to cod. By 1960, Soviet scouting trawlers had located the main overwintering and spawning aggregations of cod from the Grand Banks to Labrador.[151] In 1962, ICNAF reported that "the effort by trawlers for cod in the southern part of this subarea [ICNAF Subarea 2] has been considerably increased over the past two years with the development of a great new deepwater winter and spring fishery for this species. At the same time there has been a corresponding decrease in the effort for redfish."[152]

With the collapse of the haddock and redfish stocks, the opportunity to learn several important lessons about marine fisheries passed with little notice. The first was that severe reductions in the numbers of adults in a fish stock greatly reduced the likelihood that sufficient offspring would survive to maintain the stock, which could lead to virtual stock elimination. This confirmed what Lankester and Neilsen, among others, had concluded in the late 1800s. A second lesson should have been learned about the potentially devastating fishing power of the new fleets of trawlers. In the coming decades, not learning these lessons would prove costly to other stocks, including the overwintering and spawning aggregations of the greatest cod stocks in the world, once thought to be indestructible.

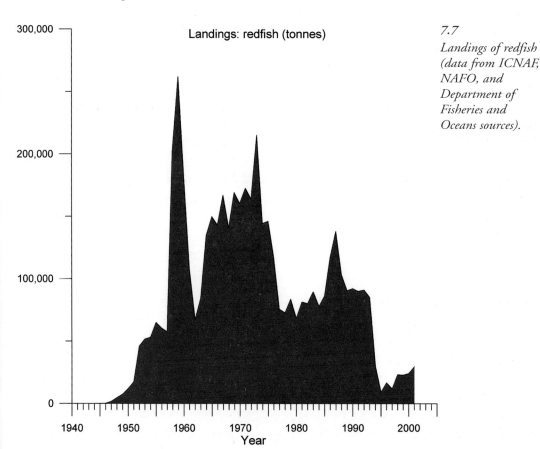

7.7
Landings of redfish (data from ICNAF, NAFO, and Department of Fisheries and Oceans sources).

Cod

It took only two decades to devastate the cod stocks. From 1960 to 1972, over 14 million tonnes of cod (perhaps 4-5 billion fish) were taken, including over 8.6 million tonnes of the northern stock, most of it by foreign fleets. The chief offenders were France, Portugal, Spain, and the Soviet Union.[153] The Grand Banks were lit up at night by cities of hundreds of trawlers and their mother ships – huge floating factories that could carry thousands of tonnes of fish. Colonel Paul Drover, who flew surveillance missions in the mid-1970s, could hardly believe what he saw: "It was quite incredible...on a night flight you'd fly about an hour offshore and all of a sudden the horizon would light up with a city of lights. It wasn't something you'd expect. The Russians would have over 200 ships there, just raking the sea bottom and the Grand Banks."[154] Cod biomass declined to less than 1.5 million tonnes, the lowest in recorded history. No stock was spared.

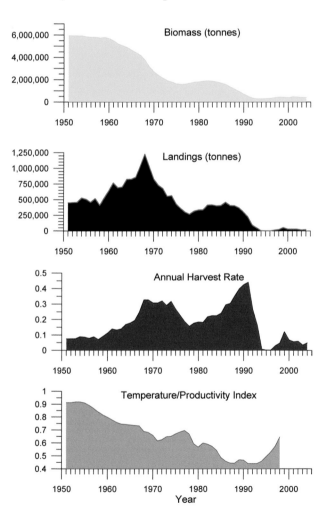

7.8
Cod biomass, landings, annual harvest rate, and the temperature-productivity index based on tree-ring analyses.

The relatively small cod stocks in the Gulf of St. Lawrence and on the Nova Scotia Shelf were the first to suffer. In the early 1960s, Canadian scientists dryly described the symptoms of overfishing:[155] "Canadian data indicate that cod are being more and more heavily fished. Line fishing is becoming less important and trawling more important. Because of heavy fishing the catch per unit effort is decreasing, large fish are becoming scarce, and Canadians are now using smaller fish than formerly." It was a trend that would be repeated over and over.

There were indications as early as 1961 that even the massive northern cod stock was declining. By then, foreign trawler landings around the Hamilton Bank, focussing on the overwintering and spawning aggregations, had increased to nearly 250,000 tonnes each year.[156] The effects on the Newfoundland traditional inshore fishery were felt almost immediately, and cod trap and longline catches went down.[157] Foreign trawler catches exceeded those of the Newfoundland fishery for the first time in the early 1960s, and it was clear to scientists in Newfoundland that the large foreign catches were damaging the Newfoundland fishery. Hodder reported in 1965 that "cod landings by trawlers from the [northern cod] stock complex…have increased substantially since 1958…coincidentally the [Newfoundland] landings in the offshore line fishery in [ICNAF area] 3L and from the inshore fishery in [ICNAF areas] 3K and 3L have declined."[158] The trawlers were fishing the same fish stocks that had supported coastal fisheries for centuries.

In 1967, an ICNAF scientific subcommittee assessing the state of the North Atlantic cod and haddock fisheries concluded that overfishing was occurring, and that mesh size restrictions, which were the primary ICNAF management tool, would not change this state. In answer to the common complaint that reduced rates of fishing would kill the industry, the subcommittee fired back that "the fishing mortality rate in most stocks is so high that, even with the largest practicable meshes, a moderate decrease in fishing would not result in a decrease in sustained catch and might well result in a slight increase."[159] The subcommittee recommended a major reduction in the fishery and concluded that "the need for taking positive steps toward effort limitation in the North Atlantic demersal fisheries before the problems become still more complex is therefore *urgent* [emphasis added]."[160] Although this advice was primarily economic and did not foresee the extent of the biological damage that was being done, even this moderate call for restraint was not heeded.

By 1968, when an almost unbelievable 1.2 million tonnes of cod were taken from Newfoundland stocks, including over 800,000 tonnes from the northern stock alone,[161] foreign catches were six times higher than the total Newfoundland fishery. This would be the peak – the highest catches of cod in Newfoundland and Labrador waters in over 450 years – and, in 1969, the ICNAF stock assessment committee concluded that:[162]

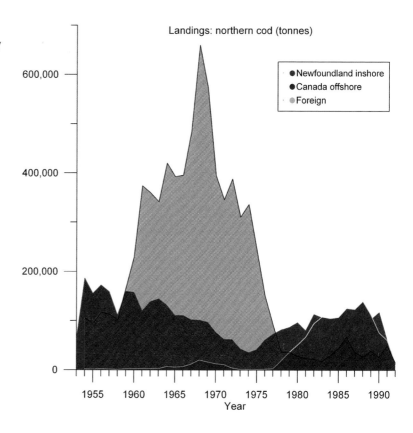

7.9 Landings of northern cod (NAFO Subdivisions 2J3KL) from 1953 to 1992, the period of great stock decline.

It is unlikely that the record catches of cod in the Convention Area in 1968 (about 1,850,000 tons) can be sustained. The high 1967-68 catches are due partly to good year-classes and partly to the removal of an accumulated stock of older age-groups. Prospects for recruitment in the near future are poor compared to the size of the year-classes that sustained the fishery in 1967-68. The effect of weak year-classes on the catches of the next few years will be made worse in some areas by the heavy fishing on the preceding good year-classes. This heavy fishing has i) reduced, to some extent, the total yield from the good year-classes...ii) concentrated the yield of a year-class into fewer years, and thereby, iii) caused higher year to year variability in catches...[and] iv) reduced spawning abundance of most of the haddock and cod stocks. This last feature is possibly the most serious...[because] there seems to be increasing evidence that a substantial reduction in spawning stock may lead to a progressive decline in average recruitment.

In simpler language, science knew well by the late 1960s that most of the haddock and cod stocks had been overfished to the point where their productivity had been reduced, even to the point of not being able to sustain themselves through reproduction.

Offshore cod catches remained high in 1969, especially on the overwintering aggregations of northern cod on the banks off Labrador, but the inshore fishery was collapsing. About 438,000 tonnes of groundfish were taken in ICNAF Subarea 2 in 1969, 91% (409,000 tonnes) of which was cod. The Soviet Union increased their catches to 131,000 tonnes, as did West Germany (72,000 tonnes) and Portugal (66,000 tonnes). Spanish catches were stable at 33,000 tonnes. Some fleets caught less than they had the year before: Poland landed 62,000 tonnes, Norway 7000 tonnes, and France 30,000 tonnes. One fishery was severely reduced – the Newfoundland inshore fishery, which "again decreased dramatically from 18,000 [tonnes] to 5000 [tonnes], the poorest catch recorded since 1930."[163]

In June 1970, the twentieth annual meeting of ICNAF was held at the Memorial University of Newfoundland. With the Newfoundland fishery in tatters and warnings from scientists that overfishing had already taken firm root, there was no mention in the press release following the meeting of problems with northern cod other than a recommendation that a 130-millimetre mesh limit be implemented in trawls. The press release did, however, welcome the delegation from Japan, which joined ICNAF that year.[164]

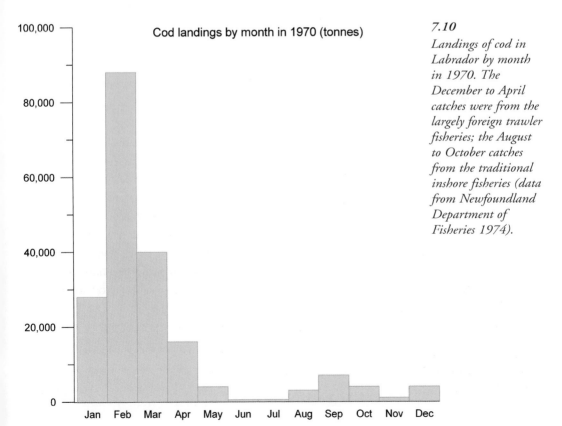

7.10
Landings of cod in Labrador by month in 1970. The December to April catches were from the largely foreign trawler fisheries; the August to October catches from the traditional inshore fisheries (data from Newfoundland Department of Fisheries 1974).

7.11

Landings of cod by gear type in 1955 and 1962 (data from Newfoundland Department of Fisheries reports). Circle diameters are proportional to landings. Trawl indicates bottom trawling and line indicates longlining.

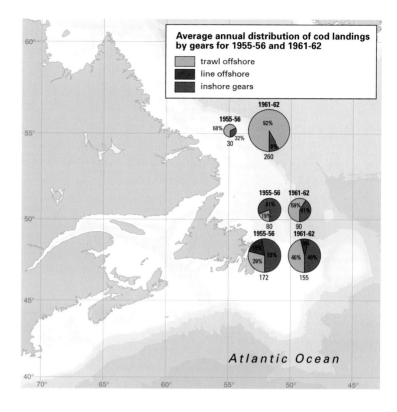

Each year in the early 1970s brought new historical lows in the coastal cod fishery in Newfoundland and Labrador. Inevitably, as the great northern cod stock declined, the size of the fish being caught became much smaller – few fish were surviving to old age.[165]

As more and more trawlers searched out the remaining fish, their catches declined as well.[166] By 1975, it was clear that strict conservation measures had to be implemented to save the fishery. At the ICNAF annual meeting that same year, Canada called for major reductions in cod fishing. Some countries, including Spain and Portugal, either denied the reality of the declines or took the position that these declines were not their fault, and stated that no reductions were acceptable.[167] There were vague commitments to conservation and sustainability, but these were largely reserved for the boardroom. At sea, where it really mattered, it remained a free-for-all.

The free fall of the cod, particularly the northern cod, brought national and then worldwide attention to the depletion of the marine ecosystems of Newfoundland and Labrador and the Grand Banks. But cod was hardly the only species impacted by the massive foreign fisheries of the 1950s and 1960s – in fact, many species were affected even more severely and more rapidly than cod. These fishes were never as abundant as cod, nor were they accessible to the historical fisheries; prior to the coming of the deep-sea trawlers, they were fished little or not at all. With the arrival of the trawlers, many of these stocks – including the flatfishes and grenadier – were literally mined to near exhaustion, and their fisheries lasted but a decade.

Flatfishes

Trawling provided an efficient means to catch flatfishes, in particular the relatively abundant American plaice, witch flounder, and yellowtail flounder. Trawling for redfish and cod would inevitably result in catches of these flatfishes and at times they outnumbered cod. A flatfish fishery developed in parallel with the foreign cod trawler fishery in the 1950s, peaking in the 1960s at about 180,000 tonnes per year. A secondary peak in flatfish landings occurred in the 1980s, similar to cod but a few years later. Much of the catch was made under ICNAF and its successor on the Grand Banks. All of these stocks in the Newfoundland and Labrador region, with the exception of yellowtail flounder on the Grand Banks, remain highly depleted at the time of this writing.

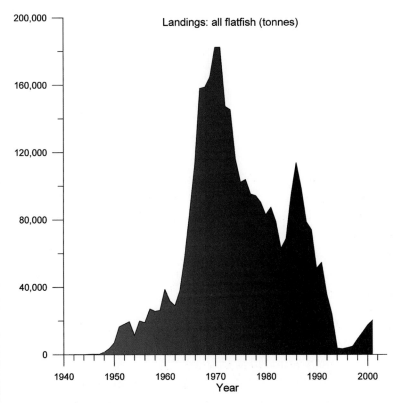

7.12
Landings of all flatfish (data from ICNAF, NAFO, and Department of Fisheries and Oceans sources).

Grenadier

As cod catches declined in the late 1960s, the Soviet scouting fleet found aggregations of the deepwater grenadier off the margins of the Grand Banks and northeast shelf. These stocks had never been fished. The Soviets developed a method to fish at depths up to and exceeding 1000 metres where the grenadier were concentrated, and within a

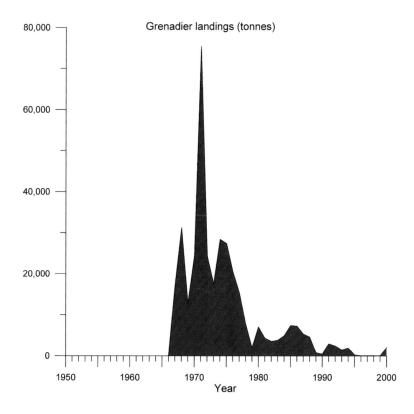

7.13
Landings of grenadier (data from ICNAF, NAFO, and Department of Fisheries and Oceans sources).

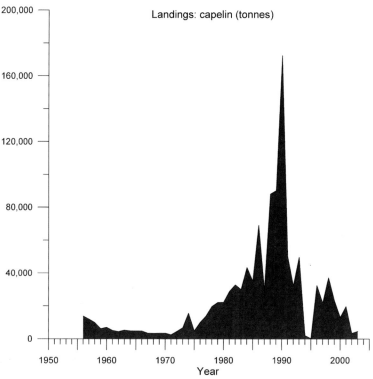

7.14
Landings of capelin (data from ICNAF, NAFO, and Department of Fisheries and Oceans sources).

few years annual catches peaked at about 75,000 tonnes. Within a few more years, the grenadier were so depleted that the fishery was no longer worth pursuing. These stocks have not recovered.

Pelagics

By the mid-1970s, with most groundfish stocks depleted, some countries turned to the pelagic fishes, the main food of the groundfish.[168] Pelagic fishes and invertebrates were typically highly migratory, moving seasonally with the changing temperatures and currents. Even more than with groundfish, fishing pelagics involved understanding their movement patterns and how these related to the environment: they had to be hunted.

Herring were considered to be nearly inexhaustible, with a world catch that approached 20 million tonnes per year in the late 1960s.[169] By the early 1970s, herring numbers were already diminishing in the northeast Atlantic, with the Norwegian spring-spawning stock reduced from perhaps 20 million tonnes to a scattering of juveniles within two decades. Off the east coast of the United States, the stronghold of northwest Atlantic herring, a veteran U.S. enforcement official recalled that "often you could count as many as 200 Communist-bloc trawlers within a 20-mile area off Hatteras Island [North Carolina]. Every one of them would be wallowing – filled to the gunwales, you might say – with herring."[170] By 1968, total catches of herring off New England had increased to 339,000 tonnes. At this point, the United States was not having any better success than Canada in restricting foreign fisheries. Despite the cold war and Cuban missile crisis of the 1960s, joint United States–Soviet Union groundfish surveys were undertaken in 1967 and expanded in 1968.[171]

For pelagic fisheries, it was now the Grand Banks' turn. Capelin were the dominant pelagic fish in most waters off Newfoundland and Labrador, and became the next main target. Capelin, like herring, were caught for reduction into fishmeal, to be used as animal feed in agriculture and aquaculture industries. Ground-up capelin fed the chickens, cattle, pigs, and, increasingly, the farmed salmon of Europe. During the mid-1970s, Norwegian and Soviet factory ships reduced about 200,000 tonnes of capelin each year, but stock abundance declined by the late 1970s, and this fishery was effectively closed in 1980.[172] The Newfoundland shore fishery for capelin increased through the 1980s: nearly 70,000 tonnes were taken in 1986 and the fishery was second only to cod in value.[173] In the late 1980s, capelin stocks were thought to have rebounded to millions of tonnes, and the capelin fishery did well – there was no hint of the rapid declines that would come within a few years.

Foreign vessels caught almost 2.5 million tonnes of Atlantic mackerel in the ICNAF area over the decade between 1968 and 1977 until they were restricted from American and Canadian waters. Canadian and American catches of mackerel during this time were minimal. The effect that this massive fishery had on the mackerel stock is not fully

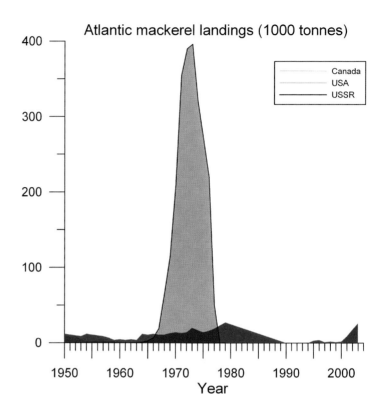

7.15
Landings of Atlantic mackerel in northwest Atlantic (data from ICNAF, NAFO, and Department of Fisheries and Oceans sources). The dominant Russian catch stands out.

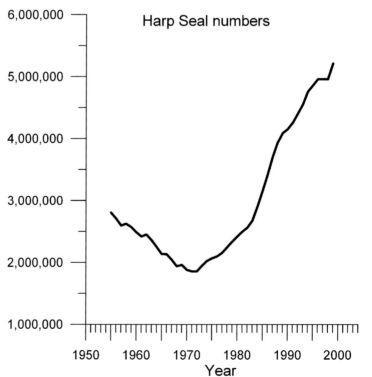

7.16
Harp seal population numbers, 1955-2000 (data from Department of Fisheries and Oceans sources).

understood, but catch rates appear to have been excessive. At the time of this writing, mackerel appear to have once again attained high abundance in the northwest Atlantic, and with the warming ocean temperatures many migrated to the northeast coast of Newfoundland in 2005 and 2006, making the fishery very successful.

Salmon

In the 1960s, Danish fishermen discovered large concentrations of Atlantic salmon off southern Greenland, and immediately began fishing for them with massive driftnets – huge walls of nets strung across the ocean for tens of kilometres. Many of these salmon were spawned in the rivers of Newfoundland and Labrador.[174] The numbers of salmon returning home to spawn declined quickly, and catches in Newfoundland and Labrador dropped from about 3000 tonnes in the early 1960s to just over 1000 tonnes in the early 1970s. In 1969, ICNAF recommended a prohibition of salmon fishing outside national fishery limits. This recommendation was neither followed nor effective, and it was weakened the next year to allow fishing outside national limits, with restrictions limited to tonnage of vessels or 1969 catch levels.[175] The issue would not be settled for another two decades, during which time salmon stocks would be reduced to very low levels in both the northwest and northeast Atlantic. The changes came too late for the salmon fishery in Newfoundland and Labrador. In 1992, a moratorium on commercial salmon fishing was declared in Newfoundland, which remains in place at time of this writing, and a voluntary buyout program of commercial fishing licences was implemented. Out of 581 fishermen, 363 volunteered to sell, leaving 218 active licences.[176] It was the end of the commercial salmon fishery in Newfoundland.

Seals

In 1971, ICNAF recommended a quota of 245,000 animals for the harp seal hunt in the Gulf of St. Lawrence and off the northeast coast of Newfoundland.[177] The following year, the hunt was shut down after massive protests by animal rights groups, which included an on-ice appearance by French actress Brigitte Bardot nuzzling a rather puzzled-looking whitecoat harp seal. It made good public relations, but poor ecology. A cruel irony was that many of the animal rights advocates were from European countries; however, they said little about the devastation their fishing fleets had inflicted on the Grand Banks, the fish stocks, and the fishing communities of Newfoundland and Labrador. With the closing of the hunt, the harp and hooded seal herds began a steady growth that would increase their numbers from about 2 million in the early 1970s to over 5 million in the early 1990s.[178] The increased seal population added additional mortality to fish stocks that were already under extreme pressure from overfishing.

Underutilized Species

With the decline in traditional commercial species, interest in harvesting other "underutilized" species strengthened. By the early 1970s, the logic of harvesting mussels, whelks, sea urchins, and anything else that had been formerly overlooked but that might be sold took hold. There was little or no knowledge of the numbers of these species in Newfoundland waters, but lack of knowledge had been no impediment to previous fisheries, so attempts were made to provide information on which a fishery might be based.[179] As the cod fishery rebounded slightly during the late 1970s to mid-1980s, interest in these species declined.

The demise of most of the fish stocks on the Grand Banks and adjacent waters was more or less complete within two decades following the *Fairtry I*'s first voyage in 1954 – the numbers and biomass of many stocks were reduced to their lowest in history. Twenty years after the Soviet side trawlers *Sevastopol* and *Odessa* first coursed the Grand Banks, the Soviets and Europeans had taken 33 million tonnes of fish from North American waters.[180] In 1974, 1076 vessels from Europe and the Soviet Union were fishing in North American waters, catching over 2 million tonnes of fish each year – 10 times more than the Americans and three time more than the Canadians.[181] The ecosystems were about to implode.

THE ECOSYSTEMS IN THE MID-1970s

In 1969, John Ryther predicted that worldwide fish catches above about 100 million tonnes were unlikely to be sustainable, based on primary production limits, and opined that "most of the existing fisheries of the world are probably incapable of [being increased, and]…many are already overexploited." His prediction was nearly perfect, and the Newfoundland and Labrador stocks were illustrative of his prognosis. By the mid-1970s, the Newfoundland and Labrador ecosystems were badly depleted. Many species had been fished to commercial extinction: haddock, redfish, grenadier, herring, and cod were down to perhaps 20%-25% of their former abundance. Harp and hooded seals were also at low levels, likely less than 2 million animals.[182]

In the mid- to late 1970s, fortune smiled briefly on these ecosystems, and productivity increased for a few years despite the general downward trend since the 1950s. Sea temperatures were moderately warm, and capelin, among the quickest to respond to changes in ocean climate,[183] were relatively abundant. The strong showing of capelin gave whales, seabirds, cod, and other fish species an abundant food supply. Cod stocks were at very low levels in the 1970s, but the key northern spawning group on Hamilton Bank, although much reduced, was still there.[184] With the coming of favourable environmental conditions, spawning groups were able to stage a relatively quick turnaround in terms of growth and reproductive capability. The numbers of young fish born in the late 1970s were better than might have been expected considering the low numbers of adults, and the stocks grew, although at a much slower rate than would have been predicted based on 1960s levels of reproduction. In parallel with the short-term improvement in ocean conditions in the mid- to late 1970s, the survival of salmon at sea rose dramatically, countering a general decline that had started at the beginning of the decade. By the early 1980s, however, ocean productivity again decreased and salmon survival would do the same.

THE POLITICAL SEASCAPE

Resettlement

Individuals and families had been moving in and out of Newfoundland outport communities to Canada and the United States and beyond ever since the first winter crews stayed ashore in the 1500s. People sought better lives or adventure outside the context of what was to many the near-indenture and lack of opportunity within the inshore fishery. By the mid-1800s, there were far more people in many communities, particularly on the northeast coast, than could make a living in the local fisheries. The Labrador fishery and the seal hunt became Newfoundland migrant fisheries that kept the economy afloat for another 60-75 years, but both of these were in steep decline by the 1920s – the seal hunt for biological reasons and the Labrador fishery for mostly economic ones.

Attempts to diversify the employment base by industrializing Newfoundland with logging, mining, and a railway had failed to alleviate the problems, and as reported by the Amulree Royal Commission in the 1930s, the fishery remained the most likely engine of economic growth in rural Newfoundland. The foreign incursions in the 1950s and 1960s effectively ended any hope of a full revival of the fisheries to historical levels and the potential for growth. In the 1950s and 1960s, the Canadian vision of the industry appeared to be a reduced,[185] corporate-organized, frozen-fish business supplying the fish stick trade in the United States, with the fishery shared with the international community. Such a fishery would employ far fewer people than it had previously, and by the early 1950s, it was increasingly evident that the 15,000 inshore fishermen and their families scattered around the hundreds of coves of Newfoundland and Labrador could not be supported at any reasonable standard. Smallwood's response was to try to get as many outport families out of the fishery as possible.

The demise of the outports played out in several ways. By the early 1950s, some outport Newfoundlanders had been exposed to better jobs and education. In particular, the American military bases in Newfoundland provided high-paying jobs during World War II and many Newfoundlanders learned valuable trades. Few of these people would go back to fishing or to the outports. In 1953, to facilitate the ability of outport families to move to better opportunities, the Newfoundland Centralization Program was initiated, administered by the provincial Department of Public Welfare. It was hardly a new idea: the Commissioner of Resources had called for a concentration of the population and fish processing in about 15 centres in 1944.[186] Only the cost of moving was covered at first, but this was soon supplemented by small cash payments of $100 to $600 to families who relocated. The catch was that all households in a community had to move before any payments were made. By 1965 about 110 communities, comprising about 7500 people, had been resettled. As the foreign fishery escalated and inshore catches plummeted, this was not enough to relieve the increasing vacuum left as the inshore fishery shrank. Smallwood issued a press release in 1957 stating that as many as 50,000 outport people "with no great future" should be resettled,[187] but this would require more funding than Newfoundland alone could muster.

In 1965, a new federal-provincial resettlement program was introduced with federal funding and more focussed goals. The new program was supposed to enable better services and prospects for resettled people in new growth centres, create a labour supply for a new industrialized society, and reduce the numbers of outport fishermen. Each household was eligible for a grant of $1000, plus $200 per person, to cover moving costs and up to $3000 for a new building lot. This was a lot of money to fishing families in the 1960s. Many saw resettlement as a chance to better the lot of their children, although few had a clear idea to what or where they were moving.[188] An additional program was implemented in 1970 under federal and provincial control.

While the foreign fleets were stripping the Grand Banks and most of the continental shelf of the fish that had supported the Newfoundland fishery for over 400 years, many families put everything they owned into a boat and sailed to a government community, some floating their houses with them. From 1965 to 1975, 4168 households totalling 20,656 people from 148 designated communities and 312 nondesignated coves were relocated. The days of the Newfoundland outport were over. A few individuals refused to leave and still remain today in communities accessible only by water, but most had little choice but to move. Around the island of Newfoundland, thousands of coves contain the rotting ruins of former outports; many graveyards, some hundreds of years old, are tended to regularly by boat.

The policy of resettlement was abandoned after 1975 as it had by then become politically distasteful. The politics of resettlement were no doubt influenced by the plans to extend jurisdiction of the fishing zone to 200 nautical miles – perhaps additional resettlement would not be necessary. With the Unemployment Insurance Program now in place, perhaps more instead of fewer people could be involved in the fishery. By the late 1970s, it was assumed that the stocks, especially the northern cod, would grow nearly exponentially, so the prospects of too much fishing were seldom considered.

7.17
The resettled and abandoned community of Ireland's Eye, Trinity Bay, showing the remains of the church in the summer of 2005 (photo by author).

With support from governments and industry, the numbers of people involved in the fishery grew quickly to levels above what they had been prior to resettlement.

The merits of resettlement have been debated for over 50 years. On one hand, contraction of the many isolated outports into fewer, more centrally located, and better-served centres was bound to happen – and, in fact, had happened prior to becoming government policy. Historically, the outports had been linked by sea and a few overland trails. In the 1960s, new roads linked the growth communities, which left the outports more isolated than ever. Many people welcomed opportunities for the better services and futures for their children thought to exist in the centres. Some of these benefits materialized and the education level of children was undoubtedly enhanced. Many people found new jobs. On the other hand, the "promised land" never materialized for some people. They had simply traded a good fishing ground for a poor or already-occupied one, with few employment alternatives. For these people, resettlement became the first step out of the fishery and out of Newfoundland.

However the resettlements of the 1950s to mid-1970s are judged, the massive movement of people was unlike any in Newfoundland's history. This was a government-sponsored, one-way ticket with no chance of turning back: at the same time that houses floated across almost every Newfoundland bay, foreign fleets were decimating the fish stocks and robbing Newfoundlanders of any chance of maintaining a sufficiently large fishery-based economy to support the outports. The growing number of resettled fishermen in the late 1970s and early 1980s would never attain the catches of their fathers – the fish were simply not there.

The notion that a smaller and more productive fishery with largely U.S. markets should be Newfoundland's lot is at best puzzling. Economist Scott Gordon thought that the heartland of colonial-minded, post–World War II Canada could not shake its old traditions and "never seriously tried to step out on her own in the developing world trading community."[189] David Alexander was blunter: "A national government dominated by central Canadians was technically incapable and mentally disinclined to assist Newfoundland to restore her international trading position. A government that could see and was interested in no other market than the United States, at a time when foreign nations were busily building vessels to come to fish in Canadian waters, was either unbelievably myopic or terribly timid about venturing out into the world."[190]

The Law of the Sea and 200-mile Limit

In 1970, with the cod and other stocks in steep decline,[191] Canada followed the lead of 57 other coastal states and unilaterally declared a true 12-mile territorial limit.[192] Bilateral agreements were made the next year with all foreign nations (with the exception of France, which was a special case as a result of their possession of the islands of St. Pierre and Miquelon) fishing within the Canadian zone to phase out their fisheries, albeit slowly and with concessions for some until 1978.[193] In 1972, the Gulf of St. Lawrence and Bay of Fundy were made exclusive fishing zones under Canadian management.

In the late 1960s, the mineral, oil, and gas wealth of many continental shelves was becoming well known, and the prospects of riches beneath the seabed were beginning to colour all concerns about zone extension. These issues, along with continued dissatisfaction with fisheries zones, led to the third United Nations Law of the Sea Conference (UNCLOS III), a convention that ran from 1973 until 1982. In the years leading up the conference, Canada and other countries jockeyed with one another to stake out positions and legal arguments. For the most part, Canadian provinces and industries wanted full control over the fisheries to the edge of the continental shelf. In Newfoundland, the Save Our Fisheries Association, headed by fisheries executive Gus Etchegary and made up of processors, the fishermen's union, and fishing communities, argued that "anything less than full management control over the main resources on the continental shelf" was unacceptable.[194] Similar views were held in Nova Scotia and British Columbia.

Canada had two central and differing situations to consider in the UNCLOS III fisheries negotiations: the anadromous Pacific and Atlantic salmon, and the Grand Banks groundfish, particularly cod. Both were fished beyond the proposed 200-mile exclusive economic zone (EEZ) that had become the cornerstone of the negotiating position of many countries. Canada's initial position advanced an ecologically based approach designed to protect species that migrated beyond 200 miles or even the continental shelf, so that species like salmon would have been protected wherever they migrated, but this approach found little favour among many states that wanted fixed geographic boundaries. In 1974, the issue came to a head. More than 100 states supported a fixed 200-mile exclusive economic zone, which covered the continental shelves of all but a few nations, and Canada's more expansive views again found little support. In 1975, Canada reluctantly abandoned its initial position, but found support for an article that could limit high-seas salmon fishing and recognized ownership of the coastal state where spawning occurred. The issue of groundfish on the Grand Banks was dropped, perhaps because these fisheries were perceived to be of less importance than salmon.[195] As pointed out by Scott Parsons, this was a "short-sighted view."[196] In reality, Canada had little support for extension beyond 200 miles, because all but a very few of the major continental shelf fishing grounds of the world were within this limit. On the other hand, there was good support for more ecological salmon management as it also impacted the United States, Japan, the Soviet Union, the United Kingdom, and other powerful nations.[197] As Otto von Bismarck once stated, "politics is the art of the possible."[198]

In the two years that followed, Canada attempted to negotiate fisheries agreements with the major foreign fishing nations through ICNAF. These were not always amicable discussions – there were confrontations between the Canadian navy and the Soviet capelin fleet in 1974 due to persistent overfishing of quotas by the Soviets. Despite growing domestic pressure to act unilaterally, Canada remained cagey if not equivocal. In December 1974, Don Jamieson, a Newfoundlander and the senior East Coast cabinet minister, argued against unilateral action because "we would find it extremely difficult and enormously costly to enforce…against countries that might…challenge Canada's action."[199] However, behind the scenes, the stage was being set to move unilaterally.

Canada's strategy was to defuse countries that might challenge a 200-mile extension through bilateral agreements. In November 1976, Canada unilaterally extended its exclusive economic zone to 200 miles by an Order in Council. A month later, ICNAF agreed to recognize Canada's right to manage all fisheries within a 200-mile zone. Unfortunately, by this time many fisheries were near commercial extinction.

The Law of the Sea Convention had achieved the custom, if not the legality, of coastal states taking exclusive ownership of the continental shelf adjacent to their shores and the rights to manage their fisheries within 200 miles. These were major accomplishments, but the convention also contained some questionable guidelines with regard to the fisheries. Article 62 stated that if coastal states did not fully exploit the fish resources within their zone, they should give access to other countries. The intent of the article was to ensure utilization of the ocean fisheries so that no nation could hoard, but not use, fish that could be exploited by others. The article was based on the theory that all species could simultaneously be harvested at maximal rates, and if there were more fish in the water than necessary to sustain a maximal harvest rate, these fish were deemed "surplus." The truth was that harvesting any stock at the maximum sustainable yield was risky enough,[200] but to harvest all stocks in an ecosystem at this level (determined for each species separately) was a recipe for disaster. In addition, the entire notion of fish being "surplus" took little account of the roles of species as both predators and prey in the marine ecosystem – all fish eat and all are eaten. Nevertheless, Article 62 achieved consensus,[201] so Canada signed 13 bilateral fisheries agreements that gave various countries access to the so-called "underutilized" species within the Canadian zone.[202] The highly valued capelin were allocated to Cuba, Japan, Norway, Poland, and the Soviet Union, and catches rose to several hundred thousand tonnes in capelin reduction fisheries to make fish meal. The Soviet Union reported catches of about 24,000-26,000 tonnes, mostly in ICNAF Divisions 2J3KL, as late as 1989 to 1990.[203] The allocation of quotas on stocks such as northern Labrador cod (ICNAF Divisions 2GH), as offered by Canada to several countries in the late 1980s, was strange politics as there had not been any commercial amounts of cod there for decades.[204] These agreements also contained trade arrangements for fish and other commodities, including those from other parts of Canada. In hindsight, it seems odd that up until the groundfish moratorium in 1992, when few fish could even be located, quotas of so-called "underutilized" and "surplus" fish were allocated to many foreign countries.

Ironically, the biggest beneficiaries of Article 62 were the Soviet bloc countries, key perpetrators of the 20-year assault on the Grand Banks ecosystems: they were given 52% of the allocations within Canadian waters. Other countries, in particular Spain and Portugal, lost out even though they were far less at fault in the destruction of the fisheries. Allocations were made on the basis of a formula that took catch histories into account, a system that in effect rewarded the guilty and penalized the innocent. As Art May opined in 1973, this type of allocation "tends to preserve a distant-fishing presence [foreign trawlers] that otherwise might be more difficult to maintain, and the greatest benefits go to the nations with the highest recent catches, the ones which caused the problem in the first place."[205] Countries with much longer histories of fishing the Grand Banks, particularly Spain and Portugal, and that had received much smaller allocations than the Soviets, moved their fisheries outside the 200-mile zone to the Nose and Tail of the Grand Banks and the Flemish Cap. There they dug in.

Limiting Canada's fishing jurisdiction to 200 miles and the allocations of fishing rights thereafter were to have major repercussions. ICNAF, which had formerly provided science and regulatory advice on most commercial fish stocks in the northwest Atlantic, now had little purpose and would be disbanded within two years. In its place, a new organization was needed to regulate the still-increasing numbers of countries wanting the fish the northwest Atlantic. The United States politely bowed out, as all of its continental shelf was within 200 miles of its shore and now within its EEZ. The real issue was the Grand Banks of Newfoundland.

Northwest Atlantic Fisheries Organization (NAFO)

Several conferences held in Ottawa in 1977 and 1978 resulted in the formation of the Northwest Atlantic Fisheries Organization (NAFO) on January 1, 1979. The inaugural NAFO meeting was held in March 1979 in Montreal and the first annual meeting in June in Halifax, home to the headquarters of ICNAF since 1949. NAFO initially had 13 members, including Bulgaria, Canada, Cuba, Denmark (for Greenland), the European Economic Community, East Germany, Iceland, Japan, Norway, Poland, Portugal, Romania, and the Soviet Union. Spain joined in 1983, but along with Portugal, joined the European Economic Community in 1987 and withdrew as an individual nation. East Germany withdrew after the reunification of Germany in 1990. After the dissolution of the Soviet Union, Russia, Lithuania, Latvia, and Estonia became members, and the Republic of Korea joined in 1993. In 1994, the European Economic Community formally became the European Union. The United States joined in 1995 and France in 1996.[206] By 1997, NAFO had 17 members.

NAFO kept the ICNAF divisions, but as its mandate was limited to the seas outside the 200-mile EEZs, its focus was, for all intents and purposes, on the outer Grand Banks – more specifically, the ancient Flemish Cap and eastern slopes that came to be known as the Nose and Tail of the Grand Banks.

NAFO was plagued with difficulties right from the start. The Grand Banks inside and outside the 200-mile line were the same marine ecosystems. For most species of fish, a single stock was now subjected to two sets of rules. Immediately after Canada implemented its EEZ, Spain threatened to withdraw from ICNAF and fish indiscriminately if its fleet was not given higher quotas.[207] Spanish vessels were reflagged – registered in Mexico, Venezuela, or Panama – in attempts to avoid compliance with Canadian or NAFO rules. Spain, Portugal, and the European Economic Community routinely objected to quotas and overfished them, at times by factors of three or four, despite the advice of NAFO science. In 1985, with an air of condescension seldom surpassed since the colonial era, the European Economic Community challenged Canada's authority to manage the northern cod, Greenland halibut, grenadier, and capelin within its EEZ because these stocks at times occupied part of the ecosystem outside the EEZ. The European Economic Community and other NAFO members wanted to fish much harder than Canada did,[208] and under NAFO, as under ICNAF, they had the right to set their own elevated quotas,[209] which they routinely exceeded. In addition, non-NAFO countries fished the same waters as NAFO countries. Gear regulations were ineffective and the use of illegal small mesh nets was not uncommon. In effect, NAFO's ability to manage, let alone conserve the fisheries, was very limited.

By the late 1980s, the few remaining commercial fish stocks in the NAFO-regulated zone of the Grand Banks continued to decline. In 1986, the NAFO-recommended quota for European Economic Community vessels was about 26,000 tonnes, but they caught more than 110,000 tonnes. NAFO science recommended that groundfish quotas be reduced to about 13,000 tonnes by 1989, but NAFO countries, particularly Spain and Portugal, unilaterally assigned themselves quotas of nearly 160,000 tonnes – about twelve times as high as the scientific advice. In addition, non-NAFO vessels, which were typically reflagged European vessels, caught an additional 20,000-30,000 tonnes. All together, with most stocks declining, foreign vessels took an average of about 135,000 tonnes of groundfish from the Grand Banks each year in the late 1980s.

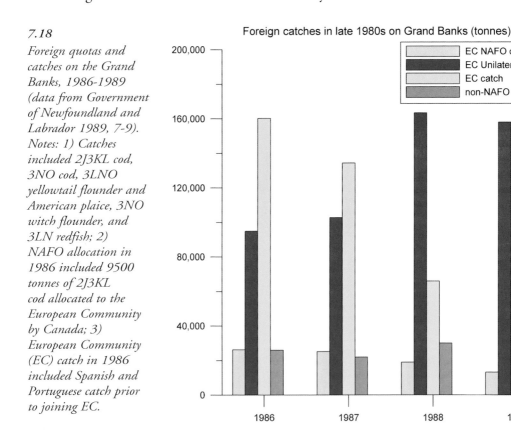

7.18
Foreign quotas and catches on the Grand Banks, 1986-1989 (data from Government of Newfoundland and Labrador 1989, 7-9). Notes: 1) Catches included 2J3KL cod, 3NO cod, 3LNO yellowtail flounder and American plaice, 3NO witch flounder, and 3LN redfish; 2) NAFO allocation in 1986 included 9500 tonnes of 2J3KL cod allocated to the European Community by Canada; 3) European Community (EC) catch in 1986 included Spanish and Portuguese catch prior to joining EC.

The European Economic Community fleets did not spare even the most threatened stocks. As the northern cod stock collapsed in 1991, Portuguese vessels and a newly enlarged Spanish fleet[210] scooped up over 50,000 tonnes of cod on the Nose of the Grand Banks. Nearly a year after Canada had declared a complete moratorium on fishing northern cod within the Canadian zone (NAFO parties were supposedly not fishing this stock after 1986), foreign trawlers still swarmed over the Nose where the northern cod were last observed in substantial aggregations. Many of these vessels were from Spain, joined by vessels registered in Panama, Korea, Honduras, Venezuela, Morocco, and Sierra Leone.[211]

Who Owned the Fish?

Prior to the establishment of territorial waters, no one owned the fish and anyone with the capacity to fish could take them. Even after the imposition of 3-, 6-, and 12-mile limits, the only stocks for which ownership might be claimed were those few that lived out their lives within those confines. None of these were of great commercial importance in Newfoundland, with the possible exception of lobster and the small coastal cod stocks. After 1949, these coastal stocks belonged to Canada, whose responsibility it was to manage the fisheries for the benefit of both Canada and Newfoundland and Labrador. It is important to note that Newfoundland did not bring the larger Grand Banks and Labrador stocks into Confederation with Canada, as according to international law, they were not theirs to bring. These stocks became "owned" by Canada only after negotiation of the 200-mile limit.[212]

How fish owned by the state would be managed and allocated to various fisheries was another matter. Unlike the ICNAF allocation policy, which was based on catch history, the Canadian approach was a mix of adjacency (those closest to the fish), historical attachment, social justice (sometimes called equity), and economics. Under these principles, Newfoundland and Labrador would gain the lion's share of the fish in adjacent waters and a priority was given to the traditional inshore fishery.[213]

Allocation issues have always been coloured by politics. In the late 1970s, a dubious perception developed with respect to the northern cod that "this resource was surplus to Newfoundland's need and, due to the size of the stock, new user groups…[would be]…granted allocations despite their non-compliance with the [established] principles of allocation."[214] Behind this, no doubt, were the interests of the moneyed fish merchants of Nova Scotia, who not only lobbied Ottawa effectively for access to the northern cod, but also for funds to build trawlers to fish it and plants to process it. In short, as deduced by Cabot Martin, "National Sea [Products] set out with federal subsidies to build a new 'distant water' dragger fleet,"[215] that would in effect replace the now displaced foreigners.

Allocation always required tradeoffs between various types of fisheries, which in Newfoundland and Labrador pitted the inshore against the offshore fisheries. The history of this animosity, well-rooted in the foreign trawler depletion of so many stocks, was transferred to all Canadian trawlers. To be fair, each fishery had advantages and disadvantages. All of the cod fisheries were seasonal. Winter trawler fisheries produced good-quality frozen fish, but were expensive to operate and maintain, interrupted overwintering and spawning behaviour,[216] and risked environmental damage. An economic analysis conducted in the late 1970s indicated that the cheapest method of catching cod was with the cod trap, but the trap fishery often produced a glut of fish that resulted in a poor quality product.[217] The same analysis showed that, when processing costs were taken into account, the most economical fishing method was a 55-foot (17-metre) longliner, which could fish for perhaps six to seven months of the year. Factory trawlers were the least economical method.[218]

Social arguments were often made to justify allocation. The inshore fishery was at times assumed to be a social or state-dependent fishery, in contrast to the profitable and rational trawler fishery.[219] This was a continuation of Stewart Bates' logic. In reality, both types of fisheries received government subsidies in one form or another, and the trawler fishery was no more economic than were the inshore fisheries – and, at times, less so. Three years before the implementation of the 200-mile zone, a $20 million federal program was put in place which largely assisted the trawler fisheries. An additional $51 million of "bridging funding" was put in place for fishermen of various species the next year. At this time, a policy of no increase in the number of vessels fishing groundfish was also set down.[220] However, industry could – and did – replace old vessels with newer models with much greater fishing power.

Newfoundland was frustrated by the applications of the adjacency and "historical dependence and attachment" principles in determining allocations of fisheries. Gulf of St. Lawrence redfish were traditionally caught by Nova Scotia and south coast Newfoundland fisheries, and like many stocks, were overfished by the mid-1970s. When the stock briefly rebounded and the fishery reopened in the late 1970s, many traditional users, including large Nova Scotia interests and the plants at Burgeo, Ramea, Harbour Breton, Gaultois, and Fortune in Newfoundland, were left out of the allocations. The principle of adjacency favoured companies situated within the Gulf of St. Lawrence in Quebec, New Brunswick, and Prince Edward Island. In this case, adjacency trumped historical attachment. However, the federal government had an ace up its sleeve: in return for losing the Gulf of St. Lawrence redfish, companies – especially those in Nova Scotia – would be subsidized to fish northern cod. As Aiden Maloney stated, the federal subsidizing of new entrants was simply "paving the way for the presence of many Canadian [Nova Scotian] offshore vessels in the northern cod fishery"[221] that had neither "historical dependence and attachment" nor adjacency to this stock. Giving up redfish for allocations of northern cod, and subsidies to boot – it must have seemed like a good trade to the trawler companies.

The complexities of allocations continued to plague the Newfoundland and Labrador fisheries after 1977 and would remain one of the chief areas of contention well into the twenty-first century. At the time of this writing, it remains unclear to many in the fishery just who owns the fish and for whom the management tolls.[222]

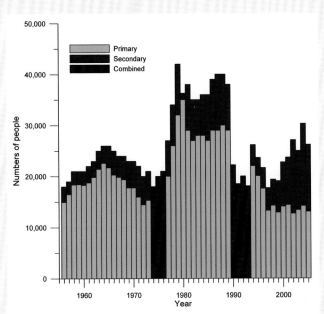

7.19 Employment in the Newfoundland and Labrador fishery (data from Newfoundland Department of Fisheries and Aquaculture and Department of Fisheries and Oceans sources). Primary employment refers to fishermen, secondary to plant workers.

UI

Beginning in 1957, the Unemployment Insurance Program – now euphemistically called the Employment Insurance Program, but in Newfoundland universally referred to as "stamps"[223] – was granted to fishermen. Workers in Canada pay into the federally run program as insurance against loss of employment. Self-employed persons, including farmers and seasonal small business operators, are not eligible to pay into or draw from this program. In 1957, Jack Pickersgill, a Manitoba farm boy turned Newfoundland bayman and federal politician for the Bonavista-Twillingate riding, succeeded in having the *Unemployment Insurance Act* amended to cover fishermen. Despite the good intentions of Pickersgill in finding ways to combat poverty in rural Newfoundland and Labrador, unemployment insurance traded the proper development and protection of one of the world's premier fisheries for what amounted to a lifestyle on the dole.[224]

Unemployment insurance had unintended consequences: the number of "fishermen" in Newfoundland and Labrador increased by 25% within five years of its implementation. Buoyed up by unemployment insurance, fishing became a much more attractive occupation rather than the so-called "employment of last resort." Unemployment insurance quickly became a one-way transfer of funds instead of an insurance plan. From 1957 to 1962, "fishermen's contributions have totalled $4.6 million while payments have amounted to $49.9 million, making a net cost of over $45 million. Benefits paid to Newfoundland fishermen in the year 1961-62 alone amounted to $3,639,000."[225] Many of the new fishermen were "ill-equipped in boats, gear, and techniques to make a satisfactory income."[226] Under the new rules, fishermen and fish plant workers needed to work only 11-14 weeks during a year to qualify for unemployment insurance payments. Plant workers – many of whom were wives of fishermen – could work for about four months, get their stamps, then draw unemployment insurance for the balance of the

> year.[227] Fishermen could only draw unemployment insurance from January to May; their pay was based on their catches, which meant that they might not qualify for unemployment insurance in a bad season.
>
> In rural Newfoundland, full employment came to mean everyone having their stamps for the year, rather than working for the year. Plants "stamped up" their workers, then laid them off and hired others. Many families in rural Newfoundland became dependent on unemployment insurance payments, their only alternative being welfare.
>
> Dependence upon unemployment insurance has affected fisheries management and conservation. Some fisheries were carried out not because they were profitable ventures in their own right and within conservation limits, but to generate stamps. The capelin fishery of 2003, with a going price of a few cents per kilogram, would not pay for the fuel used to catch the fish, but would generate stamps. Such fisheries may not have mattered when stocks were abundant, but were harmful when they were not.

Irrational Rationalization after 1977

In the late 1970s, another race for the fish occurred, based on a self-generating euphoria about rebuilding fish stocks[228] and attempts to rationalize the fishery. What followed was an irrational binge of overcapacity and overfishing. Both the provincial and federal governments participated in the binge, as did banks and industry. Newfoundland's rationalization was based on developing its own industrial capacity to revitalize the fishery and to ensure that the underutilization argument could not be used against the province: this meant a trawler fleet and new and modernized plants. Despite federal cautions that adequate, if not surplus, capacity already existed to harvest the growing stocks,[229] Newfoundland and Nova Scotia were working on fleet development plans that would expend $90 million on new vessels.[230] In 1978, Minister of Fisheries and Oceans Romeo LeBlanc addressed the Fisheries Council of Canada, a group representing industry, pleading that "the present groundfish fleet of larger vessels has the capacity to take half again its present catch, and provide better incomes – if we increase the fish…and catch rates. If we do it the other way around – increase the fleet first – we are like a man with an exhausted woodlot, who instead of planting more trees…spends all his money on more chain saws." LeBlanc continued: "I would like to see you join me in resisting suggestions that fleets should be vastly expanded, that plants be vastly enlarged – in other words, to resist the temptation of exaggerated expectations. I see no faster road to disaster than forgetting the very simple lesson that biology cannot keep up with the technology – that the wealth of the oceans cannot yet match the greed of man."[231]

Few people in industry joined LeBlanc; in fact, some outside companies re-entered the Newfoundland fishing industry during this period. However, the biggest changes in Newfoundland were in two of the old merchant companies that had survived to the 1970s: Fishery Products and the Lake Group. The provincial government invested heavily in Fishery Products, and the Lake Group was similarly, if less aggressively, assisted. The groundfish freezer capacity in Atlantic Canada increased by 150% from 1974 to 1980, and the number of licensed fish plants in Newfoundland grew from 147 to 225 from 1977 to 1981 – representing a 60% increase in four years.

The federal government also encouraged new capacity. Although Minister LeBlanc's more cautious approach was supported by science, it was neither universally accepted nor followed within his government. The federal Department of Regional Economic Development contributed over $46 million to assist 260 enterprises in developing new capacity between 1977 and 1982; Canada Works also contributed to the increase in fish processing capacity.[232] Nevertheless, the bulk of the motivation for expansion came from industry, funded by provincial governments and banks.

The expansion was breathtaking and occurred in both offshore and inshore sectors. As summarized by Bill Shrank, the era saw:[233]

> the number of registered fishermen rising between 1976 and 1981 from 14,000 to 34,000, a doubling of the number of inshore vessels, a doubling of the value of offshore vessels (despite a freeze on entry into the groundfishery of vessels greater than 65 [feet] [20 metres], the fishing capacity of this segment of the fleet increased considerably), [and] a tripling of the freezing capacity of the fish processing plants. Much, but not all, of this expansion was financed by the federal and provincial governments. The outstanding loans of the Provincial Fishermen's Loan Board, which financed inshore vessels, quadrupled. At precisely the wrong moment, in 1977, the fishermen's unemployment insurance system was made remarkably generous, thus inducing fishermen to enter and remain in the industry. Net fishermen's unemployment benefits during the expansion increased fourfold…certainly, former fishermen were not the only "new" entrants in the late 1970s since the number of fishermen in 1968 was only 19,000, while by 1980 there were 35,000 registered fishermen.

Despite the increases in groundfish plant capacity and the intentions to spread the fishery over the year, the new trawler fleet initially produced a glut similar to that of the trap fishery, although this one occurred in the winter. In 1981, the annual quota had been caught by the end of February and the fishery was closed. The 45,000 tonnes of cod caught in the first two months of 1981 were processed into cheap cod blocks, not the more valuable fillets, forfeiting any economic advantage of an extended season. This situation gave rise to the implementation of Enterprise Allocations program the next year.[234]

The Nova Scotia Connection

The rise of the Nova Scotia connection to the northern cod sealed the fate of this once great fish stock – time would prove that there was more than enough fishing capacity in the adjacent and historically dependent Newfoundland-based fisheries. In the late 1970s, H.B. Nickerson & Sons Limited, a one-plant operation, implemented a takeover plan that would make it Canada's largest fish processor.[235] With the assistance of the powerful Sobey family and other influential Nova Scotians, as well as funding from Nova Scotia government sources,[236] Nickerson & Sons gained control of the much larger National Sea Products in the same year that Canada extended its jurisdiction to 200 nautical miles. The motive was clear: they wanted access to the northern cod. The next year, the Nickerson–National Sea Products group launched a major lobbying effort to discredit what they deemed to be an overly cautious federal fisheries strategy, especially for northern cod.[237] Regrettably, Nickerson–National Sea Products gained access to northern cod shortly afterwards and had plans to build an ice-strengthened trawler fleet that was to be based in Nova Scotia and Newfoundland. In 1979, the Nova Scotia Department of Development loaned Nickerson & Sons C$25 million for trawler construction, and by the early 1980s, Nickerson & Sons were in hock to the Nova Scotia government for about C$57 million. Restrictive federal policies prevented additional vessels from being added to the fleet, but larger and more powerful replacement vessels were allowed. Newfoundland fishing interests were vehemently opposed to this breathtaking expansion, but their voices fell on deaf ears.[238]

Within a very few years, the expanded trawler fleets in both Newfoundland and Nova Scotia were in big trouble. The finger of blame was pointed in many directions: poor markets, a deepening recession in the United States, the high-valued Canadian dollar, inflation and high interest rates, insufficient fish supplies, and too many fisheries workers. In a *Globe and Mail* article, the president of National Sea Products stated that poor economic conditions were causing the closure of fish plants (primarily in Nova Scotia), but low quotas assigned by government were creating these conditions.[239] Pressure for more fish to support the newly expanded industry grew. Fisheries industry leaders in Newfoundland were of a different mind, and with a flair seldom mustered by their mainland counterparts, were more critical of governments. Subsidization of the fleet expansion in Nova Scotia was considered to be "an absolute menace to the Newfoundland fishery,"[240] and provincial Minister of Fisheries Jim Morgan was said to be acting "more and more like a bionic mouth which is unattached to any intelligent form of life."[241] For his part, Morgan fired back that the trawler companies would not be allowed to run roughshod over the Newfoundland fisheries: "If you want to see someone get tough with these companies, you'll see it in the next little while."[242] Even bankers waded into the fray, with a Royal Bank representative asserting that the real problem was "[Newfoundlanders] who continue to demand a better standard of living than we can afford."[243] It was a questionable start to the brave new post-EEZ world.

In the mid-1980s, the issue of factory-freezer trawlers resurfaced when National Sea Products, once again backed by the Nova Scotia government, requested licences to operate these vessels in the northern cod fishery. Newfoundland was dead set against this, as it meant fewer jobs in fish plants and increased the risk of overfishing. In the end, the federal government allowed three licences: one for National Sea Products, one for Fishery Products International (who had not applied for one), and one for other companies. John Crosbie, whose family had a long history in the Newfoundland fishery and who was a senior federal minister from Newfoundland at the time, opposed the introduction of factory trawlers into the fishery, but was outvoted in cabinet.[244]

It was not just Nova Scotia (and, unofficially, Quebec[245]) that was allowed to fish northern cod and other stocks in Canadian waters off Newfoundland and Labrador. Foreign fisheries continued to be granted access to several stocks within the 200-mile zone on the basis that the fish were surplus to Canada's needs. In addition, many stocks continued to be fished heavily just outside the boundary. The Government of Newfoundland and Labrador routinely protested these allocations during the mid-1980s. As late as 1985, Ray Andrews, then Deputy Minister of Fisheries, fumed to Minister Tom Rideout that[246]

> substantial portions of the contiguous fishery resources are not being utilized fully to the benefit of Newfoundland…foreign fishing outside 200 miles has prevented the full restoration of the Grand Banks cod stock, and stocks in the Flemish Cap area. The Province has stressed the necessity of extending the Canadian exclusive economic zone as a means of ensuring the conservation of fish stocks both inside and outside 200 miles. The Province continued to oppose the foreign allocation of non-surplus stocks within 200 miles…it is estimated that in 1985, foreign fleets took approximately 86,000 tons [87,376 tonnes] of cod and flounder [estimated to be worth about 2500 jobs and C$100 million] from the Grand Banks compared to approximately 145,000 tons [147,320 tonnes] of Canadian harvest in the same area.

Licensing

In 1973, Canada announced a plan to license all east coast fishermen and their boats. Licensing was imposed in the mid-1970s in Newfoundland and Labrador, but there was no restriction on numbers of licences granted, so it became more a registration system than a method of limiting access. However, it set the stage for reduction: in announcing this program, federal Minister of Fisheries Jack Davis mentioned that the government "meant to encourage expansion [additional licences] in the offshore sector."[247] Although this did not happen, this seemingly innocent comment is important because it was made several years before biological projections were issued for the badly depleted fish stocks.

The licensing issue was not received well in Newfoundland and Labrador. In 1978, at the national First Ministers Conference, Premier Frank Moores argued that either all licensing and quota allocations should fall to the province, or that at minimum, there should be a form of co-management. Moores rued the fact that the fishery had been given away at the time of Confederation: "In the case of the fisheries, we as a Province have neither ownership nor control over a resource which is of vital importance, socially, and economically."[248] Moores must have known that Canadian interests, particularly in Nova Scotia, were rubbing their hands in glee at the prospect of getting in on the Newfoundland fishery, especially for northern cod. Newfoundland established a royal commission to investigate all aspects of licensing and how this might influence the evolution of the fishery. However, the commission was upstaged by Minister LeBlanc, who called a press conference to announce a federal licensing program for all provinces: one that would recognize different types of fishermen and, to placate local concerns, be administered locally, "not by some committee of officials in some…ivory tower."[249] LeBlanc, who in general was against offshore expansion,[250] had an unlikely ally in Newfoundland's Fishermen's Union.

The Fisheries, Food, and Allied Workers Union (FFAW)

In 1970, with cod selling for 2.5 cents per pound (5.5 cents per kilogram) and plant workers excluded from the provincial minimum wage law, Father Desmond McGrath called Richard Cashin to suggest formation of a new union. Attitudes towards fishermen and plant workers had changed little in over 100 years: "The prevailing attitude towards people in the fishery was that they were not capable of running their own affairs."[251] McGrath and Cashin organized the first Northern Fishermen's Union (NFU) at Port-au-Choix soon after; by the fall of 1971, they had joined forces with locals in many fish plants around the province as well as the United Food and Commercial Workers International Union (UFCW). In 1987, the locals withdrew from the UFCW to form the Fisheries, Food, and Allied Workers Union (FFAW).

Soon afterwards, fishermen and plant workers, through their union, made it known that they would no longer be dictated to by industry and governments. The first strike was against the Lake Group operations in Burgeo. Spencer Lake was an easy target: he had said that he would never operate under a union contract, and while "he was fond of the people of Burgeo,"[252] Lake thought them incapable of running anything themselves. In the end, Lake left Burgeo and the Lake Group later folded. The provincial government nationalized the Burgeo plant and signed a union contract. The same year, the provincial House of Assembly passed the *Fishing Industry Collective Bargaining Act*, which for the first time gave fishermen the right to bargain for the price of their fish.[253]

More was to come. A strike against Fishery Products in Port-au-Choix spread to the growing trawler fleet on the south coast. Although the Port-au-Choix trouble was resolved quickly, the trawler strike was not. The main problem was that trawlermen were considered by the companies to be "co-adventurers" in the fishing enterprise, not employees. Hence, if a trip did not pay, neither did the company. This dispute went to arbitration under Leslie Harris of Memorial University, and the companies eventually conceded that trawlermen would be paid for their work.

The troubles of the early 1980s revived many of the disputes between the union and the fish processors. In 1980, facing severe economic difficulties, the companies "declared war"[254] on fishermen by unilaterally dropping the price of fish – and that summer the inshore fishery came to a standstill.

The FFAW became a powerful force in the Newfoundland and Labrador fisheries, with considerable political influence. In the late 1970s, when the licensing debate plagued Minister LeBlanc, he turned to the FFAW, with their 8000 full-time and 15,000 part-time members, both fishermen and plant workers. The FFAW had argued for licensing since 1977, and that licensing should bring rights of some sort and not be simply a registration tool. The FFAW knew that their part-time members, who often had other jobs, could interfere with full-time fishermen's ability to make a living. The inshore season was short – when cod were abundant, full-time fishermen might not be able to sell their catch because part-timers glutted the fish plants. The FFAW took the initiative to change this situation in 1977, with bona fide fishermen, as determined by local committees, given priority to sell their catch: "There's an old saying in Newfoundland that 'everyone has a right to fish.' But what the new collective agreement establishes is that full-time fishermen enjoy special rights in selling their catch."[255] The FFAW argued for local control over licensing, licences for fishermen rather than for boats, and that, upon retirement, licences should not be sold but should revert to the government. There is little doubt that federal Minister LeBlanc's actions were strongly influenced by FFAW policies.

In the mid-1980s, Newfoundland government policy remained "to acquire a greater degree of Provincial control in the area of fisheries management."[256] Newfoundland

Premier Brian Peckford rejected outright the federal licensing plan, and his government issued a pamphlet arguing that the classification of fishermen would rank two-thirds of Newfoundland and Labrador fishermen as part-timers. In the pamphlet were arguments opposing the federal initiative to put the inshore fishery on a quota and recommendations that any decisions on licensing should wait for the coming report of the Newfoundland royal commission.

Both union leaders and politicians were caught between what LeBlanc described as "the rationalizers wanting to develop and consolidate everything in sight, even if it means doing away with small fishing villages…[and] the rural romantics arguing that we must preserve the coastal way of life at all costs, even when it means preserving poverty."[257] The dichotomy of these views was most extreme in Newfoundland and Labrador, where fishing was much more firmly rooted in the culture and a tradition of egalitarianism existed among fishermen. Moreover, rationalizing was made difficult because alternative forms of employment were in short supply. In the end, few could argue against some form of rationalization, although the form it would take was much disputed, and policies were implemented that tried to satisfy both the romantic and rationalist points of view. A variety of licences were created, a trawler fishery was developed, and the traditional inshore fishery was continued, with an "allocation," not a quota, supported by unemployment insurance.

During this era of changing fisheries policy, the government of Newfoundland and Labrador lost on most counts. In the early 1980s, the federal government had ignored provincial requests and demands for more control over the licensing of fishermen, and instead granted that significant authority to the FFAW. Of greater importance, a blend of economic rationalization and patronizing romanticism triumphed, neither of which was consistent with Newfoundlanders' notions of themselves. The argument that the resources in the oceans off Newfoundland and Labrador should be used to benefit Newfoundlanders and Labradorians became a cliché, but in reality cut little ice. The federal government would make few concessions and Newfoundland would be granted no say in the management of these major resources.

It may be argued that Newfoundland's demands for management authority necessitated constitutional change, or at the least a change in law, which may not have been politically possible. However, Minister LeBlanc believed that increased provincial responsibilities were not advisable and that a mishmash of regulations might result, making management particularly difficult in waters adjacent to more than a single province. It is impossible to know if increased provincial involvement would have bettered the track record of the fisheries after confederation. All that can be said is that various forms of co-management, involving both federal and local authorities, have now become the norm, thus far with good results.

WATERSHED – THE KIRBY REPORT

> The cod of 2J, 3KL,
> produces excitement pell-mell,
> It's a fast growing stock,
> Which all want to dock,
> Who will get it? Well, we're here to tell.
> — A Kirby Task Force wit[258]

By 1980, the Newfoundland fishery was dominated by three large industrial complexes, each propped up by provincial government funds and bank loans: Fishery Products (Newfoundland), Nickerson–National Sea Products (Nova Scotia), and the Lake Group (Newfoundland). Together they controlled over 70% of fish production.[259] The balance was largely in the hands of a smattering of traditional and small Newfoundland-based, family-run companies; Smallwood's imported Canadian and U.S. companies had by this time largely fled.[260]

By 1981, all of the large fish companies were in crisis. Their problems were economic – the fish stocks, although depleted, were growing. The main problems were weakened U.S. markets and quickly rising interest rates (the large companies, especially Nickerson–National Sea Products, had expanded on borrowed money). The south coast of Newfoundland was in particular trouble, as the Lake Group, which operated many of the plants there, was nearly bankrupt. The Bank of Nova Scotia, which held most of the debt, wanted to close several plants permanently.[261]

The federal government formed a panel in 1982 to investigate the problems of the fishery and find solutions to them. Under the chairmanship of Dr. Michael Kirby, the panel began work immediately to save the Lake Group, and the federal government put up C$13 million to keep its plants open for one year.[262] This was the beginning of a massive restructuring and subsidization of the large companies in Newfoundland and Nova Scotia. The panel submitted its report later that same year and a formal release, titled *Navigating Troubled Waters: A New Policy for the Atlantic Fisheries, Highlights and Recommendations, Report of the Task Force on Atlantic Fisheries*, was made in December.[263] The Kirby report was pivotal to the future of all Canadian Atlantic fisheries. By 1982, Nickerson-National Sea Products was following the Lake Group into bankruptcy, with the Bank of Nova Scotia on the hook for nearly C$40 million. Nickerson-National Sea Products submitted several requests to the federal government for assistance. Later that year, Fishery Products in Newfoundland received refinancing assistance from the federal government and the Bank of Nova Scotia to prevent insolvency.[264]

Given the state of the fishery, economics and social aspects of the fishery became the Kirby Task Force's central mandate – it was charged with making recommendations to enable "an Atlantic fishing industry [to develop] that does not require regular or periodic government subsidies in order to survive. Indeed, the Task Force was established because, in late 1981, it became clear that once again the industry would probably require a substantial infusion of public funds – as it had in 1968 and again in 1974-76 – if it was to avoid almost total collapse. The federal government wanted to break this cycle of periodic bail-outs…[and] to employ as many people as possible."[265] It is ironic that a "substantial infusion of public funds" had been made in 1968, the very year that foreign fishery catches peaked at over 1.2 million tonnes of cod from Newfoundland and Labrador, and in 1974-76, when as a result of the foreign fisheries, the cod were virtually gone.

The economic problems identified by the Kirby Task Force included common property access, interest rates, operating costs, sluggish markets, quality, poor management of the processing sector, excessive regulation of fishing, with the priority placed on social objectives, particularly by the Newfoundland and Labrador government. There was also an emphasis on reducing the cost of Canadian fish. To address these problems and to revitalize the fishing industry, it was necessary to assume that there would be continued growth in the cod stocks, particularly the northern stock. At the time, few doubted that this would occur – the long-term damage already done by foreign overfishing was not recognized.[266] There is scant mention of the biological-ecological perspectives of fisheries science in the Kirby report, especially of the large uncertainty about the growth projections of the northern cod as outlined in Pinhorn's report[267] a few years earlier. Despite these scientific cautions, the Kirby Task Force *assumed* that an abundance of cod was coming and was perplexed about what to do with it all: "The fishery confronts us with a disturbing paradox. On our doorstep, we have one of the world's great natural fisheries resource bases, one that has made dramatic improvements since the extension of fisheries jurisdiction to 200 miles on January 1, 1977." The Kirby report was correct up to this point, but what follows is off the mark: "Although the industry has many problems, a shortage of fish is not one of them. By 1987, the groundfish harvest should reach 1.1 million tonnes, an increase of about 370,000 tonnes over 1981…Almost all the increase will be confined to one species – cod. And about 70 per cent of the growth in the harvest will take place off the northeast coast of Newfoundland and Labrador."[268]

The Kirby Task Force saw their main job as divvying up an ever-increasing pie and using this growth, combined with the export of frozen fish to the U.S. market, to solve the economic problems of the fishery. Unlike the Dawe Report or that of the Amulree Royal Commission, there was little emphasis on small cottage industries or on other markets, such as for high-quality salt fish in Europe. In hindsight, this was a mistake.

The Kirby report left little doubt that all hopes were pinned on the northern cod to save the Atlantic fishery, not just the fishery of Newfoundland. The aforementioned wit's poem compressed thoughts that "the allocation of northern cod reserved for delivery to resource-short plants offers a major opportunity for further Canadian development of a freezer trawler fleet."[269] In practice, this meant increasing access to northern cod not just by industry in Newfoundland and Labrador, but also by Nova Scotia and Quebec firms (although most of their plants were in Newfoundland). The Newfoundland government opposed much of this development, as did Newfoundland fish companies.[270] However, it was never clear that the Newfoundland position was based on a concern for the stocks, rather than on a baser jingoism, and this weakened their position.

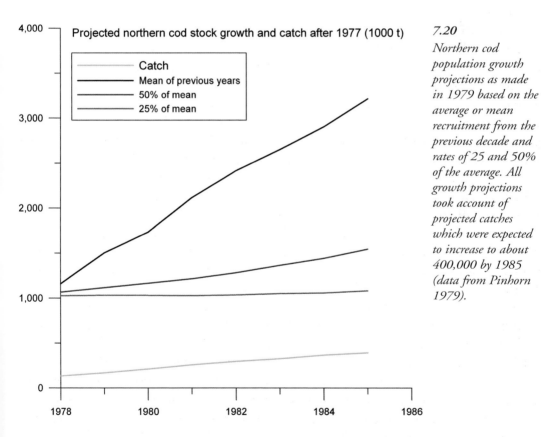

7.20 Northern cod population growth projections as made in 1979 based on the average or mean recruitment from the previous decade and rates of 25 and 50% of the average. All growth projections took account of projected catches which were expected to increase to about 400,000 by 1985 (data from Pinhorn 1979).

The Kirby Task Force was not overly concerned with the foreign fisheries. NAFO countries were thought to be allies; if there was any problem at all, it was the non-NAFO countries:[271]

Despite the best efforts on international stock management by NAFO, the activities of states that are not members of NAFO have resulted in cod stocks beyond 200 miles not being rebuilt. This has led to pressure from foreign governments to be allowed to fish, particularly for cod, within the Canadian zone…restrictive measures taken by Canada inside 200 miles to improve Canadian catch rates results in charges that there is a "surplus" that is not being allocated as required under the Law of the Sea. If it were allocated [to the foreign fleet], it would make the results of the first problem even worse, because it removes the incentive to rebuild stocks in the international area and, at the same time, makes available to our competitors raw material to compete with Canadian products on world markets.

History has shown that both NAFO and non-NAFO members were overfishing at rates sufficient to deplete the stocks. The fear of having to allow foreign fleets into Canada's EEZ if there were any surplus fish, as stipulated under the Law of the Sea, was instrumental in the rationale of the Kirby Task Force to ensure that there would be none.

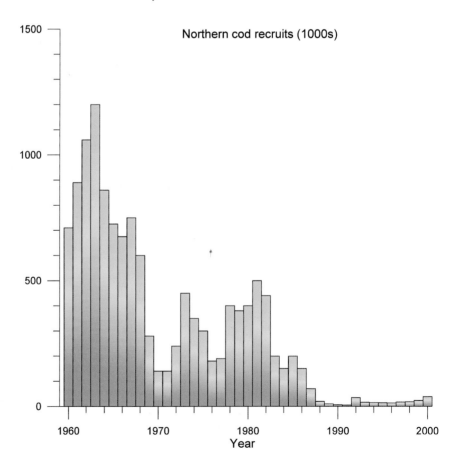

7.21 Northern cod recruitment (data from Anderson and Gregory 2000).

The Kirby report made 57 recommendations, most of them in the areas of economics and marketing as per its mandate. The Kirby Task Force outlined the problem:[272]

> The extremely strong price performance of fish products between 1969 and 1978…compounded by the declaration of the 200-mile limit and the rapid increase in Atlantic coast landings, created extraordinary optimism in the industry, in governments, and in the financial community. The optimism together with the inevitable competition among stake holders to be the first to take advantage of the 200-mile limit, accounts for the surge in investment and employment in the fishery…the financial crisis in the processing sector was the problem that precipitated the latest round of special government attention to the fishery and led to the establishment of the Task Force…The most important [problem]…in the immediate future is the near collapse of several major processors with extensive operations in Newfoundland.

Scott Parsons put it more bluntly: "More simply, the problem was unbridled greed which led to debt-financed overexpansion."[273]

In the year following the release of the Kirby report, many solutions and sources of financial salvation were put forth in consultations among the federal government, the governments of Newfoundland and Nova Scotia, banks, and private investors. For its part, the government of Newfoundland was not entirely in agreement with the corporate approach. In 1983, a detailed alternative plan was presented by the provincial to the federal government. The document stressed that "all parties must recognize that the scope of the issue extends beyond reconciling the future of these large companies. The industry is more complex than that. It involves the future of communities, the future of small operators who must compete with their larger cousins both for raw materials and to market their product, and it involves a resolution of international problems to fully access available resources."[274]

The province was against the federal notion of "vertical integration" of the fishing industry, in which processing companies operated their own trawlers and marketing arms. It argued for the separation of these levels of industry, with consolidation of the existing trawler fleet into a single and separate company and the same to apply to marketing. The province also wanted the "Resource-Short Plant" quotas[275] to be caught by this expanded trawler operation. Processors would remain much as they were.[276]

After a year of negotiations, the *Atlantic Fisheries Restructuring Act* became law in November 1983, empowering the Minister of Fisheries and Oceans to buy into the Atlantic fishery. Arrangements were made to form two large and vertically integrated trawling companies, consolidated into National Sea Products in Nova Scotia and Fishery Products International in Newfoundland.[277] Both received large amounts of public funding, and the *Fishery Products International Act* was passed in the Newfoundland legislature to focus the company on the needs of the provincial fishery. Fishery Products International was initially largely owned by the federal government and would buy out the existing capital of Fisheries Products, the Lake Group, and John Penney and Sons.[278] National Sea Products would remain largely privately owned through further consolidations of Nickerson & Sons and National Sea Products.[279] Together, Fishery Products International and National Sea Products formed the twin towers of the Canadian east coast fishing industry. Both depended heavily on northern cod.

Two fish allocation policies implemented in the early 1980s were Enterprise Allocations Program and the Resource-Short Plant Program. Under the Enterprise Allocations Program, offshore companies were given entitlement to a portion of the quota, to be fished at their convenience. The intention was to give companies assurance of their quota and reduce the race for the fish typical of a competitive fishery.[280] As federal Minister de Bané proclaimed: "The establishment of quasi-property rights for Canada's common property offshore groundfish stocks is a pivotal step in the future of the fishing industry."[281] Company quotas were first tested on a trial basis in 1982 and became policy by 1984. Most of the allocations went to National Sea Products in Nova Scotia and Fishery Products International in Newfoundland. Under the Resource-Short Plant Program, plants operating at less than full capacity would be allocated fish.

Post-Kirby

By the mid-1980s, after the dust surrounding the Kirby report and the recession of the early 1980s had settled, the processing sector in Newfoundland was comprised of about 75 freezer and 150 addition plants. Of these, 12 large trawler plants operated year round under Fishery Products International and National Sea Products; most of the plants were on the ice-free south coast, but one large National Sea Products plant was located in St. John's. Cod, flatfish, and redfish were sold as frozen fillets or frozen in blocks. The non–Fishery Products International and National Sea Products plants were mostly family businesses that operated seasonally when fish were available. About 35 registered plants specialized in salt cod production. Approximately 23,000 people in total were employed in the plants.[282]

The future looked bright – so bright that Newfoundland invested heavily in a new class of fishing vessel, often called Scandinavian longliners, as recommended by the Kirby report. These were large vessels, mostly over 100 feet (30.5 metres), that could fish year round with baited longlines or gillnets, and were more economical to operate than trawlers. It was thought in 1986 that the longliners would play an "increasingly important role in the Newfoundland fishery."[283]

These were buoyant days and it all seemed too good to be true. By the mid-1980s, Canada had become the world's largest exporter of fish. Even the old salt fish industry had experienced a mini-revival under the Canada Saltfish Corporation, which became the country's sole marketer of salt fish in 1972 and achieved marked success in improving quality and markets. There had been consolidation of marketing efforts among the major companies, and with the recession of the early 1980s now behind them and the Canadian dollar and interest rates declining, exports to the main markets in the United States and Japan looked good for the foreseeable future. The total annual value of Newfoundland fish products surpassed half a billion dollars in the early 1980s and was climbing.[284]

Storm Clouds Gathering

Amid such optimism, dark clouds began to gather. After a few years of relative prosperity, by 1986 the inshore cod fishery was failing again – even when catches were relatively good, the fish were becoming smaller. The Newfoundland Inshore Fishermen's Association (NIFA) pointed the finger squarely at trawling on the overwintering and spawning aggregations of cod. A report was commissioned by NIFA; its lead author was Derek Keats, a young Newfoundlander with a doctorate from the Memorial University of Newfoundland. Keats lambasted the NAFO stock assessments, claiming they were far too optimistic. The NIFA report focused on the known retrospective problem, and the questionable use of trawler catch rates as an index of abundance.[285]

For the most part, the federal government put aside concerns coming from inshore fishermen and scientists about the state of the stock, shifting the explanation for declining catches to cold water temperatures and ocean phenomena, which prevented the growing stocks of northern cod from reaching shore, and "out-of-date technology and inappropriate scale of operations [used in the inshore fishery]…[which was] a technological and economic anachronism." At best, the federal government realized that if the problems of the inshore fishery "cannot be resolved, [they] at least must be accommodated."[286]

NIFA took the federal Department of Fisheries and Oceans to court in 1989 over the continued trawling on the spawning grounds of the northern cod. The Department of Fisheries and Oceans advanced arguments that "a fish caught is a fish caught," and that there was no evidence that catching cod with bottom trawls during spawning hurt the population or its ability to reproduce. Closing fishing during spawning was one of the oldest fisheries management measures, but in this case the belief that the high fecundity of the cod and their overwhelming abilities to reproduce won the day. In any event, NIFA could not prove, in the legal sense, that trawling during spawning was damaging the fish stocks, and so lost their case.

The federal government clung to its belief that the trawler fleet, even though it was experiencing difficulties, was the future, and stated that "tough decisions will have to be taken regarding processing plants [mostly inshore] which are out of date, inefficient, too labor intensive and, therefore, no longer viable…[to develop] a modern, competitive, world-class fishing industry."[287] But the federal government realized that it was in a jam: if the northern cod really was declining, there was nowhere to direct the newly minted trawler fleet. Other fish stocks were already fully exploited. Most stocks classified as underutilized or unwanted by Canada were being fished by foreign fleets, which in the mid-1980s were still allocated over 20% of the fish off the east coast, despite the fact that Canadian fisheries were being closed down. As a federal report stated in 1985, "Non/Underutilized-species, we…allow to be exploited by foreign fleets. While this has been necessary in order to gain access to foreign markets, it is a strategy which, in part, comes back to haunt us."[288]

The shrinking quotas for northern cod in the late 1980s and early 1990s sharpened the political knives, and Newfoundland took more than its share of the cuts.[289] The provincial investment in a middle-distance longliner fleet in the late 1980s – an attempt to harvest top quality fish year round without using trawlers – had its miserly quota allocation of about 4000 tonnes cut in 1989 and reduced to zero in 1990.[290] The longliner fleet fell to the "last in–first out" principle, a bureaucratic invention based on conservatism that worked against improvements to a fishery.

In hindsight, it is clear that in the post-Kirby world, both Canada and Newfoundland tried to do far too much with the fisheries, especially the northern cod, based on the faulty notion that there would be plenty for everyone. The premise was that the Newfoundland and Labrador inshore fishery would be given a base allocation.[291] At the same time, an efficient, year-round offshore fishery would be developed, and after that, there would be some left over to trade for foreign favours. Art May, who had cut his scientific teeth on the cod spawning grounds of Hamilton Bank and who had witnessed the massive destruction caused by the foreign trawlers, was not so sure this would work. In 1982, May, then highly placed in Ottawa, sent a memo to the Kirby Task Force questioning whether having both an inshore and offshore fishery was possible and insisting that "someone must decide on whether or not there shall be a [human] population on the northern part of the east coast of Canada."[292] May's concerns, which turned out to be nearly prophetic, were not even noted in the Kirby report.

Some years later, Cabot Martin, the fiery Newfoundland lawyer and columnist and former head of NIFA, noted that: "In the sixties and seventies the foreigners raped our fish stocks and reduced our inshore fishery to poverty levels. But for some strange reason this did not generate a determination to ensure that such a tragedy would never be repeated." The long and the short of it was that the policies of both the federal and provincial governments were to modernize the fishery; however, the federal government position became that the northern cod was too much for Newfoundland to handle and must be shared with other provinces, because Canada must use all surplus fish within its EEZ for fear of having to say no to foreign fishing interests. Earlier programs to subsidize the construction of a Canadian migratory trawling fleet in Nova Scotia (and a much smaller longliner fleet in Quebec)[293] were replaced in the mid-1970s by provincially and bank-funded loans to accomplish the same thing. Federal policies by then attempted to restrict the numbers of trawlers, but not their size or fishing power.

The Newfoundland position – which in the 1950s and 1960s was to get people out of the fishery – changed to maximizing employment in the fishery, largely through federal supplements such as unemployment insurance. This change followed the failure of its industrialization strategies. The revised strategy was to corner as much of the northern cod as possible – or even all of it – for Newfoundland fisheries, using the principles of adjacency and historical attachment.[294] The inshore fishery had historically caught up to 400,000 tonnes of northern cod each year and it was thought that there was more than enough processing capacity to harvest this stock at maximum yields.[295] Unfortunately, the inshore fishery became dependent on unemployment insurance and various other subsidies, but was given little real encouragement in the form of investment, the development of quality fish products, and marketing. With both the trawlers and the inshore fisheries on a growth track, increases in quota were hotly contested. Aiden Maloney later concluded that the inshore fishery "has not benefited from the increases in northern cod quota during the 1980s which went totally to the offshore [trawler fleets]."[296]

As the Canadian trawler fleets increased in capacity in the early 1980s, Newfoundland initially wanted an 85%/15% split in the allocations to the inshore and offshore fisheries. Although not intended, such a strategy would have added a level of precaution to the quota-setting exercise because, unlike trawlers, the largely sedentary trap fisheries tended not to catch as much fish when abundances were declining.[297] This very situation occurred in the early 1980s, when the inshore fishery could not catch its allocation of 120,000 tonnes. The reaction of the federal and provincial governments, based on the belief that the stocks were growing, was to allocate a higher proportion of the catch to the trawlers, a more-or-less 55%/45% split.[298] This allocation lessened the inherent precaution of the inshore fishery and increased the potential for overfishing when quotas were set too high, as indeed they would be.

The tug-of-war that developed between the federal and provincial governments kept politics, not science, at the forefront of fisheries management. The federal government enabled increased capacity in the trawler fleets, especially those of the new fish barons in Nova Scotia, while the provincial government attempted to maximize the numbers of people in Newfoundland working in the fishery. These combined actions proved fatal to the fish, especially the northern cod, Newfoundland's – and arguably Canada's – most important fish stock.

SCIENCE

> The Ph.D. may become as callous as an undertaker to the mysteries at which he officiates. — ALDO LEOPOLD, "Conservation Esthetic," 1948

Developments in Fisheries Science in the 1950s

There were several developments in scientific theory in the 1950s that would have profound influences on the Newfoundland and Labrador fisheries. First, with the publication of Beverton and Holt's treatise on the population dynamics of fished stocks,[299] fisheries science, particularly as practiced in the North Atlantic, took on a much more mathematical persuasion and focussed on abstract models of fish stocks. In Canada, the mathematical approaches of William Ricker loomed large on the national fisheries stage. The knowledge of fishermen and the more empirical ecological approach that had characterized fisheries research in the first half of the 1900s were placed on the backburner.[300] With politics dictating that harvests from the sea be as large as was possible, the need to count fish and to determine how many fish could be caught became pressing.[301] As for the most part fish could not be counted directly, the mathematical approach appeared promising. Many new fisheries scientists were trained as statisticians rather than as biologists or marine scientists. They knew numbers – not fish.

The Rise of Fisheries Economics

A second and parallel development saw the rise of socioeconomic theories of fisheries management. Scott Gordon put forth the argument that the fishery was essentially a common property – fishermen would not refrain from overfishing because such investment in saving fish for the future would be nullified by others. Gordon speculated that "everybody's property is nobody's property. Wealth that is free for all is valued by none…The fish [left] in the sea are valueless to the fisherman, because there is no assurance that they will be there for him tomorrow."[302] The notion that common-property resources could lead only to tragedy was firmly planted in the late 1960s with the publication of Hardin's "The Tragedy of the Commons."[303]

Socioeconomic theories on fisheries had a large effect on fisheries management – beyond the obvious economics of selling fish. Most analyses concluded that in the 1950s and 1960s, "the focus [of Canadian fisheries policy] shifted from the purely biological aspects of fish-stock conservation to a broader consideration of the social and economic aspects of the fishery."[304] After implementation of the 200-mile EEZ in 1977, fisheries management developed in lockstep with fisheries economics, not biology. David Matthews concluded that "the regulation of the fishery came to be understood not so much as a biological problem relating to conservation and natural selection, but as a social problem relating to human greed and exploitation. Although, in some ways, the models inherent in these two metaphors are not all that different, it was nonetheless economics rather than biology that ultimately formed the basis of resource regulation in the fishery."[305] Matthews, in commenting on this shift in emphasis from biological to economic theory, was compelled in an endnote to state that "our discussion of this point is not intended to suggest in any way that fishery biologists ceased to play a significant role in their own right in the resource management of Canada's East Coast fishery."[306] Nevertheless, there is no doubt that basic biology was given short shrift in favour of economic theory in driving fisheries management.

The economic emphasis of fisheries management paralleled – and in some cases may have influenced – the more mathematical treatments of fisheries biology. Two particular areas stand out. First, fisheries biologists were compelled to make longer-term projections of fish stock performance than were biologically defensible, although easy enough to do mathematically. Second, near-paranoia about the "tragedy of the commons" led policy towards elimination of fishermen rather than a true rationalization of the fishery by encouraging a sustainable level of fishing power.[307]

Open access was a problem in the Newfoundland and Labrador fisheries, but the timing of the imposed restrictions was off. According to all scientific evidence, overfishing and stock collapses – the first and second apocalyptic horsemen of open access[308] – did not exist to any serious extent in Newfoundland and Labrador until the arrival of the foreign fleets in the 1950s. Attempts to rein in the local fishery, after having spared the foreign fleets for three decades, firmly closed the barn door with horses long gone. Economic poverty, the third horseman of open access, was perhaps a more realistic target for elimination. Unfortunately, the fourth horseman – the saviour – was a no-show.

In 1976, a federal fisheries policy document got serious about eliminating open access fisheries. According to it, "in an open-access, free-for-all fishery, competing fishermen try to catch all the fish available to them, regardless of the consequences. Unless they are checked, the usual consequence is a collapse of the fishery: that is, resource extinction in the commercial sense, repeating in a fishery context 'the tragedy of the commons.'"[309] Economist Parzival Copes picked up on this theme and applied it to the Newfoundland fisheries – in particular, the need to "reduce significantly the excessive manpower of the inshore fishery, and to rationalize the dispersed and fragmented processing industry."[310]

In 1977, with licensing in place and extension of the EEZ to 200 nautical miles on the horizon, a major report on the northern cod was released by the Economics and Intelligence Branch of the federal Department of Fisheries and the Environment.[311] The report recognized that the economic potential of the fisheries "hinge[d] on the state-of-recovery of the cod stocks." Several options were outlined as economic strategies for future fisheries, all based on the assumption that stocks would grow, including:

subsidizing a Canadian trawler fleet to fish northern cod in winter and during the spawning season (as had been the European and Soviet habit); leaving the overwintering spawning aggregations alone (as historically practiced), thus concentrating development efforts on the inshore fishery; and something called a scientific approach. Although the report was vague about what a scientific approach would entail, a restriction on fishing the overwintering and spawning aggregations appeared to have been a component. The report concluded that "a more reasonable solution is to fish the post-spawning…2J-3KL cod."[312] These biologically tempered strategies appear not to have found much favour and were never acted upon. The more economical road of fishing the highly concentrated overwintering and spawning aggregations was taken.

By 1979, the common-property perspective of fisheries economists had taken firm hold. That year, the influential *Journal of the Fisheries Research Board of Canada* published an entire issue devoted to fisheries economics that summarized the views held two decades after Gordon's seminal publication. There was general agreement on several points. The first was that open access to fisheries leads to overcrowding and poverty. Second, excess labour must be removed from the fishery and no new entrants permitted. However, there was little agreement on what "excess" meant, or rather, who it meant. The real problem was twofold: the growing dichotomy between the more traditional small-boat inshore fishery and the (mostly winter) trawler fishery, and that between the full-time and part-time inshore fishermen. Some economists, concerned with financial status and productivity, felt that fishing had become a bottom-of-the-barrel job and coined the infamous phrase "employer of last resort." Others felt differently, but everyone agreed that some form of limited entry must be imposed on all fisheries – they could not be open to everyone.[313] Another expression that gained widespread usage was "rational fisheries management." As pointed out by Matthews, this usage was intended not only to imply that any adversarial positions were irrational, but was also code for "tragedy of the commons" perspectives on fisheries management.[314]

In hindsight, it is clear that the economists missed the boat in the late 1970s: they failed to understand that the most important part of a fishery is the fish.[315] Without fish, no amount of theory or economic rationalization and no amount of social engineering or sugar coating really matters. With some exceptions, the real tragedy of fisheries was depleted stocks, with poverty its by-product – not the opposite, as some claimed.[316] For Newfoundland and Labrador fisheries, the limits of fish productivity had already been vastly exceeded, but not by the traditional inshore fishery. If there was a point to be made, it was that the "too many fishermen" notion should have been applied to the foreign fleets when it mattered, not to Newfoundland fishing communities after the fact.

THE END OF THE FISHERIES RESEARCH BOARD

The Fisheries Research Board of Canada, which had subsumed the Newfoundland Fisheries Research Board after Confederation, was put to rest in the 1970s. The Fisheries Research Board had provided 75 years of distinguished service to the fisheries, directing and implementing fisheries science in Canada at arm's length from direct government control. The Fisheries Research Board had developed an international reputation for innovation, independent thinking, and debate about the world's wild fisheries. If it had a weakness, it was that it offered no advice on managing fish stocks – it had no stock assessment or management capacity because these were the purview of ICNAF and, later, NAFO. In any case, in the 1970s, the Fisheries Research Board was out of step with the new centralist thinking of the Trudeau era and far too independent for the new breed of Ottawa mandarins.

In 1973, the Fisheries Research Board was relieved of direct control of research programs and facilities, and was reduced to an advisory body. As noted by its chairman Ron Hayes, "the [Fisheries Research Board] lost its independent status and was brought into line authority, reporting to the new assistant deputy minister for marine and fisheries…the Honorary Board [had become] a dead duck."[317] From then on, fisheries science would be conducted under a national government department that would later be renamed the Department of Fisheries and Oceans.

In the 1960s and early 1970s, the Fisheries Research Board had gone out of its way to solicit and encourage interest in fisheries research in Canadian universities. In the golden days of the Fisheries Research Board laboratory at St. Andrew's, New Brunswick, when scientific activity was arguably at its best, Hayes recalled that "there was little distinction between [federal Fisheries Research] Board people and university people…[many] came down and worked in the stations for the summer. That was deliberately removed by [previous Director] A.T. Cameron, who thought it was a waste of money. I was interested in trying to restore some of the working relationships between…governments and the universities, and I still believe this is a very important thing to do."[318] Hayes implemented a system of grants and wanted to offer more ship time to university researchers. Hayes' views were similar to those in many fishing nations of the world, such as Japan, Norway, Iceland, and the United States, where the nearly parallel development of government and academic fisheries science – as both collaborators and counter-balances to each other – was a cornerstone of the knowledge base required to manage fisheries.

Biological science was down, but not out, as economic theory took hold of fisheries management. By the late 1970s, clear-thinking fisheries scientists at the federal laboratory in Newfoundland were well aware of the devastation that overfishing had caused, especially on the northern cod stocks, based on two decades of observation. In a review in mid-1979, Allenby Pinhorn summarized the view of science:[319]

> The dramatic increase in catches by the foreign nations sustained over a period of years had a severe impact on this cod stock. The initial increase in catch resulted from fishing on hitherto inaccessible pre-spawning and spawning concentrations of cod in the spring in the northern areas of Hamilton Inlet Bank, Belle Isle Bank and Funk Island Bank. These concentrations were inaccessible prior to 1959 because of the heavy ice coverage in these areas during the winter and spring months, but during the late 1950s fishing technology was developed which enabled boats to fish in the ice. Thus, they commenced fishing on essentially a virgin stock of cod which had accumulated a very large biomass…This heavy fishing resulted in a dramatic decline in the…stock. For example, it is estimated that there were over 4 million tons of biomass in the fishable stock in the early 1960s whereas in 1978 it is estimated there were only slightly more than 1 million tons of biomass, a decline of 75%…There seems little doubt that the heavy foreign fishing offshore resulted in a severe reduction in the cod stock and, therefore, the offshore and inshore catches.

In brief, the foreign offshore fishery and the Newfoundland and Labrador inshore fishery had competed for the same fish – and the foreigners had won, although it was a Pyrrhic victory at best. There was no reason to expect that substitution of Canadian for foreign trawlers would have had a different result. The inshore fishery had always been subject to variation in cod movements to each bay, but overall, with less trawling in winter and more fish, the traditional fishery would have had higher catches.[320]

Science in the 1970s and 1980s did not realize just how bad the aftershocks of foreign overfishing in the 1960s would be.[321] Early projections of stock growth were based on the assumption that recruitment rates – the rates at which young fish were added to the stock each year – would be similar to those that had been observed in the 1960s. However, the numbers of spawning adults were considerably reduced in the 1970s, perhaps to 20%-25% of what they had been in the 1950s and 1960s; to make matters worse, the environment for cod was deteriorating and most stocks were becoming less productive. In his summary report in 1979, Pinhorn attempted to qualify his optimistic predictions, cautioning that they depended on several factors, particularly recruitment: "*It cannot be stressed too strongly* [emphasis added] that if recruitment in future is better than this average then obviously the stock will rebuild faster than projected...and, conversely, if recruitment in future is lower than average then the stock will rebuild more slowly."³²² Over 100 years of research had shown that fish reproductive success was highly variable and nearly impossible to predict. The decline in recruitment and growth that occurred in the 1980s would make these early predictions groundless.

If the authors of the Kirby report read Pinhorn's report, they either missed the qualifying remarks or did not understand their implications. With an emphasis now on economic growth, few stopped to ask what would happen to the grand plan to expand the fishery if the stocks did not grow quite as fast as expected. The Kirby report's assumption of a coming abundance of northern cod undoubtedly became political fodder. For their part, scientists had become increasingly frustrated that the "projections were gradually translated into expectations [and that science]...was expected to deliver when 1985 and 1987 arrived. However, the stock did not grow at the projected rate and this was translated as a 'mistake' in the stock assessments."³²³

Stock Assessments of Northern Cod

Stock assessments are one of the main endeavours of fisheries science. They attempt to answer the basic questions about the state of the stock: How many fish are in the ocean/stock? What will the effects of fishing be? These are not easy questions to answer. The information from stock assessments is provided by scientific agencies to political authorities and management, who consider economic and social implications, then make the regulations that govern the fishery. Scientists do not make these decisions. The goals of regulation vary widely, but the sustainability of both the fish and the fishery is one of the most crucial.

Prior to 1973, the fishery for northern cod was unregulated – there was no attempt to determine how many fish there were in the stock or how many could be caught without depleting the stock, and no catch quotas. Stock assessments began in 1972 under ICNAF with the first assessments in 1972-1975 attempting to determine the fishing harvest rate that would result in the maximum sustainable yield from the stock, a quantity known to fisheries scientists as F_{max}.³²⁴ Initial quotas based on these calculations were set at over 650,000 tonnes, a level now recognized to have been much too high as nowhere near this amount of fish could be caught sustainably. Canada was allocated only 17%-25% of this quota from 1973-1976. Canada argued each year for a reduction of the quota, but it was not until 1977, after Canada announced its intention to extend its fisheries jurisdiction to 200 miles, that a different target fishing rate was used. The new target came to be known as $F_{0.1}$, which resulted in a substantial reduction in the quota to 160,000 tonnes.³²⁵

F_{MAX} AND $F_{0.1}$

The fishing mortality rate (F) that will maximize yield from a fish stock is termed F_{max}. F_{max} occurs when the stock is at a medium level of abundance, a point at which the stock is the most productive. When the stock is very abundant, nearing the carrying capacity of the ecosystem, it is not as productive, nor is it if the stock abundance is too low. The basic calculation to obtain F_{max} is to apply various fishing rates (e.g., 10%, 15%, 20%, etc.) to fish of each age caught from a stock, and taking account fish weights as they grow, finding the F at which the yield, or total weight of catch, is highest. If this relationship is indexed to 1 virtual recruit to the stock, it is known as a yield-per-recruit curve, which quantifies the relative yield to the fishery at various levels of F. The $F_{0.1}$ (pronounced "F zero point one") fishing rate will invariably be lower than F_{max} and has no particular rationale other than that. $F_{0.1}$ is defined as the fishing rate on the yield-per-recruit curve that is 10% of the fishing rate near the origin.

In the late 1970s, stock assessment methods changed dramatically and attempts were made to estimate the numbers of fish of each age. These methods were derived from virtual population analyses, a modeling approach that amounted to a body count of fish caught over the lifetime of each year's new recruits, scaled up by the numbers that die naturally each year.

VIRTUAL POPULATION ANALYSIS

Virtual population analysis and its variants – sequential population analysis or cohort analysis – are all similar. They use the logic that in a heavily fished stock, where the ages of the fish caught are well established, the numbers of fish caught from an age-class (fish born in a single year) must have been at least as great as the numbers caught over the lifespan of these fish. For example, if 100 fish born in a given year were caught each year for ten years, then there must have been at least 1000 fish from that age-class to begin with. Allowance must be made for fish that die of other causes – typically, about 20% per year. It is relatively simple to add up the fish within an age-class and then to sum all the age-classes to reach an estimate of the stock total.

This method works well in hindsight, after all the fish of the age-classes present in the stock have been caught or have died. It is less well suited for estimating present numbers of fish and for making projections for the simple reason that all of the fish are not yet dead (or so we hope). For present numbers, various "tuning" methods are used, based on

> average historical proportions of how much fish of each age were in a stock, or on a comparison of the relationship between the numbers of dead fish in various age-classes with an index of abundance from the fishery or a scientific survey. Many methods have been devised to make these comparisons: all are useful and all are suspect.[326]
>
> The weaknesses of the virtual population analysis techniques will be exacerbated if the body count is not accurate – i.e., if the number of fish actually killed and those recorded for scientific uses differ. The techniques will also be unreliable if the independent index does not accurately represent the trends in stock abundance. Both these problems surfaced in the stock assessments of northern cod.

The virtual population analyses of the 1980s underestimated the fishing rates and overestimated the growth and resultant abundance of the northern cod. The reasons for this are worth exploring. Even after the extension of its jurisdiction to 200 miles, Canada initially allocated only 44% of the northern cod quota to itself, and this made it difficult to estimate how many fish had been caught. It was not until 1978 that Canada claimed the major proportion, but by no means all, of the quota. The foreign fishery was still active inside the Canadian zone until the closure of the fishery in 1992, and was active outside the 200-mile limit both before and after the closure. There was persistent fishing well in excess of allocated quotas – and at times without any quota – throughout this period. Spanish and Portuguese vessels, in particular, ignored the supposed moratorium in NAFO Division 3L. To add to the difficulties, millions of small cod were thrown back to the sea – dead, but not recorded as caught. An accurate virtual population analysis required information representative of the full fishery, but with as many as 10 foreign countries fishing, sampling of the catch was never consistent and some major fishing fleets submitted no information at all.[327] Hence, science had faulty data right from the start. Fisheries models are no different from any other attempt to predict something – the results cannot be better than the information on which they are based. As Canadian trawlers entered the fray in the 1980s, they apparently picked up the discarding habit and threw back an unrecorded number of fish right up until the moratorium in 1992. The Canadian policy of enterprise allocation, which allowed fishing companies to bring in their assigned quota whenever they liked, may have increased the likelihood that smaller and less valuable fish would be discarded, because each vessel could land its allocation at any time and there was no threat of losing the quota to another vessel who landed fish more quickly, as would happen with an unallocated quota.

The virtual population analysis technique works best if its track record can be compared with an independent index of abundance. The first indices for northern cod were the catch rates (the amount of fish caught in a fishing-based time interval) from the trawlers. Until 1980, pooled average catch per hour of fishing was used without any standardization for the type of vessel, the time of the fishing, or other factors known to influence catches rates – this additional information was simply not provided. As the Canadian fishery became more dominant, the quality of the data was easier to determine and a new method of standardization was implemented in 1980.

Scientific surveys typically make better indices than trawler catch rates. The first comprehensive surveys of the northern cod were carried out by the Soviet Union and West Germany in 1972. These followed the scouting surveys conducted by both countries over the previous two decades. France conducted its own surveys from 1976 to 1982.[328] Canadian surveys were not initiated until 1976-1978. All of these surveys used similar methods: a modified commercial trawl fished for a set time at randomly selected sites within subareas of the stock range, then recorded and sampled the catch. The Canadian survey has evolved somewhat since that time, with changes in vessels and nets, but has basically used the same methods since its inception.

During the 1980s, the computational methods of the stock assessment changed considerably. In the early years, measures needed for each age of fish were impossible to determine; until 1988, there was much greater uncertainty about the status of the stock in the most recent years – which, for management, were the most important years. In addition, the indices of abundance became much more reliable over time. It was suspected years ago that the trawler catch rates might not correctly represent stock abundance because the vessels hunted aggregated schools of fish whose densities were determined by fish behaviour rather than stock abundance, and because the fleets were improving their fish-finding technology and learning how to use it. These factors, as well as the inappropriate nature of the trawler catch rates, were not fully appreciated until the early 1990s.[329] The misinterpretation of the catch rates of the trawler fleet was one of the main causes of overestimates of the stock in the mid-1980s. In a nutshell, the trawler catch rates kept increasing in the 1980s – this was interpreted to mean that stock size was also increasing, but this was not true.

A third problem which impacted the stock projections, especially those made in the late 1970s and early 1980s, was uncertainty regarding recruitment and growth. These factors needed to be known in order for projections to be made. Unfortunately, ICNAF used estimates of average recruitment and growth for earlier years when there were larger fish, many more spawning fish, and environmental conditions were better. ICNAF also believed that the number of spawners was not vitally important and that "poor recruitment may result from a large spawning stock and very good recruitment from a small spawning stock."[330] While this may be technically correct, further analysis of the relationship between stock size (i.e., reproductive potential) and recruitment in most cod stocks has shown that there is a particularly strong relationship in the northern cod.[331]

Despite wide recognition that these early stock assessments were of uncertain accuracy and that the projections made from them were even more so, a pattern developed of emphasizing an optimistic view of the stock, which resulted in subsequent assessments nearly always having to readjust the numbers downward. This was termed the retrospective problem and first came to light in 1979. Most of this occurred innocently enough as a consequence of lack of knowledge about recruitment and growth to come, as well as an inability to estimate the fishing rates on each age group. The retrospective problem became acute in the mid-1980s: the spawning biomass estimated in 1985 was 30%-40% lower than that estimated the year before. By this time, it was clear that fishing rates had been much higher, and recruitment and growth rates lower, than previously thought.

Was there a preference for optimistic views of stock status?[332] There is little convincing evidence of this, although French surveys in the late 1970s that indicated much higher fishing rates than the official model used were discounted without an entirely convincing reason.[333] In addition, the use of the trawler catch rates, which inflated the apparent status of the stock, continued despite knowledge that their data were suspect. Overall, however, these are minor transgressions of the scientific method. The main conclusion that should be taken from a historical review of the science is how uncertain it was, and not that there was a preference for good or bad results.

Stock assessments of northern cod were conducted by international committees of ICNAF and NAFO using a variety of foreign and Canadian data until 1986. In the fall of 1986, the Canadian Atlantic Fisheries Scientific Advisory Committee (CAFSAC) faced a huge dilemma. Confidence in the NAFO stock assessments was evaporating in both fisheries and some scientific realms, and CAFSAC reviewed the NAFO assessments in the late fall of 1986 for the first time.[334] George Winters, a Department of Fisheries and Oceans scientist in St. John's who had studied catch rates in herring fisheries,[335] argued to CAFSAC that the NAFO northern cod assessments were badly flawed and that the use of trawler catch rates was resulting in a gross overestimation of stock size.[336] The NIFA report authored by Keats had made similar arguments and had focussed on the retrospective problem.[337] Unfortunately, the revisions suggested by Keats and Winters, although not ignored, were too easily sidestepped and the NAFO assessments were supported. Inshore catches and fish size were on the decline, but NAFO had discounted the inshore data and CAFSAC could find no fault with NAFO.

From 1987 onwards, CAFSAC conducted the assessments of the northern cod, and used only Canadian surveys and trawler catch rates as indices of abundance. But that posed a further dilemma – the survey index was flat, indicating little to no growth in the stock, but the commercial catch rate index was increasing rapidly. After considerable debate and expressed unhappiness with both indices, a middle approach was adopted. This proved to be the undoing of the assessments.

Some of the misfortune that came to the northern cod stock assessments was just bad luck. In the fall of 1986, with the state of the stock highly disputed, the survey gave an anomalously high index which supported the view that the stock was growing and that the trawler catch rates were correctly reflecting that growth. The reason for this high survey result has never been adequately explained, but it could not have come at a worse time:[338] it made the arguments of Keats and Winters appear specious. In the stock assessment of 1987, with the most recent survey so optimistic, the result was sealed. In the meantime, a federal government–sponsored independent report, *Task Force on the Newfoundland Inshore Fishery*, commonly called the Alverson report, was tasked to explain why the inshore fishery was failing when the stock assessments showed that the stock was growing. The Alverson report was generally supportive of the stock assessments, although it indicated a somewhat higher fishing rate than was generally accepted.

Industry was apparently unaware of or dubious about the controversy regarding the state of the northern cod. The big companies retained an overly optimistic view of the stock as late as 1988 when Fishery Products International "saw the first signs that there was a real possibility the TAC [total allowable catch] for northern cod would *increase significantly* [emphasis added] from 266,000 tonnes to 293,000 tonnes…Published reports by the Department of Fisheries and Oceans suggested a TAC as high as 331,000 tonnes."[339] Even after the science was too obvious to ignore, National Sea Products wrote to the Department of Fisheries and Oceans in 1989 arguing that northern cod was still abundant.[340]

In the fall of 1987, the survey result was down to less than it had been in the early 1980s, and the assessment in 1988 gave radically different results from the previous year. Fishing rates were double previous estimates and the numbers of remaining fish much lower – perhaps half. Much of the difference stemmed from a new focus on the survey data rather than the trawler catch rates. The 1988 assessment "suggested a large change in perception of stock size" – a considerable understatement.[341]

In hindsight, all that can be said about stock assessments in early to mid-1980s is that errors were made – there is little evidence of collusion or the favouring of any particular result.[342] After 1977, the growth of the northern cod was much lower than initially estimated and the recommended quotas were much too high.[343] Over this period, well over a million tonnes of northern cod were caught in excess of what would have been caught if stock assessments and quotas had better reflected reality.

While the science was uncertain, scientists did not make the decisions that brought down the northern cod. Primary responsibility must lie with governments and industry for consistently emphasizing the most optimistic view of the stock (and its management) that could be fit to the scientific advice, or ignoring the advice entirely. Governments and industry were also the source of the short-sighted notion that quotas could not be brought down because doing so would negatively impact industry.[344] One wonders how else it might have been done.

In 1989, the clock was ticking for the northern cod. Inshore fishermen were convinced the stock was in free fall, but offshore trawler companies still claimed otherwise. The 1989 memo from National Sea Products to the Department of Fisheries and Oceans stated that "our skippers have no problem locating large schools of northern cod at all times of year."[345] Science was struggling with the stark realization that stock abundance had been badly overestimated in previous years and that a quota of no more than 125,000 tonnes was indicated. If this quota had been implemented, it would have helped the fish, but gutted the trawler fisheries: the inshore fishery allocation was about 115,000 tonnes. Some options were suggested to raise the recommended quota, including an executive decision – a decision not supported by CAFSAC scientists – that allowed for a quota of up to 252,000 tonnes.[346] In the end, federal Minister of Fisheries Thomas Siddon set the quota at 235,000 tonnes, which was nearly double the quota recommended by science. In response to "alarming advice"[347] coming from scientific quarters about the state of the stock, Siddon also set up another task force, under Memorial University of Newfoundland president Leslie Harris, to examine the state of the northern cod.

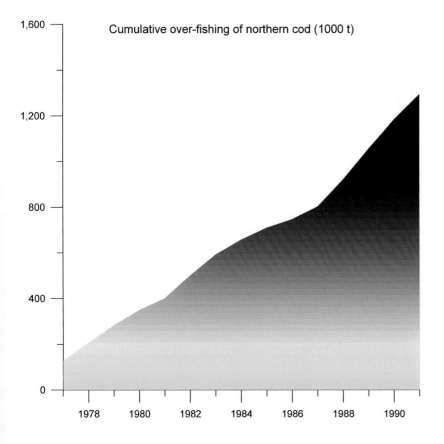

7.22 Cumulative over-fishing of northern cod from 1977 to 1991, based on what was allocated and what should have been allocated if the stock status had been better understood. This does not take into account any discarding or misreporting, which would increase the level of overfishing.

The Harris Report

The much-anticipated *Independent Review of the State of the Northern Cod Stock*, also known as the Harris report, was released in March 1990. The Harris report recommended an immediate reduction in the fishery. Unfortunately, the only immediate reaction to the report was rejection by Bernard Valcourt, the federal fisheries minister, because of the hardship such a reduction would impose on industry.[348]

In describing the background to the present state of the northern cod, the Harris report recognized that the current troubles were a product of decades of mismanagement. While respecting the many accomplishments made by ICNAF, the Harris report considered that "as an agency for conservation…ICNAF was a total failure." NAFO was ridiculed by the Harris report: "Like ICNAF before…[NAFO] is totally devoid of teeth…its moral authority counts for less than nothing in the world of realpolitik and nations of the European Community…in practice, such nations as Spain and Portugal habitually ignore scientific advice, flaunt their defiance of conservational strategies, and limit their catches only by the capacity of their fishing fleets."[349]

The Harris report, like the Kirby report before it, was a landmark in the history of the northern cod and to the future of the fishing communities adjacent to it. Harris pulled no punches and stated that the federal government, whose constitutional responsibility was to manage and protect the stock and the fishery for the people, had failed to act effectively to that end. The report was generally supportive of science – although critical of the methods and resultant decisions made during stock assessments in the 1980s – but bemoaned an overreliance on mathematical models to realistically capture the dynamics of complex marine ecosystems and the state of fish stocks, while at times ignoring obvious signs from the fishery. Vic Young, the chief executive officer of Fishery Products International, lamented to the Harris panel that "industry has to bear its share of the blame in terms of our willingness to accept, without factual foundation, that the northern cod stock was one of the best managed anywhere in the world."[350] In turn, the Harris report stressed that the concerns of fishermen and industry, some of which had been downplayed or disregarded by the Department of Fisheries and Oceans, had to be taken more seriously.

It must be said that the Harris panel, like others before it, did not fully appreciate the coming crisis. Although they knew that few fish had been born in the cold years of 1984 and 1985, and that fishing must be reduced if the stock was to be maintained,[351] they concluded that "there is not an immediate threat to the survival of the northern cod stock…[and that] the state of the stock measured by the biomass trends does not support a conclusion that anything drastic or threatening has occurred to the northern cod stock to date."[352] In hindsight, there was much to be concerned about and it would become evident within a year.

In 1990, politicians had taken the fate of the northern cod firmly in hand. The quota was set at 200,000 tonnes in advance of any scientific support or evidence. The scientific recommendation, which appeared to have been ignored, was much lower – about half the quota – and both the recommendations of the CAFSAC stock assessment and the Harris panel were ignored. The Dunne report, tasked to determine how to implement the recommendations of the Harris report, suggested that a substantial reduction in fishing would assist in the rebuilding of the stock to perhaps 1,000,000 tonnes, with a spawning stock of 450,000 tonnes, in four years. Several options for reduced catches were given, with an $F_{0.1}$ catch of about 100,000 tonnes. In response, a multiyear management plan was announced by federal Minister Valcourt in December 1990 with allocations for 1991, 1992, and 1993 of 190,000 tonnes, 185,000 tonnes, and 180,000 tonnes, respectively.[353] Where these numbers came from is unclear – they were not determined by science.[354]

The 1991 assessment was a bit more optimistic than the 1990 assessment, largely due to evidence that large numbers of fish had been born in 1986 and 1987. One Department of Fisheries and Oceans scientist referred to it as a coming "bonanza" of fish.[355] Canadian fisheries could not catch their allocations in 1991, but NAFO countries reported that over 50,000 tonnes were caught outside the EEZ (after NAFO imposed a supposed moratorium on cod fishing on the Nose of Grand Bank in 1986, about 250,000 tonnes had been caught there).[356] By 1992, the catches of both inshore and offshore Canadian fisheries had collapsed, and the euphoria over coming year-classes was shown to be misplaced. It was clear that the stock was declining rapidly, with a preliminary stock estimate of only 780,000 tonnes and 130,000 tonnes of spawners. Federal Minister of Fisheries John Crosbie, his back to the wall, reduced the quota to 120,000 tonnes in 1992, with only 20,000 tonnes to be caught by trawlers in the first six months. In March 1992, a European Economic Community group of experts conducted their own examination of the state of the northern cod and concluded that the decrease in abundance was not caused by overfishing, although they did acknowledge a change in the distribution of the stock that made it more available to their fisheries outside the Canadian EEZ.[357] At least they were right on that. At the request of Canada, NAFO also reviewed the stock status in 1992. There was agreement that the stock had declined at a rate that was difficult to reconcile with fishing effects alone; however, whatever the cause of the decline, it was recommended that any fishery be limited to 50,000 tonnes.

For the northern cod fishery, it was all over. The 1992 fishery had barely begun when it was finally realized that there were few fish left to catch. On July 2, 1992, Minister Crosbie closed the Canadian fishery for two years. The last complete stock assessment of the northern cod stock (NAFO Division 2J-3KL), based on data up to 1992 and completed in 1993, included information from regular research surveys and also from new research on the changing distribution, migration, and aggregation patterns of the stock.[358] It painted a dismal picture of a stock that was nearly gone and that was still declining. Shortly afterwards, CAFSAC was disbanded. NAFO continued – as did the foreign fishery outside the Canadian EEZ, despite a supposed moratorium on cod fishing in that region.

Newfoundland Science

Reflecting the terms of union of Confederation with Canada, the Newfoundland Fisheries Research Station in St. John's continued to operate and publish its work until 1955. Reports up until 1954 are titled *Report of the Newfoundland Fisheries Research Station* under the banner of the Fisheries Research Board of Canada. From 1955 on, they are *Reports of the Biological Station*, St. John's, Newfoundland. It seemed an easy transition. Dr. Wilfred Templeman led the Newfoundland Fisheries Research Station and stayed as Director of the Canadian Biological Station after 1955. Templeman also served with ICNAF and tried to inject some concern about the fish stocks into their proceedings on many occasions.

The Newfoundland Fisheries Research Station conducted many fishing explorations in the early 1950s. These were years of discovery. The research trawler *Investigator II* worked throughout the year on surveys of the haddock on the southern Grand Bank and in experimental fishing for cod and other species in the deep waters off the northeast coast and northern Grand Bank. Tagging programs on cod were started, and research surveys on redfish, American plaice, squid, and capelin were begun. In September 1951 and 1952, the *Investigator II* made some of the first explorations of the outer and southern Hamilton Bank. The catches of cod were enormous: up to almost 10 tonnes per hour in a relatively small research trawl. The highest concentrations of cod were found on the slopes of the bank at depths between 160 metres and 240 metres. Most of the cod were between 40-70 centimetres in length. They had struck the forming winter aggregations of cod on the slopes of Hamilton Bank and the edge of the Hawke Channel. Other valuable species were also caught. At one location in 180 metres of water and temperatures of -0.3°C, about three tonnes of cod and five tonnes of American plaice were caught in a single fishing set.[359] This report was marked confidential in 1952, but the news would soon be out. In October and November 1957, Portuguese trawlers fished the Hawke Channel and confirmed the presence of large abundances of the typically small Labrador cod.[360] At the same time, just to the north on the southeast slopes of the Hamilton Bank, the Spanish trawler *Abrego* made experimental fishing sets for cod with similar results.[361] Within a few years, hundreds of ice-breaking stern trawlers from Europe and the Soviet Union would devastate the formerly unfished overwintering and prespawning concentrations of cod around Hamilton Bank.

In most modern fishing nations, such as Iceland, New Zealand, Japan, and Norway, it is not only government departments that contribute to fisheries science: industry and academia also play major roles. Contributions from these sectors did not develop easily in Newfoundland. In the late 1950s, Jack Pickersgill, using the federal experimental farms system as a model, was instrumental in establishing the Valleyfield Experimental Station, whose mandate was to pursue the development of new salt fish products. Industry support for this initiative was weak. As David Alexander asserted, the lack of industry support for science, in particular production research that would have benefited them, "affirmed their traditional role as traders pure and simple, and [the choosing of] a conservative path...[which] was to lead them to virtual extinction."[362] With the salt fish industry failing, the station closed in the mid-1960s.

Memorial University of Newfoundland

The Memorial University of Newfoundland is the only university in Newfoundland and Labrador, and it is mandated to serve the special needs of this marine province and its fisheries.[363] The requirement for a strong academic presence in fisheries was evident from the beginnings of Memorial University College in the 1920s, with J.L. Paton establishing a biology department to do just that. Despite these early initiatives, a lack of interest in the fisheries appeared to overtake Memorial at the critical time during the late 1950s and 1960s. A growing and dominant federal presence in fisheries science after Confederation was not matched within Newfoundland, either in government itself or at the university. Unlike Japan, Norway, Iceland, or the United States, which were Newfoundland's main competitors and markets and which had major capacities for fisheries research both at the governmental and university level, Newfoundland developed no academic department, college, or institute dedicated to the study or management of its wild fisheries.

The shortcomings of Memorial's programs with respect to the *Memorial University Act* of 1949 were pointed out by the Walsh Commission soon after Confederation.[364] The Walsh Commission recommended that groupings of courses be offered that would lead to a degree in fisheries science, that postgraduate work should be undertaken "in subjects relating to the fisheries and particularly marine biology and bio-chemistry," and that "there should be generous provision for scholarships and other forms of student assistance in under-graduate and post-graduate years"[365] to attract young people into the field. In the following years, although programs in marine biology were developed, fisheries science was not.

In the 1960s, there were two major initiatives in the development of marine education and research in Newfoundland. The Fisheries College was opened in 1965; it later evolved into the Fisheries and Marine Institute.[366] The focus of the Institute was very applied and nonacademic, with diverse programs to train sea crews and seafood processors, and to build boats and nets for industry, but there was little study of the fish stocks or their management.

In 1967, the Marine Sciences Research Laboratory opened at Logy Bay as part of the Memorial University of Newfoundland. Perched on a bare rock open to the North Atlantic above a relentless pounding sea, its location had a rugged beauty; however, there was no useable harbour for either small or large vessels, and the location therefore was not well suited for field studies of wild fisheries.[367] The first director of the Marine Sciences Research Laboratory was Fred Aldrich, a giant squid expert, who divided his time between the Department of Biology and the laboratory. In 1971, D.R. Idler became the full-time director, and the laboratory turned to the reproductive physiology of aquatic organisms, biological oceanography (centred on the study of plankton), and later, aquaculture.

Fisheries science – the study of the impacts of fishing and natural phenomena on wild fish populations – was not well served at Memorial. In 1972, Wilfred Templeman retired as director of the St. John's Biological Station of the Fisheries Research Board of Canada and became the J.L Paton Professor of Marine Biology and Fisheries in the biology department at Memorial. Templeman ended his career there, but after another retirement, returned to the new federal laboratory,[368] where he worked until he died in 1990. A few faculty members with interests in fisheries were hired in the 1970s, but little was done to foster an academic program in fisheries. No degree programs in fisheries or fisheries management were developed, and no faculty, department, or academic institute was formed to tackle the many-knotted problems of the fisheries and related sciences. Faculty members interested in fisheries research were on their own and received little institutional support.

In 1978, a renewed interest in the Newfoundland and Labrador fisheries – due to the 1977 extension of jurisdiction and improved opportunities for funding – led to a workshop at Memorial to discuss the formation of a fisheries institute. Kevin Keough of the biochemistry department, a Newfoundlander with an affinity for the fisheries, stated flatly that "the only renewable natural resource which can be expected to provide employment to a significant proportion of our population [is the fishery]…it seems that the province will come to rely more heavily on the fishery than on any other industry to provide provincial wealth."[369] Leslie Harris, then vice-president academic, gave the lead presentation on the need for fisheries and marine studies at Memorial, followed by enthusiastic inputs from the departments of biology, biochemistry, geography, anthropology, mathematics, and political science. However, it all came to nothing. The main problem seemed to be an inability to define what fisheries science was, as discrete from existing departments and university structures. Some, including G.R. Peters, appeared to deny its existence: "I think it can be fairly said that 'fisheries science' is a misnomer. Science in fisheries is a collection of disciplines – pure and applied – with a common area of application."[370] The American Fisheries Society has no such problem, offering a certification program for fisheries scientists,[371] nor does Japan, where science and the supply of seafood is taken so seriously that an entire university – the Tokyo University of Fisheries – is devoted to the subject. Nor does Norway, where academic departments of fisheries at the University of Bergen complement the federal Institute of Marine Research and the academic and technical colleges at Tromso. In the East African country of Tanzania, the University of Dar es Salaam has an entire Faculty of Fisheries and Aquatic Sciences in recognition of the importance of its fisheries. There are many faculties, departments, and colleges devoted to fisheries and related disciplines in the United States.

The other problem identified at the workshop was that Memorial had no faculty in the core area of fisheries science.[372] In 1983, five years later, there was no mention of fisheries in the graduate pamphlet for the Department of Biology.

A revealing comparison can be made between the Memorial University of Newfoundland and the University of Washington in the United States. The University of Washington has a College of Ocean and Fishery Sciences comprised of four schools: Aquatic and Fishery Science, Applied Physics, Marine Affairs, and Oceanography. Each school offers degree programs. The School of Aquatic and Fisheries Science has approximately 30 faculty members, around 10 postdoctoral fellows, and dozens of graduate students. Most faculty members work closely with scientists at the nearby United States National Marine Fisheries Service laboratories and at their field facilities in the Pacific Northwest and in Alaska. In contrast, Memorial does not have a single permanent academic entity in which the name "fisheries" appears, nor any department or institute that focuses on marine or freshwater fisheries and related subjects.

The lack of a significant strength in fisheries science at Memorial likely contributed to the mismanagement of the Newfoundland and Labrador fisheries by default.[373] A single example may illustrate the point. Although no amount of research would have changed the results of the anomalous stock assessment survey of 1986, a weight of informed scientific opinion at Memorial might have added credence to the view of some scientists at the Department of Fisheries and Oceans that there was something dreadfully wrong with the survey numbers and resultant stock assessments. Such opinion may well have added support for a more cautious approach to the setting of the quotas, and thus the management of the fishery, during the critical years of the mid-1980s. In 1992, with the fishery closed and the rural economy in ruins, Cabot Martin wrote: "Memorial University…still does not teach, let alone specialize in, fisheries management. With all of our impressive local expertise in so many other areas of marine research, why has this critical area been so neglected?"[374]

Some initiatives were taken in the 1990s to attempt to upgrade the fisheries science and management capacities at Memorial. A National Science and Engineering Research Council Chair in Fisheries Oceanography was put in place in the early 1990s, and in 1992, the Fisheries and Marine Institute became part of the Memorial University of Newfoundland. Initially, there were plans to expand the fisheries science capacities of the Institute, and in 1996 a coalition of federal and provincial fisheries departments, NSERC and Fishery Products International came together to fund three new faculty members and a fisheries research program under the banner of the Fisheries Conservation Chair. Ambitious new research programs were implemented, and the new faculty put together and taught specialized courses in fisheries ecology and quantitative methods in fisheries in the new master of marine science program. In its early years, the Chair held a senior[375] and two junior chairs, a postdoctoral fellow, and ten graduate students. After an initial five-year term, the Fisheries Conservation Chair was favourably assessed by independent science reviewers and extended for an additional five years. Despite these successes, the emphasis on external funding and a perceived lack of commitment from the university led the junior chairs to seek better opportunities in other universities and departments. Nevertheless, by 2006, five postdoctoral fellows and more than twenty-five graduate students had studied with the Fisheries Conservation Chair and had produced more than 150 scientific publications in fisheries science and management, having significant influence on fisheries management in Newfoundland and Labrador and internationally. At the time of this writing, the future of these initiatives remains uncertain, as does the future of fisheries science at Memorial.

THE "PERFECT" DEMISE OF THE NORTHERN COD

> It was "the perfect storm" – a tempest...created by so rare a combination of factors that it could not possibly have been worse.
> – Rear cover of *The Perfect Storm* by Sebastian Junger

The primary depletion of the northern cod was caused by foreign overfishing during a time of high ocean productivity in the 1960s.[376] The Canadian catch represented less than 20% of the total take during that era. In 1968, Canadian and foreign fisheries reported an almost unbelievable catch of over 800,000 tonnes of northern cod.[377] Over 1,200,000 tonnes of cod from all Newfoundland and Labrador cod stocks was caught in that year. These catches were not even close to being sustainable, even during a period of relatively high productivity. Ironically, the key concentrations of cod on the edges of Hamilton Bank were discovered by Newfoundland Fisheries Research Board scientists in the early 1950s[378] and then exploited unmercifully by unprecedented numbers of trawlers from France, Poland, Spain, Portugal, and the Soviet Union. The trawlers fished during the winter and spring when the fish were highly aggregated prior to and during spawning.[379] The northern cod stock never recovered from this onslaught.

In the late 1970s, with the foreign fleets on the run, Canada was ebullient about the potential for the northern cod and, eager to flex its newfound muscles, released a report titled *Northern Cod: A Fisheries Success Story* in 1980. In hindsight, the report begins with a statement worthy of a Shakespearean tragedy: "There are few success stories in the Canadian fishery to match the recovery of the northern cod stock off Newfoundland and Labrador since Canada imposed its 200-mile economic management zone on January 1, 1977. Under management of the Department of Fisheries and Oceans, catches of northern cod increased from 79,581 metric tonnes in 1977 to 130,705 metric tonnes in 1979."[380]

There was a limited rebuilding of the northern cod under Canadian management until the late 1970s and early 1980s, but this was not to be sustained. The damage already done to the stocks by the foreign fisheries of the 1960s was not appreciated at the time, and a more complex secondary depletion began in the mid-1980s. Productivity in the northwest Atlantic had been declining since the 1960s, despite some respites from time to time. Many stocks across the northwest Atlantic were affected, not just those in Newfoundland and Labrador waters. West Greenland cod, which had boomed and been massively overfished by European and Soviet fleets in the 1960s, also collapsed by the early 1990s. The Greenland cod could not replenish itself either by local reproduction or with migrant fish from Iceland, largely because of unfavourable environmental conditions.[381]

Northern cod were subject to a rare and unlikely mix of factors that, as with *The Perfect Storm*, could hardly have been worse: it was a "perfect demise." It started with a decline in productivity after the 1960s that would result in both lower growth rates in individual fish and reduced breeding potential. Recruitment was further reduced because there were fewer and smaller adults.[382] The lack of detailed understanding that growth and recruitment were falling led to overly optimistic projections of stock increases, and fishing quotas that were set far too high. To make matters worse, after initial, relatively strong stock growth in the late 1970s and early 1980s when productivity surged briefly, the following years of 1984 and 1985 turned bitterly cold and unproductive. These were pivotal years when the stock could either have found its stride and continued to grow or fallen into a terminal decline. In the mid-1980s, fishing removals were more or less balanced with growth and reproduction. As fate would have it, however, the cold conditions of 1984 and 1985 resulted in few young fish being added to the stock. In 1986, for reasons that are still not understood, the stock assessment survey index was sky high, several times what the stock should have indicated, leading to another round of overfishing. Additions to the stock through growth and reproduction were now well below what was being removed. It was not just the Canadian-set quotas that were at fault – outside the 200-mile EEZ, both NAFO (primarily European Union countries) and non-NAFO fleets continued to hammer the northern cod.

This was only the beginning: changes in the ecosystem occurred fast and furious in the late 1980s.[383] The North Atlantic Oscillation went strongly positive in the late 1980s and brought a series of the coldest years in recorded history to Newfoundland and Labrador waters.[384] Capelin, the cod's chief food, disappeared from their northern ranges in 1990[385] during a period of the coldest weather ever observed (1990-1992). Capelin population estimates fell from about 6,000,000 tonnes to 200,000 tonnes in one year, representing an astounding drop of 97%.

By 1990, the remaining cod were moving south, likely following capelin and attempting not to lose contact with them.[386] In June of that year, the present author was chief scientist on the Department of Fisheries and Oceans research vessel *Gadus Atlantica*, and using the latest scientific echo sounders, located a huge aggregation of spawning and migrating cod in the Bonavista Corridor, the funnel-like and most southerly of the stock's main migration routes. The aggregation covered an area of about 30 kilometres by 20 kilometres, was on average about 25 metres thick, and was estimated to hold about 450,000 tonnes of cod (about 570,000,000 fish), which represented about 80% of the stock.[387] The mega-aggregation first headed south and then towards shore. Where the corridor narrowed, the aggregation compacted laterally but increased in height, with some of the fish up to 150 metres above the bottom.

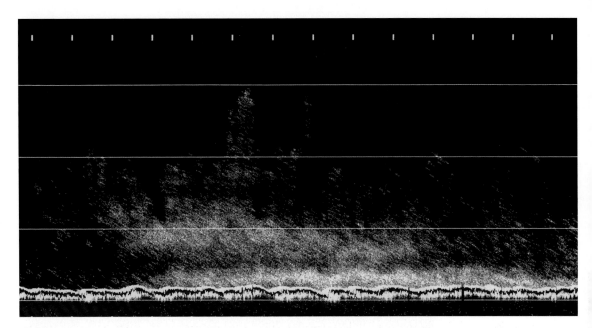

7.23
Migrating cod in the Bonavista Corridor in June 1990. The span of the echogram is about 10 kilometres and the bottom depth about 300 metres. About 450,000 tonnes of cod made this migration that year.

7.24
RV Gadus Atlantica, *the pride of the Department of Fisheries and Oceans research fleet, at sea in the late 1980s.*

It would be the *coup de grace*: the culmination of forces that would spell the end for the northern cod. The fish were becoming "hyperaggregated"[388] into a few large and very dense schools – and these superconcentrations of the remaining fish played right into the hands of both the Canadian fisheries and foreign trawler fleets.[389] In the winter and spring of 1990 and 1991, catch rates on these concentrated fish were the highest ever recorded[390] and over 300,000 tonnes of cod were taken. Many more small fish were likely discarded.[391] The inshore trap fishery was the best that it had been in decades, landing over 100,000 tonnes in 1990, despite the declining state of the stock.

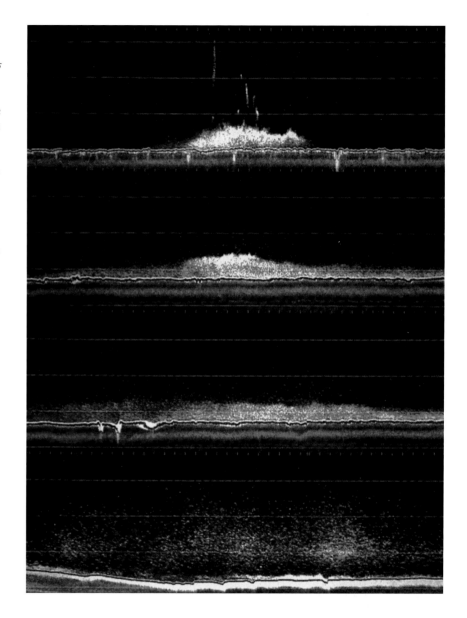

7.25
Spawning aggregation (top panel), beginnings of break up of the aggregation (middle panels), and migration (lower panel) of cod in the Bonavista Corridor in June 1992. The four panels show the same fish over a two-week time period (a different version of these data originally published in Rose 1993).

In the spring of 1991, the migration was delayed and only half as many cod moved through the Bonavista Corridor. Foreign catches on the Nose of the Grand Banks, which was the southern limit of the range of the northern cod and an area supposedly under moratorium, were among the highest recorded since the implementation of the 200-mile EEZ. In the spring of 1992, the fish were even more delayed and spawned heavily in June instead of at their normal times in April and early May. In the spring of 1992, the complete cycle of spawning through to the beginning of the migration was observed for the first time. As a final irony, it would be the last of the great migrations.

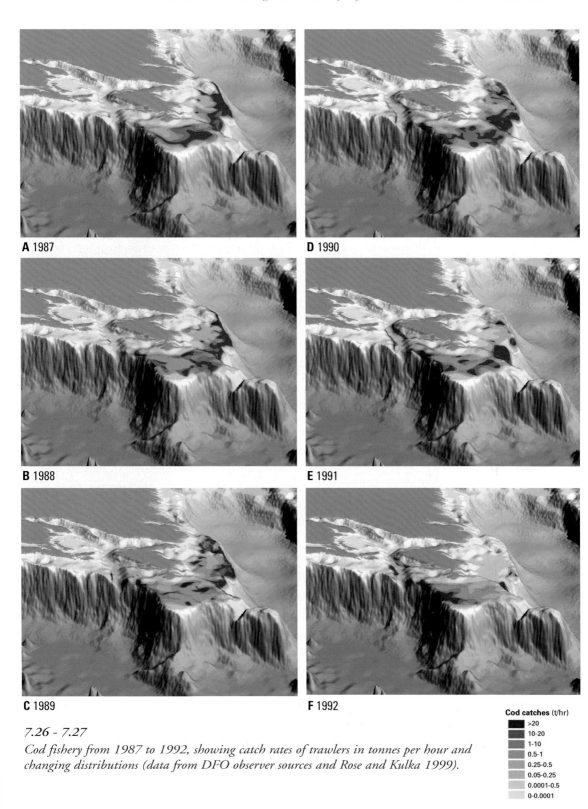

7.26 - 7.27
Cod fishery from 1987 to 1992, showing catch rates of trawlers in tonnes per hour and changing distributions (data from DFO observer sources and Rose and Kulka 1999).

The declining numbers of adult fish, their location well to the south, and lateness of the spawning in the chilled waters meant that almost no new fish were produced in 1991 and 1992.[392] In 1993, few fish were found in the Bonavista Corridor. An expanded search for them located only one large aggregation – well to the south and outside the Canadian EEZ – in the centre of the foreign fleet. As chief scientist of this research, the present author instructed the vessel to make measurements (shown here for the first time), keep radio silence, then discreetly leave the area so as not to draw attention to the fish.

In the spring of 1994, no aggregations could be found anywhere.[393]

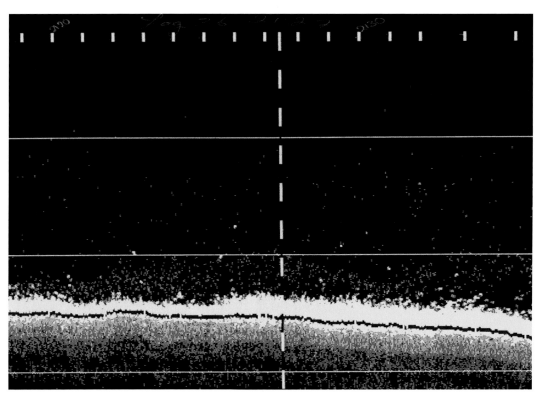

7.28 *The last known large aggregation of northern cod on the Grand Banks at approximately 47° north latitude and 47° west longitude, outside the Canadian 200 mile zone, in May 1993. The echogram spans about 5 kilometres and the depth is about 250 metres.*

Overfishing or Environmental Effects?

In the mid-1990s, debate raged over whether the decline of the cod was caused by overfishing or the environment. In reality, this was a political and not a scientific argument – one more comprised of "spin" than substance.[394] Overfishing can only be defined in terms of the productivity of the fish stock – if more are taken than are produced, or worse, if catches impact the stock's ability to produce – then that is overfishing. The whole idea of scientific fishing is based on retaining a standing stock of fish in the water that serves as a breeding group, and taking only the surplus – i.e., the production of the standing stock that is above what the stock needs to replenish itself. This notion cannot be separated easily from the environment. Fish live in a dynamic ecosystem and stock production depends heavily upon environmental conditions: what constitutes overfishing under one set of environmental conditions would not under others. A stock can be overfished even under the best environmental conditions, as occurred in the 1960s.[395] Moreover, if a stock is declining under poor environmental conditions, such as occurred with cod and several other species (e.g., capelin and American plaice) in the late 1980s and early 1990s, then fishing must be reduced or it will almost automatically become overfishing. The important factor is not whether overfishing or environmental effects caused these declines, but that fishing must be controlled in accordance with fish production, which means cutting back during environmentally unfavourable periods. For the Newfoundland and Labrador fisheries, this was not done and stock collapses were the inevitable result.

THE ECOSYSTEMS IN THE EARLY 1990s

The marine ecosystems of the Grand Banks and adjacent continental shelf in the early 1990s bore little resemblance to the postglacial abundance that had sustained fisheries for over 400 years. The millions of tonnes of cod and capelin were virtually gone. Also at low numbers were the formerly abundant redfish, haddock, and American plaice and other flatfishes.[396] Forty years of overfishing had done its work, but environmental changes were also taking a toll. As if to signal the change, massive ice floes blocked the harbour at St. John's as late as June in the spring of 1991. Ice sheets covered most of the northern parts of the Grand Banks into July, and there were few warm days, even in the heart of summer.

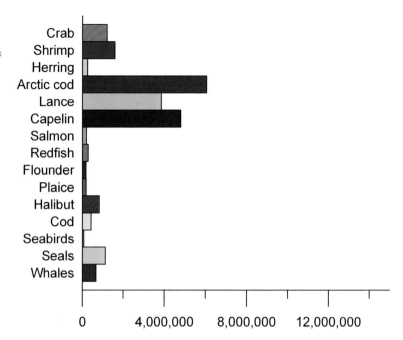

7.29
Newfoundland and Labrador ecosystem species abundance in early AD 1990s.

In the cooling waters of the early 1990s, vast schools of Arctic cod migrated far south of their normal range and were the dominant biomass recorded in acoustic surveys across the Bonavista Corridor in the early 1990s. Seabirds were forced to seek alternative foods.[397] Atlantic salmon changed their migration patterns.[398] Arctic char, like cod, found few capelin off Labrador, but unlike cod, stayed home and fed on something else.[399] The southern warm-water migrants mackerel and saury were not seen in Newfoundland waters in these years. Harp seal populations, perhaps buoyed by the increase in Arctic cod, one of their main foods, increased from about 2 million in the 1970s to nearly 6 million in the early 1990s, and came to southern Newfoundland in numbers seldom recalled. Even the humble and typically more northerly distributed *Oikopleura* (slub) were abundant far to the south. Shrimp and snow crab numbers grew – in part because they do well in cold water, and in part because their dominant predator, the cod, was virtually gone.[400]

Changes in components of an ecosystem seldom occur in isolation – one part cannot be greatly impacted without ripple effects on another. But all species are not created equal, at least in that sense. Some species are more important to an ecosystem than others because effects on them cause wider ripples throughout the system. Capelin are the key prey species in the marine ecosystems of Newfoundland and Labrador; they form the food base on which the rich predatory fish ecosystem developed and was sustained. Some of the connections are more obvious than others, for example, their importance to cod. How could 6-7 million tonnes of cod exist in this ecosystem without capelin. What would feed them? By 1991, the northern areas off Labrador had been abandoned by capelin. In the following years, capelin schools were found far to the east and south of their typical range, including the Flemish Cap and the Nova Scotia Banks where they were previously uncommon visitors. The decline, redistribution, and lack of growth in capelin in the Newfoundland and Labrador ecosystems of the early 1990s was striking and would last over a decade. Notably, there was little evidence that these changes in capelin were caused primarily by overfishing.[401] Untangling such complex relationships and their dependence on the environment, ocean climate, natural biological changes, and human-induced changes (e.g., fishing, pollution, and habitat destruction) is presently beyond our capacities. Ecosystems are simply too complicated for our mostly simple conjectures and models.

We must fall back on some simple truths. There is little doubt that the marine ecosystems of Newfoundland and Labrador – once some of the most productive in the world – have been reduced to an impoverished state. Their productivity remains well below their potential. The most dominant factor was far too much fishing, especially the foreign fishery of 1960, and destructive fishing practices, with environmental change – both physical and biological – a significant second which amplified secondary declines in the 1980s and early 1990s.

On July 2, 1992, as fishermen pounded on the outer door – some in tears, others in a tearing rage – a tense federal Minister Crosbie announced the closure of the Canadian fisheries for northern cod and Grand Banks cod at a hotel in St. John's. Foreign fisheries still operated on the Nose and Tail of the Grand Banks. It would be the first summer in 500 years with no coastal fishery for northern cod.

It had taken only forty-odd years to reduce the formerly vital Grand Banks and adjacent continental shelf – once a marvel of the world and a centre of human food production – to a virtual desert. On that fateful day in July 1992, about 40,000 Newfoundlanders and Labradorians became unemployed, and their oldest and largest industry came to an abrupt halt.

NOTES

1. Stewart Bates' view that the salt fish industry was a backwards industry and a drag on modern development deeply influenced economic thinking and public perceptions of Newfoundland in the 1950s and 1960s. In recent years, Newfoundlanders and Labradorians have often been portrayed in pejorative terms in the national media. For example, Margaret Wente (*Globe and Mail*, January 6, 2005) described "Newfs" as "surly" occupants of "the most vast and scenic welfare ghetto in the world." More viciously, animal rights activist Paul Watson spoke of Newfoundland as "the damnable rock" inhabited by "club-swinging…barbarians" who are "rapacious killers who call themselves both Christians and civilized" (Sea Shepherd Conservation Society, "Tales from the Crypt – Welcome to the Beaches of Newfoundland," April 26, 2005, http://shepherd.textamerica.com).

2. See Alexander (1977, ix).

3. Alexander (1976, 34) acknowledged that "many of the problems of the Newfoundland fishery, and through it the province's economy and society, have their roots in the deterioration of the resource base through overfishing or, at least, the excessive competition which Newfoundland fishermen must face for available stocks."

4. Alexander (1977, 16) stated that "Canadian trade policy failed miserably in maintaining and expanding markets for the highly export-oriented regional industries and economies [of Canada's east coast maritime economies]."

5. See Alexander (1977, 112-125) for a detailed description of post-Confederation economic troubles.

6. NAFEL was established in 1947 and operated for 23 years, well after Confederation, despite some opinion from Dalhousie University in Nova Scotia suggesting that NAFEL's membership structure was illegal in Canada. In 1970, federal and provincial legislation established the Canadian Saltfish Corporation, which replaced NAFEL (Maritime History Archive, "Newfoundland Fish Marketing Agencies," http://www.mun.ca/mha/holdings/findingaids/nafel_10.php).

7. See Alexander (1977, 120).

8. NAFEL (1950), cited in Alexander (1977, 121).

9. See Alexander (1977, 124).

10. Alexander (1977, 133), regarding last ditch attempts to revive the European markets in the mid-1950s, stated: "This was the last concentrated trade effort on behalf of the industry's European markets and it was too late. Demoralization had taken a firm hold, and the volume and quantity of production was deteriorating so rapidly that by the mid- to late '50s, NAFEL lacked suitable fish to offer its old European customers."

11. See Department of Fisheries (1968, 13).

12. See Alexander (1977, 139).

13. Cited in Alexander (1977, 138).

14. See Alexander (1977, 128).

15. See Gushue (1951, 69).

16. See Dawe (1963, 69).

17. In 1968, Canadians consumed an average of 5.6 kilograms of fish annually; consumption in the United States was 4.8 kilograms. Fish-eating nations consumed much more: the average annual consumption was over 27 kilograms in Japan, 23 kilograms in Portugal, 21 kilograms in Denmark and Sweden, 20 kilograms in Norway, and about 10 kilograms in the United Kingdom (Johnstone 1968, 20).

18. From 1800 to the end of World War I, the only competitors to Newfoundland-based fisheries were those of France, the United States, Canada, and a few Portuguese vessels. After World War I, growing

fleets from Portugal and Spain fished the Grand Banks.

[19] See May (1964, 5).

[20] See Beverton (1965).

[21] See May (1966, 7).

[22] Rose (2004) provided data on historical landings, estimated biomass, and harvest rates for Newfoundland and Labrador cod from the early 1500s to the present.

[23] In June 1969, the nineteenth annual meeting of ICNAF was held in Warsaw, Poland, chaired by V.M. Kamentsev of the Soviet Union. The Polish Minister of Shipping pleaded Poland's case for expanding its high seas fisheries to the northwest Atlantic (ICNAF 1968, 16): "Poland is especially interested in developing its own fisheries since we belong to those countries which suffered greatest destruction during World War II, that caused particular devastation to the Polish fisheries. The Government of the Polish People's Republic pays great attention to the problem of developing its fishing industry in order to give us the possibility of filling up the deficiency in protein for human consumption. For, as it is known, the consumption of protein per capita in Poland is below the average standard among the countries belonging to this organization." These were fine words and noble sentiments, but they gave short shrift to the Newfoundland and Labrador fisheries.

[24] See Blake (1997, 210).

[25] The *Fairtry I* was built in Aberdeen based on earlier experiment work with the *H.M.S. Felicity*, an old Navy minesweeper that was converted and renamed the *Fairfree*. Following the construction of *Fairtry I*, the *Fairtry II* and *Fairtry III* were built. Warner (1997) provides details of these developments.

[26] See Warner (1997, 235).

[27] Travin and Pechenik (1963, 1).

[28] It was not necessary to be a member of ICNAF to fish the high seas of the northwest Atlantic, but it did legitimize a fishery.

[29] Access to markets was traded for access to Canadian fisheries at least until the 1980s.

[30] Cited in Warner (1997, 236-237).

[31] See Warner (1997, 237).

[32] Marti (1958, 74).

[33] See Warner (1997, 237).

[34] Marti (1958, 74).

[35] See Travin and Pechenik (1963, 1).

[36] Travin and Pechenik (1963, 1) summarized the Soviet motivation to fish Newfoundland waters: "For several centuries the Newfoundland banks have been known to fishermen of many countries as a region with large stocks of benthic fishes, cod especially. However, intensive exploitation of the fishery resources of this region began only recently, 25-30 years ago. The decrease in catches of benthic fish in the coastal waters of northern Europe and the development of fishing boats equipped for extended sea trips abetted fishery development in this region."

[37] The first theory of sustainable fishing was advanced by the Russian F.I. Baranov prior to 1920.

[38] See Dragensund et al. (1980).

[39] The first Soviet searching trawlers explored the Grand Banks in 1954, and in the next few years located large concentrations of redfish, their preferred target. Marti (1958, 74) stated that: "In October 1957 the *Odessa* found heavy concentrations of redfish *Sebastes mentella* on the northeastern slope of the Newfoundland Bank. During the period October-December a group of big refrigerating trawlers operated in this region." Also, "in 1958 [the year the USSR joined ICNAF] the extent of Soviet investigations in the Convention area is expected to increase considerably. The three searching trawlers,

Odessa, Kreml and *Novorossiysk*, operating there will carry out the oceanographical and ichthyological research program in the regions of the Newfoundland Bank, the Flemish Cap, Labrador, and along the west coast of Greenland." The Russian trawler fleet would leave few stones unturned.

[40] See Travin and Pechenik (1963, 1).

[41] Pechenik and Troyanovskii (1969), cited in Marshall and Cohen (1973), wrote: "By 1967-1968, the survey vessels mastered the technique of trawling for *M. rupestris* to a depth of 1300 [metres]. Thus, catches per trawling hours reached 150 centners [15 tonnes] at a depth of 600-900 metres off Baffin Island, 120 centners [12 tonnes] off Labrador; near the North Newfoundland Bank the catch was as high as 200 centners [20 tonnes] per 15-minute trawling."

[42] Templeman (1959) described the large redfish concentrations in the Hawke Channel area that were first located by the Canadian research vessel *Investigator* in 1951 and were fished commercially beginning in 1958. By 1958, the fishery was prosecuted by "Canadian, United States, Icelandic, German, Belgian, and Russian trawlers." The best fishery was "on the fringes of the Hawke Channel."

[43] See Blake (1997, 211).

[44] See Parsons (1993, 224).

[45] See Parsons (1993) for details of the Santiago Declaration of 1952.

[46] Karlsson (2000, 343) provided details of Iceland's extension of limits and stated: "During this period no international law on the extent of economic jurisdiction on the seas was valid. The United Nations repeatedly tried to reach agreement on new rules, but all attempts proved fruitless because no single rule could obtain sufficient majority support. Therefore Iceland extended its fishing limits unilaterally – as other countries did in these years – and justified the action by the need to preserve its vital natural resource."

[47] Newfoundland had exclusive management of fisheries up to three miles offshore, with the baseline not following the shore but across the mouths of major bays.

[48] After the formation of ICNAF in 1950, Canada refused to concede its rights to further extend jurisdiction over fisheries beyond the current territorial limit. See Parsons (1993, 225) for further details.

[49] Government of Newfoundland (1963, 45).

[50] See Parsons (1993, 226) and Alexander (1977, 164) for somewhat different interpretations.

[51] The limits of the continental shelf were somewhat vague, but were defined as a depth of 200 metres or greater where exploitation was feasible. See Parsons (1993, 226).

[52] See Parsons (1993, 227).

[53] See Wright (1995, 400).

[54] See Parsons (1993, 227).

[55] See Alexander (1977, 164).

[56] See Alexander (1977, 164).

[57] The initial regulatory area included five divisions and extended south only to include Georges Bank (north of 39° north latitude). In 1967, the southern boundary was extended to 35° north latitude, forming the sixth division (Anderson 1998).

[58] See ICNAF (1967, 13).

[59] In the legal jargon, the members were contracting parties.

[60] See Anderson (1998, 92).

[61] Fisheries scientists have always been faced with the dichotomy of being purely disinterested observers or advocates for conservation and sustainability. In addition, scientists at the time did not know how many fish were in these stocks or how much fishing they could sustain.

62. Early fisheries conservation theories were based on the notion that if only fish above a certain size were caught, the stock would be sustained – no matter how hard it was fished. This simplistic notion has been largely discredited. It matters how much is caught and how many young fish are produced as a result.
63. Wright (1995) summarized earlier records.
64. See May (1964, 1966).
65. See Ricker (1961).
66. In 1965, Wilfred Templeman (Newfoundland) and John Gulland (United Kingdom) reviewed possible methods of conservation that could be applied to the ICNAF regions. These included: a) establishing open and closed seasons; b) closing spawning or juvenile areas; c) establishing size limits; d) prohibiting certain fishing gear; and e) prescribing catch limits. They cited some general principles of fisheries management, as true today as they were then, including: "action may have to be taken before absolute certainty in understanding the state of the stocks is reached – otherwise events may have gone too far." They related that, for whales, "some years ago the scientists pointed out that the stocks of whales in Antarctica were becoming depleted. No effective action was taken because it was felt that the scientists (who were not in complete agreement…) could be mistaken…[and] it was claimed, [that] the economic state of the industry was already too precarious" (Templeman and Gulland 1965, 56).
67. Blake (1997, 212) related how the Soviets went after silver hake on the Scotian Shelf: "When the Soviets began to develop a fishery for small-bodied silver hake, they introduced nets with meshes of 40 [millimetres] (ICNAF regulations called for large mesh nets of at least 114 [millimetres] for the groundfish sector). Young cod and other traditional species were taken, incidentally, in the small-mesh-gear fisheries, which no doubt contributed to the decline of traditional fish stocks."
68. The colonial nations of Denmark (Greenland) and France (St. Pierre and Miquelon) could also claim to be coastal states in the northwest Atlantic.
69. Anderson (1998, 82) stated: "Heavy fishing pressure on Georges Bank haddock by distant-water fleets (particularly from the USSR) in 1965 and 1966 resulted in very high catches of the very strong 1963 year-class and led to a sharp reduction in stock size."
70. An overall total quota of 12,000 tonnes for Georges Bank haddock was recommended by ICNAF scientists for 1970, "with the understanding that this was to be reserved primarily for the two Coastal States and with only a small by-catch for other countries" (Anderson 1998, 82).
71. Templeman and Gulland made the case for quotas in 1965, concluding that "there must therefore be some direct control of the amount of fishing. All methods of doing this raise difficulties, but that presenting the least…is…catch quotas, preferably allocated to each section of the industry" (Templeman and Gulland 1965, 56).
72. ICNAF was reasonably bold in setting quotas, even for stocks for which little information was available. Unfortunately, the strategy appears to have been to set the quota above the current catches, so they were largely ineffective. Nevertheless, by 1977, quotas with allocations by nation were in place for about 70 stocks, which was a major accomplishment (Anderson 1998).
73. See Needler (1970, 15-16).
74. See ICNAF (1970, 41).
75. See Needler (1971, 18).
76. See Templeman (1971, 48).
77. See Needler (1971, 19).
78. Wilfred Templeman in 1973, as reported by Johnstone (1977, 277-278).
79. See Wright (1995).
80. See Templeman and Fleming (1963).

81 See Templeman (1966) and Rose and Leggett (1988, 1989, 1990).

82 See Templeman (1959, 21; 1962, 8-27). Templeman (1959, 21) stated: "When the [Newfoundland longline fishery off Bonavista] began in 1952, the deep-water stocks of cod off Bonavista had not been fished previously, and until 1956 the Bonavista longliners were the only boats fishing the area. In 1956 a few large European trawlers began fishing in the area, and in 1957 and 1958 a much greater concentration of effort by a larger number of trawlers occurred. In addition, a fleet of longliners from the Faroe Islands and from Norway fished off Bonavista and in neighbouring deep-water areas in 1957 and 1958." This fishery collapsed within a few years.

83 Templeman (1958, 21) stated: "In 1952, the first year of commercial longlining, 30% by number of the cod landed were 81 centimetres and larger, and the average length was 76.0 [centimetres]. Each year that followed, the proportion of these larger fish in the catch was reduced and by the end of July 1958, their contribution to the catch was only 6%."

84 Memo from Colin Story to Deputy Minister Eric Gosse, November 19, 1959, Provincial Archives of Newfoundland and Labrador, GN 34/2, file 11/80, volume 2.

85 In 1959, the Federal-Provincial Atlantic Fisheries Committee (FPAFC) was implemented as a forum for provincial representatives to raise issues with the federal fisheries authorities.

86 See Wright (1995, 399).

87 The empirical determination of the amount that can be taken from a fish stock relies on first taking too much so that you know you have exceeded the maximum – then fishing is cut back. ICNAF apparently was waiting for this to occur. But quick expansions of fishing technology and power, as were occurring in the 1950s, will make the peak difficult to determine (see Hilborn and Walters (1992) for details). In essence, the maximum yield was exceeded so quickly that it could not be recognized as it occurred – by the time it was, the stocks were severely depleted (as Eric Gosse suspected they would be).

88 See Crutchfield (1961).

89 See Ricker (1961).

90 See Ricker (1961, 7). Ricker was also optimistic about herring, the flatfishes and some other species, but thought that redfish yields had little potential to increase. Ricker's theories were state-of-the-art at the time, but his notions that fish stocks could reach an equilibrium in production over a wide range of stock sizes has since been shown to be simplistic.

91 A draft presentation to the conference by C.J. Morrow, President of National Sea Products, Lunenberg, N.S., the lead off speaker to the session on efficient operations, stated that "one would be brave indeed, as a layman, to quarrel with Dr. Ricker and other scientists on future levels of fish stocks" (Newfoundland Archives, File GN 34/2 file 11/89).

92 PANL GN 34/2, file 11/89.

93 PANL GN 34/2, file 11/89.

94 See May (1964, 1966).

95 Select Committee on Inshore Fishery (1975, 9).

96 Select Committee on Inshore Fishery (1975, 10).

97 Select Committee on Inshore Fishery (1975, 32).

98 Select Committee on Inshore Fishery (1975, 34-35). Details of provincial and federal expenditures are given in Table 14 and 15, respectively, on pages 78-79.

99 Anonymous (1986, 8).

100 Anonymous (1985, 6).

101 Data compiled by Lear and Parsons (1993, 77).

[102] The full text of this letter may be found in Newfoundland Department of Fisheries (1974, 54-56).
[103] See Shepard (1973).
[104] See Shepard (1973, 148).
[105] See Neill (2006).
[106] See Government of Newfoundland (1963, Appendix A,V, 53, and Appendix B,V, 61). Expenditures per farmer were C$392 compared to C$244 per fisherman.
[107] See Government of Newfoundland (1963, preface by Joseph Smallwood).
[108] Gwyn (1999, 205) related how Smallwood's passion for Newfoundland seemed to stop at the shore: "For twenty years, Smallwood has sought to turn his people away from the sea." It was a poor strategy for an essentially marine nation.
[109] See Gwyn (1999, 204-205).
[110] See Walsh Report (1953, 24).
[111] The Walsh Report (1953, 24) gave the following example of how many people the Newfoundland fishery might support: "If the productivity of the primary fishing industry as a whole were raised to that of the most efficient current operations, i.e., dragging and long-lining, the present annual catch of cod (approximately 500,000,000 pounds [200,000 tonnes] could be landed by 2000 fishermen. A fish-processing industry capable of handling this catch would absorb some 3500 men and 2500 women (in addition to those now employed), assuming the application in all branches of a level of technique comparable to that prevailing in filleting and freezing plants." At the time there were about 20,000 fishermen in Newfoundland, a substantial decrease from the estimated 40,000 fishermen prior to WWI (Walsh Report 1953, 9).
[112] See Walsh Report (1953, 112).
[113] See Gwyn (1999, 206).
[114] See Alexander (1977, 15).
[115] See Dawe (1963, 43).
[116] See Dawe (1963, 10).
[117] Alexander (1977, 144) gave a longer version of this quote.
[118] See Dawe (1963, 16).
[119] See Alexander (1977, 143).
[120] See Dawe (1963, 17).
[121] From a scientific perspective, it had been clear since the early 1960s that the traditional inshore fishery and the new trawler fleets were fishing the same northern cod at different times of year and that the winter trawler fisheries had near-immediate effects on the summer inshore fishery (May 1964, 1966; Pinhorn 1979). The same was true on the south coast and in the northern Gulf of St. Lawrence. Only the Grand Banks cod stock did not regularly migrate to shore. Even these fish at times would come to shore, however, especially along the eastern shore of the Avalon Peninsula to Cape St. Mary's. French hooks would at times be found in cod captured on the Avalon and as far north as Conception Bay (*The Newfoundlander*, February 22, 1849; Doug Howlett, Petty Harbour, personal communication). The origin of the French hooks is not clear. They could have come from the Grand Banks or from the south coast as French fisheries operated in both regions. Recent tagging research has shown that south coast fish migrate at times to the northeast coast in summer (Lawson and Rose 2000).
[122] See Dawe (1963, 20).
[123] See Dawe (1963, 63).
[124] Government of Newfoundland (1963, 6).

[125] See Walsh Report (1953, 30).

[126] See Dawe (1963, 63-64).

[127] Dawe (1963, 83) recommended improved education on Newfoundland history and the fishery because it is "a training that should be enjoyed by all Newfoundlanders whether or not they intend to join the fishing industry, for knowledge of one's own country has traditionally been the best starting point for an intelligent appreciation of wider horizons."

[128] See Dawe (1963, 76).

[129] See Dawe (1963, 96-97).

[130] See Alexander (1977, 147).

[131] See Gwyn (1972, 384-387).

[132] See Gwyn (1972, 387).

[133] Haddock are highly vulnerable to bottom trawling in the winter and early spring because of their aggregative habits near the ocean floor prior to and during spawning.

[134] These values were estimated from historical catches using a stock reconstruction approach.

[135] St. Pierre Bank haddock were likely members of a separate and smaller population than haddock on the southern Grand Banks. St. Pierre Bank haddock had their last strong year-class in 1949. Templeman (1964, 5) stated: "On St. Pierre Bank an otter trawl survey in June once again produced very little haddock. In most of the drags no haddock or just a few specimens were caught, the best catch being 170 [pounds] (77 kilograms). Haddock fishing has been insignificant on this bank since the fishery on the very abundant 1949 year-class came to an end in 1957."

[136] See Travin and Pechenik (1963, 17).

[137] See Warner (1997, 241).

[138] See Templeman (1964) and Martin (1959, 28).

[139] Thompson (1939, 7) stated: "Within little more than a decade a greatly increased knowledge of the biology of the haddock of the Western Atlantic (east North America) region has been obtained. Part of the urge to study this fish, which, if minor variation in racial characteristics be ignored, appears to be identical with the haddock of eastern and northern Atlantic waters, has arisen from the rather sudden realization in the eastern portion of the United States of its high quality, and especially of its suitability for filleting and sharp-freezing. Over-exploitation has resulted in a remarkably short time, and conservation measures have been proposed."

[140] See Templeman (1961, 28).

[141] See Travin and Janulov (1961, 84).

[142] See Pavlov and Travin (1961, 87).

[143] See Templeman (1962, 44).

[144] See Templeman (1962, 26-27).

[145] Templeman (1964, 6) stated: "Although heavy exploitation has played its part in the present scarcity of haddock, the basic cause is the lack of successful survival of young. The last very successful year-class was that of 1955, which dominated the catches during 1960-62. Since 1955, survival has been relatively poor and consequently the haddock fishery must inevitably continue to be at a low level until a good brood occurs and is recruited to the stock."

[146] *CCGS Teleost* research trips in early 2000s, with the author as chief scientist.

[147] Pavlov (1960, 104) stated that: "In 1959, like in previous years, the Soviet trawler fleet in the North-West Atlantic fished primarily for redfish. The fishing activities were concentrated mainly on the northern and north-eastern slopes of the Newfoundland Bank and off the southern part of Labrador."

[148] The Icelandic scouting trawler *Fylkir* searched for redfish from Funk Island Bank to Hamilton Bank in 1958. New names were given to good fishing sites. The bank south of Hawke Channel proved very productive for redfish and the new fishing ground was named Sundáll. Just to the south, in what would now be called southern Belle Isle Bank, another new fishing ground was charted and given the name Ritubanki (Magnusson 1959, 57-58).

[149] Buckmann et al. (1959, 52-53) stated: "After the Icelanders had observed a rich occurrence of redfish south-east of Hamilton Inlet Bank [Hawke Channel], German trawlers moved into Subarea 2 in August. Due to the extraordinary large daily catches (33.9 tons) [34.4 tonnes], this fishery was very profitable in spite of the long trips of 2,150 nautical miles (15.9 days to and from)." In the same report, Meyer gave a German perspective on the Labrador cod: "The very slow growth of the Labrador cod is striking. In December, at the end of the feeding period, the three most abundant year-classes 1952, 1950 and 1948 had only attained average lengths of 55.1, 60.6, and 65.0 centimetres, respectively. This explains why the cod landed from Labrador in spite of its high average of 9.8 years only measure 63.4 centimetres in mean [length]. This small size causes the Labrador cod to be not especially suitable for the fresh-fish trawlers." If only it had been true.

[150] The total catches may have been much greater. Magnusson (1959, 59) stated: "The discovery of these new fishing grounds was of a very great importance for the Icelandic fishery in the year 1958. The total amount of fish caught by Icelandic vessels in 1958 was 505 thousand tons [513,000 tonnes]. The total catch of redfish was 110 thousand tons [111,760 tonnes] or 22% of the total catch, which is the highest yearly catch of redfish up to the present time. The total quantity of fish caught by Icelandic trawlers in 1958 was 199 thousand tons [202,184 tonnes], of which the redfish caught by Icelandic trawlers in [ICNAF] Subdivisions 2J and 3K was 80 thousand tons [81,280 tonnes] or nearly 77% of the total Icelandic catch of redfish in 1958." These numbers do not jibe with other reports of total redfish removals by all counties and suggest considerably higher total catches.

[151] Travin et al. (1960, 105) stated: "Cod were certainly of commercial sizes in all the Newfoundland subdivisions investigated by the scouting trawlers. In most cases their average size exceeded 50 [centimetres], the average weight being about 1.5 [kilograms]. The largest cod (average length about 60 [centimetres]) were encountered in 3L and 3N, especially early in the year."

[152] See Templeman (1962, 25 and 1963).

[153] See Hodder (1965, 34).

[154] *St. John's Independent*, February 19, 2006.

[155] See ICNAF (1963, 36).

[156] Hodder (1965, 33) stated that: "In Subarea 2 (mostly in the Hamilton Inlet Bank area, [ICNAF] Division 2J) a very great offshore cod fishery developed between 1959 and 1961 when trawler landings reached a level of almost 250 thousand tons from less than 30 thousand tons annually in 1954 to 1958."

[157] See May (1964, 1966).

[158] See Hodder (1965, 41).

[159] See ICNAF (1967, 50). Overfishing was defined as a rate of harvest which was higher than indicated to obtain the maximum sustainable yield from the stock. Scientific theory held that a stock could be sustained under such overfishing, the problem being solely lower yield to the fishery. Actual damage to the stock's abilities to sustain itself would occur if the stock was further reduced, but the stock level at which that might occur was seldom specified.

[160] See ICNAF (1967, 52).

[161] ICNAF (1969, 42) reported that the "[Federal Republic of Germany] fishery conducted from February to mid-April in Div. 2J took 35.7 tons [36.3 tonnes] per day (previous highest was 31.4 tons [31.9 tonnes] in 1965)…[but that] cod were scarce in inshore waters."

[162] See ICNAF (1969, 22).

163 See ICNAF (1970, 41).

164 See ICNAF (1970, 37-38).

165 In the 1950s and early 1960s, cod older than 10 years of age were common and older than 15 years not unusual off Labrador (Subarea 2J) (ICNAF 1957, 70). These fish were slow growing and few exceeded 80 centimetres in length (ICNAF 1961, 65).

166 Between 1969 and 1970, total cod catches in Subarea 2J declined by nearly 50%, from 412,000 tonnes to 210,000 tonnes (ICNAF 1971, 47).

167 See ICNAF (1975).

168 Pauly et al. (1998) referred to this as "fishing down the food chain."

169 See Johnstone (1968, 42).

170 Warner (1997, 240) quoted Charles Philbrook, a veteran U.S. fisheries enforcement officer.

171 See ICNAF (1969, 46-47). This was a tense time in United States–Soviet Union relations, and the survey may have had wider political implications.

172 Small allowances of 10,000 tonnes were given to allow a joint Canada–Soviet Union research program in 1980 and 1981. The inshore fishery remained largely unregulated as it took less than 25,000 tonnes in most years (DFO 1982).

173 Resource Management Division (1986).

174 "Studies of the origin and destination of salmon at West Greenland were continued in 1970 and confirm previous evidence that the stock originates mainly from rivers in Canada and the [United Kingdom]. Studies suggest that approximately equal proportions of the North American and European salmon make up the exploited stock at West Greenland" (ICNAF 1971, 46).

175 See ICNAF (1970, 37).

176 See DFO (2002).

177 See ICNAF (1970, 38).

178 See Hammill and Stenson (2003).

179 Fletcher et al. (1975) provided an early survey of some invertebrates in coastal Labrador. They reported the finding of Icelandic scallops, whelks, and mussels in good numbers in some bays, and the presence of slow-growing green sea urchins which were also thought to have market value: "Interest in the inshore resources of Newfoundland and Labrador has been increasing during the past few years, particularly with reference to unexploited or under-exploited species such as shellfish, crustaceans, seaweeds and sea urchins. The present diving survey was conducted to gain some insight into the resource potential…little or nothing is known about the shore life resources of Labrador."

180 See Warner (1997, 242) for ICNAF data.

181 Warner (1997, 241) stated: "Huge as the total catch might seem, the catch per vessel was down and the fish were running generally smaller than before, even though the foreign vessels fished longer hours with improved methods over a larger range for a greater part of the year. In fact, the foreign catch had been better for five of the preceding six years – slowly declining from a peak of 2,400,000 tons [2,438,400 tonnes] in 1968 – with fleets of equal size…Everywhere the trawlers went…more were fishing for less."

182 See Hammill and Stenson (2003).

183 See Rose (2005).

184 See deYoung and Rose (1993).

185 As pointed out by Alexander (1977, 159), it might be argued that the Canadian (i.e., Nova Scotian) and Newfoundland views of the modernization required for the fishery were not incompatible. That may well have been true; however, the difference was that Newfoundland knew that its fishery must be large in terms of employment – there were few alternatives and a small fishery was not a realistic alternative. This is why the failure of Canada to protect the entire fishery, which required gaining control over the entire continental shelf, ended any chance of maintaining a large fishery in Newfoundland and Labrador.

186 Carter (1982), citing "Fisheries Reorganized in Newfoundland," a radio program broadcast in 1944.

187 Maritime History Archive, "No Great Future: Government Sponsored Resettlement in Newfoundland and Labrador since Confederation," http://www.mun.ca/mha/resettlement/rs_intro.php.

188 Placentia: The Early Years, "Resettlement," http://collections.ic.gc.ca/placentia/newpag.htm.

189 See Gordon (1961, 15).

190 See Alexander (1977, 164).

191 Gus Etchegary, Fishery Products executive and President of the Save Our Fisheries Association, informed a House of Commons committee on fisheries in 1971 that the "Newfoundland fishery is going downhill on a toboggan" (Gwyn 1972, 389).

192 See Parsons (1993, 232-235).

193 See Blake (1997, 216).

194 *St. John's Evening Telegram*, October 12, 1971.

195 See Johnson (1977).

196 See Parsons (1993, 238-239).

197 Art May, personal communication, 2005.

198 Bismarck, a German politician, made this statement on August 11, 1867; it has since been used in a translated form by many others. See http://www.quotationspage.com/quote/24903.html.

199 See Parsons (1993, 240).

200 Canada argued vigorously for harvesting at less than the maximum theoretical rate. The Canadian recommendation was to fish at what became known as the $F_{0.1}$ level, well below the maximum rate. This recommendation was accepted by ICNAF in 1976.

201 Leonard Legault, a chief Canadian negotiator of extended jurisdiction, stated in a speech that "one of the most basic elements of this consensus (UNCLOS) is the principle of optimum utilization, which means simply that coastal states will not play dog in the manger – will not allow fish stocks under their jurisdiction to die of old age and go to waste – but rather will allow other countries access to such portions of the stocks as may be surplus to the coastal state's harvesting capacity" (Parsons (1993, 259)).

202 Maloney (1990) described how under international law and through a series of bilateral treaties between the government of Canada and the governments of various nations, including France, Cuba, the Faroe Islands (Denmark), East Germany, Japan, Norway, Poland, and the Soviet Union, each of these countries was permitted to fish within Canada's 200-mile economic zone in the waters off Newfoundland for certain allocations of fish that were deemed surplus to Canada's needs.

203 See Maloney (1990, 152-154).

204 An allocation of about 11,000 tonnes of cod in NAFO Division 2GH in the late 1980s was shared among France, the Faroe Islands, the German Democratic Republic, Norway, and the Soviet Union. These were paper fish, and none was caught. The only wealth created by this was among clerks in capital cities (Maloney 1990, 152-154).

205 DOE (1973), cited in Parsons (1993, 239).
206 For additional detail, see Anderson (1998).
207 See Blake (1997, 218).
208 The European Union wanted to fish at F_{max}, which was about twice as high as the Canadian-preferred strategy of $F_{0.1}$.
209 Under NAFO, the so-called "objection rule" allowed any member to file opposition to a quota or restriction within 60 days. They were then not bound by that limit and were free to set their own. The European Union did this regularly in the 1980s.
210 The Spanish trawler fleet was banished from Namibian waters off southwest Africa in 1990, and much of this fleet was reassigned to the Grand Banks (Blake 1997).
211 See Blake (1997).
212 Government of Newfoundland (1985, 7) stated that "Newfoundland is proud to have brought the marine and mineral resources of its Continental Shelf into Canada in 1949." This is not technically correct, as these areas in 1949 were still high seas.
213 Romeo LeBlanc, federal Minister of Fisheries in the late 1970s, was an unabashed supporter of the small boat inshore fishery and spoke of helping "the small man, the inshore man, with his own boat" (Parsons 1993, 253). In the post-extension fishery in Newfoundland, the inshore fishery was given an allocation, which was not a quota in the sense that it was constant.
214 See Maloney (1990, 115).
215 See Martin (1992, xiii).
216 See Morgan et al. (1997).
217 A perennial problem with the inshore trap fishery was the glut of fish which would occur over a period of six to eight weeks when the cod were close to shore and caught by traps. Historically, these fish would be taken from the traps and dried and salted by family enterprises; therefore, fisherman would not extract more than could be processed. However, with the change to fish plants, especially freezer plants, fishermen wanted to sell as much fish as possible, which resulted in more fish being brought in than could be handled in a short period. Delays in processing resulted in lower quality. For their part, freezing plants did not favour trap cod, as they tended to be small. In 1977, about 2558 traps were still in use in Newfoundland and Labrador, concentrated on the Labrador and northeast coasts. (Government of Newfoundland 1977) made several recommendations on freezing and storing cod during the glut. These did not overcome the fundamental problem that plant capacities were insufficient to handle the glut over a few weeks, then lay idle for the rest of the year. The reality was that traditional drying methods capable of handling more than 100,000 tonnes were no longer practicable, and nothing had been introduced to replace them that could handle so much fish in the short summer season.
218 Doucet (1979, 2) stated: "The cheapest method of catching fish is the cod trap – less than half that of any other type of gear…Much is often said about 'economic rent' when considering different options for the use of a common property resource. Economic rent or surplus is what is left when all private costs of harvesting…and processing are covered. From that point of view, the 55-foot long-liner still leads the list. The factory trawler trails far behind."
219 See Parsons (1993, 395) for a summary of these arguments in the 1970s and 1980s.
220 See Parsons (1993, 355).
221 See Maloney (1990, 40).
222 Alistair Hann, Mayor of Burgeo, personal communication, 2005.

[223] Credit for work was given in the form of stamps of differing value, literally pasted into a book. Higher value stamps would return higher payments under the *Unemployment Insurance Act*.

[224] The notion that Newfoundland started milking an innocent federal cow immediately after Confederation is a mainland myth. In fact, throughout the 1950s, of all the provinces of Canada, Newfoundland actually received the second least money from Ottawa, with only Quebec receiving less. Newfoundland had proportionately fewer older people and almost no unemployment claimants until 1957. In addition, Newfoundland veterans received their basic pension from Britain, with only a top-up from Canada (Gwyn 1999, 208). The dependence of rural Newfoundland communities on federal money is a new thing, created largely by the Unemployment Insurance Program being extended to fishermen in 1957.

[225] Government of Newfoundland (1963, 61).

[226] See Dawe (1963, 43).

[227] See Matthews (1993, 36).

[228] See Blackwood (1996).

[229] Both the federal Department of Regional Economic Expansion and Department of the Environment issued documents in 1976 suggesting that no increase in employment should be expected after the coming 200-mile extension was implemented (Shrank 2005, 408). Federal Minister Romeo LeBlanc responded to notions that Newfoundland and Nova Scotia should expand their fleets: "Do we want to double a fleet that is getting half loads? I hardly think so" (Parsons 1993, 253).

[230] See Parsons (1993, 253).

[231] See Parsons (1993, 359).

[232] Parsons (1993, 359), citing government reports.

[233] See Shrank (2005, 408).

[234] See Parsons (1993, 131-132).

[235] Shrank (2005, 408) stated of H.B. Nickerson & Sons Limited and the major part they played in the overcapacity of the Newfoundland cod fishery: "A Nova Scotia company with delusions of grandeur formed joint ventures with Newfoundland companies to gain access to, process and sell much of the fish newly under Canadian jurisdiction. One result was that substantial new processing capacity was added in an industry that already had excess capacity." Shrank stuck mainly to the economics, but there is little doubt that these delusions of grandeur also contributed to ignoring science when stock assessments indicated declines and the eventual destruction of several stocks.

[236] Carter (undated) gave details of the takeover of National Sea Products by Nickerson & Sons Limited.

[237] Parsons (1993, 359) gives details of the Nova Scotian-based position.

[238] Gus Etchegary, personal communication, 2005.

[239] William Morrow, *Globe and Mail*, August 29, 1981.

[240] See Etchegary (1981, 59).

[241] Bill Wells, CBC radio, July 11, 1982.

[242] Jim Morgan, *St. John's Evening Telegram*, June 26, 1982.

[243] John Macpherson, *St. John's Evening Telegram*, March 26, 1982.

[244] See Crosbie (1997, 357-358).

[245] Maloney (1990, 73) described how dozens of Quebec boats fished northern cod off Black Tickle, Labrador, in the mid-1980s until the stock collapsed. This may not have been legal, as the vessels were not technically licensed for that stock, but the Department of Fisheries and Oceans management

authorized it anyway to placate Quebec interests. In the mid-1980s, the author was conducting research at Blanc Sablon, and saw many of these vessels heading north. Reports of gillnets criss-crossed across the inshore grounds near Black Tickle became common lore among Labradorians. Black humour joked that only the last person to set their nets would catch fish. Much gear must have been lost and many fish destroyed but not harvested. This undoubtedly contributed to the collapse of the stock.

[246] See DFO (1986, 8).

[247] See Matthews (1993, 49).

[248] Moores (1978), cited in Matthews (1993, 63).

[249] LeBlanc (1980, 5-6), cited in Matthews (1993, 54).

[250] Art May, personal communication, 2005.

[251] See FFAW (1996, 3).

[252] See FFAW (1996, 3).

[253] This was the last piece of legislation passed by a Smallwood administration. It is ironic in that Smallwood was at best equivocal about unions and workers' rights.

[254] See FFAW (1996, 6).

[255] See FFAW (1996, 6).

[256] See DFO (1986, 6).

[257] LeBlanc (1980, 5), cited in Matthews (1993, 52).

[258] See Kirby (1982, 91). This gives some idea of what the commissioners thought of northern cod. The problem, as they saw it, was what to do with all the cod to come. The statement that NAFO Division 2J3KL cod was a fast growing stock was at odds with 100 years of scientific observation.

[259] Government of Newfoundland (1978, 225-226).

[260] In the 1950s and 1960s, the frozen-fish trade was given all encouragement, with Premier Smallwood attracting several large non-Newfoundland firms into this business, including British Columbia Packers Limited (owned by George Weston Limited of Toronto) and Birds Eye Foods (owned by the giant multi-national Unilever). Both these companies were not strongly committed to the Newfoundland fishery and sold their operations as stocks declined in the 1970s.

[261] See Parsons (1993, 361).

[262] See Parsons (1993, 361) for additional details.

[263] "Why the need for such haste? Because the fishery, viewed from either an economic or a social perspective, is in serious trouble. Change in the way a great many things are now done is essential and must be undertaken as quickly as possible if the fishery is to survive" (Kirby 1982, preface).

[264] See Parsons (1993, 362).

[265] See Kirby (1982, 60-61).

[266] Similar mistakes were made in the 1990s in projections of northern cod growth by Roughgarden and Smith (1976), Myers et al. (1997), and Hutchings (2000).

[267] Pinhorn (1979, 15), concluded that "the biomass of cod is projected to rebuild to a level capable of supporting 350,000-400,000 tons [355,600-406,400 tonnes] by 1985 under present management strategies *if levels of recruitment in the future are average*…Inshore and offshore fisheries are competing for the same resources essentially at different seasons so that an increase in one fishery will be at the expense of the other" [emphasis added].

[268] See Kirby (1982, 7-9).

[269] See Kirby (1982, 140).

[270] Gus Etchegary, former chief executive officer of Fishery Products, personal communication.

[271] See Kirby (1982, 72).

[272] See Kirby (1982, 122).

[273] See Parsons (1993, 363).

[274] Government of Newfoundland (1983, 1).

[275] Under this program, fish plants with unused capacity, being resource-short, would be given fish to process.

[276] Government of Newfoundland (1983) questioned the logic of the rationalization of the industry suggested in a federal report undertaken by the accounting firm of Price Waterhouse that specified that at least five plants would close, including Grand Bank, Burin, Gaultois, Fermeuse, and St. Lawrence.

[277] Parsons (1993, 364-376) gave details of the negotiations that led up to the *Atlantic Fisheries Restructuring Act* in 1983.

[278] The initial agreement provided for a C$75 million federal cash injection to buy the companies, a conversion of C$32 million of debt to equity by Newfoundland, and C$44 million by the Bank of Nova Scotia, giving ownership as follows: federal government (60%), Newfoundland (25%), Bank of Nova Scotia (12%), and employees (3%). This was modified somewhat as Fishery Products International was formed (Parsons 1993, 370).

[279] The federal government put C$44 million into the formation of National Sea Products (Parsons 1993, 375). Pêcheries Cartier in Quebec received about C$32 million.

[280] In the late 1970s, Canadian trawler companies fishing under a competitive quota system caught their entire quota within about six weeks, resulting in a closure of the fishery and poor fish quality.

[281] See DFO (1983). This was also the year in which the quota for northern cod was increased from 200,000 tonnes to 266,000 tonnes, largely to service the new allocations. The quotas for the Scotian Shelf and southern Newfoundland cod stocks were both reduced in the same announcement, from 64,000 tonnes to 55,000 tonnes, and from 33,000 tonnes to 25,000 tonnes, respectively.

[282] See Government of Newfoundland (1986, 16-17).

[283] See Government of Newfoundland (1986, 7).

[284] See Government of Newfoundland (1986, 20).

[285] See Keats et al. (1986).

[286] DFO (1985, 5).

[287] DFO (1985, 5-7).

[288] DFO (1985, 13).

[289] See Maloney (1990).

[290] In Newfoundland and Labrador, these vessels were referred to as "the bankers" because they had been named *Funk Island Banker*, *Hamilton Banker*, *Nain Banker*, etc. These vessels were sold off by the provincial government at a great loss in the 1990s. Bureaucracy can be amusing: access to groundfish was determined by vessel size – the first of the longliner fleet, the *Grand Banker*, was a bit shorter in length than the other longliners, and so was in a different vessel category than the others and retained her quota (Maloney 1990).

291 A fishing quota was defined as a percentage of the total allowable catch, while an allocation was an absolute number. If the overall catch limit for the stock declined, quotas also declined. Allocations, however, did not. The belief of federal officials that the northern cod would grow made this an easy succession; as the total allowable catch increased, most of it was given to the offshore companies.

292 Memo from Dr. Arthur May to the Kirby Task Force, April 1982, cited in Shrank (1995). Dr. May confirmed that this was one of the battles not won (personal communication 2005).

293 A new shipbuilding subsidy program was instituted on May 12, 1961. This program included two types of assistance on fishing vessels: 1) a 50% subsidy on steel trawlers over 75 feet [22.5 metres] and effective in the Atlantic provinces, including Quebec; and 2) a 40% subsidy on fishing vessels over 100 gross tons (a gross ton is 2240 pounds or 1016 kilograms) [midsized vessels generally made of wood or fibreglass] until March 31, 1963, after which the subsidy declined to 35% (Government of Newfoundland 1963, 61).

294 As late as the mid-1980s, Newfoundland argued against allocations of northern cod to non-Newfoundland vessels or plants. Government of Newfoundland and Labrador (1985) stated that "existing Canadian fisheries policy allocates approximately 50,000 tons [50,800 tonnes] of northern cod each year to fishing interests outside Newfoundland, and in recent years this level has been increasing. Allocations have been granted to fleets with no historic presence on the stock, at a time when the raw material needs of Newfoundland plants historically dependent on northern cod are far from being met."

295 Whether or not this was true is hard to say. Even in good years the inshore fishery caught too much too fast – the glut – and the problems in processing this fish were not resolved.

296 See Maloney (1990, 89).

297 See Rose (1992).

298 See Parsons (1993, 132).

299 Beverton and Holt (1957).

300 See Rose (1996) and Hutchings (2000).

301 Maximum sustainable yield became the holy grail of fisheries science in the 1950s.

302 See Gordon (1954).

303 See Hardin (1968).

304 See Matthews (1993, 40).

305 See Matthews (1993, 239).

306 See Matthews (1993, 253).

307 Fishing power is the ability to kill fish: it is not a measure of how many fishermen are in a fishery. One trawler with 20 men has much more fishing power than 200 small-boat fishermen.

308 The Four Horsemen of the Apocalypse were war, famine, death, and goodness (signified in Christian tradition as Christ). The Horsemen of open access fisheries by analogy are overfishing, stock collapse, and poverty, all of which have occurred. The white horse (goodness) is the hope that sustainable management may follow.

309 DOE (1976), cited in Matthews (1993, 50).

310 See Copes (1980, 142).

311 See Dawe (1977).

312 See Dawe (1977, 78)

313 A more detailed discussion of these economic papers is given by Matthews (1993, 43-47).

314 See Matthews (1993).

315 This statement is made with full realization that this is not all that is needed for a successful fishery. Economics are important, as are social concerns – a successful fishery must make the most of the resource. However, first you must have the fish. That is the bottom line.

316 MacKenzie (1979, 816) argued that the fishery is an employer of last resort: "He is a fisherman because he is poor, not the other way around." A case may be made that the Newfoundland and Labrador inshore fishery had suffered from the tragedy-of-the-commons effect and resulting lack of fish since the mid-1800s, when human population growth began to outstrip fish population growth. However, for the most part, the poverty of the fishery was more related to economics than lack of fish until the 1960s, and the full effect of the tragedy of the commons would not be felt until the coming of the foreign trawlers.

317 See Johnstone (1977, 307).

318 Comment by Ronald Hayes made to Kenneth Johnstone (1977, 248).

319 See Pinhorn (1979, 9-10).

320 Pinhorn (1979, 14-15) made these conclusions.

321 In the 1990s, both popular and scientific accounts of the decline of the Newfoundland cod stressed the scientific troubles in the 1980s as being of primary importance in allowing the stock declines. The much larger declines in the 1960s were considered to be of only historical interest (e.g., Harris 1998; Hutchings and Myers 1993). This view is unlikely to be correct as the large Grand Banks and northern cod stocks were reduced to levels well below those at which reproductive impairment became evident in the 1970s. The mistake of later science was not to take this into account. For example, since the mid-1970s, the remnant northern cod stock has not produced a single year-class of the magnitude estimated for the 1960s when the stock was at 3-4 million tonnes.

322 See Pinhorn (1979, 11).

323 See Bishop and Shelton (1997, 43).

324 Lear (1998) summarized these early developments in applying yield-per-recruit analyses that can be used to establish the maximum sustainable yield. See Hilborn and Walters (1992) for details of these calculations.

325 See Pinhorn (1979) and Bishop and Shelton (1997) for additional detail.

326 Ulltang (1998) discussed the weaknesses of virtual population analysis and related methods, noting that they routinely overestimate stock abundance. This was often catalogued as the "retrospective problem," as if this somehow puts it under control. Ulltang (1998, 136) referred to the "delusion of…[virtual population analysis]–tuning" and the dangers of overreliance, at worst blind faith, in abstract numbers and models.

327 Bishop and Shelton (1997, 8) stated: "Canadian catches were mainly from the inshore fixed gear fishery and sampling was generally in excess of minimum requirements. However, sampling by some countries obtaining large catches was totally lacking for most years." The main offenders were France, East Germany, West Germany, and Spain.

328 See Bishop and Shelton (1997, 11).

329 Rose and Kulka (1999) showed that the trawler catch rates continued to increase as the northern cod stock declined and were at their highest recorded levels when the stock was near total collapse.

330 See ICNAF Redbook (1977, 54).

331 See Cushing (1995)

332 See Finlayson (1994) for a sociological explanation of this phenomenon.

333 Bishop and Shelton (1997) offered no convincing explanation of why these surveys were discounted, other than they may not have been extensive enough.

334 See Bishop and Shelton (1997).

335 See Winters and Wheeler (1985).

336 George H. Winters, "Aide-Memoir on 2J3KL assessment: *Non gratum Anus Rodentum?*" Unpublished CAFSAS working paper, 1986, 13 p. (cited with permission of the author).

337 See Keats et al. (1986).

338 See Warren (1997).

339 See Young (1988, 2).

340 National Sea Products memo to Department of Fisheries and Oceans, 1989 (mimeo in possession of the author).

341 See Bishop and Shelton (1997, 27).

342 Hutchings et al. (1997) inferred that scientific mistakes had overtones of collusion, and Finlayson (1994) proposed that pressure on scientists to prefer positive results greatly influenced the stock assessments of the 1980s. I found little evidence that either of these arguments was correct.

343 Various calculations indicated that the fishing mortality (i.e., the proportion of fish killed) was likely at least double what the stock could have sustained (e.g., Bishop and Shelton 1997).

344 Crosbie (1997, 378) related how he knew that if he had accepted the scientific advice in 1989, both National Sea Products and Fishery Products International would have gone bankrupt. He believed he had no choice but to set a higher quota, because "we are dealing with thousands of human beings, who live and breathe and eat and need jobs – fishermen and the like, so we are not going to, because of the formula…immediately go to a quota of 125,000 tonnes." Crosbie's actions were understandable, and overall, his performance during those years was solid, but delaying the closure of the fisheries allowed the continual removal of fish; this, combined with continued foreign fishing on the Nose of the Grand Bank, has undoubtedly delayed stock rebuilding.

345 National Sea Products memo to the Department of Fisheries and Oceans, 1989 (mimeo in possession of the author).

346 See Bishop and Shelton (1997).

347 See Harris Report (1990, 10).

348 See Lear and Parsons (1993).

349 See Harris (1990, 8).

350 See Young (1988, 13).

351 Harris (1990, 134) argued that fishing mortality must be reduced to about 20% (it had previously been as high as 40%–50%); if this was not achieved, there was likely to be "a significant continuing decline in the spawning population."

352 See Harris Report (1990, 64).

353 See DFO (1991, 17).

354 Bishop and Shelton (1997, 34) stated of the multiyear quotas: "The rationale for these TAC [total allowable catch] steps is not known – it did not come from the Groundfish Subcommittee of CAFSAC, the scientific body responsible for carrying out the stock assessments."

355 Jake Rice, St. John's, personal communication, 1990.

356 DFO (1991, 17) stated that "the moratorium on cod fishing in NAFO Division 3L outside Canadian fishing waters has been extended for 1991."

[357] See Bishop and Shelton (1997).

[358] See Bishop et al. (1993).

[359] See Templeman (1952).

[360] A Portuguese trawler fished from August to November 1956 inshore from the Hawke Channel off Labrador. This was a departure from typical Portuguese fishing patterns on the southern Grand Banks. Large numbers of relatively small cod (45-70 centimetres) were caught (Ruivo 1957). Twenty-five trawl samples were made, all but two in Hawke Channel, in October-November 1957. Cod of lengths 45-70 centimetres dominated the catches (Ruivo and Quartin 1958, 56). These reports contain some of the first detailed information on Labrador cod size, age, and growth. In 1958, Portuguese trawlers again scouted the Hawke Channel and southwest corner of Hamilton Bank in September and October, with reports of sizes, ages, and maturities of about 2000 sampled fish (Ruivo and Quartin 1959).

[361] Rojo (1958, 65) stated: "Samples were collected on board the trawler Abrego of the PYSBE company by special personnel instructed by me for this work in a campaign on the Grand Bank[s] on board the same boat. With this work Spain initiates its study of the Labrador cod."

[362] See Alexander (1977, 132).

[363] Section 8.1 of the *Memorial University Act* stated that the university's academic responsibilities were to provide "b) instruction, whether theoretical, technical, artistic, or otherwise, that is of special service to persons engaged or about to be engaged in the *fisheries*, manufacturing or the mining, engineering, agricultural and industrial pursuits of the province; and c) facilities for the prosecution of original research in science, literature, arts, medicine, law and *especially the application of science to the study of fisheries and forestry*" [emphasis added].

[364] The Walsh Commission referenced Section 8.1c of the *Memorial University Act*.

[365] See Walsh (1953, 109).

[366] The Fisheries and Marine Institute became part of the Memorial University of Newfoundland in 1992.

[367] See Johnstone (1977, 289).

[368] Northwest Atlantic Fisheries Centre in St. John's, Newfoundland.

[369] See Keough (1978).

[370] See Peters (1978).

[371] The American Fisheries Society (AFS) has certified over 2500 professionals since the program began in 1963, and currently has about 650 active certified fisheries scientists working in several countries, mostly in the United States and Canada (information provided by the AFS, 2006).

[372] Department of Biology, Memorial University of Newfoundland (1978) indicated that "the big gap in our staff, however, is in the area of fisheries management and population dynamics."

[373] Keats et al. (1986) and a few others made attempts to contribute from time to time, but there was no sustained and effective involvement that could have offered the counterbalance necessary to sway the federal position. Notably, Dr. Keats, a Newfoundlander, found work not at Memorial or even in Newfoundland, but achieved a successful career in South Africa.

[374] See Martin (1992, 40).

[375] The present author was the senior chair.

[376] Reasoned accounts of the effects of overfishing and environment on the fate of the northern cod have come to the same conclusion – that the declines of the 1960s were caused almost totally by overfishing in a time of relatively high ocean productivity (Rice 2002; Rose 2004). The fishery was about 80% foreign.

377 Just what they were thinking is difficult to fathom, but various Canadian and Newfoundland reports gave no hint of the level of devastation that these fisheries were causing as late as 1974. Their misplaced belief in the invincibility of the cod stocks was unshakeable. A letter from the Newfoundland Minister of Fisheries to his federal counterpart in March 1974 focussed entirely on the social, and not biological, impacts of the foreign fisheries on the Hamilton Bank.

378 See May (1959) for details of the early investigations.

379 Pinhorn (1979, 7), stated that "by this time [1968] about 12 countries were fishing the area but about 80% of the catch was taken by 5 countries (France, Poland, Portugal, Spain, USSR)."

380 See Department of Fisheries and Oceans (1980, 1).

381 See Hamilton (2003).

382 Older and larger female cod produce many more eggs (see Chapter 2).

383 See Rose et al. (2000) and references therein.

384 See Colbourne et al. (1997), Dickson and Turrell (2000), and Drinkwater (2000).

385 See Frank et al. (1996).

386 See O'Driscoll et al. (2000a, b).

387 Rose (1993) initially reported the numbers of fish in the mega-aggregation surveyed in 1990; these numbers were subsequently revised slightly to an estimate of 450,000 tonnes (Rose et al. 2000).

388 See Rose and Kulka (1999). The catch rates of the winter fishery continued to rise even as the stock collapsed. The schooling behaviour of the cod was referred to as "hyperaggregation."

389 See Rose et al. (1994).

390 See Bishop et al. (1993).

391 See Myers and Cadigan (1995).

392 See deYoung and Rose (1993) and Rose et al. (2000).

393 The sequence of events described here was supported by Department of Fisheries and Oceans autumn trawl surveys and the fish catches of the offshore Canadian fleet (Rose et al. 2000).

394 Hutchings and Myers (1994) – and many other papers by the same authors – argued that the declines of the late 1980s and early 1990s, especially of northern cod, were caused entirely by overfishing and that environmental influences were not important. Few disagreed that overfishing had taken place and was the main mortality factor in stock collapses, but many rejected the implication that productivity declines had not also taken place. Productivity is largely a function of recruitment of new fish to the stock and individual growth. deYoung and Rose (1993) had earlier provided data to show that northern cod had retreated southward in the early 1990s and speculated that this might reduce recruitment because eggs and larvae would be swept off the Grand Banks by the Labrador current. The reality of the southern shift was confirmed by Rose et al. (2000) after additional investigation. Rose et al. (1994), reinforced by Atkinson et al. (1997), argued that this southward shift would also have led to increased vulnerability to both Canadian and foreign fisheries, and increased overfishing. Drinkwater et al. (2002) showed that the growth of individual northern cod declined over this same period, thus further reducing productivity. Rice (2002) came to largely the same conclusions. Both of these recent reviews concluded that environmental changes in the late 1980s and early 1990s reduced stock productivity. Studies of capelin (Frank et al. 1996) and plankton (Sameoto 2004) confirmed this. Fish catches were not reduced in parallel with declines in productivity – this, by definition, is overfishing.

395 See Rose (2004).

[396] See Bowering et al. (1997) for an overview of the decline in American plaice and Rose et al. (2000) for a general description of ecological changes in the early 1990s.
[397] See Montevecchi and Myers (1997) and Rowe et al. (2000).
[398] See Narayanan et al. (1995).
[399] See Dempson et al. (2002).
[400] Details of this era are given in Rose et al. (2000) and Drinkwater (2002).
[401] See Frank et al. (1996).

A FISHERY WITHOUT COD

AD 1993 TO 2006

Who is going to look after the sea if the fishermen are gone?

ANGELA SANFILIPPO, *an activist with Fishermen's Wives of Gloucester, to Mark Kurlansky in the late 1990s*

If proper fisheries management does not represent a vital development opportunity, what does?

CABOT MARTIN, *No Fish and Our Lives*, 1992

Moratoria . . .	501
Trepassey . . .	501
Burgeo . . .	502
Harbour Breton . . .	502
The Ecosystem . . .	503
Cod . . .	507
Long-term Effects of Overfishing . . .	510
Climate Change – Global Warming . . .	511
Other Species . . .	518
Species Increases . . .	518
The Fisheries . . .	520
Snow Crab . . .	520
Shrimp . . .	523

Cod . . .	524
Seals . . .	526
Capelin . . .	527
Other Species . . .	527
Long-term Social Consequences . . .	529
An Age-Old Question: Must the Fishery be Poor? . . .	531
The Fishery Products International Story . . .	533
Oil and Gas . . .	535
Fisheries Science . . .	537
AD 2005: A Year of Reports . . .	538
Government Reports . . .	538
Nongovernmental Organizations . . .	539
COSEWIC and Endangered Species . . .	540
The Sea Ahead . . .	541

MORATORIA

The summer of 1992 saw the single largest layoff of workers in Canadian history, and it left the people of Newfoundland and Labrador with few options. Without a cod fishery and with few employment alternatives, tens of thousands of fishermen and plant workers had to move or seek government assistance. Almost immediately, Minister Crosbie arranged the Northern Cod Adjustment and Recovery Program (NCARP), an income supplementation plan to assist those displaced by the moratorium for two years. There was some initial hope that the cod stocks would rebound quickly and that a fishery might resume in 1994,[1] but it was not to be.

NCARP was followed in 1994 by The Atlantic Groundfish Strategy (TAGS), another income supplementation plan. TAGS was intended to support some 26,500 people, but the numbers quickly rose to about 40,000. Nearly C$2 billion were allotted to TAGS, but the money ran out in mid-1998, before the scheduled termination of the program in the spring of 1999.

NCARP and TAGS brought short-term relief to tens of thousands of fishermen and fisheries workers, but could not prevent the eventual economic decline of many fishing communities. Without cod, the economies of hundreds of communities were grounded. Even with new fisheries for other species, there were too many people attempting to make a living from an ocean depleted of fish. The hardest-hit communities were those most dependent on northern and Grand Banks cod and on the flatfish fisheries on the Grand Banks. Many of these communities were on the ice-free south coast.

Trepassey

The southern Avalon Peninsula town of Trepassey is one of the oldest European settlements in North America, first appearing on charts in the early 1600s. It was initially an inshore fishing harbour that served as a border town between the English and French fisheries, but later became a centre for the Grand Banks fishery. As late as the mid-1980s, the Fishery Products International fish plant at Trepassey employed a workforce of about 850 people and processed about 15,000 tonnes of flatfish (American plaice and yellowtail flounder) and cod from the southern Grand Banks each year. Overfishing, largely by European countries fishing under the NAFO banner, forced the closure of this plant in the early 1990s. Many people left, leaving their devalued homes boarded up. A few new manufacturing industries came to Trepassey, but they could not replace the loss of the fishery.

Burgeo

Burgeo, an isolated outport on the rugged southwest coast of Newfoundland, was settled in the 1700s. The fishery was its only industry. A fish plant was constructed in 1943 and operated by Fishery Products until it was taken over by Burgeo Fish Industries, part of the Lake Group, in 1955. Strike action by the plant workers in 1971 resulted in the plant being purchased by the provincial government and operated by National Sea Products of Nova Scotia. A new plant, jointly owned by National Sea Products and the government of Newfoundland and Labrador, was opened in 1978. The plant processed redfish and northern cod throughout the 1980s, but was closed and sold to the Barry Group (Seafreeze) in the early 1990s as the stocks declined. The Barry Group moved its redfish operations to Nova Scotia and, after processing small amounts of crab in Burgeo in the late 1990s, mothballed the plant. At present, this isolated community struggles to survive by maintaining a small fishery and a diversified economy, but many people have packed up and moved away. Mayor Allister Hann is one of the most eloquent spokesmen for the fishing communities of Newfoundland – he often inquires as to the purpose of fisheries management, asking "Who is it for?"[2] This question receives no easy answer.

Harbour Breton

The town of Harbour Breton lies about halfway between Trepassey and Burgeo on the south coast. Like them, it is an ice-free port year round, and became a trade and fishing centre in the 1800s. The Fishery Products International plant was Harbour Breton's main employer in the 1980s, processing cod and redfish. After the moratorium, Harbour Breton processed Barents Sea cod that had been harvested and frozen by Russian vessels, then resold it to American markets. By the early 2000s, however, the cost of the Russian fish had climbed, and with the increase in value of the Canadian dollar and the rise of China as a major seafood processor, the Harbour Breton plant could no longer process this fish economically. In 2005, with insufficient local fish to process, the Fishery Products International plant at Harbour Breton closed.

The future of Harbour Breton is uncertain. A proposal to convert the plant to produce fish meal from pelagic species as feed for the aquaculture industry was put forth by the Barry Group in 2006. The forecasted demand was 50,000 tonnes of capelin and various amounts of herring and mackerel. However, the current allowable quota of capelin is only 34,000 tonnes and there is little evidence that the troubled capelin stock can sustain an increase in catch, either inshore or on the Grand Banks. Similarly, it appears unlikely that herring stocks can sustain an increased level of harvest. The mackerel fishery is typically fully subscribed when this species is abundant in Newfoundland waters, an event which occurs only sporadically – as fishermen will tell you, the mackerel does not "belong to Newfoundland" and only migrates here during warm periods. The town of Harbour Breton supported a plan by the Barry Group to revitalize the plant, and to add farmed fish to its products, but at the time of this writing only a fraction of the former employment has been restored and the future remains uncertain.

The reopening of the south coast cod stock in 1997 did little to help the fish plants in Trepassey, Burgeo or Harbour Breton.[3] The small quota of 10,000 tonnes was simply not enough fish to supply these and other plants in the region.

Other regions and towns that were less dependent on cod and flatfish have fared better since the moratorium. Access to the burgeoning crab and shrimp fisheries enabled some communities to prosper, especially on the Avalon Peninsula and northeast coast where snow crab became increasingly abundant in adjacent waters in the mid-1990s. With strong markets and prices, many new crab fishermen made hundreds of thousands of dollars, and crab processing provided thousands of jobs in quickly converted plants. If there was surprise at how quickly the ecosystem could change, the rapid transition of the fishing industry was no less remarkable. It was the largest change in hundreds of years.

Despite the success stories, the increases in shrimp and crab, and increasing revenues from the new fisheries, there has been a marked decline in the prospects for most of Newfoundland and Labrador's fishing communities. This seeming paradox has occurred almost solely because of the decline in cod and other groundfish. The long-term viability of many small fishing communities ultimately depends on the productivity of the fishing grounds of the Grand Banks and continental shelf, and the abundance of cod, as it always has. With the productivity of these ecosystems reduced substantially since the 1950s – especially to the north – the ecosystems are mere shadows of their former selves, and many fishing communities live in those shadows. Many can survive only as bedroom or seasonal communities, not as working fishing villages – their fate sealed by the radical changes that have occurred in the marine ecosystems of the Northwest Atlantic.

THE ECOSYSTEM

> The ordinary citizen today assumes that science knows what makes the community clock tick; the scientist is equally sure that he does not. He knows that the biotic mechanism is so complex that its workings may never be fully understood. — ALDO LEOPOLD, "The Land Ethic," 1948

The declines in capelin and cod stocks were markers of profound changes in the marine ecosystems of the Grand Banks and adjacent continental shelves, whose origins may be traced back to the calamities of the late 1960s and early 1970s. In just a few short years in the early 1990s, changes occurred in rapid-fire succession. Capelin virtually disappeared over much of their former range north of the Grand Banks: their abundance

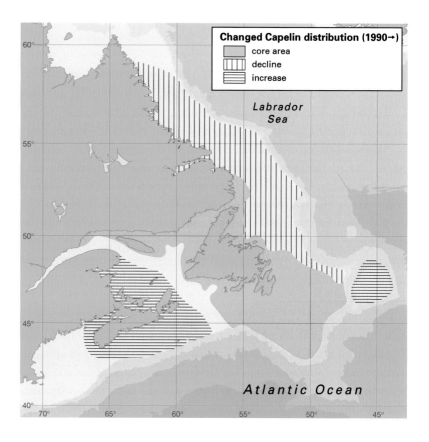

8.1 Distribution changes in capelin during the 1990s.

decreased from about 3-6 million tonnes in the late 1980s to only a few hundred thousand tonnes in the early 1990s.[4] At the same time, capelin numbers were increasing farther south on the Scotian Shelf and in the Gulf of St. Lawrence, as well as to the east on the Flemish Cap – areas where they had formerly been scarce.[5] In some inshore areas of Newfoundland, especially in Trinity and Placentia bays, estimates of capelin abundance based on catch rates indicated that they were holding their own,[6] but even there, abrupt changes had occurred within the population. Growth rates decreased: the average weight of spawning adults in the 1990s was 50% less than it had been in the 1980s.[7] Spawning was delayed: fishing communities that had once set their calendar by the arrival of capelin in late spring now waited weeks or even months for the spawning schools to appear, if they came at all.[8] Capelin historically spawned in late June; in the 1990s, however, many capelin spawned in August or even September, leaving too little time for their young to grow large enough to survive the coming winter. Capelin behaviour also changed: where once their grand schools could be found throughout much of the sea, seeming to overwhelm their many predators by sheer numbers, they were now observed in weak and inconspicuous layers deep in the ocean.[9]

Why the capelin disappeared so quickly and why their biology changed so radically are still open questions. A superficial answer is overfishing, but there is little evidence that fishing caused the decline in abundance or shift in distributions of capelin, or that fishing could have caused the changes in growth and behaviour. On the other hand, the changes coincided with rapid decreases in sea temperatures, and the decline in growth rates suggested that food was a limiting factor.[10] Had something gone wrong in the food web for capelin? Capelin feed almost exclusively on zooplankton, the tiny animals of the sea, and in the late 1980s and early 1990s, changes in the zooplankton species and abundance accompanied the more obvious changes in sea temperatures.[11] Poor feeding and colder waters would explain the decline in growth in capelin, and perhaps the population decline in the north. Another theory suggested that there were two kinds of capelin: a dominant, fast-growing, early-spawning type and a slow-growing, late-spawning type that never achieved the numerical abundance of the first type, but was more resistant to environmental change.[12] Under this theory, the numbers of fast growers could have been reduced primarily by poor survival in an unfavourable environment and, perhaps, by the fishery, while the smaller but less abundant type fared better. Alternatively, ocean conditions for capelin in the early 1990s may simply have been so poor that massive distribution shifts, decreased food availability, reductions in growth rates, and behavioural changes occurred across a single stock. Whatever the truth, it is clear that since 1990, capelin have not grown at their usual rates or maintained their high historical overall abundance levels. This suggests that there have been fundamental changes in the ecosystem that have affected capelin stocks. In turn, reduced numbers of capelin may have caused further changes in the ecosystems, because capelin are an important prey species for many marine predatory fishes and mammals.

With so much attention paid to the collapse of the cod stocks, two factors were often overlooked: capelin may be more fundamental to the functioning ecosystem than cod, and the capelin stock collapsed over most of its former range in Newfoundland and Labrador just before the final collapse of the cod stocks. Most scientific works on the so-called cod stock collapse – which was really an ecosystem collapse – did not even mention capelin, and others have misinterpreted species interactions. One modelling study claimed that an overabundance of small capelin may have been part of the reason that the cod stocks were not rebuilding.[13] There was little evidence for such a contention, and cod diet analyses in the 1990s suggested quite the opposite, that cod could find few capelin to eat north of the Grand Banks.[14]

Capelin are a keystone species in the Newfoundland and Labrador ecosystems: they form the main conduit from plankton production up the food chain to the larger fishes. The same is true in the large marine ecosystems of the Barents Sea and Iceland.[15] If capelin distribution or abundance changes, for whatever reason, many other species will be affected. The affected species include cod – historically the major predator of capelin[16] – but also many other capelin-feeders, such as Arctic char, Atlantic salmon, Greenland halibut, and winter flounder. Marine mammals and seabirds are also dependent on capelin, and some species have few alternative food sources of equivalent nutritional value. Traditional knowledge of fishermen is nearly unanimous on the importance of capelin in this ecosystem, especially to cod.

Although there have been sporadic pulses of capelin observed in various bays since the early 1990s, there has been no consistent rebuilding of the capelin stock, particularly in the northern areas, up to and including 2005, with the possible exception of the capelin in the Gulf of St. Lawrence. In 2006, however, capelin spawned widely in many areas around the island of Newfoundland, and their size was notably larger, approaching historical norms. Most of these spawning capelin were born in 2002 and 2003, and their marked growth indicates better feeding during their lifetime than capelin have experienced since the early 1990s. Increasing growth in capelin is a positive sign that the North Atlantic ecosystem may be beginning to return to a state resembling that which existed prior to the late 1980s. No attempts have been made to assess the total biomass of the stock since 1990,[17] but acoustic surveys of some of the former heartland areas for capelin on the Grand Banks – where relatively few schools were found from 1992 to 2004 – have begun to observe larger concentrations. The causes and effects of the changes in the capelin stocks in the waters of Newfoundland and Labrador after 1990 remain key unresolved scientific questions about the recent ecosystem collapse.

Another fundamental change in these ecosystems over the past decades, near the summit of the food chain, was a large increase in seal populations. Harp seal herds grew quickly after the reductions in the hunt in the early 1970s and the ban on killing pups in 1987. Harp seal numbers increased from a low of about 1.2 million animals in the early 1970s to 5.5 million animals in the early 2000s. Pup production may have been as high as 1 million in years with good ice conditions.[18] Hooded seal numbers are also believed to have increased substantially over the same period. Although historical population numbers of seals are not known, hunt statistics suggest that current seal numbers are well above any since the mid-1800s.

Both harp and hooded seals eat fish in considerable quantities. Harp seals prefer small fish such as Arctic cod and capelin as prey, but they will take larger fish such as Atlantic cod when opportunities arise.[19] In the inshore region of the northeast coast, overwintering aggregations of cod have been corralled by harp seals in very shallow and cold water on several occasions. The seals killed thousands of the helpless fish, eating only the soft internal organs and killing far more than they needed, as predators sometimes do.[20] Hooded seals eat larger fish and may prey more heavily on cod than harp seals, as do the less common grey seals.

In the 1990s, all species in the marine ecosystems of Newfoundland and Labrador had to contend with severe environmental changes. For many species, the most important changes were the decline of the capelin stocks and the increase in numbers of harp seals. For some species, these changes may have meant little, but for others the effects were extreme. Among the most affected species was the cod.

Cod

The steep decline in the northern cod stock over most of its range continued well after the moratorium on fishing imposed in 1992. Early hopes and predictions that this stock would rebound quickly proved to be folly. In reality, their levels fell to as little as one percent of their former abundance. The numbers of fish born into the population in the cold years of the early 1990s, when distributions were shifted southward,[21] were very low, so the continuing decline was perhaps not unexpected. However, the increased rate of deaths at early ages, when the fish were not being caught in any fishery, was much harder to explain. By the mid-1990s, few fish lived more than five or six years.[22]

With the decline in capelin, most northern cod were reduced to eating shrimp, a lipid-poor food source for mature cod that likely impacted their survival and reproductive potential.[23] In cod, the liver is the primary storage organ for surplus lipid (fat), the energy currency in marine ecosystems, consumed during the summer feeding season. Stored fat is slowly depleted as the fish uses the energy for survival, and especially for reproduction, during the winter and spring, when cod feed less or in the northern stocks not at all.[24] After spawning, cod livers are depleted and the fish are at their lowest seasonal condition.[25] Historically, northern cod did not reproduce before reaching six or seven years of age, and their livers recovered within weeks after spawning once schools of the lipid-rich capelin were located. Northern cod on the continental shelf north of the Grand Banks now spawn at a much younger age (three to five years) and their livers remain in poor condition for most of the year.[26] Poor livers signify weak fish, and weakened fish are more susceptible to disease and predation by seals. It is possible that the stresses of early reproduction are leading to early death. Nevertheless, as yet science cannot say with certainty why the life expectancy of cod has decreased. Fisheries science relies on body counts from fisheries, and forensic work to determine cause of death requires bodies, but seals do not record their catches and dead fish are seldom found in the open ocean. All we know for certain is that cod with poor livers are disappearing at an unprecedented rate, and that few cod older than five or six years of age can be found from the northern Grand Banks to Labrador, where, historically, cod normally lived for 15-20 years. It is impossible for this stock to increase quickly or at all with such elevated death rates – with every gain through fish born into the population, there is nearly as much loss.[27]

The cod stock that supported fisheries on the Grand Banks for 350 years has increased in size only marginally since the declines in the 1980s. Little is known about their recent feeding habits, but there is less evidence of poor condition and unusual mortality than for the northern cod. Large cod (exceeding one metre in length and masses of 10 kilograms) still inhabit the southern Grand Banks, although in much reduced numbers. There is an additional problem on the Grand Banks – the fishery there never stopped. Fishing companies registered in European Union countries, in particular Spain and Portugal, fished for cod long after the moratorium was declared and continue to do so under the guise of bycatch. From 1994 to 2001, foreign fisheries

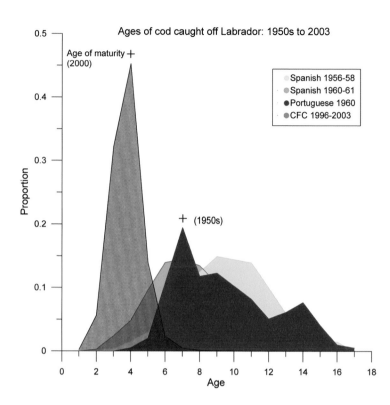

8.2 Ages of cod caught in the Labrador fishery around the Hawke Channel in the late 1950s and early 1960s as compared to research catches in the late 1990s and early 2003. In recent years, few fish have lived beyond the age of five. The Spanish and Portuguese data are from exploratory fisheries, the CFC data from research of the Chair in Fisheries Conservation at Memorial University. The ages when 50% of fish were mature are indicated by + signs.

caught over 18,000 tonnes of cod on the Nose and Tail of the Grand Banks and on the Flemish Cap.[28] In 2003 alone, almost 5000 tonnes of cod were caught by these fleets – far more than would have been allocated had there been a legal fishery. At present, the biomass of the cod spawning stock is thought be about 5500 tonnes.[29] This so-called bycatch is sufficient to further deplete and prevent recovery in these stocks.[30]

Cod fared much better in other stock areas in the 1990s, no doubt helped by the moratorium. On St. Pierre Bank and in Placentia and Fortune bays, cod made a quick comeback. Relatively large numbers of fish were born in 1989, 1990, and 1992. There was no increase in inexplicable deaths in these regions; additionally, capelin and sand lance were relatively abundant,[31] and seal numbers were modest. Rebuilding followed a similarly positive pattern on the northeast coast in Trinity and Bonavista bays, where large numbers of fish born in 1990 and 1992 formed spawning aggregations by 1995. In these stocks, productivity has been relatively high.[32]

The variable responses of cod in different areas of the fishing moratorium may hold the key to understanding why some stocks have not grown, even when they were given nearly full protection. The patterns are complex and difficult to discern, but some general features have emerged. It seems evident that early conjectures that the collapse of the cod stocks was a simple response to overfishing, without any longer-term effects or important interactions with the environment, were incorrect; predictions that the northern cod stocks would rebuild to millions of tonnes within a decade if fishing stopped have proven false.[33] Fishing does not affect species or ecosystems quite so neatly, nor is rebuilding so straightforward.

All of the cod stocks, and stocks of other species before them, were heavily overfished, and several factors appear to be important in determining how quickly a stock may rebuild. Some relate to how severe the overfishing was, and some to the environmental conditions during and after the period of overfishing: if conditions were relatively favourable – i.e., if fish growth was reasonable, if a large number of young fish were born and survived, if there was an adequate food supply and a limited number of predators, and if there was no further fishing – then the moratorium had the desired effect and the stocks grew at a predictable rate.[34] This positive scenario was played out on the south coast and in the coastal stock on the northeast coast. On the other hand, if the majority of these factors was not favourable, especially if the original overfishing was severe, then rebuilding was hindered or completely stymied, as in the case of the Grand Banks and the northern stock. Science would like to be able to offer a more quantitative description of these factors and their effects, but at present this is not possible.

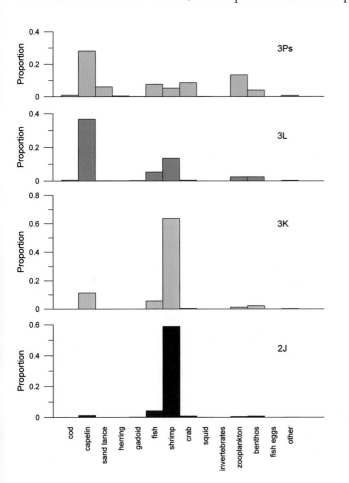

8.3

Diet of cod around Newfoundland and Labrador in the late 1990s and early 2000s. 2J, 3K, and 3L are the NAFO divisions that span the range of the northern cod, and 3Ps is the NAFO division on the south coast of Newfoundland. In the areas where capelin-feeding has been prominent, cod have rebuilt their numbers much more quickly.

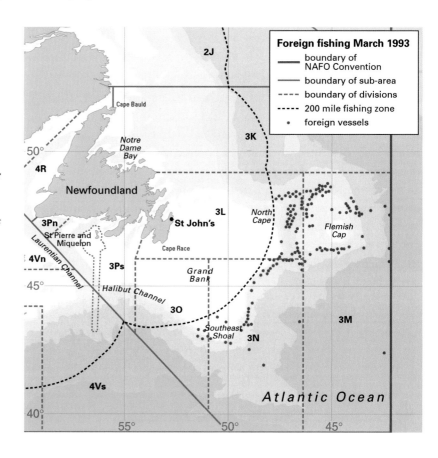

8.4

Foreign fishing in March 1993 after the imposition of the moratorium on cod. Note that the vessel positions off the North Cape correspond closely with the area where the last large cod aggregation was found a few months later (see 7.28).

Long-term Effects of Overfishing

Overfishing may not just reduce a population, it may also alter that population's abilities to regenerate itself and exploit its environment as it has in the past. This may be especially damaging to stocks that are near the edge of their geographical range, as was the case for many species in Newfoundland and Labrador waters. These effects are broader in scope than simple overfishing, and they may temporarily reduce a population's ability to sustain itself and result in much longer recovery times. For example, eliminating most of the larger and older fish can change the behaviour of a stock, particularly with respect to migration and spatial utilization of habitat.[35] Range restriction will ultimately limit population size. Overfishing may even cause changes in population genetics if it is sufficiently intense to disallow reproduction by all but a select group – for example, if all the big fish are caught first and only smaller fish are left to reproduce. Genetic effects should be regarded as more permanent. Little is known about these effects, and they are likely to manifest only under severe fishing conditions, but recent evidence indicates they should not be dismissed out of hand.[36]

Trawling may have damaged the sea floor habitat for some species, including cod. There was diverse evidence on this issue. On the one hand, it is indisputable that trawling can damage sensitive and highly featured bottom types. Sea mounts in New Zealand suffered extensive damage from trawlers; the fishing industry consequently offered to close one-third of the fishing grounds to trawling in perpetuity[37] and the government has accepted this offer. On the other hand, Iceland and Norway harvest most of their cod and other groundfish using otter trawlers, although not on the spawning grounds as was done in Newfoundland and Labrador. There have been calls for outright bans on otter trawls, but it is hard to imagine how flatfish and shrimp fisheries could be practiced without them. While there is room here for compromise among the extreme positions, there is no doubt that fisheries management in Newfoundland and Labrador has allowed far too much trawling, particularly on spawning grounds and in juvenile fish habitat, and that restrictions such as those proposed or practiced in other countries might greatly benefit ecosystem rebuilding.

The experiences of 1950-1992 have shown beyond any reasonable doubt that industrial fishing decimated many fish stocks, and some have changed almost beyond recognition. For all that, these experiences have also shown that environmental changes can greatly impact marine ecosystems and ultimately determine how much fishing can be sustained. Ignoring either is folly. Over the long term, however, the most persistent changes are likely to be related to climate.

Climate Change – Global Warming

Global warming is fact. Climate change is occurring in the northern and southern oceans, and it will affect the fisheries in these oceans.[38] Ice is disappearing and water temperatures are increasing. These changes will likely lead to warmer conditions in Newfoundland and Labrador over the next few decades, although this is far from certain.[39] Secondary effects of the general global warming trend – such as increased snow, which adds ice to the Greenland glaciers, and increased cold water flows in the Labrador Current – could lead to local cooling conditions in Newfoundland and Labrador waters, at least in some years. The ocean climate of Newfoundland and Labrador is not always in synch with the rest of the northern hemisphere.

Overall, the best guess is that warming will take place, and this should benefit the Newfoundland and Labrador cod stocks, which live near the northern limits of the range of the species.[40] In these ecosystems, cod and many other species are limited more by cold waters than warm. Hence warming should result in more widespread distribution and increased abundance. Millions of years of evolution have prepared cod to be top dog in these ecosystems, and they are particularly fit to take advantage of gross changes in ocean climate, having experienced them several times in the past. In time, if warming does occur, cod should move farther north on the Labrador coast, following capelin that will move first,[41] as has happened in modern times in Icelandic and Greenland waters and in the Barents Sea. As capelin become re-established, the remaining groups of spawning cod will have a much better chance of reproductive success and their numbers should increase; once this occurs, range expansion will become far more likely. In Newfoundland and Labrador, this scenario may not play out quickly because so few fish remain – their numbers are not great enough to spur movement to new grounds or even to migrate as once they did. Politicians and interest groups often want science to make up timetables and schedules for fish stock changes as climate change advances, but Nature confounds such simplicity. Predicting the timing and track of range expansion is not possible – it could take a century or more, or it could occur within one or two decades, or not at all if climate predictions are off. If expansion does occur, it will likely occur in steps or stages.[42] Shorter-term rebuilding is impossible.

For the time being, the Newfoundland and Labrador ecosystems are dominated by shrimp and crab. The reasons for the increases in the largely bottom-dwelling shrimp and crab populations are twofold. These are cold-water species – the cold conditions of the mid- to late 1980s and early 1990s favoured the survival of their young and allowed them to expand their range to nearly all areas of the continental shelf.[43] Additionally, predation rates decreased with the decline in the cod stocks, enhancing their life expectancy. Several strong years of shrimp and snow crab reproduction occurred during the cold years between 1991 and 1994, and by the early 1990s, both species were numerous in areas where they were formerly rare. However, their time of dominion over these ecosystems is likely to be fleeting. When the cod recover, it is likely that conditions will no longer be as favourable for shrimp and snow crab. Even if their reproduction rates are still high, many will fall prey to the growing numbers of cod. Strong cod stocks are unlikely to coexist with abundant shrimp and crabs stocks; hence, the rebuilding of cod will almost certainly be accompanied by a decline in these invertebrates.[44]

Ecosystems are complex – even the relatively simple systems in the cold waters of Newfoundland and Labrador are not easily understood. Changes in one part of the system, whether caused by natural forces or human interventions, may ripple through the system with unpredictable results. In general, marine ecosystems are best thought of as flexible rather than rigid systems: they bend with changes in the balance of species and ages of fish, but seldom break. The energy transferred from the primary producing phytoplankton through the zooplankton and upwards to the top predators flexes with changes in species, with remarkable suppleness. However, recent experiences in Newfoundland and Labrador illustrate that, despite their inherent resiliency, marine ecosystems can be pushed beyond their bending points. If that happens, as it has on the Grand Banks and in the northern cod ecosystems, predictions based on even the most realistic fisheries models are of little use.

As an example of the complexity of the North Atlantic ecosystem and how changes in one species can affect all others, consider the following. Productivity changes in the ocean may lead to shifts in zooplankton distribution and abundance that cause decreases in capelin populations. As capelin are a key prey species, this means that there will be less food for cod and other capelin-consuming species. Cod growth and reproduction will subsequently decline. If fishing continues with differential removal of the larger individuals from the stock, cod populations may crash, potentially with concomitant changes in behaviour and the genetic makeup of the population. The removal of major predators of small fishes and other species in the system gives prey species better chances of survival. The differential removal of large cod – that is currently occurring even without fishing in the northern cod stock – may result in larger populations of other large prey fish, such as yellowtail flounder on the Grand Banks. Yellowtail flounder feed on capelin eggs; increases in their numbers will lead to increased pressure on the capelin stock. The lack of cod may also increase the food available to other predators such as harp seals, which will then increase in number and prey on the remaining capelin and cod. These interactions may be described loosely in words, and some models (as in this book) attempt to balance the ups and downs of various species; however, even with the best ecosystem models, quantifying them with certainty is beyond our present abilities.[45] It is sobering to note that the most important changes in the Newfoundland and Labrador marine ecosystems that have occurred since 1992 were not and could not have been predicted. General direction perhaps, but not specific details. This should be kept in mind when making forecasts about what will happen in future ecosystems.

The Smith Sound Cod. In the post-moratorium world, after three years without fishing, northern cod numbers continued to decrease and surveys on the Grand Banks located few fish. In April 1995, researchers studying coastal cod ecology in Trinity Bay were searching for fish to tag. Having had little success, the research vessel *Shamook* moved into Smith Sound on the west side of the bay, where to the astonishment of everyone on board, the echo sounder recorded an extensive layer of what appeared to be fish. The skipper and first mate had over 60 years of experience in the coastal waters of Newfoundland between them and they were baffled. In previous years of tagging in this area nothing even remotely like this had been observed. Their first thought was that the echosounder traces must be herring. The scientists on board suggested that a few cod might be caught around the supposed herring and a fishing set was called. The net was towed briefly through the aggregation of fish, and the skipper immediately felt the weight of the catch as the net filled. Experience told him that these were not herring, as small fish are seldom caught in a typical fishing set as made from the *Shamook*, and when the net was brought to the surface, it was revealed that 100% of the catch was cod – so many fish that the net could not be pulled back onto the deck of the boat. As the net was near bursting, it was cut open as it lay on the water surface, and thousands of cod were returned to the sea.

8.5
The tranquil waters of Smith Sound, Trinity Bay, in the spring of 2005, home to the largest remaining overwintering and spawning aggregation of northern cod in the early 2000s (Tom Clenche collection).

I was in St John's that day when an enthusiastic scientist involved in the work grabbed me by the arm and said I must come quickly: cod had unexpectedly been located in Smith Sound and needed to be surveyed. The research vessel was turned over to me immediately, and the initial survey revealed a biomass of about 10,000 tonnes – the largest find of cod since 1992. Subsequent surveys conducted in May and June of the same year showed that the aggregation was holding fast.

In February 1996, I returned to Smith Sound on board the Canadian research vessel *Teleost* with a crew of experienced marine biologists to determine if the cod were still there. Just a few miles into the Sound, a dense aggregation of fish rising over 100 metres from the bottom was encountered. Few on board had ever seen an overwintering cod aggregation, and most didn't believe that this mostly pelagic echosounder reading could be coming from a supposed groundfish, the cod. As I called the bridge to request a fishing set, several colleagues jokingly suggested that we get the herring smokers ready. Shortly afterwards, our midwater research trawl briefly skimmed the surface of the aggregation and confirmed the evidence from the previous spring: the catch was 100% cod, most of which were mature fish – a classic overwintering and prespawning aggregation.

Just where these fish came from is still uncertain. The initial aggregation was comprised for the most part of fish born in 1989 and 1990. These were the "founder" year-classes of the aggregation. Two years later, they were bolstered by relatively large numbers of maturing fish born in 1992. There is little indication that fish of the founder year-classes are abundant anywhere else within the range of the northern cod. Occam's Razor might suggest that these fish were born in Trinity Bay, but there was little indication of substantial spawning in Trinity Bay between 1989 and 1992. There are two other possibilities. One is that surviving juveniles from 1989 and 1990 – the last years when large spawning aggregations existed on the adjacent banks – congregated in Trinity Bay, made Smith Sound their winter home, and stayed there. Another is that some of these fish came from the south coast, where the founder year-classes were also strong. Genetic evidence suggested that the Smith Sound and south coast cod were genetically indistinguishable and differences between these fish and fish from the adjacent Bonavista Corridor were very small.[46] There are arguments for and against each of these possibilities, and although additional work is being done on these matters, it is likely that we will never know for certain.

Wherever their origin might have been, once these cod began to use Smith Sound for overwintering and spawning, they kept doing so. They have been there every winter from 1995 to 2007, and their numbers more

8.6
Echogram of the overwintering Smith Sound cod aggregation in February 1996. The fish were highly pelagic at that time. The echogram spans about 2 kilometres and shows the bottom 150 metres (total water depths were about 300 metres). The orange band marks the bottom; the blue above the cod, with some fish up to 100 metres off bottom (blue boomerangs).

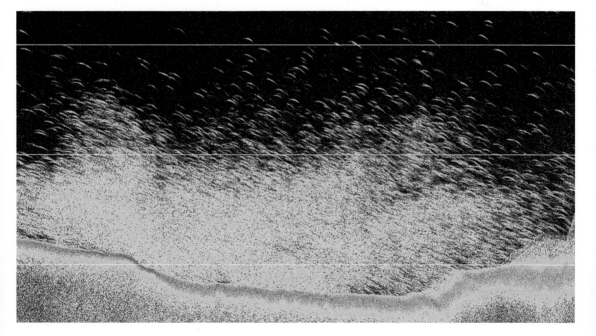

than doubled from the mid-1990s until the early 2000s, reaching a total biomass of over 25,000 tonnes by 2001. In 2003, the biomass dipped to about 20,000 tonnes, likely a result of the fishery conducted from 1999 until 2002. The fishery closed in 2003 and this stock began to rebound once again. Unfortunately, with the reopening of the fishery in 2006, against the general recommendations of science, the numbers of fish in this group are at risk of declining again.[47]

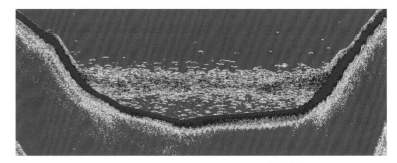

8.7
Smith Sound cod on April 18, 2000, during the spawning period. The echogram spans about 2 kilometres showing the bottom 150 metres (total water depth about 225 metres).

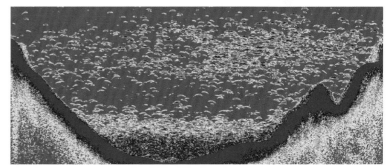

8.8
Smith Sound cod on April 17, 2006, during the spawning period. The echogram spans about 2 kilometres showing the bottom 150 metres (total water depth about 225 metres).

There may be at least two types of cod in Smith Sound. There are fish resident there year round, but the majority migrate seasonally in and out of the Sound. In a normal year, the migrating cod return to the Sound in late fall or early winter when the deep waters in the trench are warm (3-5°C). As winter passes and spring approaches, the deep waters cool, but temperatures are still above zero degrees when the cod begin to spawn in early April. By late May, temperatures in the deep waters may dip below zero degrees, and the migrants either move out or up into the warming surface waters. At this time of year, cod may be found "right to the rocks" in just a few metres of water. They are often visible from the surface or even from shore in the clear waters of the Sound. At times they flick their tails at the surface, for no known reason, maybe for feeding or temperature sensing, but maybe just for the fun of it. Spawning in the Sound has been observed as late as July in some years.[48] Recent evidence suggests that the migrating fish may spawn early, with spring storms washing their eggs far out of the Sound, while residents spawn later in the calm summer, with large retention of young within the Sound.

In the spring of 2003, something went dreadfully wrong. The normal cooling of the deep waters accelerated quickly in early April, trapping the cod in sub-zero waters deep in the Sound. The fish schooled tightly together, but the blood of many began to freeze.[49] As they lost their ability to hold their position near the bottom, many spiralled upwards to the surface, either dead or dying. In all, about 500,000 fish died, comprising about 5% of the Smith Sound cod.[50]

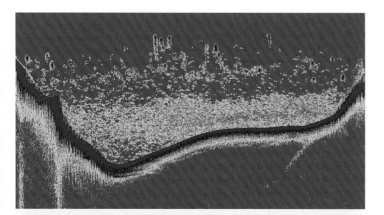

*8.9
Smith Sound cod on April 18, 2006, during the spawning period with columns forming above the mass of fish. The echogram spans about 2.5 kilometres showing the bottom 50 metres (total water depth about 225 metres).*

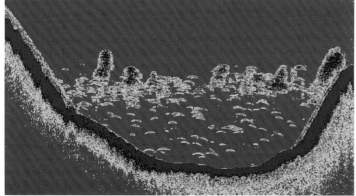

*8.10
Smith Sound cod on April 18, 2006, during the spawning period showing details of spawning columns. The echogram spans about 1.5 kilometres showing the bottom 205 metres.*

*8.11
Smith Sound cod on January 15, 2006, hunkered down to bottom in an overwintering aggregation. There were very few fish up in the water at that time. The echogram was taken from a drifting ship over about half an hour, at water depth of 225 metres. For figures 8.7-8.11 the bottom is the blue band with cod above as yellow-red-blue masses or single fish (boomerangs).*

Many questions remain about the Smith Sound cod. Of foremost importance is their role in the rebuilding of the larger northern cod stock. Genetic studies have yielded varied results, but some have suggested that the majority of the Smith Sound cod are coastal fish that will not mix with their Grand Banks cousins.[51] Nevertheless, it remains unclear whether or not these fish have the capacity to expand their range and recolonize new regions along the coast or on the offshore banks. There has been little evidence of expansion to date, at least not to the continental shelf or banks, but perhaps cod numbers are too few. Range expansion no doubt is possible given longer (i.e., geological) timescales, but whether it is likely to happen in the short term remains uncertain.

Other Species

Few of the fish species depleted during the decades of the mid- to late 1900s have rebounded to any extent. These species include haddock, redfish, American plaice, witch flounder, and grenadier. The only exceptions have been the yellowtail flounder, which has rebounded strongly on the southern Grand Banks, and to some extent the Greenland halibut, a deepwater species that was not heavily fished in the early years of the trawler fisheries. Rare and unfished species may also have declined.[52] In the mid-1990s, echo sounding and trawl surveys across the Grand Banks and the adjacent continental shelf revealed a nearly empty ecosystem.

SPECIES INCREASES

Nature has little tolerance for emptiness. As the capelin and cod stocks declined, populations of other species increased and took their place. As luck would have it, in the waters of Newfoundland and Labrador, these growing populations were the commercially valuable snow crab and shrimp stocks.

Shrimp. Shrimp numbers first began to increase in the mid-1980s. Their range also expanded at that time.[53] At first the increases were modest, but the population exploded during the coldest years of the early 1990s: from 1991 to 1994, highly successful reproduction released untold billions of young shrimp into the waters off Newfoundland and Labrador.[54] Although the total biomass cannot be known accurately, it is certain that it increased dramatically in the past two decades. It is not entirely clear, even in hindsight, why the shrimp did so well. One theory suggests that the reduced numbers of cod increased shrimp survival rates, while another suggests that environmental conditions favoured shrimp in the 1980s and 1990s.[55] It is likely that both factors were important.

The cold conditions of the late 1980s and early 1990s coincided with expanding numbers and distribution of shrimp prior to the decline in cod, suggesting that ocean conditions were important in initiating the increase.[56] It is also possible that cod numbers were sufficiently reduced by that time to decrease predation and increase shrimp survival. In any event, when shrimp populations grew rapidly in the 1990s, the cod were nearly gone and shrimp survival was almost certainly enhanced. If cold ocean conditions were a key factor in the outbreak of shrimp, it might be expected that the warmer conditions of the mid-1990s would result in reduced numbers of young shrimp surviving and joining the population, with a corresponding decline in the overall numbers of shrimp in the 2000s. This seems to have been borne out in the fishery.[57]

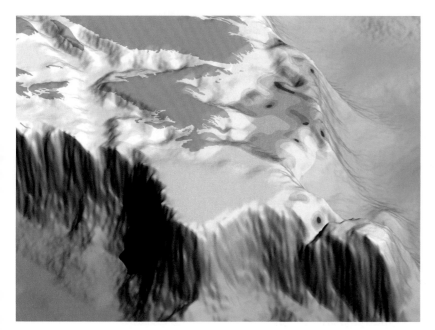

8.12
Shrimp distribution in 2000 (catch per set data from Department of Fisheries and Oceans trawl survey).

Snow crab. Snow crab stocks are given to strong fluctuations. Although there is uncertainty regarding how many crabs were present historically, snow crab stocks appear to have increased rapidly in the 1990s. The causes of the growth in snow crab populations were likely similar to those that affected shrimp. Cod are a chief predator of young and female snow crabs, and the cooler conditions of the late 1980s and early 1990s were likely favourable to snow crab reproduction and survival. With the warming ocean conditions of the late 1990s, the survival rates of young crabs sharply declined in most areas,[58] and it is highly unlikely that the high levels of catch in the fishery can be sustained.

In summary, increases in both shrimp and snow crab no doubt have been enhanced by the decline in cod, but any indication that this is the only factor should be balanced by the fact that the chief prey of cod, the capelin, has declined over the same period.

8.13
Snow crab distribution in 2000 (data from Department of Fisheries and Oceans sources).

THE FISHERIES

The overall value of the Newfoundland and Labrador fisheries increased after the decline of the cod stocks, with landed value exceeding half a billion dollars per year, and production value reaching nearly a billion dollars per year in the early 2000s.[59] Although employment in the fisheries has decreased, about 25,000 people are still involved in the industry.

Snow Crab

Snow crab fisheries began in Atlantic Canada in 1967 in the Gulf of St. Lawrence. In that year, a total of 616 tonnes were caught. Catches increased to 8000 tonnes in 1976 and to 18,000 tonnes in 1979, all from the Gulf of St. Lawrence. In the 1970s, Atlantic Canada was a minor player in the world's crab fisheries, with total landings of snow crab seldom exceeding 11,000 tonnes out of a world total of about 60,000 tonnes.[60] Newfoundland and Labrador accounted for only about 2000 tonnes per year.[61] Further increases in the snow crab fishery were not anticipated: in 1976, it was concluded that "the annual sustainable yield of snow crab on the Atlantic coast is in a range of 10 to 15 thousand metric tons."[62] In 2000, however, landings of snow crab in Newfoundland and Labrador surpassed 50,000 tonnes per year – levels that were four to five times higher than those thought to be sustainable for the entire Atlantic coast in the 1970s. What had been a minor fishery up until the mid-1980s became the saving industry for Newfoundland and Labrador.

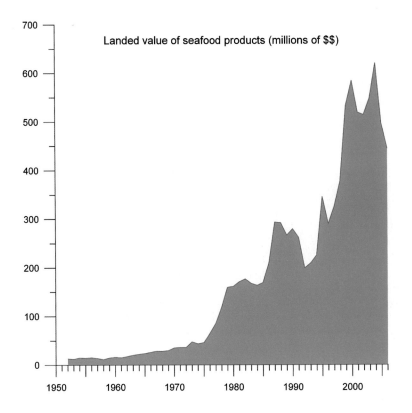

8.14
Total landed value of seafood in Newfoundland and Labrador, 1952-2006.

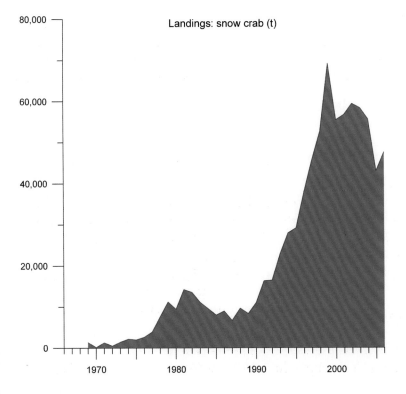

8.15
Landings of crab in Newfoundland and Labrador, 1969-2006.

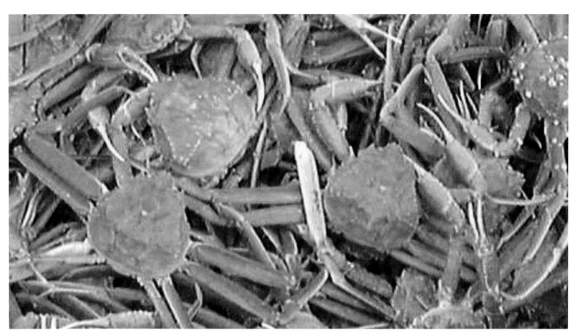

8.16 Fresh crab (Tom Clenche collection).

The snow crab fishery has a built-in conservation mechanism of sorts: females never grow large enough to be of commercial interest, so they are not fished. Only the large males are caught, after recruiting to the fishery at 7-8 years of age. Nevertheless, within a decade of the rapid rise in the crab fishery in the 1990s, the fishery began to fail, first in the north off Labrador.[63] Many crab fishermen thought that the increasing numbers of shrimp trawlers fishing the crab grounds were destroying the crab. With the crab in the Hawke Channel nearly gone, a small area was closed to trawling in 2002, then expanded to 2500 square nautical miles (about 8500 square kilometres) in 2003. By 2005, the crab fishery was failing on the south coast (NAFO Division 3Ps) and was in trouble on the northeast coast (NAFO Division 3K). Only the northern Grand Banks (NAFO Division 3L) remained strong, but by 2006 there were mixed signs, with a low survey index and poor expected recruitment. Four years after the first closure of the Hawke Channel to shrimp trawling, crab recruitment and most stock indicators were on the upswing in 2J and the adjacent area to the south in 3K – the only area where this was true.[64]

In 2005 and 2006, the high prices that characterized the snow crab market of the previous decade evaporated abruptly, crashing from record highs of C$2.50 per pound paid to fishermen (C$5.50 per kilogram) to about C$1.00 per pound (C$2.20 per kilogram). In addition, the strong Canadian dollar meant that the returns to processors were substantially reduced from the American markets, and consequently, both processors and fishermen found their profits cut to a small margin.

Heavy ice in the spring of 2007 prevented an early fishery, but prices have improved and the market is good, so at least in the short term the crab fishery should prosper.

Shrimp

Like snow crab, the pandalid shrimp were not considered to have the potential to support a large fishery in Newfoundland and Labrador waters. Historically, as far as is known, shrimp were not numerous and only patchily distributed. A shrimp fishery began in the mid-1960s in the Gulf of St. Lawrence, although it was recognized that the prospects for higher abundances would be better farther north. In the mid-1970s, a small shrimp fishery began off Labrador and was most successful in the deep cross-shelf channels. By 1978, there were 11 shrimp licences issued to companies based in Newfoundland and Labrador, Nova Scotia, New Brunswick, and Quebec. Large stern trawlers chartered from countries such as the Faroe Islands were used at first, but these were eventually phased out in favour of Canadian vessels and crews. Landings remained less than about 20,000 tonnes until the early 1990s, by which time the fleet had expanded to 13 factory trawlers that were fishing with 17 licences held by 14 companies.[65]

In 1997, the shrimp fishery expanded abruptly, with about 360 new entrants given access to the fishery on a temporary basis. Most of these entrants were smaller inshore vessels (less than 65 feet or 20 metres) displaced from the cod fishery. The quota increased from 37,600 tonnes in 1997 to 152,102 tonnes in 2003.

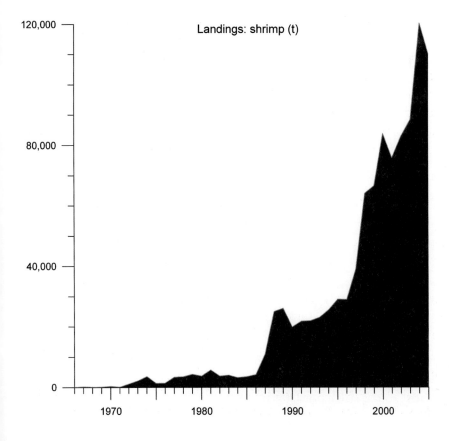

8.17 Landings of shrimp in Newfoundland and Labrador, 1965-2005.

In the early 2000s, despite continuing increases in shrimp abundance and expanded distribution,[66] the shrimp fishery became a marginal economic industry for many fishermen and plants. Prices fell, supplies remained high, and costs increased. A 20% tariff on Canadian shrimp entering the European Union limited market opportunities. In contrast, shrimp caught by Iceland and other countries in the Flemish Cap region were not subject to this tariff.

Together, crab and shrimp formed the twin pillars of a new fishery that expanded rapidly in the 1990s, mostly as a replacement for the vanished cod. High prices and strong demand for crab products resulted in the rapid conversion of several plants to crab production and increased demand for access to this fishery among many former cod fishermen. All along the coast, from Labrador to the south of Newfoundland, former cod fishermen saw crab and, to a lesser extent, shrimp as their saving grace, as indeed they were. In the late 1990s, hundreds of new temporary licences were issued, most to displaced cod fishermen and plants – but there is a saying in Newfoundland that nothing is as permanent as a temporary fishing licence. At the time of this writing, with both crab and shrimp fisheries are facing a range of new problems, the issuing of so many new licences seems at best a short-term solution to social and economic problems.

Cod

All cod stocks in Newfoundland and Labrador waters were closed to fishing after 1993, with the understanding that fisheries would resume as stocks rebuilt. The first stock to reopen was the southern stock that was concentrated in Placentia Bay and on St. Pierre Bank. By 1997, fishermen's observations and acoustic surveys of spawning biomass in Placentia Bay were sufficiently optimistic that the Fisheries Resource Conservation Council (FRCC) recommended that a small fishery of 10,000 tonnes be reopened in this economically depressed region.[67] The Department of Fisheries and Oceans followed this advice. In the following years, the harvest quota increased far too quickly to be sustained – up to 36,000 tonnes were allocated within one 12-month period. The rapid quota increases resulted from political and industry demands for more fish. Unfortunately, the highly variable stock assessments offered no counter to these demands.[68] However, if these difficulties can be viewed as start-up problems, they have been largely overcome, and the quota settled at a more realistic and sustainable 15,000 tonnes, with the stock remaining in relatively good shape based on large numbers of fish born from 1997 to 1999. However, the following years of the early 2000s were poor for young fish and the stock assessments have remained variable. Hence, this stock must be watched carefully, in part because it is the sole major cod fishery remaining in Newfoundland and Labrador. In 2006, the quota was reduced to 12,500 tonnes, and further reductions may be necessary if recruitment to the stock does not improve.[69]

On the northeast coast, the presence of the Smith Sound cod led many people to believe that the northern cod were coming back. While it was true that the rebuilding population of coastal fish was well advanced relative to the formerly dominant Grand Banks fish, the total abundance of the spawning Smith Sound cod, which was likely the bulk of the coastal fish, never exceeded 30,000 tonnes. When compared to the historical levels of the northern cod stock (3,500,000 tonnes) and migrating numbers of cod through the Bonavista Corridor in 1990 (450,000 tonnes), the Smith Sound cod stock seems quite small. However, 30,000 tonnes of cod can give the appearance of great abundance when they are crammed into a few small coves or bays. The FRCC recommended in 1999 that a small index fishery be implemented on the entire northeast coast. This recommendation was implemented by the Department of Fisheries and Oceans. At first, the resulting fishery appeared to do well and took an average of about 5000 tonnes per year over a three-year period. By 2002, however, it was evident that the coastal fish were again declining, and the fishery was closed in 2003. The FRCC regarded this fishery as an experiment to determine if a small, sustainable fishery could be conducted on the northeast coast, thereby providing employment and information about the state of the stock. The results of the experiment were not encouraging. This was to be one of the last recommendations on groundfish to be made by the FRCC, as its mandate was changed shortly thereafter and its former advisory role in groundfish management dropped. The Department of Fisheries and Oceans reopened both the commercial index fishery and a recreational fishery in 2006.[70] Although the rationale given for the reopening was to test the stock, the commercial quota (2500 tonnes) was more than twice that necessary to develop an index fishery, and the recreational fishery had little or no scientific value. The value of these fisheries was almost entirely social and political, with some economic benefits at the local level. These are legitimate benefits and goals, but politicians chose to sidestep them and base their case to reopen the fisheries on questionable benefits to science. As the dust settled on the 2006 fishery, a model constructed by the Department of Fisheries and Oceans, the same type of model that overestimated cod stocks in the 1980s, and estimates of harvest rates from tagging, suggested that there had been no reduction in the spawning biomass, but directed surveys of the Smith Sound cod, the majority of this stock, were down, as they had been after the earlier reopening. The fishery is scheduled to proceed with a few modifications in 2007.

In the northern Gulf of St. Lawrence, cod growth and condition fell to low levels during the cold period of the early 1990s, but both have since improved markedly.[71] Fishermen's reports suggest that capelin and mackerel have returned in large numbers, providing a good food base for the recovery of the cod (although there is no information on the stock sizes of the pelagic fish in the Gulf). Mortality inflicted by seals on cod likely remains high. The science on this cod stock is not well accepted by fishermen or industry. Stock assessments paint a rather dismal picture, a view not accepted by most fishermen and industry. Fishermen are seeing large numbers of cod inshore along the full range of the stock, which seem to indicate that the stock is rebuilding. At present, a small fishery (3000-5000 tonnes) has been reopened, with an increased quota in 2006 of 5500 tonnes. Optimism about this stock has led to an increase of the quota to 7000 tonnes for the 2007 fishery.

There is little doubt that the reopened fisheries will slow, or in worst case prevent, the rebuilding of cod stocks.[72] But there is another side to that story. While allowing too much fishing risks eliminating the fish, allowing no fishing risks eliminating entire communities.[73] Social and economic goals are not irrelevant, and a balance must be found. Current Federal Minister of Fisheries and Oceans Loyola Hearn, a Newfoundlander, has attempted to find that balance, particularly in the reopening of the small commercial and recreational fisheries on the northern coastal cod in 2006-2007. Unfortunately, the low stock levels do not permit much error in setting quotas – too much fishing will quickly remove the gains made in the past decade and could even prevent recovery altogether.

The small re-emerging commercial cod fisheries have faced the recurring problem of poor processing facilities and limited markets. The small numbers of cod caught in the early 2000s were of little interest to the large plants and fetched low prices. Some of the fish were frozen into blocks to be further processed elsewhere.[74] In 2005, northern Gulf of St. Lawrence cod was sold for as little as 40 cents per pound (C$1 per kilogram). At the same time, fresh cod in Gloucester, Massachusetts, sold for US$2.50 to US$3.00 per pound right off the boat (US$5.50-US$6.50 per kilogram, about C$6.50-C$7.50 per kilogram) – six to seven times the price paid to Newfoundland fishermen. The Newfoundland and Labrador cod is a superior product when it comes from the water – it has firm flesh and is mostly free of the worms that characterize more southerly stocks. It would be better to use this high quality fish to produce high-value product, such as salt or fresh fish targeted for international niche markets, than to freeze it into blocks.

Seals

From 1995 to 1999, a federal government subsidy was introduced to encourage the fullest possible use of seal meat left over from seals that had been taken for their pelts. By 1999, rising seal numbers led to an increase in quotas to 275,000 harp seals and 10,000 hooded seals, with the subsidy removed. In 2000, a five-year management plan was introduced for 2001-2005, with the quotas set as in 1999.

The seal fishery has resumed an important place in the small boat fishery of the northeast and west coasts. In the early 2000s, regulations were introduced to ensure that the hunt was carried out as humanely as possible and that maximum use was made of the carcasses. Currently, either the pelt or the meat or both must be brought in – killing whitecoats for their pelts or any seal for specific body parts is illegal. In the early 2000s, the harp seal population numbered over 5 million animals and annual pup production was as high as 1 million animals.[75] The demand for seal pelts is currently very strong, and the hunt has once again become a sustainable economy for many fishing and aboriginal communities in Newfoundland and Labrador.

The seal hunt faces continued harassment and negative international lobbying in Europe and the U.S. from animal rights groups. These actions have in the past raised legitimate issues about cruelty to animals, which have been addressed, but concerns about conservation and ecosystem issues misplaced and would be better directed towards the overfishing and pollution being caused, to a large extent, by industries in those same countries.

Capelin

The decline in the size and availability of capelin over most of its former range in the early 1990s led to a collapse of the capelin fishery. The small fish that remained were devalued and brought very poor prices, if they could be sold at all. In the late 1990s, a small inshore fishery resumed with a quota of about 40,000 tonnes; by 2005, it had been reduced to 34,000 tonnes. Prices have remained poor, at times as low as a few cents per kilogram.

In 2006, a proposal was put forth by the Barry Group to reopen the fish plant in Harbour Breton to process pelagics into fish meal for Barry's developing aquaculture interests in the same region. Unfortunately, the proposal to harvest 50,000 tonnes of capelin far exceeds the current quota, which is fully allocated, and there is no basis on which to expand it. Many fishermen share the views of scientific work that suggest that there is no surplus of capelin and that no such expansion should take place in the capelin fishery.[76] A corollary of this argument is that the cod, formerly the largest consumer of capelin, need the capelin more than we do, and the best strategy for the rebuilding of the ecosystem is to leave them in the water.

Other Species

Salmon. The moratorium on commercial salmon fishing was extended to all of Labrador in 1998.[77] The current salmon fishery is limited to primarily catch-and-release recreational fishing and some aboriginal fishing under communal licences.[78] Until there is a substantial recovery in the stock status of Atlantic salmon, it is unlikely that the commercial fishery will be reinstated.

Lobster. The lobster fishery is currently a small but important coastal fishery. There have been no substantial declines in this fishery for many years, and it is one of the best examples of a fishery that has been sustainable. The lobster fishery is managed by effort limits rather than by quota; perhaps most importantly, management involves considerable input at the local level.

Underutilized species. The decline of the cod fishery in the 1990s spurred renewed interest in what were previously considered underutilized fisheries. The green sea urchin roe fishery was advocated as a means of supplementing the fishing economy of rural Newfoundland, as it had been in the United States, British Columbia, and Chile for some time.[79] However, the strength of stocks was unknown, and although the urchins were thought to be locally abundant, it was likely that they would be a slow-growing, low-productivity species in the cold waters of coastal Newfoundland and Labrador. These traits would make them vulnerable to rapid depletion as had occurred in some other areas of the world, such as France and Korea.[80] Nevertheless, with other fisheries failing, the prospect of harvesting so-called underutilized species was compelling, and management plans based on trial-and-error effort control were put into place.[81] A 1993 report judged this fishery to be economically unviable as it was then practiced. The report was only partly correct: processing in Newfoundland and Labrador turned out to be uneconomical, but harvesting was not.[82]

New or expanded fisheries also developed for nontraditional species such as Greenland halibut, lumpfish, hagfish, winter flounder, sea anemones, and sea cucumbers. Little to nothing is known about the biology or abundance of most of these species.

8.18 Lobster boat belonging to fisherman Denis Ivany of Petley, Newfoundland (Tom Clenche collection).

LONG-TERM SOCIAL CONSEQUENCES

While there is optimism that the marine ecosystems of Newfoundland and Labrador can recover in time, there is less certainty about the future of the fisheries. Nevertheless, with some fundamental changes, the Newfoundland and Labrador fishery can have a bright future. Of prime importance, the fish stocks and the marine environment must be protected, including the parts of the Grand Banks that are currently outside Canadian jurisdiction. In addition, the proportion of the Newfoundland and Labrador population expecting to make a living from the fisheries must decrease, and so too their absolute numbers. The decline in production per person in the fisheries began in the early 1800s; using the population of the time as a guideline, no more than about 25,000-30,000 people should be employed in the fisheries in a fully rebuilt ecosystem. Perhaps 10,000-15,000 people can be employed with present production levels. This is about the same number directly employed in Iceland in a fishery that produces about 500,000-1,000,000 tonnes of product annually.[83] In Newfoundland and Labrador, about 40,000 people claimed attachment to the cod fishery alone during the years of the TAGS subsidy program.

Changing market and labour conditions are likely to dictate how the shift in employment happens. After the resettlements of the 1960s, Newfoundlanders and Labradorians have little taste for social engineering, and the prevailing sentiment is that people must decide for themselves. Many young people do not want to be involved in the fishery, preferring to pursue education, jobs, and urban lifestyles in St. John's or other Canadian and American cities. They will vote with their feet.[84] Those that stay

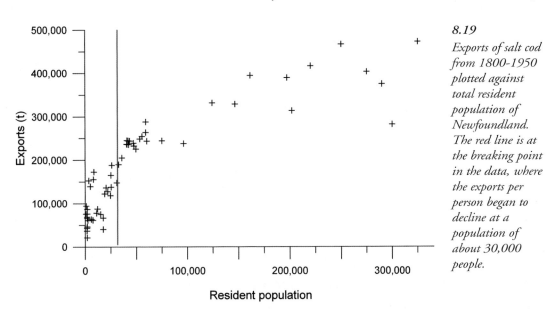

8.19 Exports of salt cod from 1800-1950 plotted against total resident population of Newfoundland. The red line is at the breaking point in the data, where the exports per person began to decline at a population of about 30,000 people.

in the fishing industry will likely benefit from improved working conditions and higher earnings. The value of seafood fluctuates, but overall has never been higher, and the potential for increased incomes from more productive ecosystems is real. Improved fisheries resources can be expected to translate directly into better livelihoods for all involved in the fisheries. The wharf at St. John's is lined with multimillion dollar vessels whose owners have resettled on their own initiative and made fortunes fishing shrimp and crab. The processing part of the industry is less certain, as low-wage economies such as China can process fish caught anywhere in the world more cheaply than most Newfoundland and Labrador plants. Low-grade and low-priced markets cannot be serviced by relatively high-cost plants in Newfoundland and Labrador. Nevertheless, high-end and high-value seafood products and markets, such as fresh and niche markets, have a bright future in Newfoundland and Labrador and should be aggressively pursued.

Many small settlements will change considerably, as they cannot sustain an economy capable of supporting schools, hospitals, and the services expected by most people, but those that are accessible will not be abandoned. Many people will choose to live there, at least seasonally. Some communities will become gentrified, such as the old settlement of Trinity, or bring in new industries and ecotourism, and with this reap additional economic benefits.

The fishery needs to reinvent itself with new markets and products, and a new philosophy based on a conservation ethic. A campaign to encourage consumption of Newfoundland fish – both in Newfoundland and Canada, where the traditional diet has been beef, pork, and chicken – could be beneficial, with the widely researched health benefits of fish eating emphasized. Old products, like salt fish, may find niche markets. Prior to the 1900s, salt fish were produced by cottage industries in which each family processed their own fish. By the mid-1900s, fish plants processed most of the fish caught in a community. This economic transformation led to the fish plant becoming the centre of a wage economy for many rural communities, which later became linked with government subsidy programs such as employment insurance. As these plants failed, the underpinnings of the economies faltered. A return to smaller and more independent cottage industries that produce high-value products such as light-salted, air-dried cod – as was the historical practice on the northeast and south coasts – could bolster the economies of some communities. There is a strong demand for such product, and it commands much higher prices than the plant-produced frozen blocks or fillets. Both Norway and Iceland, Newfoundland and Labrador's main counterparts and competitors in the international cod trade for centuries, did not ditch their salt fish production; they saw the markets, and today produce over 200,000 tonnes of salt cod in most years for the rich European markets in Italy, France, Spain and Portugal.

A change in philosophy, in attitude, would greatly benefit future fisheries in Newfoundland and Labrador, and elsewhere. If fisheries and conservation are incompatible, then neither has much future. Conservation requires that fishing practices change. It is wrong to think of the historical salt fish product as an anachronism, but not so the unrestrained fish killers of old. A new breed of fishermen is needed, a breed that will adopt a sea ethic, as Aldo Leopold described for the land almost 60 years ago.[85] A sea ethic means thinking, and acting, as part of the ocean ecosystems that we harvest, and not as dominators or controllers, something we may pine for but can never achieve. A sea ethic means caring for the harvest as food, and eating it, in preference to the unhealthy salted fat meat that has been consumed by generations of Newfoundlanders and Labradorians for centuries. On the political side, the fisheries in Newfoundland and Labrador would be better off if politicians and industry leaders got out in front of the many changes that are almost certain to come to marine fisheries worldwide. Quality standards and ecologically friendly fishing are going to dictate markets and value in the twenty-first century. Bottom trawling will face increasing restrictions and outright bans to conserve habitat and fish.[86] There will be worldwide initiatives to protect perhaps 10-20% of the ocean from all human intervention. These are positive steps for marine conservation and for the fisheries that adapt to these new realities. Whether the fisheries lead or follow these developments, or simply founder in their wake, will largely determine their place in future world economies.

An Age-Old Question: Must the Fishery be Poor?

The question of why the fishery resulted in poverty at so many times in the history of Newfoundland and Labrador has been posed by many people. Most economic analysts – even those who were poles apart in their thinking, such as Stewart Bates and David Alexander – have agreed that poverty was an imposed and not a necessary condition for fishermen or fisheries workers. The fishing industry starts with an inherently valuable resource which is caught, processed, and then exported to markets in a fish-hungry world. What in this process has brought poverty to the fishery as a recurring theme?

The ultimate limit to a fishery is fish. This lesson has been learned in the hardest of ways in Newfoundland and Labrador. Historically, the coastal fisheries depended on cod migration patterns which were always variable. The Newfoundland and Labrador cod stocks inhabit waters with temperatures at or near their limits of tolerance; hence, changes in the ocean environment translated into year-to-year and place-to-place variability in the shore fishery. Shore fisheries always had very good years, very bad years, and many in between. Most of the variation in cod landings up until the 1950s can be attributed to changes in availability of fish, not stock depletion. An economic system that buffers against the inevitable bad years, with some sort of self-sustaining insurance plan, might help to alleviate short-term poverty, but no such system has ever existed in Newfoundland and Labrador.

Mismanagement and stock depletion became more serious factors causing poverty in the fishery only in the final 50 years of a 450-year-old fishery. Although conflict between competing interests began with the earliest fisheries, and haggles between Newfoundland and foreign interests were well established by the late 1700s,[87] serious stock depletion did not begin until the 1950s. With the rapid buildup of the trans-Atlantic European fleets after World War II, the lack of protection of the stocks from overfishing became the single most important factor in the decline of the fisheries, the poverty of fishermen, and the failure of fisheries management.

In the most recent decades, with the foreign fisheries pushed outside the 200-mile EEZ, there has been far too much fishing effort, including vessels, people, and processors within the Canadian zone. Having too many people dependent upon the fisheries is not simply a social problem: it affects fish stocks and the marine ecosystems on which the fisheries depend. Too many people caps individual productivity and leads to poverty. Poverty inevitably leads to desperate fishing practices and lack of concern for conservation.[88] In Newfoundland and Labrador, those who became wealthy from the fishery were driven by the equally destructive motive of greed. Taken together, a greedy few controlling a poor and sometimes desperate majority does not lead to much concern for the oceans or to much investment back into the fisheries. This issue is in no way unique to Newfoundland and Labrador – it is played out daily in fisheries around the world.

Another reason for poverty in the fisheries has been the lack of development of markets and new seafood products. Even when fish were abundant the fishery could be poor. In the 1800s, unchanging prices for salt cod for nearly a century effectively devalued the industry and guaranteed a rise in poverty. In the 1900s, while other fishing nations were developing new products, improving old ones, and enhancing the value of their seafood products, Newfoundland fell far behind. The failures of the commercial system contributed greatly to poverty in the outports. Attempts to modernize the fishery in the 1950s might have worked, but by then the fish were being rapidly depleted by foreign interests. The total lack of protection of the stocks quickly undid any attempts to modernize, or to keep ahead, or at least abreast of, changing trends in world seafood markets. It is noteworthy that in recent years, Norway and the Faroe Islands put about half of their total cod production of 270,000 tonnes into salt fish (the same product that was discarded as antiquated and obsolete in Newfoundland and Labrador in the 1950s and 1960s). Iceland puts about one-third of its 225,000 tonne catch into salt fish (apparently when it comes down to it, much of life is still salt fish).[89]

A final factor underlying the poverty of the Newfoundland fisheries can be put under the heading of political and economic intrigues. Newfoundland history is littered with self-serving initiatives and individuals: belligerent admirals, aloof governors, greedy merchants and politicians who imposed unfair truck trading practices,[90] and power-hungry clergy.[91] Newfoundland is not unique for having had these features, but it has endured more than its share. The development of a "fish-ocracy" ensured that the tremendous wealth of the fishery was channelled into few hands, oblivious to endemic poverty and lack of education. There have been few democratic heroes in Newfoundland's history, although many – from Whitbourne to the more enlightened governors such as Berry and Cochrane, to politicians such as Bond, Coaker, and even Smallwood – undoubtedly did their best. There have been many cynics. Perhaps the place is not kind to heroes or idealists. There have been no Washingtons, Jeffersons or Mandelas in Newfoundland. Some of the least heroic people in this cast of characters have been the fish merchants.

THE FISHERY PRODUCTS INTERNATIONAL STORY

The processing and marketing of fish products is an essential part of any fishery, and the lack of effective policies and practices to do this has hampered the Newfoundland and Labrador fisheries for centuries. With the trawler fleets and companies near bankruptcy in the late 1970s, Fisheries Products International was formed in 1983 with a large infusion of government money in order to provide a financially sound and socially responsive company to process and market Newfoundland fish products. The provincial *Fishery Products International Act* guided its formation and ownership, and provided a legal basis for its operating principles. Based in St. John's, Fishery Products International became the largest fish processor in Canada and the flagship of the Newfoundland fishing industry. During the 1980s, Fishery Products International processed cod and other groundfish caught both by their own trawlers and by the inshore fishery, and expanded their seafood domain in lucrative United States and world markets.

In 1992, with the closure of the northern cod fishery, it might have been expected that Fishery Products International would be in deep financial trouble. Although it was forced to close some operations and sell off most of its trawler fleet, which now had little to catch, these troubles did not sink the company. Under the leadership of chief executive officer Victor Young, Fishery Products International processed redfish and yellowtail flounder from the Grand Banks, imported Barents Sea cod, and most importantly, turned their hand quickly to the new fisheries for shrimp and crab. Fishery Products International remained profitable as well as a good corporate citizen, as best as it could recognizing its social responsibilities to the one-industry towns that its fisheries and plants supported and a commitment to fisheries science.[92]

In 2000, outside interests, led by Nova Scotian John Risley, put in place a scheme to take over Fishery Products International. The plan was to have Fishery Products International buy Clearwater Fine Foods, Risley's company, for about $500 million and then make Risley the chief executive officer of the whole lot. The first attempt didn't succeed. Undaunted, Risley returned the next year, this time with a few Newfoundland

8.20 Fishery Products International flagship Newfoundland Otter *engaged in the shrimp fishery in the ice of Labrador (photo by author).*

businessmen, including fish processor Bill Barry,[93] former politician and federal fisheries minister John Crosbie, and Derrick Rowe, as allies and launched a hostile takeover of Fishery Products International. Their claim to stockholders was that Young's conservative management of Fishery Products International was not making enough money and that they would do better. This was a smokescreen: under Young's direction, the company had just recorded its best profits in many years. However, promises were made regarding increased labour forces and increased profits, and a majority of the stockholders bought it. The Liberal government of Roger Grimes might have intervened, based on the *Fishery Products International Act*, but chose not to; Young and most of his team, who had piloted Fishery Products International through the treacherous waters of the 1980s and 1990s, were out. Young sensed that Risley's claims were not what they seemed and that the new Fishery Products International would fail to be a good corporate citizen. After his ousting, his parting words were from the "Ode to Newfoundland": "God guard thee, Newfoundland." Perhaps he should have said, "God help thee, Newfoundland."

By late 2005, Fishery Products International stock had plunged from C$10-C$12 per share to about C$5 per share,[94] and plants were closing or being downsized. Few of the promises made by the new directors were kept. The once-successful plant at Harbour Breton was closed, with only a few days' notice, just before Christmas 2004. Closure of the plants at Bonavista, Catalina, Fortune, and Marystown were threatened. Some fish were sent to China for processing. The company proposed raising a trust fund through the sale of some of its hard-won U.S. assets to save the Newfoundland operations, but none of this came to pass. Soon after these events, CEO Derrick Rowe resigned. The company had incurred an unmanageable debt load, something Young carefully avoided. Some people believe that the takeover was a well-thought-out plan by hostile competitors to founder Fishery Products International and then salvage its most lucrative parts at bargain prices.

With FPI in ruins, in 2007 its assets have been sold off: its lucrative U.S. marketing branch and secondary processing plant in Burin to High Liner of Nova Scotia (a successor to National Sea Products), and its other plants to Ocean Choice International, a firm owned by Newfoundland and international interests.

Provincial governments in the 1990s and early 2000s have for the most part sat on the sidelines during the dismantling of FPI; despite the existence of the *Fishery Products International Act*, they have seemed either unwilling or unable to protect what was once Newfoundland and Labrador's flagship industry from hostile control or dissolution. One thing is clear: the takeover of Fishery Products International did little to benefit the Newfoundland and Labrador fishery or economy, and that alone should have been enough to justify government intervention – what else is a government for?

OIL AND GAS

No story of the Grand Banks would be complete without mention of the oil and gas resources that formed under accumulated sediments many millions of years ago. The rise of the oil industry took place as the fisheries were being depleted. By the mid-1990s, with the development of the Hibernia oilfields on the northeastern part of the Jean D'Arc basin, oil had become not only a source of a growing economy in Newfoundland and Labrador, but also a cause for concern about the marine environment.

The first concern is the effect that seismic surveys may have on fish stocks and other marine life. Seismic surveys are used to study the structure of the earth's crust and to search for oil and gas beneath the sea. These surveys use underwater air guns to transmit high-energy sound waves through the water to penetrate deeply into the sea floor. Changes in the density of the subsea sediments result in echoes that register on a series of listening devices that can be interpreted by geologists. Seismic surveys have been conducted over much of the Grand Banks and the continental shelf, with little regard for potential effects on marine life. Research in Norway and Australia has indicated that seismic surveys at the very least may disturb various fish species, disrupt the functioning of marine mammals' sensitive sonar, and can even cause the death of marine organisms under some circumstances.[95] Little research has been conducted on the effects of seismic surveys in Newfoundland and Labrador waters. At times, survey planning has ignored even the most basic biological considerations: surveys were planned over spawning grounds and migration routes of fish, including cod and capelin, and in sensitive habitats of marine mammals and crustaceans. While some of these surveys were stopped or modified after protests, many others have gone ahead.[96]

A second concern is the direct environmental damage incurred from the dumping of waste material around oil rigs, and from oil spills from rigs and oil tankers. The large oil spill that occurred in Prince William Sound, Alaska, killed thousands of marine animals, polluted coastal habitats for decades, and caused long-term damage to fisheries and fish stocks.[97] Exxon continues to deny responsibility for much of the damage this oil spill caused. The potential for damage from an oil spill is high in coastal Newfoundland, particularly in Placentia Bay, where oil from the Grand Banks is shipped and where an older oil refinery receives heavier oil from distant sources.[98] Tanker traffic in Placentia Bay continues to increase. Placentia Bay is home to the largest coastal cod stock in Newfoundland, one of the largest seabird colonies in the North Atlantic at Cape St. Mary's, and is prime habitat for juvenile capelin, herring, whales, and variety of other species. An oil spill in Placentia Bay would likely do damage equivalent to that which occurred in Prince William Sound.[99]

A third concern that is applicable to all shipping industries is the release of oil at sea, primarily by the pumping out of bilge water. This illegal practice occurs regularly, especially on the major sea routes across the southern Grand Banks from Europe. One estimate is that as many as 300,000 seabirds are killed each year by oil in Newfoundland waters.[100] Even a drop of oil on the feathers of a seabird can result in loss of insulation and death. This cruel fate awaits far more birds than are taken by hunters and appears to result at least in part because of minor penalties to offenders.

Oil money, primarily through the Pew Charitable Trusts,[101] has funded considerable recent research condemning the influence of fishing on the world's oceans. Far less has been directed towards research on the impacts of oil exploration and spills on marine life, or the longer-term impacts of fossil fuel production on climate change and ocean life.[102] The most recent areas to be explored for oil and gas are in the ancient Orphan Basin, of which the famous cod highway of the Bonavista Corridor is part, and also off Labrador, both key areas for northern cod spawning and migration. The low cod numbers currently present in those regions draw little public attention to the seismic surveys and proposed drilling – as in the past, Newfoundland and Labrador has turned all too easily towards promises of riches through industrialization, but the returns are non-renewable and likely to be short-term.

Despite most of the cash from Grand Banks oil going elsewhere, a deal hammered out by Premier Danny Williams has ensured that Newfoundland and Labrador at least gets a share. The money from non-renewable oil could do much for Newfoundland and Labrador, and one of the most important goals would be to lay the groundwork for longer-term sustainability in renewable ocean resource industries, the foremost of which is the fishery. At the time of this writing, Williams is locked in a battle with the Canadian government over revenues from the oil industry – with central Canadian interests compelling the federal government to renege on a promise made to Newfoundland and Labrador, and other provinces, to enable provinces to keep their portion of these revenues. It might be pointed out that foreign and federal government interests are the chief beneficiaries of the Grand Banks oil revenues, whereas potential harm to the marine ecosystems and fisheries will be borne almost exclusively by Newfoundland and Labrador.

FISHERIES SCIENCE

> Politicians, fisheries management officials, union leaders and even fishermen were far more experienced in public relations and media manipulation than scientists were…when they came under attack from all quarters, they never had a prayer.   – JIM WELLMAN, "The Broadcast," 1997

Fisheries science fared badly in the years following the moratorium, serving as a scapegoat to a smattering of politicians, fishermen, and industry people who needed someone to blame. As pointed out by John Crosbie to angry fishermen, he didn't catch the fish – but neither did scientists. Scientists also never pushed for increased quotas, as did industry and many politicians, and their pleas for moderation in catches and warnings that model predictions were based on uncertain assumptions often went unheeded. Nevertheless, and not without some justification, public confidence in science and its models was shattered when the cod stocks collapsed. Scientists themselves contributed to the confusion by pointing fingers at each other over mistakes that had been made decades earlier – harangues that looked more like opportunism and exploitation of a crisis than attempts to correct past mistakes.[103] In the end, fisheries scientists became subject to public ridicule. One host of the Canadian Broadcasting Corporation's *Fisheries Broadcast* often referred to fishermen as the "real scientists." In this air of dissention and controversy, politicians saw an opportunity to relieve science of much of its funding without public protest and did just that.

Two large-scale science programs initiated around 1990 to address cod ecology were not renewed following the moratoria: the Ocean Production Enhancement Network, which was a Natural Sciences and Engineering Research Council Network of Excellence that included fisheries scientists and oceanographers from Newfoundland to British Columbia; and the Department of Fisheries and Oceans–sponsored Northern Cod Science Program launched in the wake of the Harris report. The collapse of the fishery influenced decisions not to continue these programs. The prevailing attitude was that if there was no fishery, why was research needed? Even a substantive fishery did not ensure that science would be well supported: no major new research initiatives were put in place to study the snow crab and shrimp fisheries as they developed in the 1990s.

Cutbacks in most areas of government-funded fisheries science became the order of the day in the 1990s and early 2000s. Fish surveys were cut.[104] Many scientists left government and many more retired with few replacements. Ships were no longer available for researchers, a situation that was made worse when the former fisheries research vessels were reallocated to the Coast Guard. Under the Coast Guard, research became a secondary concern after search and rescue and other activities. Even top priority science projects, such as annual multi-species surveys, were badly compromised by lack of availability of serviceable vessels.[105]

One positive scientific outcome of the decline in the cod fisheries was the implementation of what came to be known as sentinel fisheries. Fishermen were paid to fish at set sites and at sites of their own choosing around the coast. Sentinel fishermen tagged fish and made measures of water temperatures and salinities. Catches from sentinel fisheries eventually formed an index of stock status. Although there was at first much scepticism about the usefulness of this information, the sentinel fishery has proven to be a very useful indicator of stock status, providing information on the age and size of fish as well as a means to conduct research on coastal fish. The main problem with the sentinel fishery is its cost: it consumes a considerable portion of the total fisheries science budget.

AD 2005: A YEAR OF REPORTS

> Conservation is a state of harmony between men and land [sea]…[it] proceeds at a snail's pace: progress still consists largely of letterhead pieties and convention oratory. — ALDO LEOPOLD, "The Ecological Conscience," 1948

Government Reports

There may have been a shortage of fish in the early 2000s, but there was no shortage of government and nongovernment committees and reports. The federal government commissioned its Eminent Panel on Seal Management, the province set up a royal commission on the Future of Newfoundland within Canada, and the FRCC studied the crab fishery. These reports mostly gathered dust, and were neither widely read nor implemented. By 2005, with the problems in the fishery and the degradation of the Grand Banks ecosystems persisting, a plethora of additional reports were commissioned or written directly by governments and nongovernmental organizations.

In the spring of 2005, the federal government commissioned the Advisory Panel on Straddling Stocks to examine the rebuilding potential of the Grand Banks ecosystems. The Canadian panel was led by Dr. Arthur May, and included the chief executive officer of Fishery Products International and a law professor from Dalhousie University. The report of the panel was titled *Breaking New Ground: An Action Plan for Rebuilding the Grand Banks Ecosystems.* The report and its externally commissioned reviews described the destruction of the Grand Banks fisheries and the potential benefits of rebuilding this ecosystem and its fisheries. The report also stated unequivocally that the present Canadian strategy of reforming NAFO was unlikely to achieve that goal, although it stopped short of recommending unilateral Canadian action to take control of the Grand Banks (often referred to under the general but ill-defined heading of custodial management). The report recommended that Canada form a new regional fisheries management organization in partnership with a few NAFO members committed to the Grand Banks fisheries, and that new fishing rules be based on rebuilding the ecosystem. By early 2006, the federal government had yet to respond to the report or to release it to the public, and continues its long-standing approach of attempts to reform NAFO.

In the fall of 2005, the all-party, federal Standing Committee on Fisheries and Oceans released a report about the future of the Newfoundland and Labrador fisheries based on a number of interviews and consultations. This report was targeted at northern cod and was titled *Northern Cod: A Failure of Canadian Fisheries Management*. The report rehashes much of the frustration about the decline of the northern cod and, based on a scattered and selective mix of science, makes many recommendations, including a call for immediate actions by the federal government to declare custodial management of the Grand Banks, the rejection of a recommendation to declare cod an endangered species, reinvestment in fisheries science, and the opening of a small fishery on the northeast coast of Newfoundland, on a bay-by-bay basis.

The Canada-Newfoundland and Labrador Action Team for Cod Recovery published its report titled *A Strategy for the Recovery and Management of Cod Stocks in Newfoundland and Labrador* in late 2005. This report summarized the history of the cod stocks around Newfoundland and Labrador and described the various measures taken since the moratorium to protect the remaining cod. The importance of seals and capelin to the fate of the reduced cod stocks was emphasized, and new directions towards smaller scale management and reinvestment in science were suggested. There were few hard-hitting recommendations in this report, underlining a tone of frustration that despite the closures, the protection of spawning aggregations, and other measures, little rebuilding has occurred in the major stocks.

Nongovernmental Organizations

Nongovernment organizations have played a role in the Newfoundland and Labrador fisheries for several decades, but for the most part their actions were seen as meddlesome attempts by well-heeled outsiders from Europe and the United States to harass and exploit poor fishing communities. At times their motivation appeared to be purely financial and their strident methods lacked ethics.[106] Too many movie and pop stars, such as Brigitte Bardot, Richard Dean Anderson, and Paul McCartney, have parachuted in and just as quickly departed, with little time to do much except pose for numerous photographers and gossip about the morality of the traditional fishing and hunting societies of Newfoundland and Labrador. This was particularly offensive behaviour at a time when European trawlers were still ravaging the Grand Banks for the remaining fish with little regard for the conservation of many species whose very existence may be threatened. In the 1970s and 1980s, pictures of whitecoat harp seals were used in fundraising campaigns of various animal rights groups to protest the hunt; these campaigns are still used in the present day despite the fact that whitecoats have been protected by law for decades. Old pictures of clubbed seals are still trotted out, even though seals are now shot as humanely as possible in the Newfoundland and Labrador hunt. It is not surprising that few people in Newfoundland and Labrador have much time for these groups and their methods.

The World Wildlife Fund, however, is widely respected in Newfoundland and Labrador as an organization whose primary purpose is the conservation of wildlife, including fish, and ecosystems around the world. The World Wildlife Fund does not oppose fishing and hunting societies, recognizing that such activities are essential to conservation, often forming the front line against habitat destruction and overexploitation. In the early 2000s, the World Wildlife Fund turned its considerable influence towards the world's oceans, in particular the Grand Banks, and a report titled *Bycatch on the High Seas: A Review of the Effectiveness of the Northwest Atlantic Fisheries Organization* was released in 2005. This report summarized the available assessment data for Newfoundland and Labrador for stocks currently under the management of NAFO. The general assumption of the report was that NAFO will continue to exist, and the report focussed on reducing the substantial bycatch in the fisheries on the Grand Banks. How this might be achieved under NAFO remains problematic: the minor scrutiny that the foreign fleet has been subjected to over the past 20 years resulted in over 500 citations for fishing infractions.[107]

COSEWIC and Endangered Species

The Committee for the Status of Endangered Wildlife in Canada (COSEWIC) is a federally funded group tasked to identify species or local populations of species that are in danger of extinction in Canada. For many years, COSEWIC dealt with and made recommendations only on terrestrial species, and there was no requirement for provincial or federal governments to act on their recommendations. However, when the *Species at Risk Act* became law in Canada in 2003, COSEWIC became a much more powerful group: their recommendations, once accepted, were to be followed up with rebuilding plans and various restrictions.

COSEWIC's venture into the world of marine fishes in Canada was a strange tale, considering that the Department of Fisheries and Oceans employed most of Canada's marine scientists, conducted extensive assessments of commercial stocks of fish, and had jurisdiction for the management of all species. COSEWIC had little of this expertise or experience, and relied to a large extent on university students to conduct reviews and reinterpret Department of Fisheries and Oceans' material. Consultations with industry and fishing communities were not thought to be necessary.

The first Atlantic fish to be listed as threatened[108] were two wolffish species. Research funding was diverted to support this designation and create a rebuilding plan. Whether or not these actions will help the wolffish will take years to assess. In 2005,

COSEWIC recommended that the northern and Grand Banks cod be listed as endangered, and the south coast and northern Gulf of St. Lawrence stocks as threatened. The recommendation was not well received in Newfoundland and Labrador, especially in the case of the south coast cod which both science and industry believed to be in relatively good shape. Even though recruitment in this stock has waned in recent years, it was difficult for anyone with knowledge of these fisheries to accept that this stock was threatened with extinction. Not surprisingly, the Department of Fisheries and Oceans did not accept COSEWIC's recommendation on cod.

All of the above reports and initiatives contain useful descriptions and guidance about the current state of the fish stocks and fisheries. Despite their different approaches, several points of convergence, if not outright consensus, stand out, and they form a straightforward approach to rebuilding the cod stocks and marine ecosystems of Newfoundland and Labrador. The approach includes: a large reduction in fishing effort, landings, and bycatch in keeping with scientific guidelines; protection of habitat from destructive fishing practices, especially bottom trawling; reducing pollution; closing sensitive spawning and juvenile areas, the "hotspots" of the ocean, to all fishing and industrial activities; and re-establishment of the pelagic forage fishes, especially capelin. Points for which convergence is lacking but that appear to have validity are Canadian custody of the Grand Banks and a reduction in the seal herd.

THE SEA AHEAD

> For a start, our university could start teaching fisheries management. For another, the province could, for small money, build up a competent team of fisheries managers. These are not frills but the essential preconditions of ensuring that we, Newfoundlanders, regain full control over our fisheries...If we do not do this, we can expect to become nothing more than a pathetic shadow of ourselves; no longer a Newfoundland people; no longer a Newfoundland society. — CABOT MARTIN, *No Fish and Our Lives*, 1992

In the early 2000s, the marine ecosystems of the Grand Banks and adjacent continental shelves remain degraded in terms of overall production and diversity of species. As go the Grand Banks, so goes Newfoundland. The list of species that are no longer abundant enough to be fished is long: haddock, redfish, Atlantic cod, American plaice, witch flounder, grenadier, and capelin. The list of profitable fisheries grows shorter and shorter, down now to a few crustaceans. While some people in the industry have relished the rich catches of crab and shrimp, in most of fishing communities there is at best a begrudging acceptance of this new ecosystem, a hunch that it won't last long, and a deep longing to see cod return. For all its foibles, cod were trusted and are deeply rooted in Newfoundland's culture. There is something vaguely suspicious about shrimp and crab – like finding money. A day on the water with fishermen is instructive: many become more excited by catching cod worth a few dollars than by harvesting a load of shrimp or crab worth thousands.

In the final pages of this book, I feel compelled to try to explain why the fisheries of Newfoundland and Labrador have not prospered as they might. There are two recurring themes that span hundreds of years of history: lack of support for fisheries as an industry and outside control. The lack of support has been expressed in several ways: repeated attempts to turn away from fisheries, lack of protection of the fish, wavering support for science in government and academic circles, poor marketing, and the disorganization of a productive industry. There are many symptoms: the lack of fish in the typical Newfoundland diet; the low social standing and educational level of most fishermen; and the paucity of seafood restaurants on the St. John's waterfront (there is a steak house).

For the past 130 years, the governments of Newfoundland and Labrador attempted to turn away from the fisheries towards new industries. These attempts survived several forms of government: responsible, commission, and post-Confederation. They began with the railroad (which no longer exists), worked their way through C$26 million worth of rubber boot and chocolate factories (also gone),[109] and reached their peak with the hydroelectric power giveaway at Churchill Falls (the profits from which go to Quebec). Grasping for new industries did little to resolve the economic problems of fishing communities: pulse industries offered no more solution than did pulse fisheries.

Attempts to diversify the economy are unquestionably necessary, but have proven insufficient to reduce the numbers of people dependent on the fishery, and intended or not, have diverted attention and resources away from managing the fisheries. The present focus on oil and gas extraction provides few jobs after the construction phase, and the wells will be exhausted in a very few decades. What then? If oil and gas developments are harming marine ecosystems, are they really worth the benefits, the majority of which do not accrue to Newfoundland and Labrador anyway? Some think that the degradation of marine ecosystems can be countered with aquaculture, by relieving the demand for wild fish. But aquaculture is farming: whatever its other benefits, it is more likely to exacerbate than help a weakened ecosystem. Farmed fish may consume 5-10 kilograms of fishmeal, comprised mostly of pelagics like capelin, herring, and sand lance, to produce one kilogram of marketable farmed fish. Is it sensible to deplete wild stocks to feed farmed animals and fish? There are also problems associated with the use of pesticides and antibiotics to keep fish penned in close quarters, trouble with diseases and parasites, and unknown risks with genetically modified fish. In the coming decades, high-end food products will become more organic, harvested as naturally as possible and grown without the use of chemicals. This will almost certainly be repeated in international fish markets. Newfoundland and Labrador, with its coldwater environment, has wild organic seafood in abundance at its doorstep – why would it not utilize this natural advantage?

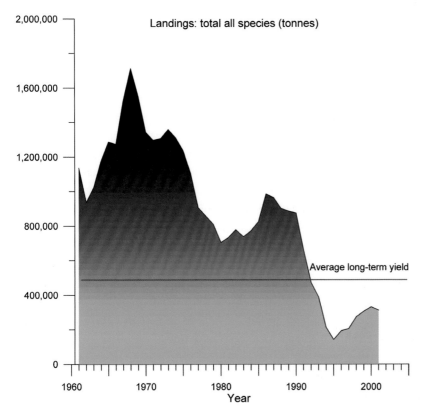

8.21
Total landings of all species by all countries, including those from the Flemish Cap, 1960-2001 (data from ICNAF, NAFO and Department of Fisheries and Oceans sources).

Over the centuries, much has been taken from the Northwest Atlantic, but little returned. Since the fisheries began, about 100 million tonnes of cod have been taken from the waters of Newfoundland and Labrador. If this amount of fish were sold at today's prices in St. John's, it would be worth some C$650 billion[110] – yet only paltry amounts have been reinvested into fisheries research or improving cod products and markets. Underlying this disparity is the age-old negative social attitude towards the fishery and its support services, including science. The scientific study of the fisheries, like the fishery itself, has not taken the prominent position that it should have in Newfoundland and Labrador institutions: in the late 1800s, a short-sighted politician berated science for throwing bottles in the water. In February 2006, influential fishing industry executive Bill Barry derided science, apparently inferring his own, and much higher, cod abundance estimate for the south coast cod stock.[111] Just what Barry's estimate was based on is not known. Newfoundland and Labrador should not blame the negative attitude towards the fisheries on others: while it undoubtedly had foreign roots, it has been a homegrown phenomenon for over 100 years.

A related thread through history is outside control. Historical dependence on Britain and, for nearly 60 years, Canada has not resulted in appropriate or successful management of the oceans or the fisheries. If the oceans are to be better managed, the people and institutions of Newfoundland and Labrador must take a greater interest in and ownership of the fishing industry and of fisheries science and management. Doing so will require a greater commitment from the Memorial University of Newfoundland and the provincial government.[112] The lack of capacity in these organizations, given the importance of the fisheries to the economy and history of Newfoundland and Labrador, is nearly inexplicable and has undoubtedly contributed to the degradation of the fisheries and marine ecosystems and limited the ability of Newfoundland and Labrador to play a larger role in the management of its fisheries.[113]

Local involvement does not stop at academic and government doorways: it stretches to fishing communities and to small and large industries. If Fishery Products International had been owned by widespread Newfoundland interests and by employees of the company, rather than by diverse and disinterested groups such as the teachers of Ontario and industrial interests in Iceland, New Zealand, and Nova Scotia, the hostile takeover engineered by largely outside interests might not have happened. No one prevented local ownership – it was forfeited by default.

In 2007, as I finish this book, a new wave of pessimism has engulfed the Newfoundland and Labrador fishery. The rising Canadian dollar, low prices for product, competition from China, European tariffs on shrimp, the lack of rebuilding of the cod stocks, the decline in snow crab, spring ice, the never ending beat-ups at the hands of the Canadian media and the annual animal rights circus, the dismantling of Fisheries Products International, and the continuing emigration of Newfoundlanders to seek employment elsewhere all contribute to a pervasive melancholy. Despite these problems, the underlying cause of the gloom may be none of these, but instead a deep apprehension that societal interests in maintaining a fishing economy have waned. This apprehension, oft unspoken, begs an answer but fears one, and is mostly met with silence.

In short, Newfoundland and Labrador and Canada have a choice to make. They can continue with the historical status quo of downplaying the potential of the fisheries and concentrate on industrial-urban economic developments, leaving the fisheries to do what they can, or they can put policies in place to protect the marine environment, rebuild the great fish stocks of the Grand Banks, and recognize that the fisheries are a key food production system of global importance. The only long-term solution to the problems of the fisheries is to allow the marine ecosystems to restore themselves. Even if no specific historical state can be targeted,[114] their productivity in terms of desirable commercial and non-commercial species can almost certainly be increased. To do so requires a new ethic for the ocean: if the Grand Banks fisheries have been an icon for abuse and mismanagement, they can become an icon for restoration and rebuilding. The cultural symbolism of a rebuilt fishery in a rebuilt Newfoundland and Labrador is powerful, a chance to re-embed a remarkable and adaptable culture within a marvelous marine world. It comes down to political will; a lack of it underlies every major obstacle in rebuilding marine ecosystems.

Whichever path is taken, the fishery will write the future of Newfoundland and Labrador as it has written its history. Ten thousand years after the Wisconsin ice sheets retreated, and 500 years after the first export fisheries took salt cod to Europe, the Grand Banks still anchor one of the world's most latently productive marine ecosystems – capable of employing tens of thousands and feeding millions. For all its transient problems, the seafood industry is still highly productive and is not in the general state of decline and crisis that many people believe.[115] Change it must, but arguably no more so than other industries already have over the past century. Newfoundland and Labrador is blessed with ample and lucrative energy resources, but its most valuable assets are clean water, fish, and its unique culture and people – its natural advantage with the sea. Some might claim that the next generation will forget their heritage as they become more urbanized and, as Cabot Martin wrote, "a pathetic shadow" of their past. Perhaps, but Newfoundland culture, like "longliner tea,"[116] is steeped in the sea and the fisheries. Better fisheries management represents Newfoundland and Labrador's most vital development – the only way to sustain its essentially rural culture and heritage. The head of Newfoundland and Labrador may be in the offices and warehouses of St. John's, as it has been for centuries, but its heart and soul will remain in the outports and in fishing boats by the ocean, where they were born.

Like the ever-changing ecosystems of the North Atlantic, there is no true conclusion to this story…sometimes, as in the spring of 2006 in Smith Sound, cod rise from the depths in massive columns like apparitions of past abundance, flooding the waters with billions of eggs each holding a tiny embryo, and waves of capelin swarm the beaches, all revealing that Nature never forgets. But unlike the memories of the ocean, our story must end…

NOTES

[1] See Lear and Parsons (1994).

[2] Allister Hann, personal communication, December 2005.

[3] The reopening of the south coast stock was criticized by several academics outside of Newfoundland who based their opinions on obsolete information, but it has been followed by a decade of a small but relatively prosperous fishery. This fishery could not supply sufficient fish for all of the former plants that depended on Grand Banks flatfish, redfish, and cod.

[4] See Narayanan et al. (1995).

[5] See Frank et al. (1996).

[6] See Narayanan et al. (1995).

[7] Spawning capelin averaged over 17 centimetres in length in the mid-1980s, but only about 15 centimetres in the 1990s (Narayanan et al. 1995). In most fish, weight is related to length by a power of about three, so 17-centimetre capelin would be about 50% heavier than 15-centimetre capelin.

[8] At Brador Bay on the Labrador coast inside the Strait of Belle Isle (now part of Quebec), capelin historically spawned at the large beach on June 24, give or take a few days. They rarely failed. In the early 1990s, spawning was delayed. On the northeastern coast of Newfoundland, capelin did not show up at their regular time and spawning occurred weeks late – some fish spawned as late as August or September. The spawning capelin were tiny, perhaps half their former weight. The oldest people on the coast have no memory of similar events happening in the past.

[9] See Mowbray et al. (2002).

[10] Shannon Obradovich, unpublished MSc research.

[11] See Sameoto (2004) and Edwards et al. (2005).

[12] Jim Carscadden, personal communication, 2005.

[13] See DeRoos and Persson (2002).

[14] See Rose and O'Driscoll (2002) and Sherwood et al. (2006).

[15] See Orlova et al. (2005) and Vilhjálmsson (2002).

[16] Changes in capelin distribution or abundance mainly affected cod, their major predator (Bundy 2001).

[17] Brian Nakashima, Department of Fisheries and Oceans, CBC *Fisheries Broadcast*, February 10, 2006.

[18] See Hamill and Stenson (2003).

[19] See Bundy (2001).

[20] "Belly-feeding" occurs when seals take large cod that they cannot eat whole. They literally rip the bellies out of the fish and eat only the liver and internal organs, throwing the rest of the dying fish back to the sea. When cod are in overwintering aggregations, they are slow to respond to any stimulus, and many fish may be killed in quick succession by even a few seals. Underwater videos have been made of cod bodies littering the bottom after one of these events.

[21] See deYoung and Rose (1993).

[22] See Lilly et al. (2003) and Shelton et al. (2006).

[23] See Rose and O'Driscoll (2002).

[24] See Rose and O'Driscoll (2002).

[25] See Mello and Rose (2005).

26 See Rose and O'Driscoll (2002) and Sherwood et al. (2007). Recent unpublished research shows that early maturation, elevated egg production and early death now characterize the life history of offshore northern cod.

27 Shelton et al. (2006) claimed that northern cod would grow without a fishery; however, this analysis lumped coastal and bank fish together, an unreasonable choice given the very different mortality rates in the two regions. Only the coastal fish have had a fishery since 1992, and these fish have grown considerably in number and have not suffered any unknown mortality as have the bank fish.

28 Northwest Atlantic Fisheries Organization, NAFO Annual Fisheries Statistics Database. http://www.nafo.int/publications/frames/fisheries.html.

29 See Power et al. (2005).

30 The World Wildlife Fund recently launched a program to attempt to reduce bycatch on the southern Grand Banks as a step towards rebuilding this fishery.

31 See Mello and Rose (2005).

32 Shelton et al. (2006) showed relatively high productivity in the south coast cod population, but the failure to separate the coastal from bank northern cod precluded showing high productivity in the coastal stock.

33 Roughgarden and Smith (1996) and Myers et al. (1997) predicted that the northern cod would rebuild to over a million tonnes within a decade after fishing was stopped in 1992. Other later predictions (e.g., Hutchings et al. (1999)) have also proven to be far too optimistic. The predictions made by deYoung and Rose (1993) shortly after the start the moratorium were the only ones to stand the test of time. These authors stated that ecosystem change made any predictions using simple fisheries models baseless, and that "predictions of future performance of the northern cod stock based on our conceptual model [that took perceived changes in performance of the stock into account] suggest a longer rebuilding time."

34 The northeast coastal cod grew at a rate close to 20 percent per year in the late 1990s prior to the start of a fishery (Rose 2003b). This growth rate was close to that determined historically for northern cod (e.g., Myers et al. (1997)). Unfortunately, Myers et al. (1997) applied this rate of growth to the entire northern cod complex, which resulted in an overly optimistic prediction of stock rebuilding.

35 See Rose (1993) and McQuinn (1997).

36 See Olsen et al. (2004) and Hutchings (2005).

37 New Zealand Minister of Fisheries Jim Anderton, CBC *Fisheries Broadcast*, March 3, 2006.

38 See McGinn (2002) and Vilhjálmsson and Rose (2007).

39 See Rose (2005).

40 See Drinkwater (2005).

41 See Rose (2005).

42 See deYoung and Rose (1993).

43 Lilly et al. (2000) showed that the increase in shrimp began prior to the decline of cod in the late 1980s. Although the authors agreed that the lack of cod predation almost certainly increased the survival of shrimp, they believed that a change in ocean conditions initiated the "outbreak" of shrimp in the 1990s.

44 Worm and Myers (2003) demonstrated an inverse relationship between the abundances of cod and shrimp in several northern marine ecosystems. The same type of inversion appears to occur in the North Pacific with walleye pollock and shrimp. Worm and Myers (2003) interpreted the relationship as what is referred to as top-down control: the predator (cod) controls the numbers of the prey (shrimp). This may be true, but as Lilly et al. (2000) pointed out, may not be the whole story. Not enough is known of marine ecosystem interactions to make definitive conclusions.

45 The most widely used models are of the Ecopath family developed at the University of British Columbia Fisheries Centre (http://www.ecopath.org/).

46 See Ruzzante et al. (1999) and Beacham et al. (2002).

47 The most recent surveys were conducted from January to June 2007.

48 In the spring of 2006, cod spawning in Smith Sound was observed, with high egg densities, from early April until mid-July. In previous years cod have left the Sound by early June. This variability likely depends on ocean conditions and is presently being studied.

49 Cod can develop blood antifreeze if they are exposed slowly to sub-zero waters. However, the Smith Sound cod were not exposed to such cold waters in winter. They avoided rather than endured the coldest waters. It was no surprise that analysis of the blood of the dead fish in 2003 showed no trace of antifreeze.

50 A report in *Nature* (April 24, 2003, news) titled "Atlantic cod meet icy death" inferred that the Smith Sound cod had done something unusual in 2003 and that the population had been decimated: "Some environmentalists are describing the mass kill as an environmental disaster, as it destroyed one of the last remnants of the region's few cod stocks." Neither was true.

51 See Ruzzante et al. (1999) and Beacham et al. (2002).

52 See Devine et al. (2006).

53 Drinkwater and Mountain (1997) reported the link between decreasing temperatures and increasing range and abundance of shrimp from the mid-1980s to early 1990s.

54 Koeller et al. (2000) reported the shrimp population was comprised of about 100 billion animals from 1996 to 1998.

55 Lilly et al. (2000, 58) argued that increases in shrimp biomass began in the early 1980s when cod stocks were still relatively strong, and stated that "if the initial increase in shrimp biomass resulted from a decrease in predation by cod, then the decrease in predation should have started in the early 1980s. This is not consistent with the trend in cod biomass." This implied an ecosystem change that favoured shrimp, beginning in the early to mid-1980s. Lilly et al. (2000) also concluded that the cod decline enabled enhanced shrimp survival in the 1990s. On the other hand, Worm and Myers (2003), using a many-population modelling approach, concluded that the predation of shrimp by cod was sufficient to explain the shrimp explosion.

56 See Lilly et al. (2000).

57 Koeller et al. (2000) reported that, for the northern shrimp, "available information suggests a decline in recruitment from 2000+, so it is uncertain if the current TAC [total allowable catch] can be sustained."

58 See Dawe et al. (2004).

59 See DFA (2007).

60 New Brunswick took about half of the total Atlantic Canadian catch, but in the same years the United States took on average about 35,000 tonnes of king crab and 20,000 tonnes of snow crab, mostly from Alaska. Japan had by far the largest landings of snow crab, averaging about 40,000 tonnes in the early 1970s, but declining thereafter (Industry, Trade and Commerce Canada 1976).

61 See DFO (1980).

62 See DFO (1980, 4).

63 See Dawe et al. (2004).

64 See DFO (2007a).

65 The ownership of these companies remains somewhat obscure in some cases.

66 See DFO (2006).

67 At the time, the Department of Fisheries and Oceans survey and population model showed few fish and a much lower population, but this result was not thought to be accurate by the FRCC. Well-known university scientists in Nova Scotia, in particular Dr. Ransom Myers, went public to deride the FRCC for advising a reopening of this fishery. However, the fish were there, and the fishery, despite several problems, in particular poor recruitment in recent years, continues to produce and support the local economy 10 years after being reopened. The population of this stock is slowly increasing and some consider it to be nearly fully rebuilt. History has proven the FRCC to have been correct.

68 The stock assessments of NAFO Subarea 3Ps cod have been highly variable as a result of the research survey. If taken at face value by the FRCC, which they were not, the recommended quota would have been much higher. Nevertheless, the stock assessment had to be considered, and gave no rebuttal to incessant demands for higher quotas.

69 See Brattey et al. (2004). In 2006, the quota was lowered from 15,000 to 12,500 tonnes.

70 Recreational and food fisheries are generally banned during fishing moratoria. The issue of whether or not such fisheries should be allowed to continue is highly controversial. Science and the FRCC have not supported them for stocks in dire straights or under moratoria, but there is a widespread demand among the general public to reinstate these fisheries, which has occurred for most coastal stocks.

71 See Dutil et al. (1999) and Dutil and Lambert (2000).

72 See Shelton et al. (2006).

73 See Wroblewski et al. (2005).

74 Fishermen in the northern Gulf of St. Lawrence received as little as 40 Canadian cents per pound (C$1 per kilogram) for fish in the summer of 2005. Fishermen in Gloucester, Massachusetts, received the equivalent of C$2.50-C$3.50 per pound in the fall of 2005 for their fish.

75 See Hamill and Stenson (2003).

76 Rose and O'Driscoll (2002) showed that the northern cod could find few capelin on which to feed in the 1990s. Sherwood et al. (2006) confirmed that this had not changed up until 2004.

77 See DFO (2002).

78 See DFO (2003).

79 In 1993, about 70% of the sea urchin roe sold to the key Japanese market came from the United States. Canada had about 20% of the market, mostly in British Columbia (Gillingham and Penney 1993).

80 See Gillingham and Penney (1993).

81 Gillingham and Penney (1993, iv) stated: "With the current resource problems in the Canadian Atlantic fishery, the harvesting of under-utilized fish species such as sea urchins, is desirable from both a business and an environmental perspective. The diversion of dependence away from traditional fish species (e.g., cod) allows for the continued survival of this resource, while at the same time causing attention to be focused on other lucrative, under-utilized fish species – an example being sea urchins."

82 Gillingham and Penney (1993) concluded that present labour-intensive methods in the sea urchin fishery were uneconomical. This has proven to be incorrect.

83 Iceland fisheries data (Statistics Iceland), http://www.statice.is.

84 There is some concern that, in the future, there may be insufficient labour to support the Newfoundland and Labrador fisheries. This has happened in Iceland, where many processors hire foreign workers (Alastair O'Reilly, personal communication).

85 See Leopold (1948).

86 The Canadian government announced in October 2006 that it would not support an international call led by Australia, New Zealand, the U.K., Norway and the U.S.A. to ban deep sea otter trawling (*The Telegram*, St. John's, Sat. Oct. 7). The influence of Canada's involvement in NAFO in this decision is all too obvious.

87 Newfoundland's prosperity in the early 1800s during the brief period when both French and American fisheries were excluded from the fishery suggests that competition may have been limiting to the fisheries. A downturn came swiftly when both foreign fleets were allowed back in from 1815 to 1818. O'Flaherty (1999, 126) pointed out that the return of an international fishery – in particular the fleets of France, with even greater rights than granted previously, and the Americans, with new rights on the south coast from Ramea to Cape Ray (the American Shore) and also on the west coast and in Labrador – coincided with increasing difficulties in the Newfoundland fishery.

88 The link between poverty and destructive fishing practices can be seen in many places in the world. Coral reefs in tropical countries are routinely dynamited by poor fishermen to kill fish, and wildlife poaching is often exacerbated if people are poor and not benefiting from the wildlife.

89 Data from International Groundfish Forum on World Product Flow in 2005 (courtesy of the Icelandic Consul General to Newfoundland and Labrador and Gus Etchegary). Icelandic comment on salt fish is a reference to their historical notion that life was salt fish (see Chapter 6).

90 See Candow (1997).

91 O'Flaherty (1999, 203-205) argued that dirty politics formed a large part of the reason why a systematic development of an independent political position for Newfoundland was so slow to develop and why Newfoundland failed as a nation. He was not shy about who was primarily responsible, stating that in the 1830s, and for many decades thereafter, "[Catholic Bishop] Fleming, his priests, and the new Irish bear much of the responsibility for the coarsening of political life." Fleming was politically ambitious, and "the Irish and their leaders would not wait to be given their rightful place in the society [which was largely made up of Protestants prior to the Irish immigration]. They would take it, and quickly." O'Flaherty also cast blame on self-serving merchants for stifling the development of democratic institutions and the state of mind in Newfoundlanders and their communities. All in all, Newfoundland politics came to be known as "dirty" and at times violent, with no coherent vision for the future or for its main industry, the fishery.

92 Fishery Products International became the principal industrial sponsor of Natural Sciences and Engineering Research Council Industrial Research Chair at the Memorial University of Newfoundland, which enabled the author to come to Memorial from the Department of Fisheries and Oceans.

93 Barry bought voting shares in Fishery Products International to help the takeover. He subsequently sold them when Fishery Products International did not perform well (CBC *Here and Now*, February 22, 2006).

94 *The Telegram* stock quotes, St. John's, December 17, 2005.

95 See Sverdrup et al. (1994) and Engas et al. (1996).

96 See Canadian Council of Professional Fish Harvesters, "The fishing and oil and gas industries: Is a peaceful co-existence possible?" Presentation by Earle McCurdy, http://www.ccpfhccpp.org/e_dbViewer.asp?cs=policy&id=34. See also Rose (2003a).

97 See Peterson et al. (2003).

98 See Carew (2000).

99 See Spies et al. (1996) and Peterson et al. (2003).

100 See Weise (2002).

[101] The Pew Charitable Trusts is a multibillion dollar fund based on the oil fortunes of Joseph N. Pew (Sun Oil Company) and his heirs that has funded several works to study the effects of marine fisheries, including the Sea Around Us Project (http://www.seaaroundus.org) which studied the North Atlantic (Pauly and Maclean 2003).

[102] See FishNet USA, "Pew, SeaWeb shrug off oil to target fishing," http://www.fishingnj.org/netusa17.htm.

[103] At its worst point, personal attacks on then DFO senior scientists Drs. Scott Parsons and Bill Doubleday by Dr. Ransom Myers of Dalhousie University, who had been a DFO modeller during the final collapse of the cod, were so intense as nearly to end in court.

[104] See Anderson and Rose (2001).

[105] The survey took several months to complete in some years as a result of vessel problems, which severely compromised the notion of a survey as a synoptic image of stock states. Fish can move considerably during a three-month time period; a slow survey may measure them several times or not at all.

[106] The Fund for Animal Welfare sent cameramen with deep pockets either to hire bogus seal hunters to commit cruel practices on the ice, or mislead legitimate and honest seal hunters as to their intentions.

[107] See May et al. (2005, 7).

[108] COSEWIC has seven categories: extinct, extirpated, endangered, threatened, special concern, data deficient, and not at risk.

[109] See Letto (1998).

[110] Upgrading historical fish volumes by modern prices may not always be justified, but serves a point. This calculation is based on a 50 million tonnes yield from 100 million tonnes landings at C$13.00 per kilogram (the price for frozen cod in St. John's stores in February 2006).

[111] Bill Barry, CBC *Fisheries Broadcast*, February 22, 2006.

[112] The provincial Department of Fisheries and Aquaculture is currently one of the smallest government departments.

[113] Maloney (1990, 99) advocated that "the Province of Newfoundland approach the Government of Canada to initiate discussions with respect to the establishment of a joint Management Board or Commission so as to ensure that the Fishery objectives of both Governments can be better coordinated and implemented for the betterment of the Newfoundland fishery as a whole."

[114] There is debate about whether specific targets should be used in rebuilding fisheries. On the one hand, targets are useful to set policy, but on the other, they may be unachievable, set up unrealistic expectations, and lock in poor policy. For example, if a target biomass of the northern cod was set as 500,000 tonnes, which is historically reasonable, before any fishery could occur, in effect no fishery could ever take place, at least not in any living person's lifetime. Such a policy would have tremendous social, economic and psychological repercussions in Newfoundland and Labrador well beyond the fishery.

[115] Beaudin (2001, 195-197) stated: "It is not true, as many people believe, that the [seafood] industry as a whole is in a crisis, even though the moratoriums are still in effect…Canada's East Coast fisheries and the fish-processing industry as a whole have indeed survived the unprecedented groundfish crisis that paralyzed those segments of the industry with the greatest value-added potential. Some may say that the expansion of the snow crab and northern shrimp fisheries contributed extensively to the unexpected turnaround…[but] that alone does not explain the renewed activity in the seafood industry…over the last ten years."

[116] Tea made on small fishing boats, literally boiled for hours.

APPENDIX

Species of Northwest Atlantic Fishes

FAMILY	COMMON NAME	SCIENTIFIC NAME	FAMILY ORIGIN
Acipenseridae	Shortnose sturgeon	*Acipenser brevirostris*	Ancient
Acipenseridae	Atlantic sturgeon	*Acipenser sturio*	Ancient
Agonidae	Alligatorfish	*Aspidophoroides monopterygius*	Pacific
Agonidae	N. Alligatorfish	*Leptagonus decagonus*	Pacific
Alepisauridae	Lancetfish	*Alepisaurus ferox*	Atlantic
Alepisauridae	Lancetfish	*Bathytroctes homopterus*	Atlantic
Alepisauridae	Lancetfish	*Xenodermichthys copei*	Atlantic
Ammodytidae	Northern sandlance	*Ammodytes hexapterus*	Ancient
Ammodytidae	American sandlance	*Ammodytes americanus*	Ancient
Anarhichadidae	Broadhead wolffish	*Anarhichas denticulatus*	Atlantic
Anarhichadidae	Striped wolffish	*Anarhichas lupus*	Atlantic
Anarhichadidae	Spotted wolffish	*Anarhichas minor*	Atlantic
Anguillidae	American eel	*Anguilla rostrata*	Ancient
Bothidae	Summer flounder	*Paralichthys dentatus*	Ancient
Bothidae	4spot flounder	*Paralichthys oblongus*	Ancient
Bothidae	Windowpane	*Rhombus aquosus*	Ancient
Clupeidae	Alewife	*Alosa pseudoharengus*	Atlantic
Clupeidae	Shad	*Alosa sapidissima*	Atlantic
Clupeidae	Herring	*Clupea harengus*	Atlantic
Cottidae	Hookeared sculpin	*Artediellus atlanticus*	Pacific
Cottidae	Arctic sculpin	*Cottunculus microps*	Pacific
Cottidae	Staghorn sculpin	*Gymnocanthus tricuspis*	Pacific
Cottidae	Sea raven	*Hemitripterus americanus*	Pacific
Cottidae	Grubby sculpin	*Myoxocephalus aeneus*	Pacific
Cottidae	Longhorn sculpin	*Myoxocephalus octodedemspinosus*	Pacific
Cottidae	Seascorpion	*Myoxocephalus scorpius*	Pacific
Cottidae	Mailed sculpin	*Troglops nybelini*	Pacific
Cottidae	Ribbed sculpin	*Troglops pingelii*	Pacific
Cryptacanthodidae	Wrymouth	*Gryptacanthodes maculatus*	Pacific
Cyclopteridae	Sea tadpole	*Careproctus longipennis*	Pacific
Cyclopteridae	Sea tadpole	*Careproctus ranulus*	Pacific
Cyclopteridae	Sea tadpole	*Careproctus reinhardi*	Pacific
Cyclopteridae	Lumpfish	*Cyclopterus lumpus*	Pacific
Cyclopteridae	Atl. Spiny lumpsucker	*Eumicrotremus spinosus*	Pacific
Cyclopteridae	NF Spiny lumpsucker	*Eumicrotremus terraenovae*	Pacific
Gadidae	Arctic cod	*Boreogadus saida*	Atlantic
Gadidae	Cusk	*Brosme brosme*	Atlantic
Gadidae	4beard rockling	*Enchelyopus cimbrius*	Atlantic
Gadidae	Longfinned hake	*Ficus chesteri*	Atlantic
Gadidae	Atlantic cod	*Gadus morhua*	Atlantic
Gadidae	Greenland cod	*Gadus ogac*	Atlantic
Gadidae	3beard rockling	*Gaidropsarus ensis*	Atlantic

FAMILY	COMMON NAME	SCIENTIFIC NAME	FAMILY ORIGIN
Gadidae	Haddock	*Melanogrammus aeglefinus*	Atlantic
Gadidae	Silver hake	*Merluccius bilinearis*	Atlantic
Gadidae	Atlantic tomcod	*Microgadus tomcod*	Atlantic
Gadidae	Hake	*Micromesistius poutassou*	Atlantic
Gadidae	European ling	*Molva molva*	Atlantic
Gadidae	Pollack	*Pollachius virens*	Atlantic
Gadidae	Squirrel hake	*Urophycis chuss*	Atlantic
Gadidae	Spotted hake	*Urophycis regius*	Atlantic
Gadidae	Mud hake	*Urophycis tenuis*	Atlantic
Gasterosteidae	Bloody stickleback	*Apeltes quadracus*	Ancient
Gasterosteidae	3spine stickleback	*Gasterosteus aculaeatus*	Ancient
Gasterosteidae	Stickleback	*Gasterosteus wheatlandi*	Ancient
Gasterosteidae	9spine stickleback	*Pungitius pungitius*	Ancient
Labridae	Cunner	*Tautogolabrus adsperus*	Tropical
Liparidae	Seasnail	*Liparis atlanticus*	Pacific
Liparidae	Northern seasnail	*Liparis liparis*	Pacific
Liparidae	Liparid	*Paraliparis copei*	Pacific
Lophiidae	Monkfish	*Lophius americanus*	Ancient
Macrouridae	Grenadier	*Chalinura murrayi*	Ancient
Macrouridae	Straptailed grenadier	*Chalinura occidentalis*	Ancient
Macrouridae	Longnose grenadier	*Coelorhynchus carminatus*	Ancient
Macrouridae	Rock grenadier	*Coryphaenoides rupestris*	Ancient
Macrouridae	Marlin spike	*Macrourus bairdii*	Ancient
Macrouridae	Rough head grenadier	*Macrourus berglax*	Ancient
Macrouridae	Grenadier	*Macrourus holotrachys*	Ancient
Myctophidae	Glacier lanternfish	*Benthosema glaciale*	Ancient
Myctophidae	Lanternfish	*Ceratoscopelus madeirensis*	Ancient
Myctophidae	Headlight lanternfish	*Diaphus effulgens*	Ancient
Myctophidae	Lanternfish	*Gonichthys cocco*	Ancient
Myctophidae	Lanternfish	*Lampadena braueri*	Ancient
Myctophidae	Jewel lanternfish	*Lampanyctus crocodilus*	Ancient
Myctophidae	Lanternfish	*Myctophum affine*	Ancient
Myctophidae	Lanternfish	*Myctophum humboltii*	Ancient
Myctophidae	Common lanternfish	*Myctophum punctatum*	Ancient
Myctophidae	Lanternfish	*Notoscopelus castaneus*	Ancient
Myctophidae	Pearly lanternfish	*Notoscopelus margaritifer*	Ancient
Osmeridae	Capelin	*Mallotus villosus*	Pacific
Osmeridae	Rainbow smelt	*Osmerus mordax*	Pacific
Paralepididae	Barracudina	*Paralepis brevis*	Atlantic
Paralepididae	Barracudina	*Paralepis rissoi*	Atlantic
Pholidae	Blenny	*Pholis gunnellus*	Pacific
Pleuronectidae	Witch flounder	*Glyptocephalus cynoglossus*	Pacific
Pleuronectidae	Atlantic halibut	*Hippoglossoides hippoglossoides*	Pacific
Pleuronectidae	American plaice	*Hippoglossoides platesoides*	Pacific

FAMILY	COMMON NAME	SCIENTIFIC NAME	FAMILY ORIGIN
Pleuronectidae	Yellowtail flounder	*Limanda ferruginea*	Pacific
Pleuronectidae	Smooth flounder	*Liopsetta putnami*	Pacific
Pleuronectidae	Winter flounder	*Pseudopleuronectes americanus*	Pacific
Pleuronectidae	Greenland halibut	*Reinhardtius hippoglossoides*	Pacific
Rajidae	Little skate	*Raja erinacia*	Ancient
Rajidae	Skate	*Raja fyllae*	Ancient
Rajidae	Jensen's skate	*Raja jenseni*	Ancient
Rajidae	Barndoor skate	*Raja laevis*	Ancient
Rajidae	Skate	*Raja mollis*	Ancient
Rajidae	Eyed skate	*Raja ocellata*	Ancient
Rajidae	Thorny skate	*Raja radiata*	Ancient
Rajidae	Smooth skate	*Raja senta*	Ancient
Rajidae	Spinytailed skate	*Raja spinicauda*	Ancient
Salmonidae	Atlantic salmon	*Salmo salar*	Pacific
Salmonidae	Arctic charr	*Salvelinus alpinus*	Pacific
Salmonidae	Brook charr	*Salvelinus fontinalis*	Pacific
Scomberesocidae	Atlantic saury	*Scomberesox saurus*	Ancient
Scombridae	Frigate mackerel	*Auxis thazard*	Ancient
Scombridae	Atlantic mackerel	*Scomber scombrus*	Ancient
Scombridae	Bluefin tuna	*Thunnus thynnus*	Ancient
Scorpaenidae	Redfish	*Sebastes fasciatus*	Pacific
Scorpaenidae	Rosefish	*Sebastes marinus*	Pacific
Scorpaenidae	Redfish	*Sebastes mentella*	Pacific
Stichaeidae	Blenny	*Chirolophis ascanii*	Atlantic
Stichaeidae	Fourlined snakeblenny	*Eumesogrammus praecisus*	Atlantic
Stichaeidae	Shanny	*Leptoclinus maculatus*	Atlantic
Stichaeidae	Greenland blenny	*Lumpenus fabricii*	Atlantic
Stichaeidae	Snakeblenny	*Lumpenus lampretaeformis*	Atlantic
Stichaeidae	Arctic shanny	*Stichaeus punctatus*	Atlantic
Stichaeidae	Radiated shanny	*Ulvaria subbifurcata*	Atlantic
Zoarcidae	Unernak	*Gymnelis viridis*	Pacific
Zoarcidae	Eelpout	*Lycenchelys sarsii*	Pacific
Zoarcidae	Jimmy Durante eelpout	*Lycenchelys verrillii*	Pacific
Zoarcidae	Eelpout	*Lycodes agnostus*	Pacific
Zoarcidae	Eelpout	*Lycodes atratus*	Pacific
Zoarcidae	Esmark's eelpout	*Lycodes esmarki*	Pacific
Zoarcidae	Eelpout	*Lycodes frigidus*	Pacific
Zoarcidae	Eelpout	*Lycodes lavalaei*	Pacific
Zoarcidae	Reticulated eelpout	*Lycodes reticulatus*	Pacific
Zoarcidae	Newfoundland eelpout	*Lycodes terraenovae*	Pacific
Zoarcidae	Polar eelpout	*Lycodes turneri*	Pacific
Zoarcidae	Eelpout	*Lycodes vachonii*	Pacific
Zoarcidae	Vahl's eelpout	*Lycodes vahlii*	Pacific
Zoarcidae	Ocean pout	*Macrozoarces americanus*	Pacific
Zoarcidae	Eelpout	*Melanostigma gelatinosum*	Pacific

REFERENCES

Abele, L.G. 1982. Biogeography. *In* The biology of crustacea: systematics, the fossil record, and biogeography. *Edited by* L.G. Abele. Academic Press, New York, pp. 242-304.

Abreu-Ferreira, D. 1995. The cod trade in early-modern Portugal: deregulation, English domination, and the decline of female cod merchants. PhD dissertation, Memorial University of Newfoundland, Department of History.

Alexander, D. 1976. The political economy of fishing in Newfoundland. Journal of Canadian Studies 11: 32-40.

Alexander, D. 1977. The decay of trade: an economic history of the Newfoundland saltfish trade, 1935-1965. Institute of Social and Economic Research, Memorial University of Newfoundland, 173 pp.

Altringham, J.D. and R.E. Shadwick. 2001. Swimming and muscle function. In Tuna physiology, ecology, and evolution. *Edited by* B.A. Block and E.D. Stevens. Academic Press, San Diego, pp. 314-341.

Amulree Royal Commission. 1933. Newfoundland royal commission report, 1933, presented by the Secretary of State for Dominion Affairs to Parliament by command of His Majesty, November 1933. His Majesty's Stationary Office, London.

Anderson, E.D. 1998. The history of fisheries management and scientific advice – The ICNAF/NAFO history from the end of World War II to the present. Journal of Northwest Atlantic Fisheries Science 23: 75-94.

Anderson, J.T. and Dalley, E.L. 2000. Interannual differences in hatching times and growth rates of pelagic juvenile cod in Newfoundland waters. Fisheries Research 46:227-238.

Anderson, J.T. and R.S. Gregory. 2000. Factors regulating survival of northern cod (NAFO 2J3KL) during their first three years of life. ICES Journal of Marine Science 57: 349-359.

Anderson, J.T. and G.A. Rose. 2001. Offshore spawning and year-class strength of northern cod (2J3KL) during the fishing moratorium, 1994-1996. Canadian Journal of Fisheries and Aquatic Sciences 58: 1386-1394.

Anderson, R.N. 1986. Marine geology: a planet earth perspective. John Wiley & Sons, New York, 328 pp.

Andrews, J.T. 1987. The late Wisconsin glaciation and deglaciation of the Laurentide ice sheet. *In* North America and adjacent oceans during the last glaciation. *Edited by* W.F. Ruddiman and H.E. Wright, Jr. Geological Society of America. Boulder, Co., pp.13-38.

Anglin, D.G. 1970. The St. Pierre and Miquelon affaire of 1941. University of Toronto Press, Toronto, 9 pp.

Anspach, L.A.1819. A history of the Island of Newfoundland, containing a description of the island, the banks, the fisheries and trade of Newfoundland, and the coast of Labrador. Allman, Richardson, London, 512 pp.

Appleton, L. 1891. The foreign policy of Europe. Simpkin, Marshall, Hamilton, Kent, London, 175 pp.

Arnason, R. 1995. The Icelandic fisheries. Fishing News Books, Oxford, UK, 177 pp.

Astthorsson, O.S. and A. Gislason. 1999. Inter-annual variation in abundance and development of *Calanus finmarchicus* in Flaxaflói, West Iceland. Rit Fiskideildar 16: 131-140.

Astthorsson, O.S. and H. Vilhjálmsson. 2002. Iceland shelf LME: Decadal assessment and resource sustainability. *In* Large marine ecosystems of the North Atlantic. *Edited by* K. Sherman and H.R. Skoldal. Elsevier, Amsterdam, pp. 219-243.

Atkinson, D.B., G.A. Rose, E.F. Murphy and C.A. Bishop. 1997. Distribution changes and abundance of northern cod (*Gadus mohua*), 1981-1993. Canadian Journal of Fisheries and Aquatic Sciences 54:132-138.

Backus, R.H., J.E. Craddock, R.L. Haedrich, and B.H. Robison. 1997. Atlantic mesopelagic zoogeography. *In* Fishes of the Western North Atlantic. Sears Foundation for Marine Research, Yale University, New Haven, CT, pp. 266-287.

Bailey, R. 1982. The Atlantic snow crab. Communications Branch, Department of Fisheries and Oceans, Ottawa.

Baker, M. and S. Ryan. 1997. The Newfoundland fishery research commission, 1930-1934. *In* How deep is the ocean? Essays on Canada's Atlantic fisheries. *Edited by* J.E. Candow and C. Corbin. University College of Cape Breton Press, Sydney, NS, pp. 161-173.

Baldwin, J.D., A.L. Bass, B.W. Bowen, and W.H. Clark. 1998. Molecular phylogeny and biogeography of the marine shrimp *Penaeus*. Molecular Phylogenetics and Evolution 10: 399-407.

Bannister, J. 1997. Reform era, 1815-1832. Newfoundland and Labrador Heritage, 1-5. Available: http://www.heritage.nf.ca/law/reform.html.

Bannister, J. 2003. The rule of admirals: law, custom and naval government in Newfoundland, 1699-1832. University of Toronto Press, Toronto, 423 pp.

Barry, R.G. 1989. The present climate of the Arctic Ocean and possible past and future states. *In* The Arctic seas: climatology, oceanography, geology, and biology. *Edited by* Y. Herman. Van Nostrand Reinhold, New York, pp. 1-46.

Barsukov, V.V. 1959. The wolffish (Anarhichadidae). Novaya Seriya No. 73, Vol. 5., Smithsonian Institution and the National Science Foundation, Washington, DC.

Bates, S. 1944. The report on the Canadian Atlantic Sea-Fishery. Royal Commission on Reconstruction and Rehabilitation. Government of Nova Scotia, Halifax, NS.

Beacham, T.D., J. Brattey, K.M. Miller, K.D. Le, and R.E. Withler. 2002. Multiple stock structure of Atlantic cod (*Gadus morhua*) off Newfoundland and Labrador determined from genetic variation. ICES Journal of Marine Science 59: 650-665.

Beaudin, M. 2001. Towards greater value: enhancing Eastern Canada's seafood industry. Canadian Institute for Research on Regional Development: Moncton, N.B.

Bell, M.V. and J.R. Sargent. 1996. Lipid nutrition and fish recruitment. Marine Ecology Progress Series 134: 315-316.

Bergstrom, B. I. 2000. The biology of Pandalus. In Advances in Marine Biology, Vol. 38. *Edited by* A.J. Southward, P.A. Tyler, C.M. Young, and L.A. Fuiman. Academic Press, San Diego, pp. 55-256.

Bertin, L. 1956. Eels: A biological study. Cleaver-Hume Press, London, 192 pp.

Beverton, R.J.H. 1965. Catch/effort assessment in some ICNAF fisheries. ICNAF Research Bulletin 2: 59-72.

Beverton, R.J.H. and S.J. Holt. 1957. On the dynamics of exploited fish populations. Fishery investigations (Great Britain Ministry of Agriculture, Fisheries, and Food), Series 2, Vol. 19. Her Majesty's Stationary Office, London.

Bishop, C.A., E.F. Murphy, M.B. Davis, J.W. Baird, and G.A. Rose. 1993. An assessment of the cod stock in NAFO divisions 2J+3KL. Northwest Atlantic Fisheries Organization Scientific Council Research Document 93/86.

Bishop, C.A. and P.A. Shelton. 1997. A narrative of NAFO 2J3KL cod assessments from extension of jurisdiction to moratorium. Canadian Technical Report of Fisheries and Aquatic Sciences No. 2199.

Blackwood, G. 1983. Lumpfish Roe Fishery Development in Newfoundland, 1982-83. Newfoundland Department of Fisheries, St. John's, report No. 31.

Blackwood, G. 1996. Past and future goals and objectives in the allocation of the northern cod resource. MSc dissertation, Memorial University of Newfoundland, Department of Geography.

Blake, R.B. 1997. The international fishery off Canada's east coast in the 20th century. *In* How deep is the ocean? Historical essays on Canada's Atlantic fisheries. *Edited by* J.E. Candow and C. Corbin. University College of Cape Breton Press, Sydney, NS, pp. 207-221.

Boag, D. and M. Alexander. 1986. The Atlantic puffin. Blandford Press, New York, 129 pp.

Bowering, W.R. 1983. Age, growth, and sexual maturity of Greenland halibut, *Reinhardtius hippoglossoides* (Walbaum), in the Canadian northwest Atlantic. Fishery Bulletin 81: 599-611.

Bowering, W.R. 1989. Witch flounder distribution off southern Newfoundland, and changes in age, growth and sexual maturity patterns with commercial exploitation. Transactions of the American Fisheries Society 6: 659-669.

Bowering, W.R. 1990. Spawning of witch flounder (*Glyptocephalus cynoglossus*) in the Newfoundland-Labrador area of the northwest Atlantic as a function of depth and water temperature. Fisheries Research 9: 23-29.

Bowering, W.R., M.J. Morgan, and W.B. Brodie. 1997. Changes in the population of American Plaice (*Hippoglossoides platessoides*) off Labrador and northeastern Newfoundland: A collapsing stock with low exploitation. Fisheries Research 30: 199-216.

Bowman, T.E. and L.G. Abele. 1982. Classification of the recent crustacea. *In* The biology of the crustacea: systematics, the fossil record, and biogeography. *Edited by* L.G. Abele. Academic Press, New York, pp. 1-27.

Boyd, C. 1997. Come on all the crowd, on the beach: The working lives of beachwomen in Grand Bank, Newfoundland, 1900-1940. *In* How deep in the ocean? Historical essays on Canada's Atlantic fishery. *Edited by* J.E. Candow and C. Corbin. University of Cape Breton Press, Sydney, NS, pp. 175-184.

Brattey, J., N.G. Cadigan, B.P. Healey, G.R. Lilly, E.F. Murphy, P.A. Shelton, and J.-C. Mahé. 2004. Assessment of the cod (*Gadus morhua*) stock in NAFO Subdivision 3Ps in October 2004. Department of Fisheries and Oceans Science Advisory Secretariat Report No. 2004/083.

Brawn, V.M. 1961. Aggressive behaviour in the cod (*Gadus callarias* L.). Behaviour 18: 107-143.

Breummer, F. 1971. St. Pierre and Miquelon. Canadian Geographical Journal 83: 138.

Brière, J.-F. 1997. The French fishery in North America in the 18th century. *In* How deep is the ocean? Historical essays on Canada's Atlantic fisheries. *Edited by* J.E. Candow and C. Corbin. University College of Cape Breton Press, Sydney, NS, pp. 47-60.

Briggs, J.C. 1974. Marine zoogeography. McGraw-Hill Book Company, New York, 475 pp.

Briggs, J.C. 2003. Marine centres of origin as evolutionary engines. Journal of Biogeography 30:1-18.

Briggs, J.C. 2006. Proximate sources of marine biodiversity. Journal of Biogeography 18:595-622.

Brown, C. 1972. Death on the ice: The great Newfoundland sealing disaster of 1914. Doubleday Canada, Toronto, 217 pp.

Buckmann, A., G. Dietrich, A. Meyer, and A. von Brandt. 1959. German research report, E. Fishing activities. IV. ICNAF Annual Proceedings for the Year 1958-59, Vol. 9. Dartmouth, NS.

Bundy, A. 2001. Fishing on ecosystems: The interplay of fishing and predation in Newfoundland-Labrador. Canadian Journal of Fisheries and Aquatic Sciences 58: 1153-1167.

Bundy.A., G.R. Lilly, and P.A. Shelton. 2000. A mass balance model of the Newfoundland-Labrador shelf. Canadian Technical Report of Fisheries and Aquatic Sciences No. 2310.

Burger, A.E. and M. Simpson. 1986. Diving depths of Atlantic puffins and common murres. The Auk 103: 828-830.

Cairns, D.K., K. Bredin, and W.A. Montevecchi. 1987. Activity budgets and foraging ranges of breeding common murres. The Auk 104: 218-224.

Candow, J.E. 1997. Recurring visitations of pauperism: change and continuity in the Newfoundland fishery. *In* How deep is the ocean? Historical essays on Canada's Atlantic fisheries. *Edited by* J.E. Candow and C. Corbin. University College of Cape Breton Press, Sydney, NS, pp. 139-160.

Carew, A.M.E. 2000. Oil pollution and the Newfoundland and Labrador fishery: Current and potential threats for the conservation of commercial fisheries resources in Placentia Bay. MMS major project, Memorial University of Newfoundland.

Carlson, C.C. 1996. The (in)significance of Atlantic salmon. Federal Archaeology VIII, No. 3/4: 22-30.

Carr, S.M., D.S. Kivlichan, P. Pepin, and D.C Crutcher. 1999. Molecular systematics of gadid fishes: Implications for the biogeographic origins of Pacific species. Canadian Journal of Zoology 77: 19-26.

Carscadden, J.E., K.T. Frank, and D.S. Miller. 1989. Capelin (*Mallotus villosus*) spawning on the southeast shoal: influence of physical factors past and present. Canadian Journal of Fisheries and Aquatic Sciences 46: 1743-1754.

Carscadden, J.E., K.T. Frank, and W.C. Leggett. 2001. Ecosystem changes and the effects on capelin (*Mallotus villosus*), a major forage species. Canadian Journal of Fisheries and Aquatic Sciences 58: 73-85.

Carter, R. 1982. Something's fishy: policy and private corporations in the Newfoundland fishery. St. John's Oxfam Committee.

Castonguay, M., G.A. Rose, and W.C. Leggett. 1992. Onshore movements of Atlantic mackerel (*Scomber scombrus*) in the northern Gulf of St Lawrence: associations with wind-forced advections of warmed surface waters. Canadian Journal of Fisheries and Aquatic Sciences 11: 2232-2241.

Cell, G.T. 1969. English enterprise in Newfoundland, 1577-1660. University of Toronto Press, Toronto, 181 pp.

Cell, G.T. 1982. Newfoundland discovered: English attempt at colonization, 1610-1630. Hakuyt Society, London, 310 pp.

Chadwick, J. 1967. Newfoundland: Island into Province. Cambridge University Press, London, 298 pp.

Cisne, J.L. 1982. Origin of the crustacea. *In* The biology of crustacea: systematics, the fossil record, and biogeography. *Edited by* L.G. Abele. Academic Press, New York, pp. 65-92.

Cleghorn, J. 1854. On the fluctuations in the herring fisheries. British Association for Advancement in Science 24: 124.

Cochrane, J.A. 1938. The story of Newfoundland. Ginn and Company, Boston, 256 pp.

Cohen, D.M., T. Inada, T. Iwamoto, and N. Scialabba. 1990. Gadiform fishes of the world (Order Gadiformes). An annotated and illustrated catalogue of cods, hakes, grenadiers, and other gadiform fishes known to date. FAO Species Catalogue, Vol. 10: 125.

Colbert, E.H. and M. Morales. 1991. Evolution of the vertebrates: a history of backboned animals through time. Wiley-Liss, New York, 470 pp.

Colbourne, E., B. deYoung, S. Narayanan, and J.A. Helbig. 1997. Comparison of hydrography and circulation on the Newfoundland shelf during 1990-1993 with the long-term mean. Canadian Journal of Fisheries and Aquatic Sciences 54: 68-80.

Collette, B.B., C. Reeb, and B.A. Block. 2001. Systematics of the tunas and mackerels (Scombridae). *In* Tuna physiology, ecology, and evolution. *Edited by* B.A. Block and E.D. Stevens. Academic Press, San Diego, pp. 5-30.

Collins, M.A.J. 1976. The lumpfish (*Cyclopterus lumpus L.*) in Newfoundland waters. The Canadian Field-Naturalist 90: 64-67.

Colonial Office. 1886. Minutes of June 7, 1886. St. John's.

Colonial Office. 1701. Record 194/2. 46xii, 175-178. St. John's.

Conan, G.Y., M. Starr, M. Comeau, J.-C. Therriault, F.X.M. Hernandez, and G. Robichaud. 1996. Life history strategies, recruitment fluctuations, and management of the Bonne Bay fjord Atlantic snow crab (*Chionoecetes opilio*). In High latitude crabs: biology, management and economics, Alaska Sea Grant College Program report No. 96-02. University of Alaska, Fairbanks, AK.

Cooper, R.A. and J.R. Uzmann. 1980. Ecology of juvenile and adult Homarus. *In* The biology and management of lobsters, Vol. 2. *Edited by* J.S. Cobb and B.F. Phillips. Academic Press, New York, pp. 97-139.

Copes, P. 1980. The evolution of marine fisheries policy in Canada. *In* Resource policy: international perspectives. *Edited by* P.N. Nemetz. Institute for Research on Public Policy, Montreal, pp. 125-148.

Cormack, W.E. 1928. Narrative of a Journey across the Island of Newfoundland in 1822. *Edited by* F.A. Bruton. Longmans, Green and Co., London, 138 pp (originally published by Edinburgh Philosophical Society, 1824, and reprinted in St. John's, 1856).

Corten, A. 2002. The role of "conservatism" in herring migrations. Reviews in Fish Biology and Fisheries 11: 339-361.

Cox, P. and M. Anderson. 1924. A study of the lumpfish (*Cyclopterus lumpus L.*). Contributions to Canadian Biology 1: 1-20.

Craig, P.C., W.B. Griffiths, L. Haldorson, and H. McElderry. 1982. Ecological studies of Arctic cod (*Boreogadus saida*) in Beaufort Sea coastal water, Alaska. Canadian Journal of Fisheries and Aquatic Sciences 39: 395-406.

Crawford, R.E., C. Hudon, and D.G. Parsons. 1992. An acoustic study of shrimp (*Pandalus montagui*) distribution near Resolution Island (eastern Hudson Strait). Canadian Journal of Fisheries and Aquatic Sciences 49: 842-856.

Crosbie, J.C. 1997. No holds barred: my life in politics. McClelland and Stewart, Toronto, 505 pp.

Crutchfield, J.A. 1961. The role of the fisheries in the Canadian economy. Government of Canada pamphlet.

Cuff, R.H. 1997. New-founde-land. Harry Cuff Publications, St. John's, 175 pp.

Cumming, W.P., R.A. Skelton, and D.B. Quinn. 1971. The discovery of North America. McClelland and Stewart, Toronto, 80 pp.

Cushing, D. 1995. Population production and regulation in the sea: a fisheries perspective. Cambridge University Press, London, 354 pp.

Cushing, D.H. 1988. The cod fishery off Newfoundland. *In* The provident sea. Cambridge University Press, London, pp. 53-76.

Dalley, E.L. and J.T. Anderson. 1997. Age-dependent distribution of demersal juvenile Atlantic cod (*Gadus morhua*). Canadian Journal of Fisheries and Aquatic Sciences 54: 168-176.

Dawe, G.M. 1977. The northern cod (*Gadus morhua*). Fisheries and Environment Canada, St. John's, 82 pp.

Dawe, H.A. 1963. Report and recommendations of the Newfoundland Fisheries Commission to the Government of Newfoundland, April 1963, St. John's,123 pp.

Dawe, E.G., D. Orr, D.G. Parsons, D. Stansbury, D.M. Taylor, H.J. Drew, P.J. Veitch, P.G. O'Keefe, E. Seward, D. Ings, A. Pardy, K. Skanes and P.C. Beck. 2004. An assessment of Newfoundland and Labrador snow crab in 2003. Department of Fisheries and Oceans Stock Assessment Secretariat Research Document No. 2004/024.

de Gómara, F.L. 1552. Historia general de las Indias. Seville, Spain.

de L. Brooke, M. 2002. Seabird systematics and distribution: A review of current knowledge. *In* Biology of marine birds. *Edited by* E.A. Schreiber and J. Burger. CRC Press, Washington, pp. 57-86.

de la Morandiere, C. 1962. Histoire de la pêche Francais de la morue dans l'Amerique Sepentrionale. Maisonneuve et Larose, Paris, 1397 pp.

DeRoos, A.M. and L. Persson. 2002. Size-dependent life-history traits promote catastrophic collapses of top predators. Proceedings of the National Academy of Sciences 99: 12907-12912.

DeBlois, E.M. and G.A. Rose. 1996. Cross-shoal variability in the feeding habits of migrating Atlantic cod (*Gadus morhua*). Oecologia 108: 192-196.

Dempson, J.B. 1984. Identification of anadromous Arctic charr stocks in coastal areas of northern Labrador. *In* Biology of the Arctic charr: Proceedings of the international symposium on Arctic charr. *Edited by* L. Johnson and B. L. Burns. University of Manitoba Press, Winnipeg, pp. 143-162.

Dempson, J.B., M. Shears, and M. Bloom. 2002. Spatial and temporal variability in the diet of anadromous Arctic charr, *Salvelinus alpinus*, in northern Labrador. Environmental Biology of Fishes 64: 49-62.

Department of Biology, Memorial University of Newfoundland. 1978. Submission to the workshop on fisheries and related matters. St. John's.

Department of Environment Canada. 1976. Policy for Canada's commercial fisheries, May 1976. Ottawa.

Department of Fisheries and Aquaculture. 1986. Department of Fisheries annual report, 1985/86. St. John's.

Department of Fisheries and Aquaculture. 2007. Fishery. Economics, Research and Analysis Division. http:\\www.economics.gov.nl.ca.

Department of Fisheries and Oceans. 1980. Northern cod: a fisheries success story. Ottawa.

Department of Fisheries and Oceans. 1982. Proposed management strategies for the capelin fisheries of the Newfoundland region. St. John's.

Department of Fisheries and Oceans. 1983. Canadian Atlantic groundfish plan announced. News release NR-HQ-083-096E. Ottawa.

Department of Fisheries and Oceans. 1985. Atlantic fisheries development 1984-85. Ottawa.

Department of Fisheries and Oceans. 1986. Report on the 1986 capelin fishery. St. John's.

Department of Fisheries and Oceans. 1990. Atlantic groundfish multi-year management plan (1991-1993). Ottawa.

Department of Fisheries and Oceans. 1991. Atlantic Canada groundfish management plan. Ottawa.

Department of Fisheries and Oceans. 2002. Newfoundland and Labrador Atlantic salmon stock status for 2001.

Report D2-01. Ottawa.

Department of Fisheries and Oceans. 2003. Newfoundland and Labrador Atlantic salmon stock status update. Canadian Science Advisory Secretariat Science Advisory Report No. 2003/048.

Department of Fisheries and Oceans. 2006. Assessment of division 0B-3K northern shrimp. Canadian Science Advisory Secretariat Science Advisory Report No. 2006/007.

Department of Fisheries and Oceans. 2007a. Assessment of Newfoundland and Labrador snow crab. Canadian Science Advisory Secretariat Science Advisory Report No. 2007/008.

Department of Fisheries and Oceans. 2007b. Stock assessment of northern (2J3KL) cod in 2007. Canadian Science Advisory Secretariat Science Advisory Report No. 2007/018.

Department of Industry, Trade and Commerce. 1976. Crab review. Agriculture, Fisheries and Food Products Branch, Ottawa.

Desbarats, G.J. 1918. Deputy minister's report. Annual report of the Fisheries Branch, Department of Naval Service. Sessional paper 39, 1-45. Ottawa.

Desbarats, G.J. 1920. Deputy minister's report. Annual report of the Fisheries Branch, Department of Naval Service. Sessional paper 39, 7-26. Ottawa.

Devine, J.A., K.D. Baker, and R.L. Haedrich. 2006. Deep-sea fishes qualify as endangered. Nature 439: 29.

deYoung, B. and G.A. Rose. 1993. On recruitment and distribution of Atlantic cod (*Gadus morhua*) off Newfoundland. Canadian Journal of Fisheries and Aquatic Sciences 50: 2729-2741.

Diamond, J.M. 1975. Assembly of species communities. *In* Ecology and evolution of communities. *Edited by* M.L. Cody and J.M. Diamond. Harvard University Press, Cambridge, MA, pp. 342-444.

Dickinson, A.B. and C.W. Sanger. 2005. Twentieth-century shore-station whaling in Newfoundland and Labrador. McGill-Queen's University Press, Montreal, 254 pp.

Dickson, R.R. and W.R. Turrell. 2000. The NAO: the dominant atmospheric process affecting oceanic variability in home, middle and distant waters of European Atlantic salmon. *In* The ocean life of the Atlantic salmon. *Edited by* D. Mills. Fishing News Books, Oxford, pp. 92-115.

Doel, P.A. 1992. Port o' call: Memories of the Portuguese white fleet in St. John's, Newfoundland. Social and Economic Studies, No. 49. Institute of Social and Economic Research, Memorial University, 204 pp.

Doucet, F.J. 1979. The utilization of northern cod. Report prepared for Fisheries and Oceans Canada. Toledo, Ontario, 42 pp.

Dragesund, O., J. Hamre, and Ø. Ulltang. 1980. Biology and population dynamics of the Norwegian spring-spawning herring. Journal du Conseil Internationale pour l'exploration de la Mer 177: 43-71.

Drinkwater, K. 2000. Changes in ocean climate and its general effect on fisheries: examples from the north-west Atlantic. *In* The ocean life of the Atlantic salmon. *Edited by* D. Mills. Fishing News Books, Oxford, pp. 116-136.

Drinkwater, K. 2002. A review of the role of climate variability in the decline of northern cod. American Fisheries Society Symposium 32: 113-130.

Drinkwater, K. 2005. The response of Atlantic cod (*Gadus morhua*) to future climate change. ICES Journal of Marine Science 62: 1327-1337.

Drinkwater, K.G. and D.G. Mountain. 1997. Climate and oceanography. *In* Northwest Atlantic groundfish: perspectives on a fishery collapse. *Edited by* J. Boreman, B.S. Nakashima, J.A. Wilson, and R.L. Kendall. American Fisheries Society, Bethesda, MD, pp. 3-25.

Dudnik, Y.I., V.K. Zilanov, V.D. Kudrin, V.A. Nesvetov, and A.S. Nesterov. 1981. Distribution and biology of the Atlantic saury, *Scomberesox saurus* (Walbaum), in the northwest Atlantic. NAFO Science Council Studies, No. 1: 23-29.

Dunfield, R.W. 1985. The Atlantic salmon in the history of North America. Canadian Special Publication of Fisheries and Aquatic Sciences No. 80.

Dutil, J.D., M. Castonguay, D. Gilbert, D. Gascon. 1999. Growth, condition, and environmental relationships in Atlantic cod (*Gadus morhua*) in the northern Gulf of St. Lawrence and implications for management strategies in the Northwest Atlantic. Canadian Journal of Fisheries and Aquatic Sciences 56:1818-1831.

Dutil, J.D. and Y. Lambert. 2000. Natural mortality from poor condition in Atlantic cod (Gadus morhua). Canadian Journal of Fisheries and Aquatic Sciences 57:826-836.

Earll, R.E. 1887. The herring fishery and the sardine industry. *In* The fisheries and fishing industries of the United States. United States Commission of Fish and Fisheries, Washington, pp. 417-526.

Eastman, J.T. and L. Grande. 1991. Late Eocene gadiform (Teleostei) skull from Seymour Island, Antarctic peninsula. Antarctic Science 3: 87-95.

Edwards, M., P. Licandro, A.W.G. John, and D.G. Johns, D.G. 2005. Ecological status report: Results from the CPR survey 2003/2004. SAHFOS Technical Report, No. 2: 1-6.

Ehrenberg, R.E. 2006. Mapping of the world: an illustrated history of cartography. *Edited by* R.E. Ehrenberg. National Geographic Society, Washington, 256 pp.

Einarsson, H. 1945. Euphausiacea I. Northern Atlantic species. Bianco Luno, Copenhagen, 191 pp.

Ekman, S. 1953. Zoogeography of the sea. Sidgwick and Jackson, London, 417 pp.

Elliott, C.B. 1887. The United States and the northeastern fisheries. University of Minnesota, Minneapolis, MN.

Emery, K.O. and E. Uchupi. 1984. The geology of the Atlantic ocean. Springer-Verlag, New York, 1050 pp.

Engås, A., S. Lokkeborg, E. Ona, and A.V. Soldal. 1996. Effects of seismic shooting on local abundance and catch rates of cod (*Gadus morhua*) and haddock (*Melanogrammus aeglefinus*). Canadian Journal of Fisheries and Aquatic Sciences 53: 2238-2249.

Ennis, G.P. 1973. Food, feeding, and condition of lobsters, *Homarus americanus*, throughout the seasonal cycle in Bonavista Bay, Newfoundland. Journal of the Fisheries Research Board of Canada 30: 1905-1909.

Ennis, G.P. 1983. Observations on the behavior and activity of lobsters, *Homarus americanus*, in nature. Canadian Technical Report of Fisheries and Aquatic Sciences No. 1165.

Erickson, J. 1996. Marine geology. Facts on File, Inc., New York, 243 pp

Esselmont, P. 1893. Reports of Fishery Board of Scotland. 11 (Part 3). Edinburgh, Scotland.

Etchegary, G. 1981. Fishing News International 20: 59.

Fardy, B.D. 1994. John Cabot: The discovery of Newfoundland. Creative Publishers, St. John's, 115 pp.

Farmer, G.H. 1981. The cold ocean environment of Newfoundland. *In* The natural environment of Newfoundland, past and present. *Edited by* A.G. Macpherson and J.B. Macpherson. Department of Geography, Memorial University of Newfoundland, pp. 56-82.

Fillon, R.H. 1975. Deglaciation of the Labrador continental shelf. Nature 253: 429-431.

Fillon, R.H. 1976. Hamilton Bank, Labrador shelf: Postglacial sediment dynamics and paleo-oceanography. Marine Geology 20: 7-25.

Finlayson, A.C. 1994. Fishing for truth: a sociological analysis of northern cod stock assessments from 1977 to 1990. Institute of Social and Economic Research, Memorial University of Newfoundland, 176 pp.

Fish, Food and Allied Workers. 1996. Solidarity in adversity: 25 years of social unionism. St. John's.

Fisher, J. and H.G. Vevers. 1943. The breeding and distribution, history and population of the North Atlantic gannet (*Sula bassana*). Journal of Animal Ecology 12: 173-213.

Fitzgerald, J.E. 2002. Newfoundland at the crossroads: documents on Confederation with Canada. Terra Nova Publishing, St. John's, 180 pp.

Fletcher, G.L., L.C. Haggerty, and G. Campbell. 1975. The distribution of Iceland scallops, mussels, clams, and cockles and sea urchins in St. Lewis Bay, Alexis Bay, St. Michael's Bay and Sandwich Bay, Labrador. Unpublished report, Marine Sciences Research Laboratory, Memorial University of Newfoundland.

Flint, R.F. 1943. Growth of North American ice sheet during the Wisconsin age. Geological Society of America Bulletin 54: 352-362.

Flowers, P. 1891. The meaning of fish. Colonial Office, London.

Forest, J. and M. Saint-Laurent. 1976. Further captures off the Phillepine Islands of *Neoglyphea inopinata* (Crustacea: Decapoda: Glypheidae). C.R. Hebd. Seances Academy of Science 283: 935-938.

Fortin, P. 1852. Fisheries report. Journal of the Legislative Assembly of Canada. Ottawa.

Fortin, P. 1867. Fisheries report. Sessional papers. Commissioner for Crown Lands, Ottawa.

Frank, K.T., J.E. Carscadden, and J.E. Simon. 1996. Recent excursions of capelin (*Mallotus villosus*) to the Scotian Shelf and Flemish Cap during anomalous hydrographic conditions. Canadian Journal of Fisheries and Aquatic Sciences 53: 1473-1486.

Frost, N. 1938a. Some fishes of Newfoundland waters. Fisheries Research Bulletin No. 4. Newfoundland Department of Natural Resources.

Frost, N. 1938b. Trout and their conservation. Fisheries Service Bulletin No. 6. Newfoundland Department of Natural Resources.

Frost, N. 1938c. Newfoundland fishes: a popular account of their life histories (Parts I and II). Fisheries Service Bulletin No. 8. Newfoundland Department of Natural Resources.

Frost, N. 1940. Newfoundland flatfishes: a popular account of their life histories. Fisheries Service Bulletin No. 14. Newfoundland Department of Natural Resources.

Gaskell, J. 2000. Who killed the great auk? Oxford University Press, Toronto, 227 pp.

Gaskell, T.F. 1972. The gulf stream. Camelot Press, London, 170 pp.

Gaston, A.J. 1985. The diet of thick-billed murre chicks in the eastern Canadian Arctic. The Auk 102: 727-734.

Gillingham, J. and S. Penney. 1993. Newfoundland sea urchins: the new pearls of the sea. Unpublished report to the Canadian Centre for Fisheries Innovation, St. John's.

Gillispie, J.G., R.L. Smith, E. Barbour, and W.E. Barber. 1997. Distribution, abundance, and growth of Arctic cod in the northeastern Chukchi Sea. *In* Fish Ecology in Arctic North America. *Edited by* J.D. Reynolds. American Fisheries Society, Bethesda, MD, pp. 81-89.

Gjøsæter, H.1998. The population biology and exploitation of capelin (*Mallotus villosus*) in the Barents Sea. Sarsia 83: 453-496.

Goddard, S.V., M.H. Kao, and G.L. Fletcher. 1999. Population differences in antifreeze production cycles of juvenile Atlantic cod (*Gadus morhua*) reflect adaptations to overwintering environment. Canadian Journal of Fisheries and Aquatic Sciences 56: 1991-1999.

Goode. G.B. 1887. The fisheries and fishery industries of the United States. United States Government Printing Office, Washington.

Goode, G.B. and J.W. Collins. 1887. The fisheries and fishery industries of the United States. Part II: the cod, haddock, and hake fisheries. United States Government Printing Office, Washington.

Gordan, S. 1961. The historical perspective: nineteenth century trade theory and policy. *In* Canada and the new international economy. *Edited by* H.E. English. University of Toronto, Toronto, pp. 1-15.

Gordon, H.S. 1954. The economic theory of a common-property resource: the fishery. Journal of Political Economy 62: 124-142.

Gosling, W.G. 1910. Labrador, its discovery, exploration and development. Rivers, London, 574 pp.

Gotceitas, V., S. Fraser, and J.A. Brown. 1997. Use of eelgrass bed (*Zostera marina*) by juvenile Atlantic cod (*Gadus morhua*). Canadian Journal of Fisheries and Aquatic Sciences 54: 1306-1319.

Government of Canada. 1948. A regional forum for the discussion of problems pertaining to the fishing industry. Second East Coast Conference Contribution No. 23, 1-118. Québec City.

Government of Newfoundland. 1963. National fisheries development: a presentation to the Government of Canada by the Government of Newfoundland. St. John's.

Government of Newfoundland. Select committee on inshore fishery. 1975. Report to House of Assembly of Newfoundland. St. John's.

Government of Newfoundland. 1977. Seasonal inshore codtrap fishery glut. St. John's.

Government of Newfoundland. 1978. Setting a course. St. John's.

Government of Newfoundland. 1983. Restructuring the fishery: a detailed presentation by the Government of Newfoundland to the Government of Canada. St. John's.

Government of Newfoundland and Labrador. 1985. Strength from the sea: Newfoundland's position on northern cod. St. John's.

Government of Newfoundland. 1986. The fishing industry today: a profile of the Newfoundland fishery. St. John's.

Graham, J.B. and K.A. Dickson. 2001. Anatomical and physiological specializations for endothermy. *In* Tuna physiology, ecology, and evolution. *Edited by* B.A. Block and E. D. Stevens. Academic Press, San Diego, pp. 121-160.

Grant, W.S. and Leslie, R.W. 2001. Inter-ocean dispersal is an important mechanism in the zoogeography of hakes (Pisces: *Merluccius* spp.). Journal of Biogeography 28: 699-721.

Green, J. 1961. A biology of crustacea. H.F. & G. Witherby, London, 180 pp.

Greenberg, D.A. and B.D. Petrie. 1988. The mean barotropic circulation in the Newfoundland shelf and slope. Journal of Geographysical Research 93: 15,541-15,550.

Gushue, R. 1951. Newfoundland fisheries in 1950. Newfoundland Journal of Commerce January, 69 pp.

Gwyn, R.J. 1972. Smallwood: the unlikely revolutionary. McClelland and Stewart, Toronto, 364 pp.

Gwyn, R.J. 1999. Smallwood: the unlikely revolutionary. 2nd edition. McClelland and Stewart, Toronto, 459 pp.

Hakluyt, R. 1958. Voyages and documents. With an introduction and a glossary by Janet Hampden. Oxford University Press, New York.

Hamilton, L.C. 2003. West Greenland's cod-to-shrimp transition: local dimensions of a climatic change. Arctic 56: 271-282.

Hammill, M.O. and G.B. Stenson. 2003. Harvest simulations for 2003-2006 harp seal management plan. Department of Fisheries and Oceans Science Advisory Secretariat Report No. 2003/068.

Hampton, W.F. 1938. The dogfish and how it can be used. Fisheries Service Bulletin No. 5. Newfoundland Department of Natural Resources.

Handcock, G. 1967. Settlement. In Encyclopedia of Newfoundland and Labrador. *Edited by* J.R. Smallwood and C. Poole. Vol. 5. Newfoundland Book Publishers and Harry Cuff Publications, St. John's, p. 135.

Handcock, W.G. 1977. English migration to Newfoundland. *In* The peopling of Newfoundland: essays in historical geography. *Edited by* J.J. Mannion. Institute of Social and Economic Research, Memorial University of Newfoundland, pp. 15-48.

Hansen, L.P. and J.A. Jacobsen. 2000. Distribution and migration of Atlantic salmon, *Salmo salar L.*, in the sea. *In* The ocean life of Atlantic salmon: environmental and biological factors influencing survival. *Edited by* D. Mills. Fishing News Books, Oxford, pp. 75-87.

Hansen, P.M. 1948. Synopsis of investigations into fluctuations in the stock of cod at Greenland during the years 1930-1933. Rapport et Procès-verbaux de la Conseil Permanent International pour l'exploration de la Mer 36: 1-11.

Hansen, P.M. 1949. Studies on the biology of cod in Greenland waters. ICES Rapport et Procès-verbaux 123: 1-77.

Harden-Jones, F.R. 1968. Fish migration. Edward Arnold, London, 325 pp.

Hardin, G. 1968. The tragedy of the commons. Science 162: 1243-1248.

Harris, L. 1990. Independent review of the state of the northern cod stock. Department of Fisheries and Oceans, Ottawa, 154 pp.

Harris, M. 1998. Lament for an ocean: the collapse of the Atlantic cod fishery: a true crime story. McClelland and Stewart, Toronto, 342 pp.

Harris, M.P., D.J. Halley, and R.L. Swann. 1994. Age of first breeding in common murres. The Auk 111: 207-209.

Harvey, M. 1894. Newfoundland as it is in 1894: a handbook and tourists guide. J.W. Withers, Queen's Printers, St. John's.

Haug, T., H. Gjosaeter, U. Lindstrom, and K.T. Nilssen. 1995. Diet and food availability for north-east Atlantic minke whales (*Balaenoptera acutorostrata*) during the summer of 1992. ICES Journal of Marine Science 52: 77-86.

Hayman, R.1628. Qvodlibets, lately come over from New Britaniola, Old Newfoundland. Elizabeth All-de, London.

Head, C.G. 1976. Eighteenth century Newfoundland: a geographer's perspective. McClelland and Stewart, Toronto, 296 pp.

Helbig, J.A. and P. Pepin. 1997. Distribution and drift of Atlantic cod (*Gadus morhua*) eggs and larvae on the northeast Newfoundland shelf. Canadian Journal of Fisheries and Aquatic Sciences 54: 670-685.

Hendry, A.P., T. Bohlin, B. Jonsson, and O.K. Berg. 2004. To sea or not to sea? Anadromy versus non-anadromy in salmonids. *In* Evolution illuminated: salmon and their relatives. *Edited by* A.P. Hendry and S. C. Stearns. Oxford University Press, Oxford, pp. 92-125.

Hendry, A.P. and S.C. Stearns. 2004. Evolution illuminated: salmon and their relatives. Oxford: Oxford University Press, Oxford, 510 pp.

Hilborn, R. and C.J. Walters. 1992. Quantitative fisheries stock assessment: choice, dynamics and uncertainty. Chapman and Hall, New York, 570 pp.

Hiller, J.K. 1998. Confederation: deciding Newfoundland's future, 1934-1949. Newfoundland Historical Society, St. John's, 75 pp.

Hiller, J.K. 2004. The state and the great war. Newfoundland and Labrador heritage. Available: http://www.heritage.nf.ca/law/state_gw.html.

Hjort, J. 1914. Fluctuations in the great fisheries of northern Europe. Rapport et Procès-Verbaux de la Conseil Permanent International pour l'exploration de la Mer 20: 1-13.

Hobson, K.A. and W.A. Montevecchi. 1991. Stable isotopic determinations of trophic relationships of great auks. Oecologia 87: 528-531.

Hodder, V.M. 1965. Trends in the cod fishery off the east coast of Newfoundland and Labrador. ICNAF Research Bulletin 2: 31-41.

Hodych, J.P., A.F. King, and E.R.W. Neale. 1989. Rocks and time – geological overview of the island of Newfoundland. Newfoundland Journal of Geological Education 10: 1-16.

Hollet, S.M. 1986. Sharing the wealth or dividing the spoils? The Canada-France dispute over jurisdiction and fishing in the northwest Atlantic. MSc dissertation, University of London, London.

Hop, H., W.M. Tonn, and H.E. Welch. 1997. Bioenergetics of Arctic cod (*Boreogadus saida*) at low temperatures. Canadian Journal of Fisheries and Aquatic Sciences 54: 1772-1784.

Hopkins, D.M. and L. Marincovich. 1984. Whale biogeography and the history of the Arctic Basin. Arctic Whaling: Proceedings of the International Symposium 8: 7-24

Howes, G.J. 1991. Biogeography of gadoid fishes. Journal of Biogeography 18:595-622.

Howley, J.P. 1997. Reminiscences of James P. Howley: selected years. *Edited by* W. J. Kirwin, G.M. Story and P A. O'Flaherty. The Chaplain Society, Toronto.

Howley, J.P. 1915. The Beothuks or Red Indians. Prospero Books, Toronto, 348 pp.

Hunt, J.P. 1844. The life of Sir Hugh Palliser, Bart. admiral of the white and governor of the Greenwich Hospital. Chapman and Hall, London.

Hurrell, J.W. and R.R. Dickson. 2004. Climate variability over the North Atlantic. *In* Marine ecosystems and climate variation: the North Atlantic. *Edited by* N.C. Stenseth, G. Ottersen, J.W. Hurrel and A. Belgrano. Oxford University Press, Toronto pp.15-32.

Hutchings, J.A. 1995. Spatial and temporal variation in the exploration of the northern cod, *Gadus morhua*: a historical perspective from 1500 to present. *In* Marine resources and human societies in the North Atlantic since 1500. *Edited by* D. Vickers. Institute of Social and Economic Research Conference Paper No. 5, Memorial University of Newfoundland, pp. 41-68.

Hutchings, J.A. 1999. Influence of growth and survival costs of reproduction on Atlantic cod, *Gadus morhua*, population growth rate. Canadian Journal of Fisheries and Aquatic Sciences 56: 1612-1623.

Hutchings, J.A. 2000. Collapse and recovery of marine fishes. Nature 406: 882-885.

Hutchings, J.A. 2005. Life history consequences of overexploitation to population recovery in northwest Atlantic cod (*Gadus morhua*). Canadian Journal of Fisheries and Aquatic Sciences 62: 824-832.

Hutchings, J.A., T.D. Bishop, and C.R. McGregor-Shaw. 1999. Spawning behaviour of Atlantic cod, *Gadus morhua*: Evidence of mate competition and mate choice in a broadcast spawner. Canadian Journal of Fisheries and Aquatic Sciences 56: 97-104.

Hutchings, J.A. and R.A. Myers. 1993. Effect of age on the seasonality of maturation and spawning of Atlantic cod, *Gadus morhua*, in the northwest Atlantic. Canadian Journal of Fisheries and Aquatic Sciences 50: 2468-2474.

Hutchings, J.A. and R.A. Myers. 1994. What can be learned from the collapse of a renewable resource? Atlantic cod, *Gadus morhua*, of Newfoundland and Labrador. Canadian Journal of Fisheries and Aquatic Sciences 51: 2126-2146.

Hutchings, J.A., R.A. Myers, and G.R. Lilly. 1993. Geographic variation in the spawning of Atlantic cod, *Gadus morhua*, in the northwest Atlantic. Canadian Journal of Fisheries and Aquatic Sciences 50: 2457-2467.

Hutchings, J.A., C. Walters, and R.L. Haedrich. 1997. Is scientific inquiry incompatible with government information control? Canadian Journal of Fisheries and Aquatic Sciences 54: 1198-1210.

Hvingel, C. 2003. Data for the assessment of the shrimp (*Pandalus borealis*) stock in Denmark Strait/off East Greenland, 2003. NAFO Scientific Council Research Document No. 03/77.

Innis, H.A. 1940. The cod fisheries: the history of an international economy. Carnegie Endowment for International Peace, Division of Economics and History, New Haven CT, 520 pp.

International Convention for the Northwest Atlantic. 1957. ICNAF Annual Proceedings for the Year 1956-57, Vol. 7. Dartmouth, NS.

International Convention for the Northwest Atlantic. 1961. Portuguese research report, 1960. ICNAF Annual Proceedings for the Year 1960-61, Vol. 11. Dartmouth, NS.

International Convention for the Northwest Atlantic. 1963. Part 3, Subarea 4, 5. Cod. ICNAF Annual Proceedings for the Year 1962-63, Vol. 13. Dartmouth, NS.

International Convention for the Northwest Atlantic. 1964. Summaries of research and status of fisheries by subareas. ICNAF Annual Proceedings for the Year 1963-64, Vol. 14. Dartmouth, NS.

International Convention for the Northwest Atlantic.1965. Summaries of research and status of fisheries by subareas. ICNAF Annual Proceedings for the Year 1964-65, Vol. 14. Dartmouth, NS.

International Convention for the Northwest Atlantic. 1967a. ICNAF Annual Proceedings for the Year 1966-67. Dartmouth, NS.

International Convention for the Northwest Atlantic. 1967b. State of the North Atlantic cod and haddock fisheries. Part 4, A. ICNAF Annual Proceedings for the Year 1966-67. Dartmouth, NS.

International Convention for the Northwest Atlantic. 1969a. Part 3, Subarea 2 and 5. ICNAF Annual Proceedings for the Year 1968-69, Vol. 19. Dartmouth, NS.

International Convention for the Northwest Atlantic. 1969b. ICNAF Annual Proceedings for the Year 1968-69, Part 2, Report f the: 19th Annual Meeting of the ICNAF, Warsaw, Poland 2-7 June 1969. Dartmouth, NS.

International Convention for the Northwest Atlantic. 1970. Part 3. Subarea 2. ICNAF Annual Proceedings for the

Year 1969-70, Vol. 20. Dartmouth, NS.

International Convention for the Northwest Atlantic. 1971a. Part 3, Subarea 2. ICNAF Annual Proceedings for the Year 1970-71, Vol. 21. Dartmouth, NS.

International Convention for the Northwest Atlantic. 1971b. Atlantic salmon. Part 3, 5. ICNAF Annual Proceedings for the Year 1970-71, Vol. 21. Dartmouth, NS.

International Convention for the Northwest Atlantic. 1975. Report of the 25th annual meeting, June 2-7, 1975. ICNAF Annual Proceedings for the Year 1974-75. Dartmouth, NS.

International Convention for the Northwest Atlantic. 1977. ICNAF Redbook. Standing Committee on Research and Statistics, Dartmouth, NS.

Jamieson, D. 1989. No place for fools: the political memoirs of Don Jamieson, Vol. 1. Breakwater Books, St. John's, 213 pp.

Jansa, L.F. and J.A. Wade. 1974. Geology of the continental margin off Nova Scotia and Newfoundland. *In* Offshore geology of Eastern Canada, Vol. 2. Regional Geology. *Edited by* W.J.M. van der Linden and J.A. Wade. Geological Survey of Canada, Ottawa, pp. 51-106.

Jeffers, G.W. 1931. The life history of the capelin, *Mallotus villosus* (O.F. Muller). University of Toronto Press, Toronto, 146 pp.

Johnson, B. 1977. Canadian foreign policy and fisheries. *In* Canadian foreign policy and the law of the sea. *Edited by* B. Johnson and M.W. Zacher. University of British Columbia Press, Vancouver, pp. 53-95.

Johnstone, K. 1968. The vanishing harvest: the Canadian fishing crisis. Montreal Star Books, Montreal, 87 pp.

Johnstone, K. 1977. The aquatic explorers: a history of the Fisheries Research Board of Canada. University of Toronto Press, Toronto, 342 pp.

Joint Commission of the Newfoundland Legislative Council and House of Assembly. 1880. Report. St. John's.

Jónsson, J. 1994. Fisheries off Iceland, 1600-1900. ICES Marine Science Symposium, 198: 3-16.

Jordan, D.S. 1905. A guide to the study of fishes, Vol. 2. Henry Holt and Company, New York. 299 pp.

Karlsson, G. 2000. Iceland's 1100 years: the history of a marginal economy. Hurst and Company, London, 418 pp.

Katsarou, E. and G. Naevdal. 2001. Population genetic studies of the roughhead grenadier, *Macrourus berglax L.*, in the North Atlantic. Fisheries Research 51: 207-215.

Keats, D.W., D.H. Steele, and J.M. Green. 1986. A review of the recent status of the northern cod stock (NAFO division 2J, 3K and 3L) and the declining inshore fishery: a report to the Newfoundland Inshore Fisheries Association on scientific problems in the northern cod controversy. Unpublished report, Department of Biology, Memorial University of Newfoundland.

Keen, C.E., B.D. Loncarevic, I. Reid, J. Woodside, R.T. Haworth, and H. Williams. 1990. Tectonic and geophysical overview. *In* Geology of the continental margin of Eastern Canada. *Edited by* M.J. Keen and G.L. Williams. Geological Survey of Canada, Ottawa, pp. 31-35.

Keigwin, L.D., J-P. Sachs, and Y. Rosenthal. 2003. A 1600-year history of the Labrador Current off Nova Scotia. Climate Dynamics 21: 23-62.

Keough, K. 1978. Working paper for workshop on fisheries and related matters. Memorial University of Newfoundland, Centre for Newfoundland Studies.

King, C.A.M. 1974. Introduction to marine geology and geomorphology. Edward Arnold, London, 309 pp.

Kinnison, M.T. and A.P. Hendry. 2004. From macro- to micro-evolution: tempo and mode in salmonid evolution. *In* Evolution illuminated: salmon and their relatives. *Edited by* A.P. Hendry and S.C. Stearns. Oxford University Press, Oxford, pp. 208-231.

Kirby, M.J.L. 1982. Navigating troubled waters: a new policy for the Atlantic fisheries. Ministry of Supply and Services Canada, Ottawa.

Koeller, P., L. Savard, D.G. Parsons, and C. Fu. 2000. A precautionary approach to assessment and management of shrimp stocks in the Northwest Atlantic. Journal of Northwest Atlantic Fisheries Science 27: 235-246.

Komai, T. 1999. A revision of the genus *Pandalus* (Crustacea: Decapoda: Caridea: Pandalidae). Journal of Natural History 33: 1265-1372.

Kurlansky, M. 2002. Salt: a world history. Vintage Canada, Toronto, 484 pp.

Lajus, D.L., J.A. Lajus, Z.V. Dmitrieva, A.V. Kraikovski, and D.A. Alexandrov. 2005. The use of historical catch data to trace the influence of climate on fish populations: examples from the White and Barents Sea fisheries in the 17th and 18th centuries. ICES Journal of Marine Science 62: 1426-1435.

Lawson, G.L. and G.A. Rose. 2000. Small-scale spatial and temporal patterns in spawning of Atlantic cod (*Gadus morhua*) in coastal Newfoundland waters. Canadian Journal of Fisheries and Aquatic Sciences 57: 1011-1024.

Laxness. H.K. 1936. Salka Valka. Translated by F.H. Lyon. Houghton Mifflin, Boston.

Lazier, J.R.N. 1979. Recent oceanographic observations in the Labrador Current: proceedings of symposium on research in the Labrador coastal and offshore region. Memorial University of Newfoundland, pp. 195-204.

Lazier, J.R.N. and D.G. Wright. 1993. Annual velocity variations in the Labrador Current. Journal of Physical Oceanography 23: 659-678.

Lear, W.H. 1998. History of fisheries in northwest Atlantic: The 500-year perspective. Journal of Northwest Atlantic Fisheries Science 23: 41-73.

Lear, W.H. and L.S. Parsons. 1993. History and management of the fishery for northern cod in NAFO divisions 2J, 3K, and 3L. *In* Perspectives on Canadian marine fisheries management. *Edited by* L.S. Parsons and W.H. Lear. Canadian Bulletin of Fisheries and Aquatic Science No. 226: 55-90.

LeBlanc, R. 1980. An address by the honourable Roméo LeBlanc, Minister of Fisheries and Oceans, at the Gulf Ground Fish Seminar. Memramcook, New Brunswick, September 25, 1980.

LeDrew, L.J. 1984. Historical development of the arctic charr fishery in northern Labrador. *In* Biology of the Arctic charr: proceedings of the international symposium on Arctic Charr. *Edited by* L. Johnson and B. L. Burns. University of Manitoba Press, Winnipeg, pp. 537-548.

Leggett, W.C., K.T. Frank, and J.E. Carscadden. 1984. Meteorological and hydrographic regulation of year-class strength in capelin (*Mallotus villosus*). Canadian Journal of Fisheries and Aquatic Sciences 41: 1193-1201.

Leon-Rodriguez, L. 2004. Emergence of the Isthmus of Panama: paleoceanographic implications in the Atlantic and Pacific ocean. Department of Earth Sciences, Florida International University.

Leopold, A. 1949. A Sand County almanac and sketches here and now. Oxford University Press, New York, 228 pp.

Letto, D.M. 1998. Chocolate bars and rubber boots: the Smallwood industrialization plan. Blue Hill Publishing, Paradise, NL, 117 pp.

Liem, A.H. and W.B. Scott. 1966. Fishes of the Atlantic coast of Canada. Fisheries Research Board of Canada Bulletin 155: 1-485.

Lilly, G.R. 1987. Interactions between Atlantic cod (*Gadus morhua*) and capelin (*Mallotus villosus*) off Labrador and eastern Newfoundland: a review. Canadian Technical Report of Fisheries and Aquatic Sciences No. 1567.

Lilly, G.R., D.G. Parsons, and D.W. Kulka. 2000. Was the increase in shrimp biomass on the Northeast Newfoundland shelf a consequence of a release in predation pressure from cod? Journal of Northwest Atlantic Fisheries Science 27: 45-61.

Lilly, G.R., P.A. Shelton, J. Brattey, N. Cadigan, B.P. Healey, E.F. Murphy, D. Stanbury, and N. Chen. 2003. An assessment of the cod stock in NAFO Divisions 2J+3KL in February 2003. Department of Fisheries and Oceans Science Advisory Secretariat Report No. 2003/023.

Lindsay, S.T. and H. Thompson. 1932. Biology of the salmon (*Salmo salar*) taken in Newfoundland waters in 1931. Fisheries Research Bulletin No 1. Newfoundland Fishery Research Commission, Newfoundland Department of Natural Resources.

Lockwood, S.J. 1988. The mackerel: its biology, assessment and the management of a fishery. Fishing News Books, Farnham, Surrey, UK, 181 pp.

Lockwood, W.B. 2006. On the philology of cod and stag. Transactions of the Philological Society 104: 13-15.

Lodge, T. 1939. Dictatorship in Newfoundland. Cassell, London, 273 pp.

Longhurst, A.R. 1998. Ecological geography of the sea. Academic Press, San Diego, 398 pp.

Loughrey, A.G. 1959. Preliminary investigation of the Atlantic walrus, *Odobenus romarus romarus* (Linnaeus). Wildlife Management Bulletin Series 1, Number 14. Canadian Department of Northern Affairs and National Resources, Ottawa.

Love, M.S., M. Yoklavich, and L. Thorsteinson. 2002. The rockfishes of the Northeast Pacific. University of California Press, Berkeley, CA, 404 pp.

MacKenzie, W.C. 1979. Rational fishery management in a depressed region: the Atlantic groundfishery. Journal of the Fisheries Research Board of Canada 36: 811-854.

MacPherson. N.L. 1933. Vitamin A Concentration of Cod Liver Oil Correlated with Age of Cod. Nature 132: 26.

MacPherson, N.L. 1935. The Dried Codfish Industry. Fisheries Service Bulletin No. 1. Newfoundland Department of Natural Resources.

MacPherson, N.L. 1937. Newfoundland Cod Liver Oil Pill. Fisheries Service Bulletin No. 3. Newfoundland Department of Natural Resources.

Magnússon, J. 1959. Icelandic research report, redfish. ICNAF Annual Proceedings for the Year 1958-59, Vol. 9. Dartmouth, NS.

Magnússon, M. and H. Pálsson. 1965. The Vinland sagas: The Norse discovery of North America. Penguin, London, 124 pp.

Mahoney, E.M. and R.G. Buggeln. 1983. Seasonal variations in the concentration of *Oikopleura* spp. (Tunicata: Appendicularia) in Conception Bay, Newfoundland. Canadian Technical Report of Fisheries and Aquatic Sciences No. 1155.

Maloney, A. 1990. Report of the Commission of Enquiry into the Alleged Erosion of the Newfoundland fishery by non-Newfoundland Interests. The Executive Business Centre Limited, St. John's.

Mannion, J.J. 1977. Introduction. *In* The peopling of Newfoundland: essays in historical geography. *Edited by* J.J. Mannion. Institute of Social and Economic Research, Memorial University of Newfoundland, pp. 1-14.

Markham, C.R. 1889. A life of John Davis, the navigator, 1550-1605: discoverer of the Davis Straits. George Philip and Son, London, 301 pp.

Marshall, I. 1996. A history and ethnography of the Beothuk. McGill-Queen's University Press, Montreal, 640 pp.

Marshall, N. B. and D. M. Cohen, 1973: Order Anacanthini (Gadiformes). Characters and synopsis of families. *In* Fishes of the Western North Atlantic. *Edited by* D.M. Cohen. Sears Foundation for Marine Research, Yale University, New Haven, CT, pp. 479-495.

Marshall, N. B. and T. Iwamoto. 1973. Family Macrouridae. *In* Fishes of the Western North Atlantic *Edited by* D.M. Cohen. Sears Foundation for Marine Research, Yale University, New Haven, CT, pp. 496-665.

Marti. 1958. Soviet Union researches in recent years. ICNAF Annual Proceedings for the Year 1957-58, Vol. 8. Dartmouth, NS.

Martin, W.R. 1959. Canadian research report, 1958. B. Subareas 4 and 5. ICNAF Annual Proceedings for the Year 1958-59, Vol. 9. Dartmouth, NS.

Martin, C. 1992. No fish and our lives: some survival notes for Newfoundland. Creative Publishers, St. John's, 209 pp.

Martin, W.B.W. 1990. Random Island pioneers. Creative Publishers, St. John's, 268 pp.

Matthews, D.R. 1993. Controlling common property: regulating Canada's east coast fishery. University of Toronto Press, Toronto, 277 pp.

Matthews, K. 1968. A history of the west England–Newfoundland fishery. PhD dissertation, Oxford University.

Matthews, K. 1988. Lectures on the history of Newfoundland, 1500-1830. Breakwater Books, St. John's, 191 pp.

May, A. 1959. Cod investigations in Subarea 2-Labrador, 1950 to 1958. ICNAF Annual Proceedings for the Year 1958-59, Vol. 9. Dartmouth, NS.

May, A. 1973. A speech by Art May to the Fisheries Council of Canada. Charlottetown, Prince Edward Island.

May, A.W. 1964. New cod fishing grounds off Labrador and the northeast coast of Newfoundland. Trade News 1964: 3-6.

May, A.W. 1966. Effect of offshore fishing on the inshore Labrador cod fishery. ICNAF Research Document 66-23.

May, A.W., D.A. Russell, and D.H. Rowe. 2005. Breaking new ground: an action plan for rebuilding the Grand Banks fisheries. Report to Fisheries and Oceans Canada, 67 pp.

McAllister, D.E. 1960. List of the marine fishes of Canada. National Museum of Canada Bulletin No. 168. Department of Northern Affairs and National Resources, Ottawa, 76 pp.

McDonald, I.D.H. 1987. To each his own: William Coaker and the Fisherman's Protective Union in Newfoundland politics, 1908-1925. Institute for Social and Economic Research, Memorial University of Newfoundland, 201 pp.

McGinn, N.A. 2002. Fisheries in a changing climate. American Fisheries Society, Symposium 32, Bethesda MD, 295 pp.

McFarland, R. 1911. A history of the New England fisheries. University of Pennsylvania, D. Appleton and Co., New York, 547 pp.

McIntosh, W.C. 1921. The resources of the sea as shown in the scientific experiments to test the effects of trawling and of the closure of certain areas off the Scottish shores. Cambridge University Press (2nd edition), London, 352 pp.

McLaren, I.A. 1960. Are the pinnipedia biphyletic? Systematic Zoology 9: 18-28.

McQuinn, I.H. 1997. Metapopulations and Atlantic herring. Reviews in Fish Biology and Fisheries 7: 297-329.

McMurrich, J.P. 1884. Science in Canada. The Week, Nov. 6, 1884: 776-777.

Meek, A. 1916. The flat fishes. *In* The Migrations of Fish. Edward Arnold, London, pp. 256-283.

Mello, L.G.S. and G.A. Rose. 2005. Seasonal cycles in weight and condition in Atlantic cod (*Gadus morhua* L.) in relation to fisheries. ICES Journal of Marine Science 62: 1006-1015.

Methven, D.A., D.C. Schneider, and G.A. Rose. 2003. Spatial pattern and patchiness during ontogeny: Post-settled *Gadus morhua* from coastal Newfoundland. ICES Journal of Marine Science 60: 38-51.

Millais, J.G. 1907. Newfoundland and its untrodden ways. Longmans, Green and Co., London, 340 pp.

Minet, J.P. and J.B. Perodou. 1977. Predation of cod (*Gadus morhua*) on capelin (*Mallotus villosus*) in northern Gulf of St. Lawrence (ICNAF Div. 4R-4S-3P) and off Labrador–East Newfoundland (ICNAF Div. 2J-3K-3L). ICNAF Research Document 77/VI/22.

Mitchell, J.M. 1864. The herring: its natural history and national importance. Edmonston and Douglas, Edinburgh, 372 pp.

Montevecchi, W.A. and R.A. Myers. 1997. Centurial and decadal oceanographic influences on changes in northern gannet populations and diets in the north-west Atlantic: implications for climate change. ICES Journal of Marine Science 54: 608-614.

Montevecchi, W.A. and J.F. Piatt. 1984. Composition and energy contents of mature inshore spawning capelin (*Mallotus villosus*): Implications for seabird predators. Comparative Biochemistry and Physiology 78A: 15-20.

Montevecchi, W.A. and L.M. Tuck. 1987. Newfoundland birds: exploitation, study, conservation. Nuttall Ornithological Club, Cambridge, MA, 273 pp.

Moores, F.D. 1978. Fisheries in the future. Paper presented at the First Ministers Conference. Ottawa, February 13-15, 1978.

Moores, J.A. and G.H. Winters. 1984. Migration patterns of Newfoundland west coast herring, *Clupea harengus*, as shown by tagging studies. Journal of Northwest Atlantic Fisheries Science 5: 17-22.

Morgan, M.J. and E.B. Colbourne. 1999. Variation in maturity at age and size in three populations of American plaice. ICES Journal of Marine Science 56: 673-688.

Morgan, M.J., E.M. DeBlois, and G.A. Rose. 1997. An observation on the reaction of Atlantic cod (*Gadus morhua*) in a spawning shoal to bottom trawling. Canadian Journal of Fisheries and Aquatic Sciences 54: 217-223.

Morgan, M.J. and E.A. Trippel. 1996. Skewed sex ratios in spawning shoals of Atlantic cod (*Gadus morhua*). ICES Journal of Marine Science 53: 820-826.

Morison, S.E. 1978. The great explorers. Oxford University Press, New York, 752 pp.

Morris, C.J. and J.M. Green. 2002. Biological characteristics of a resident population of Atlantic cod (*Gadus morhua* L.). ICES Journal of Marine Science 59: 666-678.

Mowbray, F. 2002. Changes in the vertical distribution of capelin (*Mallotus villosus*) off Newfoundland. ICES Journal of Marine Science 59: 942-949.

Munn, W.A. 1922. Annual migration of codfish in Newfoundland waters. Newfoundland Trade Review December 23: 21-24.

Murray, H. 1829. Historical account of the discoveries and travels in North America. Longman, Rees, Orme, Brown, and Green, Edinburgh, 530 pp.

Murua, H., F. González, and D. Power. 2005. A review of the fishery and the investigations of roughhead grenadier (*Macrourus berglax*) in Flemish Cap and Flemish Pass. Journal of Northwest Atlantic Fisheries Science 37: 1-15.

Myers, R.A. and N.G. Cadigan. 1995. Was an increase in natural mortality responsible for the collapse of northern cod? Canadian Journal of Fisheries and Aquatic Sciences 52: 1274-1285.

Myers, R.A., G. Mertz, and P.S. Fowlow. 1997. The population growth rate of Atlantic cod (*Gadus morhua*) at low abundance. Fishery Bulletin 95: 762-772.

Myers, R.A., B.R. MacKenzie, K.G. Bowen, and N.J. Barrowman. 2001. What is the carrying capacity for fish in the ocean? A meta-analysis of population dynamics of North Atlantic cod. Canadian Journal of Fisheries and Aquatic Sciences 58: 1464-1476.

Myers, R.A. and B. Worm. 2003. Rapid worldwide depletion of predatory fish communities. Nature 423: 280-283.

Newfoundland Associated Fish Producers Limited. 1950. General Circular, No. 396, 12 June 1950. St. John's.

Nafpaktitis, B.G., R.H. Backus, J.E. Craddock, R.L. Haedrich, B.H. Robison, and C. Karnella. 1977. The families Neoscopelidae and Myctophidae. *In* Fishes of the Western North Atlantic *Edited by* D.M. Cohen. Sears Foundation for Marine Research, Yale University, New Haven, CT, pp. 1-265.

Nakashima, B.S. 1992. Patterns in coastal migration and stock structure of capelin (*Mallotus villosus*). Canadian Journal of Fisheries and Aquatic Sciences 49: 2423-2429.

Nakashima, B.S. and J.P. Wheeler. 2002. Capelin (*Mallotus villosus*) spawning behaviour in Newfoundland waters: the interaction between beach and demersal spawning. ICES Journal of Marine Science 59: 909-916.

Nakken, O. 1994. Causes of trends and fluctuations in the Arcto-Norwegian cod stock. ICES Marine Science Symposium 198: 212-228.

Narayanan, S., J. Carscadden, J.B. Dempson, M.F. O'Connell, S.J. Prinsenberg, D.G. Reddin, and N. Shackell. 1995. Marine climate off Newfoundland and its influence on Atlantic salmon (*Salmo salar*) and capelin (*Mallotus villosus*). In Climate Change and Northern Fish Populations. *Edited by* R.J. Beamish. Canadian Special Publication of Fisheries and Aquatic Sciences, Vol. 121, Ottawa, pp. 461-474.

Neary, P. and P. O'Flaherty. 1974. By great waters: a Newfoundland and Labrador anthology. University of Toronto Press, Toronto, 262 pp.

Needler, A.W.H. 1970. Chairman's report. ICNAF Annual Proceedings for the Year 1969-70, Vol. 20, Part 2. Dartmouth, NS.

Needler, A.W.H. 1971. Chairman's report. ICNAF Annual Proceedings for the Year 1970-71, Vol. 21, Part 2. Dartmouth, NS.

Neill, R. 2006. It is farming, not fishing. Why bureaucrats and environmentalists miss the point of Canadian aquaculture. *In* How to farm the seas. *Edited by* B.L. Crowley and G. Johnson, Atlantic Institute for Market Research.

Newfoundland Associated Fish Exporters Limited. 1950. General Circular 346.

Newfoundland Board of Trade. 1915. Seventh annual report of the Newfoundland Board of Trade. St. John's.

Newfoundland Board of Trade. 1925. Seventeenth annual report of the Newfoundland Board of Trade, St. John's.

Newfoundland Department of Fisheries. 1903. Annual fisheries report. St. John's, Newfoundland.

Newfoundland Department of Fisheries. 1904. Annual report. St. John's.

Newfoundland Department of Marine and Fisheries. 1917. Annual report. St. John's.

Newfoundland Department of Marine and Fisheries. 1921. Annual report. St. John's.

Newfoundland Department of Natural Resources. 1937. Annual report. St. John's.

Newfoundland Department of Fisheries. 1968. Annual report 1967/68. St. John's.

Newfoundland.Department of Fisheries. 1974. Annual report for year ending March 31, 1974. St. John's.

Newfoundland Department of Marine Fisheries. 1917. Annual report. St. John's.

Newfoundland Department of Marine Fisheries. 1921. Annual report. St. John's.

Newfoundland Department of Natural Resources. 1937. Report of the fisheries research committee. St. John's.

Newfoundland Fisheries Commission. 1915. Summary report. St. John's.

Newfoundland Inshore Fisheries Association. 1991. A preliminary report on the history of the cod trap fishery in Newfoundland. St. John's.

Newton, E. T. 1891. Vertebrata of the pliocene deposits of Britain. Memoirs of the Geological Survey of the United Kingdom. Geological Survey of the United Kingdom, London, 131 pp.

Nielsen, A. 1894. Report for the year 1893. Government of Newfoundland.

Nolf, D. 1995. Studies on fossil otoliths – the state of the art. *In* Recent developments in fish otolith research. *Edited by* D.H. Secor, J.M. Dean, and S.E. Campana. University of South Carolina Press, pp. 513-544.

Norgaard-Pedersen, N., R.F. Spielhagen, H. Erlenkeuser, P.M. Grootes, J. Heinemeier, and J. Knies. 2003. Arctic Ocean during the last glacial maximum: Atlantic and polar domains of surface water mass distribution and ice cover. Paleoceanography 18: 1063.

Norman, J. R. 1963. A history of fishes. Ernest Benn, London, 398 pp.

Nova Scotia Royal Commission on Reconstruction and Rehabilitation. 1944. The report on the Canadian Atlantic sea-fishery. Nova Scotia Government.

O'Driscoll, R.L. 2003. Determining species composition in mixed-species marks: An example from the New Zealand hoki (*Macruronus novaelandiae*) fishery. ICES Journal of Marine Science 60: 609-616.

O'Driscoll, R.L., D.C. Schneider, G.A. Rose, and G.R. Lilly. 2000a. Potential contact statistics for measuring scale-dependent spatial pattern and association: An example of northern cod (*Gadus morhua*) and capelin (*Mallotus villosus*). Canadian Journal of Fisheries and Aquatic Sciences 57: 1355-1368.

O'Driscoll, R.L., G.A. Rose, J.T. Anderson, and F. Mowbray. 2000b. Spatial association between cod and capelin: a perspective on the inshore-offshore dichotomy. Canadian Stock Assessment Secretariat Report No. 2000/083.

O'Driscoll, R.L., J.D. Parsons, and G.A. Rose. 2001. Feeding of capelin (*Mallotus villosus*) in Newfoundland waters. Sarsia 86: 165-176.

O'Flaherty, P. 1999. Old Newfoundland: a history to 1843. Long Beach Press, St. John's, 278 pp.

O'Flaherty, P. 2005. Lost country: the rise and fall of Newfoundland, 1843-1933. Long Beach Press, St. John's, 514 pp.

O'Neill, P. 2003. The oldest city: the story of St. John's Newfoundland. Boulder Publications, Portugal Cove-St. Philip's, 888 pp.

Odling-Smee, L. and V.A. Braithwaite. 2003. The role of learning in fish orientation. Fish and Fisheries 4: 235-246.

Olsen, E.M., M. Heino, G.R. Lilly, M.J. Morgan, J. Brattey, B. Ernande, and U. Dieckmann. 2004. Maturation trends indicative of rapid evolution preceded the collapse of northern cod. Nature 428: 932-935.

Orlova, E.L., V.D. Boitsov, A.V. Dolgov, G.B. Rudneva, and V.N. Nesterova. 2005. The relationship between plankton, capelin, and cod under different temperature conditions. ICES Journal of Marine Science 62: 1281-1292.

Osborn, H.L. 1887. Report on a cod-trawling trip to Grand Bank in 1879. In The fisheries and fishing industries of the United States. United States Commission of Fish and Fisheries, Washington.

Ottersen, G., N.C. Stenseth, and J.W. Hurrell. 2004. Climatic fluctuations and marine systems: A general introduction to the ecological effects. In Marine ecosystems and climate variation: the North Atlantic. Edited by N.C. Stenseth, G. Ottersen, J.W. Hurrell, and A. Belgrano. Oxford University Press, Toronto, pp 3-14.

Palliser, H. 1765. Colonial Office, December 18th, 1765. St. John's.

Parsons, D.G., G.R. Lilly, and G.J. Chaput. 1986. Age and growth of northern shrimp *Pandalus borealis* off northeastern Newfoundland and southern Labrador. Transactions of the American Fisheries Society 115: 872-881.

Parsons, D.G. and G.E. Tucker. 1986. Fecundity of northern shrimp, *Pandalus borealis* (Crustacea, Decapoda) in areas of the northwest Atlantic. Fishery Bulletin 84: 549-558.

Parsons, L.S. 1993. Management of marine fisheries in Canada. Canadian Bulletin of Fisheries and Aquatic Sciences No. 225, 762 pp.

Patterson, G. 1890. The Portuguese on the north-east coast of America and the first European attempts at colonization there: A lost chapter in American history. Transactions of the Royal Society of Canada 8, No. 2: 146.

Pauly, D., V. Christensen, J. Dalsgaard, R. Froese, and T. Francisco, Jr. 1998. Fishing down marine food webs. Science 279: 860-863.

Pauly, D. and J. Maclean. 2003. In a perfect ocean: the state of fisheries and ecosystems in the North Atlantic ocean. Island Press, Washington, 175 pp.

Pavlov, M.A. 1960. U.S.S.R. research report, 1959. D. Fishing conditions in the area of the Grand Newfoundland Bank in 1959, as reported by the Soviet trawler fleet. ICNAF Annual Proceedings for the Year 1959-60, Vol. 10. Dartmouth, NS.

Pavlov, M.A. and V.I. Travin. 1961. Soviet Union research report, 1960. B. Soviet trawl fishery in the northwest Atlantic, 1960. ICNAF Annual Proceedings for the Year 1960-61, Vol. 11. Dartmouth, NS.

Pechenik and Troyanovskii. 1969. Studies of grenadiers in Newfoundland-Labrador waters (In Russian). Ryb. Khoz. 12: 7-9.

Pepin, P., D.C. Orr, and J.T. Anderson. 1997. Time to hatch and larval size in relation to temperature and egg size in Atlantic cod (*Gadus morhua*). Canadian Journal of Fisheries and Aquatic Sciences 54: 2-10.

Peters, G.R. 1978. The nature of a programme in fisheries and related studies. Presentation at the workshop on fisheries and related matters, Memorial University of Newfoundland, St. John's, January 1978.

Peterson, C., S. Rice, J. Short, D. Esler, J. Bodkin, B. Ballachey, and D. Irons. 2003. Long-term ecosystem response to the Exxon Valdez oil spill. Science 302: 2082-2086.

Petrie, B. 1983. Circulation on the Newfoundland continental shelf. Atmosphere-Ocean 21: 207-226.

Pflaumann, U., M. Sarnthein, M. Chapman, L. d'Abreu, B. Funnell, M. Huels, T. Kiefer, M. Maslin, H. Schulz, J. Swallow, S. van Kreveld, M. Vautravers, E. Vogelsang, and M. Weinelt. 2003. Glacial North Atlantic: Sea-surface conditions reconstructed by GLAMAP 2000. Paleoceanography 18: 1065.

Phillips, B.F., J.S. Cobb, and R.W. George. 1980. General biology. In The biology and management of lobsters, Vol. 1. Academic Press, New York, pp. 1-72.

Piatt, J.F. and D.N. Nettleship. 1985. Diving depths of four alcids. The Auk 102: 293-297.

Pinhorn, A.T. 1979. The northern cod resource, 1979. A paper presented to the northern cod seminar, Corner Brook, Newfoundland, August 27, 1979.

Pinhorn, A.T. 1984. Temporal and spatial variation in fecundity of Atlantic cod (*Gadus morhua*) in Newfoundland waters. Journal of Northwest Atlantic Fisheries Science 5: 161-170.

Ponomareva, L.A. 1966. Euphausiids of the North Pacific: their distribution and ecology. Israel Program for Scientific Translations. S. Monson, Jerusalem.

Pope, P. 1995. Early estimates: assessment of catches in the Newfoundland cod fishery, 1660-1690. In Marine resources and human societies in the North Atlantic since 1500. Institute of Social and Economic Research, Memorial University of Newfoundland, pp. 9-36.

Pope, P. 1997. The many landfalls of John Cabot. University of Toronto Press, Toronto, 244 pp.

Pope, P. 2004. Fish into wine: the Newfoundland plantation in the seventeenth century. University of North Carolina Press, Chapel Hill, NC, 463 pp.

Power, D., B.P. Healey, E.F. Murphy, J. Brattey, and K. Dwyer. 2005. An Assessment of the Cod Stock in NAFO Divisions 3NO. NAFO Scientific Council Meeting, June 2005. NAFO Scientific Council Research Document No. 05/67.

Prowse, D.W. 1895. A history of Newfoundland, from the English, colonial and foreign records. MacMillan and Company, London, 742 pp.

Quinn, D.B. 1979. Newfoundland from fishery to colony. Arno Press, New York, 454 pp.

Rasmussen T.L., D.W. Oppo, E. Thomsen, and S.J. Lehman. 2003. Deep sea records from the southeast Labrador Sea: Ocean circulation changes and ice-rafting events during the last 160,000 years. Paleoceanography 18 (1):18-1-18-15.

Reddin, D.G. and K.D. Friedland. 1993. Marine environmental factors influencing the movement and survival of Atlantic salmon. *In* Salmon in the sea and new enhancement strategies. *Edited by* D.H. Mills. Fishing News Books, Oxford, pp. 79-103.

Reddin, D.G., J.A. Helbig, A. Thomas, B.G Whitehouse, and K.D. Friedland. 2000. Survival of Atlantic salmon (*Salmo salar* L.) related to marine climate. *In* The ocean life of Atlantic salmon: environmental and biological factors influencing survival. *Edited by* D. Mills. Fishing News Books, Oxford, pp. 88-91.

Reist, J.D., J.D. Johnson, and T.J. Carmichael. 1997. Variation and specific identity of char from northwestern Arctic Canada and Alaska. American Fisheries Society Symposium 19: 250-261.

Resource Management Division. 1986. Report on the 1986 Capelin Fishery. Department of Fisheries and Oceans, St. John's.

Rice, J.C. 2002. Changes to the large marine ecosystem of the Newfoundland and Labrador shelf. *In* Large Marine Ecosystems of the North Atlantic. *Edited by* K. Sherman and H.R. Skoldal. Elsevier, Amsterdam.

Richard, P.R. and R.R. Campbell. 1988. Status of the Atlantic walrus, *Obdobenus romarus romarus*, in Canada. The Canadian Field-Naturalist 102: 337-348.

Ricker, W.E. 1961. Productive capacity of Canadian fisheries. Fisheries Research Board of Canada Circular No. 64, Nanaimo, BC.

Rideout, R.M., M.P. M. Burton, and G.A. Rose. 2000. Observations on mass atresia and skipped spawning in northern Atlantic cod, from Smith Sound, Newfoundland. Journal of Fish Biology 57: 1429-1440.

Robards, M.D., M.F. Willson, R.H. Armstrong, and J.F. Piatt. 1999. Sand lance: a review of biology and predator relations and annotated bibliography. Exxon Valdez Oil Spill Restoration Project 99346. Research Paper PNW-RP-521. U.S. Department of Agriculture, Portland, OR.

Robichaud, D. and G.A. Rose. 2002. Assessing evacuation rates and spawning abundance of marine fishes using coupled telemetric and acoustic surveys. ICES Journal of Marine Science 59: 254-260.

Robichaud, D. and G.A. Rose. 2004. Stock structure and range in Atlantic cod (*Gadus morhua*): Inference from 100 years of tagging. Fish and Fisheries 5: 1-33.

Rogerson, R.J. 1989. The glacial history of Newfoundland and Labrador. Newfoundland Journal of Geological Education 11: 117-134.

Rojo, A. 1957. Researches carried out on board the trawlers Santa Ines and Santa Celia in Subdivision 3N, Grand Bank of Newfoundland, September 1956. ICNAF Annual Proceedings for the Year 1956-57, Vol. 7. Dartmouth, NS.

Rojo, A. 1958. Spanish research report 1957. B. The cod in the Labrador area. ICNAF Annual Proceedings for the Year 1957-58, Vol. 8. Dartmouth, NS.

Rose, E.R. 1952. Torbay map-area, Newfoundland. In memoir, Vol. 265. Geological Survey of Canada, Ottawa.

Rose, G.A. 1992. Indices of total stock and biomass in the "northern" and Gulf of St. Lawrence Atlantic cod (*Gadus morhua*) stocks derived from time series analyses of fixed gear (trap) catches. Canadian Journal of Fisheries and Aquatic Sciences 49: 202-209.

Rose, G.A. 1993. Cod spawning on a migration highway in the north-west Atlantic. Nature 366: 458-461.

Rose, G.A. 1997. The trouble with fisheries science! Reviews in Fish Biology and Fisheries 7: 365-370.

Rose, G.A. 2003a. Drilling and fishing the Grand Banks of Newfoundland. Newfoundland Quarterly 96: 28-29.

Rose, G.A. 2003b. Monitoring coastal northern cod: Towards an optimal survey of Smith Sound, Newfoundland. ICES Journal of Marine Science 60: 453-462.

Rose, G.A. 2004. Reconciling overfishing and climate change with stock dynamics of Atlantic cod (*Gadus morhua*) over 500 years. Canadian Journal of Fisheries and Aquatic Sciences 61: 1553-1557.

Rose, G.A. 2005. On distributional responses of North Atlantic fish to climate change. ICES Journal of Marine Science 62: 1360-1374.

Rose, G.A., D.B. Atkinson, J.W. Baird, C.A. Bishop, and D.W. Kulka. 1994. Changes in distribution of Atlantic cod and thermal variations in Newfoundland waters, 1980-1992. ICES Marine Science Symposia 198: 542-552.

Rose, G.A., B. de Young, D.W. Kulka, S.V. Goddard, and G.L. Fletcher. 2000. Distribution shifts and overfishing the northern cod (*Gadus morhua*): A view from the ocean. Canadian Journal of Fisheries and Aquatic Sciences 57: 644-664.

Rose, G.A. and D.W. Kulka. 1999. Hyperaggregation of fish and fisheries: How catch per unit effort increased as the northern cod (*Gadus morhua*) declined. Canadian Journal of Fisheries and Aquatic Sciences 56: 118-127.

Rose, G.A. and W.C. Leggett. 1988. Atmosphere-ocean coupling in the northern Gulf of St. Lawrence: Frequency-dependent wind-forced variations in nearshore sea temperatures and currents. Canadian Journal of Fisheries and Aquatic Sciences 45: 1222-1233.

Rose, G.A. and W.C. Leggett. 1989. Predicting variability in catch-per-unit effort in Atlantic cod, *Gadus morhua*, trap and gillnet fisheries. Journal of Fish Biology 35: 155-161.

Rose, G.A. and W.C. Leggett. 1990. The importance of scale to predator-prey spatial correlations: An example of Atlantic fishes. Ecology 71: 33-43.

Rose, G.A. and R.L. O'Driscoll. 2002. Capelin are good for cod: Can the northern stocks recover without them? ICES Journal of Marine Science 59: 1018-1026.

Roughgarden, J. and F. Smith. 1996. Why fisheries collapse and what to do about it. Proceedings of the National Academy of Science 93: 5078-5083.

Rowe, F.W. 1980. A History of Newfoundland and Labrador. McGraw-Hill Ryerson, Toronto, 563 pp.

Rowe, S., I.L. Jones, J.W. Chardine, R.D. Elliot, and B.G. Veitch. 2000. Recent changes in the winter diet of murres (*Uria* spp.) in coastal Newfoundland waters. Canadian Journal of Zoology 78: 495-500.

Ruddiman, W.F. 1987. Northern oceans. *In* North America and adjacent oceans during the past deglaciation. *Edited by* W.F. Ruddiman and H.E. Wright. Geological Society of America, Boulder, CO, pp. 137-154.

Ruddiman, W.F. and H.E. Wright. 1987. Introduction. *In* North America and adjacent oceans during the last glaciation. *Edited by* W.F. Ruddiman and H.E. Wright. Geological Society of America. Boulder, CO, pp. 1-12.

Ruivo, M. 1957. Portuguese research report, 1956. B. Subarea 2-Labrador, cod, 1956. ICNAF Annual Proceedings for the Year 1956-57, Vol.7. Dartmouth, NS.

Ruivo, M. and G. Quartin. 1958. Portuguese research report, 1957. Observations on cod in Subarea 2-Labrador. ICNAF Annual Proceedings for the Year 1957-58, Vol. 8. Dartmouth, NS.

Ruivo, M. and G. Quartin. 1959. Portuguese research report. Observations of cod in Subarea 2-Labrador. ICNAF Annual Proceedings for the Year 1958-59, Vol. 9. Dartmouth, NS.

Ruppert, E.E. and R.D. Barnes. 1994. Invertebrate zoology. Saunders College Publishing, Fort Worth, VA, 896 pp.

Ruzzante, D.E., C.T. Taggart, and D. Cook. 1999. A review of the evidence for genetic structure of cod (*Gadus morhua*) populations in the NW Atlantic and population affinities of larval cod off Newfoundland and the Gulf of St. Lawrence. Fisheries Research 43: 79-97.

Ryan, B. 1989. Revolution in distant ages – the precambrian geology of Labrador. Newfoundland Journal of Geological Education 10: 77-84.

Ryan, S. 1986. Fish out of water. Breakwater Books, St. John's, 320 pp.

Ryan, S. 1994. The ice hunters: a history of Newfoundland sealing to 1914. Breakwater Books, St. John's. 515 pp.

Ryther, J.H. 1969. Photosynthesis and fish production in the sea. Science 166: 72-76.

Sabine, L. 1853. Report on the principle fisheries of American seas. Report prepared for the Treasury Department of the United States. Robert Armstrong, Washington.

Sainte-Marie, B., J.-M. Sevigny, B.D. Smith, and G.A. Lovrich. 1996. Recruitment variability in snow crab (*Chionoecetes opilio*): pattern, possible causes, and implications for fishery management. *In* High latitude cabs: biology, management, and economics. University of Alaska, Fairbanks, AK, pp. 451-478.

Sameoto, D. 2004a. Northwest Atlantic plankton trends, 1959-2000. Department of Fisheries and Oceans Science Advisory Secretariat Report No. 2004/073.

Sameoto, D. 2004b. The 1990s: A unique decade for plankton change in the northwest Atlantic. Department of Fisheries and Oceans Science Advisory Secretariat Report No. 2004/074.

Sauer, C.O. 1971. Sixteenth century North America. University of California Press. Berkeley, CA.

Schram, F.R. 1982. The fossil record and evolution of crustacea. *In* The biology of crustacea: systematics, the fossil record, and biogeography. *Edited by* L.G. Abele. Academic Press, New York, pp. 94-147.

Schwarzbach, M. 1963. The geological knowledge of the North Atlantic climates of the past. *In* North Atlantic biota and their history. *Edited by* A. Love and D. Love. Pergamon Press, Oxford, pp. 11-19.

Sclater, J.G., J.G. Hellinger, and C. Tapscott. 1977. The paleobathymetry of the Atlantic Ocean from the Jurassic to the present. Journal of Geology 5: 509-552.

Sclater, J.G. and C. Tapscott. 1979. The history of the Atlantic. Scientific American 240: 156-174.

Scott, G.K., G.L. Fletcher, and P.L. Davies. 1986. Fish antifreeze proteins: recent gene evolution. Canadian Journal of Fisheries and Aquatic Sciences 43: 1028-1034.

Scott, W.B and M.G. Scott. 1988. Atlantic Fishes of Canada. University of Toronto Press, Toronto, 731 pp.

Scudder, N.P. 1887. The salt-halibut fishery. *In* History and methods of the fisheries, Volume I. *Edited by* G.B. Goode. United States Commission of Fish and Fisheries, Washington, pp. 90-119.

Seary, E.R. 1971. Place names of the Avalon Peninsula of the Island of Newfoundland. University of Toronto Press, Toronto, 383 pp.

Second East Coast Fisheries Conference. 1948. A regional forum for the discussion of problems pertaining to the fishing industry. Chateau Frontenac, Québec City, February 3-5, 1948.

Select Committee on Inshore Fishery. 1975. Report to House of Assembly of Newfoundland. Government of Newfoundland.

Serebryakov, V.P. 1967. Cod reproduction in the northwest Atlantic. Fisheries Translation Series 1133.

Sergeant, D.E. 1991. Harp seals, man and ice. In Canadian Special Publication of Fisheries and Aquatic Sciences 114: 1-153.

Shelton, P.A., A.F. Sinclair, G.A. Chouinard, R.K. Mohn, and D.E. Duplisea. 2006. Fishing under low productivity conditions is further delaying recovery of northwest Atlantic cod (*Gadus morhua*). Canadian Journal of Fisheries and Aquatic Sciences 63: 235-238.

Shepard, P. 1973. The tender carnivore and the sacred game. Charles Scribner's Sons, New York, 302 pp.

Sherwood, G.D. and G.A. Rose. 2005. Stable isotope analysis of some representative fish and invertebrates of the Newfoundland and Labrador continental shelf food web. Estuarine Coastal Shelf Science 63: 537-549.

Sherwood, G., Rideout R., Fudge, S. and G.A. Rose. 2007. Growth, condition and reproduction capacity in Newfoundland and Labrador cod under variable feeding regimes. Deep Sea Research II. In Press.

Shrank, W.E. 1995. Extending fisheries jurisdiction: origins of the current crisis in Atlantic Canada's fisheries. Marine Policy 19: 285-299.

Shrank, W.E. 2005. The Newfoundland fishery: ten years after the moratorium. Marine Policy 29: 407-420.

Shumway, S.E., H.C. Perkins, D.F. Schick, and A.P. Stickney. 1985. Synopsis of Biological Data of the Pink Shrimp *Pandalus borealis* (Kroyer, 1838). NOAA Technical Report NMFS 30/FAO Fisheries Synopsis No. 144. Food and Agriculture Organization of the United Nations, Rome.

Simpson, M. and S.J. Walsh. 2004. Changes in the spatial structure of Grand Bank yellowtail flounder: testing MacCall's basin hypothesis. Journal of Sea Research 52: 199-210.

Sleggs, G.F. 1933. Observations upon the economic biology of the capelin. Report of the Newfoundland Fishery Research Commission 1, No. 3.

Slijper, E.J. 1979. Whales. Cornell University Press, Ithaca, NY, 511 pp.

Smith, T.D. 1994. Scaling fisheries: The science of measuring the effects of fishing, 1855-1955. Cambridge University Press. Cambridge, 392 pp.

Spies, R.B., S.D. Rice, D.A. Wolfe, and B.A. Wright. 1996. The effects of the Exxon Valdez oil spill on the Alaskan coastal environment. *In* Proceedings of the Exxon Valdez Oil Spill Symposium. *Edited by* S.R. Rice, R.B. Spies, D. Wolfe, and B. Wright. American Fisheries Society, Bethesda, MD, pp. 1-16.

Springer, A.M. and S.G. Speckman. 1997. A forage fish is for what? *In* Proceedings of the Symposium on the Role of Forage Fishes in Marine Ecosystems. *Edited by* B.R. Baxter. University of Alaska, Fairbanks, AK, pp. 773-805.

Squires, H.J. 1992. Recognition of *Pandalus eous* Makarov, 1935, as a Pacific species not a variety of the Atlantic *Pandalus borealis* Kroyer, 1838 (Decapoda, Caridea). Crustaceana 63: 257-262.

Stanley, R.D., R. Kieser, K. Cooke, A.M. Surry, and B. Mose. 2000. Estimation of a widow rockfish (*Sebastes entomelas*) shoal off British Columbia, Canada, as a joint exercise between stock assessment staff and the fishing industry. ICES Journal of Marine Science 57: 1035-1049.

Stearns, S.C. and A.P. Hendry. 2004. Introduction: the salmonid contribution to key issues in evolution. *In* Evolution illuminated: salmon and their relatives. *Edited by* A.P. Hendry and S.C. Stearns. Oxford University Press, New York, pp. 3-19.

Stevenson, S.C. and J.W. Baird. 1988. The fishery for lumpfish (*Cyclopterus lumpus*) in Newfoundland waters. Canadian Technical Report of Fisheries and Aquatic Sciences Report No.1595.

Stobo, W.T., J.A. Moores, and J.J. Maguire. 1982. The herring and mackerel resources on the east coast of Canada. Canadian Technical Report of Fisheries and Aquatic Sciences No. 1081.

Stommel, H. and E. Stommel. 1979. The year without a summer. Scientific American 240: 176-186.

Sverdrup, A., E. Kjellsby, P.G. Krüger, R. Floysand, F.R. Knudsen, P.S. Enger, G. Serck-Hanssen, and K.B. Helle. 1994. Effects of experimental seismic shock on vasoactivity of arteries, integrity of the vascular endothelium and on primary stress hormones of the Atlantic salmon. ICES Journal of Marine Science 45: 973-995.

Svetovidov, A.N. 1948. Gadiformes. Israel Program for Scientific Translations, Vol. IX, No.4. S. Monson, Jersusalem, 304 pp.

Svetovidov, A.N. 1952. Herrings (Clupeidae). The USSR Fauna, Vol. 11 No. 1. AS Press, Moscow-Leningrad.

Svetovidov, A.N. 1963. Clupeidae. Israel Program for Scientific Translations, S. Monson, Jerusalem, 428 pp.

Swain, D.P. and A.F. Sinclair. 2000. Pelagic fishes and the cod recruitment dilemma in the northwest Atlantic. Canadian Journal of Fisheries and Aquatic Sciences 57: 1321-1325.

Sweeny, R.C.H. 1997. Accounting for change: Understanding merchant credit strategies in outport Newfoundland. *In* How deep is the ocean? Historical essays on Canada's Atlantic fishery. *Edited by* J.E. Candow and C. Corbin. University College of Cape Breton Press, Sydney, NS, pp. 121-138.

Taggart, C.T. and K.T. Frank. 1987. Coastal upwelling and *Oikopleura* occurence ("Slub"): A model and potential application to inshore fisheries. Canadian Journal of Fisheries and Aquatic Sciences 44: 1729-1736.

Taylor, V.R. 1985. The early Atlantic salmon fishery in Newfoundland and Labrador. Canadian Special Publications of Fisheries and Aquatic Sciences No.76.

Templeman, P. 1920. Proceedings of the Convention of Licensed Codfish Exporters. St. John's.

Templeman, W. 1940. The life history of the lobster. Research Bulletin (Fisheries) 15: 1-42. Newfoundland Department of Natural Resources.

Templeman, W. 1944. The life history of the spiny dogfish (*Squalus acanthias*) and the vitamin A values of dogfish liver oil. Newfoundland Department of Natural Resources. Research Bulletin (Fisheries) 15: 1-102.

Templeman, W. 1948. The life history of the capelin (*Mallotus villosus*) in Newfoundland waters. Bulletin of the Fisheries Research Board of Canada 17: 1-151.

Templeman, W. 1952. Report of the Newfoundland Fisheries Research Station for 1952. Fisheries Research Board of Canada, St. John's.

Templeman, W. 1959. Summaries of research, 1958. Canadian research report. A. Subareas 2 and 3. ICNAF Annual Proceedings for the Year 1959-60, Vol. 9. Dartmouth, NS.

Templeman, W. 1961. Canadian research report, 1960. A. Subareas 2 and 3. Part 3. ICNAF Annual Proceedings for the Year 1960-61, Vol. 3. Dartmouth, NS.

Templeman, W. 1962a. Summaries of research, 1961, by Subareas. Subarea 3, status of the fisheries. ICNAF Annual Proceedings for the Year 1961-62, Vol. 12. Dartmouth, NS.

Templeman, W. 1962b. Summaries of research, 1961, by Subareas. Subarea 3(6), cod. ICNAF Annual Proceedings for the Year 1961-62, Vol. 12. Dartmouth, NS.

Templeman, W. 1962c. Summaries of Research, 1961, by Subareas. Subarea 2(5), haddock. ICNAF Annual Proceedings for the Year 1961-62, Vol. 12. Dartmouth, NS.

Templeman, W. 1962d. Report of the Biological Station for 1961-62. Fisheries Research Board of Canada, St. John's.

Templeman, W. 1963. Summaries of research, 1962, by Subareas. Subarea 2(7). ICNAF Annual Proceedings for the Year 1962-63, Vol. 13. Dartmouth, NS.

Templeman, W. 1964. Report for the 1963-1964 of the Biological Station. Fisheries Research Board of Canada, St John's.

Templeman, W. 1966. Marine resources of Newfoundland. Bulletin of the Fisheries Research Board of Canada 154: 1-170.

Templeman, W. 1971. Summaries of research and status of fisheries by Subareas, 1970. Subarea 2. ICNAF Annual Proceedings for the Year 1970-71, Vol. 21, Part 3. Dartmouth, NS.

Templeman, W. and A.M. Fleming. 1953. Long-term changes in hydrographic conditions and corresponding changes in the abundance of marine animals. ICNAF Annual Proceedings 3: 79-86.

Templeman, W. and A.M. Fleming. 1962. Cod tagging in the Newfoundland area during 1947 and 1948. Fisheries Research Board of Canada 19: 446-486.

Templeman, W. and A.M. Fleming. 1963. Longlining experiments for cod off the east coast of Newfoundland and Labrador, 1950-1955. Fisheries Research Board of Canada Bulletin 141: 1-65.

Templeman, W. and J. Gulland. 1965. Review of possible conservation actions for the ICNAF area. Part 4. ICNAF Annual Proceedings for the Year 1964-65, Vol. 15. Dartmouth, NS.

Thompson, H. 1939. The occurrence and biological features of haddock in the Newfoundland area. Fisheries Research Bulletin No. 6. Newfoundland Department of Natural Resources.

Thompson, H. 1943. A biological and economic study of cod (*Gadus callarias*, L.) in the Newfoundland area, including Labrador. Fisheries Research Bulletin 14. Newfoundland Department of Natural Resources.

Travin, V.I. and K.P. Janulov. 1961. Soviet Union research report, 1960. A. Soviet Investigations in the ICNAF area in 1960. ICNAF Annual Proceedings for the Year 1960-61, Vol. 11. Dartmouth, NS.

Travin, V.I. and L.N. Pechenik. 1963. Soviet fishery investigations and fishing northwest Atlantic. In Soviet Fisheries Investigations in the Northwest Atlantic. Translated by B. Hershkovitz. U.S. Department of the Interior and the National Science Foundation, Washington.

Travin, V.I., E.I. Surkova, and E.M. Makevich. 1960. U.S.S.R. research report, 1959. C. Fishing activities and composition of the catches by areas. ICNAF Annual Proceedings for the Year 1959-60, Vol. 10. Dartmouth, NS.

Trippel, E.A. 1998. Egg size and viability and seasonal offspring production of young cod. Transactions of the American Fisheries Society 127: 339-359.

Tuck, L.M. 1960. The murres: their distribution, populations and biology. Canadian Wildlife Series 1. Department of Northern Affairs and National Resources, National Park Branch, Canadian Wildlife Service, Ottawa.

Ulltang, Ø. 1998. Where is fisheries science heading – how can fish stock assessments be improved? Journal of Northwest Atlantic Fisheries Science 23: 133-141.

Vigneris, L-A. 1956. New light on the 1497 Cabot voyage to America. Hispanic-American Historical Review XXXVI, No. 4.

Vilhjálmsson, H. 2002. Capelin (*Mallotus villosus*) in the Iceland-East Greenland-Jan Mayen ecosystem. ICES Journal of Marine Science 59: 870-883.

Vilhjálmsson, H. and G.A. Rose. Impacts of climate change on high latitude fisheries. In Climate Change Impacts, Adaptation and Vulnerability. Intergovernmental Panel on Climate Change (IPCC) (in press).

Walsh, A.J. 1953. Report of the Newfoundland Fisheries Development Committee. Government of Newfoundland.

Walsh, S. 1992. Factors influencing distribution of juvenile yellowtail flounder (*Limanda ferruginea*) on the Grand Bank of Newfoundland. Netherlands Journal of Sea Research 29: 193-203.

Walsh, S. and M.J. Morgan. 2004. Observations of natural behaviour of yellowtail flounder derived from data storage tags. ICES Journal of Marine Science 61: 1151-1156.

Warner, W.W. 1997. The fish killers. *In* How deep is the ocean? Historical essays on Canada's Atlantic fishery. *Edited by* J.E. Candow and C. Corbin. University College of Cape Breton Press, Sydney, NS, pp. 223-242.

Warren, W.G. 1997. Changes in the within-survey spatio-temporal structure of the northern cod (*Gadus morhua*) population, 1985-1992. Canadian Journal of Fisheries and Aquatic Sciences 54: 139-148.

Webb, J.A. 2001. The fishermen's protective union and politics. Newfoundland and Labrador Heritage, 1-4. Available at: http://www.heritage.nf.ca/law/fpu_politics.html.

Weise, F. 2002. Seabirds and Atlantic Canada's ship source oil pollution. World Wildlife Fund Canada, Toronto.

Wheeler, J. P. and G. H. Winters. 1984. Migrations and stock relationships of east and southeast Newfoundland herring (*Clupea harengus*) as shown by tagging studies. Journal of Northwest Atlantic Fisheries Science 5: 121-129.

Whidborne, R. 2005. Crosses and comforts: being the life and times of Captain Sir Richard Whitbourne (1561-

1635) of Exmouth in Devonshire. Great Auk Books, St. John's, 217 pp.

Whitbourne, R. 1620. A discourse and discovery of New-Found-Land. The English experience: its record in early printed books, 1971 (facsimile). Da Capo Press, Amsterdam.

Whiteley, A.S. 1977. A century on Bonne Espérance. Cheriton Graphics, Ottawa, 96 pp.

Williams, A.F. 1996. John Cabot and Newfoundland. Newfoundland Historical Society, St. John's, 64 pp.

Williamson, A.M. 1992. Historical lobster landings for Atlantic Canada, 1892-1989. Canadian Manuscript Report of Fisheries and Aquatic Sciences No. 2164.

Wilson, I. 1991. The Columbus myth: did men of Bristol reach America before Columbus? Simon and Schuster, New York, 240 pp.

Wilson, J.T. 1966. Did the Atlantic close and then re-open? Nature 211: 676-681.

Windle, M.J.S. and G.A. Rose. 2005. Migration route familiarity and homing of transplanted Atlantic cod (*Gadus morhua*). Fisheries Research 75: 193-199.

Windle, M.J.S. and G.A. Rose. 2006. Do cod form spawning leks? Evidence from a Newfoundland spawning ground. Marine Biology 150: 671-680.

Winters, G.H. 1977. Migrations and activity levels of overwintering Atlantic herring (*Clupea harengus harengus*) along southwest Newfoundland. Journal of the Fisheries Research Board of Canada 34: 2396-2401.

Winters, G.H. 1983. Analysis of the biological and demographic parameters of northern sand lance, *Ammodytes dubius*, from the Newfoundland Grand Bank. Canadian Journal of Fisheries and Aquatic Sciences 40: 409-419.

Winters, G.H. 1986. *Non gratum anus rodentum*. Unpublished working paper mimeo (courtesy of the author).

Winters, G.H. 1989. Life history parameters of sand lances, *Ammodytes* spp., from the coastal waters of eastern Newfoundland, Canada. Journal of Northwest Atlantic Fisheries Science 9: 5-12.

Winters, G.H. and E.L. Dalley. 1988. Meristic composition of sand lance (*Ammodytes* spp.) in Newfoundland waters with a review of species designations in the northwest Atlantic. Canadian Journal of Fisheries and Aquatic Sciences 45: 516-529.

Winters, G.H. and J.P. Wheeler. 1985. Interaction between stock area, stock abundance, and catchability coefficient. Canadian Journal of Fisheries and Aquatic Sciences 42: 989-998.

Winters, G.H. and J.P. Wheeler. 1987. Recruitment dynamics of spring-spawning herring in the northwest Atlantic. Canadian Journal of Fisheries and Aquatic Sciences 44: 882-900.

Winters, G.H. and J.P. Wheeler. 1996. Environmental and phenotypic factors affecting the reproductive cycle of Atlantic herring. ICES Journal of Marine Science 53: 73-88.

Worm, B. and R.A. Myers. 2003. Meta-analysis of cod-shrimp interactions reveals top-down control in oceanic food webs. Ecology 84: 162-173.

Worrall, S. 2005. Admiral Lord Nelson's fatal victory. National Geographic. October 1, 2005, Washington, pp. 54-69.

Wright, M. 1995. Fishers, scientists, the state and their responses to declines in the Bonavista fishery, 1950-1964. *In* Marine resources and human societies in the North Atlantic since 1500. *Edited by* D. Vickers. Institute of Social and Economic Research, Memorial University of Newfoundland, pp. 387-409.

Wright, M.C. 1997. Fishing in modern times: Stewart Bates and the modernization of the Canadian Atlantic fishery. *In* How deep is the ocean? Essays on Canada's Atlantic fisheries. *Edited by* J.E. Candow and C. Corbin. University College of Cape Breton Press, Sydney, NS, pp. 195-205.

Wroblewski, J., B. Neis and K. Gosse. 2005. Inshore stocks of Atlantic cod are important for rebuilding the east coast fishery. Coastal Management 33:411-432.

Young, V. 1988. A Question of Balance. Presentation to the Harris Review Panel on the State of the Northern Cod Stock.

INDEX

A

Abrego, 464
Act to Regulate the Whaling Industry, 299
Adolphus Nielsen, 299, 311, 322, 323, 324
Advisory Panel on Straddling Stocks, 538
Alaska, 25, 29, 78, 95, 113, 132, 467, 535
Alderdice, Frederick, 346, 347
Aldrich, Dr. Fred, 466
Alexander, David, 267, 325, 377, 380, 390, 403, 426, 465, 531
alligator fish, 30, 32, 144
Alverson report, 460
American Fisheries Society, 467
American fishery, 283
American lobster, 126, 129, 133, 134, 135, 221, 247, 248, 268, 279, 299, 301, 302, 323, 324, 331, 351, 403, 431, 527, 528
American Revolution, 229, 234
American Shore, 274
Amulree Royal Commission 1933, 318, 343, 346, 347, 350, 380, 403, 423, 442
anadromous fish, 122
Andrews, Ray, 437
Anglo-French commissions, 278
Ango, Jean, 185, 186, 191
animal rights, 421, 527, 539, 544
Anspach, Lewis Amadeus, 294
Antarctica, 24, 27, 29, 59, 128, 136, 140
aquaculture, 401, 419, 466, 502, 527, 542
Arctic char, 110, 123, 124, 125, 284, 307, 308, 476, 505
Arctic cod, 30, 59, 63, 83, 84, 85, 99, 142, 284, 476, 506
Arctic Ocean, 27, 30, 64, 136
Argentina, 388
Atlantic cod, 18, 24, 32, 36, 38, 44, 55, 57, 58, 59, 61, 62, 63, 64, 65, 66, 67, 68, 69, 70, 71, 72, 73, 75, 76, 77, 78, 79, 80, 81, 82, 83, 84, 85, 86, 91, 92, 95, 102, 110, 114, 115, 118, 120, 123, 126, 127, 128, 130, 135, 137, 145, 168, 171, 179, 180, 181, 183, 184, 185, 186, 187, 188, 189, 191, 193, 194, 207, 208, 209, 210, 214, 215, 219, 224, 226, 229, 230, 237, 239, 240, 241, 242, 243, 244, 245, 246, 247, 248, 249, 250, 266, 268, 269, 270, 271, 272, 274, 275, 277, 278, 280, 282, 284, 285, 286, 287, 289, 290, 291, 292, 294, 295, 299, 300, 301, 303, 304, 305, 306, 307, 308, 309, 310, 311, 312, 313, 316, 318, 319, 320, 321, 323, 324, 325, 327, 328, 330, 331, 332, 334, 335, 336, 337, 338, 340, 341, 342, 343, 344, 345, 346, 349, 351, 352, 353, 354, 378, 379, 381, 383, 384, 385, 386, 387, 393, 394, 395, 396, 397, 399, 400, 401, 403, 404, 405, 407, 410, 411, 412, 413, 414, 415, 416, 417, 419, 422, 426, 427, 428, 430, 431, 432, 435, 436, 437, 438, 439, 441, 442, 443, 444, 446, 447, 448, 449, 451, 452, 454, 455, 457, 458, 459, 460, 461, 462, 463, 464, 469, 470, 471, 472, 473, 475, 476, 477, 499, 500, 501, 502, 503, 505, 506, 507, 508, 509, 510, 511, 512, 513, 514, 515, 516, 517, 518, 519, 520, 523, 524, 525, 526, 527, 528, 529, 530, 531, 532, 533, 535, 536, 537, 538, 539, 541, 542, 543, 544, 545
Atlantic eel, 102, 125, 126, 135
Atlantic herring, 86, 87, 88, 419
Atlantic mackerel, 57, 66, 117, 118, 119, 120, 137, 186, 189, 247, 248, 282, 284, 290, 307, 321, 328, 329, 330, 352, 393, 403, 419, 420, 476, 502, 525
Atlantic salmon, 110, 122, 123, 124, 245, 307, 426, 476, 505, 527
Atlantic saury, 57, 118, 119, 121, 284, 328, 476
Atlantic walrus, 89, 116, 117, 169, 246, 247, 249, 285
Atwood, N.E., 290
Aubert, Captain Thomas, 185
Avalon Channel, 35, 43
Avalon Peninsula, 20, 23, 35, 71, 77, 123, 125, 170, 184, 186, 187, 188, 191, 193, 208, 218, 220, 221, 225, 233, 244, 271, 312, 348, 501, 503
Azores, 44, 181, 182

B

Baccalieu Island, 143, 182, 213, 215, 285
Bacon, Francis, 218
Baie Bois, 187
Baine Johnston Company, 271, 313, 318, 381
bait, 71, 99, 133, 186, 215, 249, 279, 281, 282, 283, 284, 287, 289, 290, 291, 292, 294, 303, 304, 305, 306, 308, 330, 336, 338
Bait Act, 279
Balaena, Hermitage Bay, 136, 299
Baltimore, Lord, 221
bank crash, 318, 327, 348
bank fishery, 288
Bank of Nova Scotia, 348, 441
Bannister, Jerry, 237
Bardot, Brigitte, 421, 539
Barents Sea, 18, 28, 37, 57, 77, 81, 101, 502, 505, 512, 533
Barry Group, 502, 527
Barry, Bill, 534, 543

Basques, 179, 189, 191, 193, 211, 232, 298
Bates, Stewart, 332, 349, 380, 383, 391, 432, 531
Battle Harbour, 271, 330
Bay Bulls, 71, 184, 218, 230, 235, 239, 275, 332, 333, 339, 341, 343, 345, 347
Bay of Castles, 186
Bay of Chaleur, 214
Bay of Exploits, 243, 245
Bay of Fundy, 45, 88, 232, 426
Bay of Islands, 304
Bay St. George, 73
baymen, 269
beachwomen, 288
Belles Amour, 187
Bering Strait, 29, 30, 45
Bermuda, 240, 275
Berry, Sir John, 215, 224
Bertin, Léon, 125
Bertius, 213
Beverton, Raymond, 384, 450
Biological Station of the Fisheries Research Board of Canada, 340, 466
Bird Rocks, 143, 285
Bismarck, Otto von, 427
Blackwood, Captain Baxter, 407
Blair, Arthur, 345
Blanc Sablon, 109, 186, 187, 192, 227, 244, 271, 312, 335
Bland, John, 246
blood antifreeze, 55, 57, 101
bluefin tuna, 119, 120, 121
Bonaparte, Napoleon, 235, 272
Bonaventure, 246, 270, 315
Bonavista Corridor, 24, 69, 75, 76, 108, 395, 470, 471, 472, 474, 476, 515, 525, 536
Bonavista Platform, 325, 326
Bond, Sir Robert, 281, 324, 533
Bonne Bay, 301, 304
Bonne Espérance, 185, 186, 187, 244, 245, 308, 309, 330
Boston, 229, 278, 301
Boston Tea Party, 278
bowfin, 29
Bowring Brothers, 299, 314, 318, 325, 381
Bradore, 185, 282
Brady, Major, 223
Brazil, 332, 337, 381
Brest, 185, 186, 192, 213, 308
Bristol, 179, 180, 184, 208, 209, 219
Bristols Hope, 220
British Columbia, 105, 126, 393, 402, 426, 528, 537
British Empire Marketing Board, 339

British Harbour, 270
British-Norwegian Whaling Company, 336
brook char, 125
brown trout, 125
Brown, Cassie, 298
Browns Bank, 393
Buchan, David, 243
Buckle, William, 271
bultows, 286
Burin, 187, 188, 393, 534
bye-boat keeper, 217, 240, 250, 269

C

C.F Bennett Company, 298
Cabot Steam Whaling Company, 299
Cabot, Sebastien, 180, 181, 194
Cabot, Zuan or John, 18, 179, 180, 181, 184, 185, 194, 213, 282, 299, 431, 448, 467, 499, 541, 545
Calanus, 55, 60, 61, 126, 127, 137, 145
Cameron, A.T., 453
Canada, 43, 123, 193, 221, 233, 267, 281, 282, 283, 286, 301, 319, 320, 322, 323, 332, 338, 339, 340, 341, 343, 346, 347, 348, 349, 350, 351, 352, 354, 379, 380, 381, 382, 383, 384, 385, 386, 388, 389, 390, 391, 393, 394, 395, 396, 397, 401, 402, 404, 405, 406, 416, 419, 423, 426, 427, 428, 429, 430, 431, 433, 435, 436, 437, 438, 444, 446, 448, 449, 450, 451, 453, 455, 457, 463, 464, 466, 469, 520, 530, 533, 538, 539, 540, 544
Canada Saltfish Corporation, 446
Canadian Fisheries Research Board, 322, 339, 340
Canso, 214, 303
Cape Bauld, 274
Cape Bonavista, 180, 182, 186, 192, 231, 232, 405
Cape Breton Island, 119, 171, 180, 189, 214, 231, 242, 303
Cape Broyle Whaling and Trading Company, 299
Cape Chidley, 299, 329
Cape Cod, 43, 140
Cape Degrat, 180, 192, 244
Cape Harrison, 120, 289, 292, 304, 319
Cape Pine, 213
Cape Race, 182, 213, 224, 315
Cape Ray, 234
Cape Spear, 184, 213
Cape St. Francis, 241
Cape St. John, 234, 242, 271, 274
Cape St. Mary's, 141, 143, 182, 213, 285, 536
Cape Verde, 182
capelin, 30, 36, 55, 57, 58, 59, 60, 61, 62, 63, 64, 66, 68, 71, 72, 73, 75, 77, 93, 95, 99, 101, 106, 107, 108, 109, 110, 114, 118, 120, 123, 125, 130,

136, 137, 138, 140, 141, 142, 145, 171, 186, 187, 189, 240, 244, 247, 248, 279, 287, 290, 294, 303, 305, 306, 313, 328, 330, 338, 341, 351, 352, 354, 393, 403, 407, 409, 418, 419, 422, 427, 428, 429, 434, 464, 470, 475, 476, 477, 500, 502, 503, 504, 505, 506, 507, 508, 509, 511, 512, 513, 518, 519, 525, 527, 535, 536, 539, 541, 542, 545
caravelles, 208
carbon productivity, 56
Carbonear, 187, 218, 224, 229, 230, 288
Caribbean, 23, 92, 120, 180, 215, 242, 272, 278, 332, 337, 381
caribou, 168, 169, 170, 171, 224
Carolina, 133, 191, 419
Carson, Dr. William, 275
Cartier, Jacques, 143, 183, 185, 188, 192, 246, 285
Cartwright, George, 243, 274
Cashin, Richard, 299, 438
Catadromous, 125
Catalina, 192, 534
catch rates, 18, 286, 287, 289, 393, 421, 434, 444, 447, 457, 458, 459, 460, 471, 473, 504
catfish
see eelpout, 110, 115
Cell, Gillian, 212
Centre for Newfoundland Studies, 213, 214, 220, 227, 234, 236
Cetaceans
 Atlantic right whale, 136, 189, 298
 blue whale, 99, 135, 136, 137, 139, 299, 380
 bottlenose whale, 136
 bowhead whale, 136
 fin whale, 135, 136, 139, 299, 336
 Greenland whale, 136
 humpback whale, 135, 136, 137, 299
 killer whale or orca, 136
 minke whale, 135, 136, 137, 299
 pilot whale, 136, 298, 336
 right whale, 136, 189, 298
 sei whale, 135, 136, 137
 sperm whale, 135, 136, 298
 white-sided dolphin, 136
Chain Rock, 235
Champlain Sea, 89
Charlie transform, 23, 75
Chesapeake Bay, 101
Chile, 388, 528
China, 178, 187, 502, 530, 534, 544
Chionoecete, 132
Clearwater Fine Foods, 533
Clenche, Thomas, 40
climate, 24, 25, 28, 34, 38, 42, 43, 45, 55, 61, 84, 168, 169, 172, 173, 194, 284, 285, 422, 477, 511, 512, 536
climate change, 511, 512, 536
Clutterbuck, P.A., 346
Coaker, William, 268, 272, 325, 326, 327, 338, 533
Cochrane, Sir Thomas, 275
cod biomass, 62, 63, 241
cod evolution, 32, 65
cod landings, 241, 286, 287, 384, 400, 412, 413, 414, 415, 531
cod seines, 287
cod trap, 308, 309, 310, 312, 320, 324, 331, 413, 432
Cod Trap Act, 310, 311
Collins, Captain J.W., 303
Collins, Harold, 400
Columbus, Christopher, 179, 180, 181, 282
Commission of Government, 268, 346, 347, 348, 388
Committee for the Status of Endangered Wildlife in Canada (COSEWIC), 500, 540, 541
Conception Bay, 191, 208, 218, 219, 220, 246, 248, 271, 274, 285, 288, 295, 317, 328
Confederation, 267, 268, 281, 340, 346, 347, 349, 351, 354, 377, 381, 402, 431, 438, 453, 464, 465, 542
Convention of 1857, 278
Cook, Captain James, 239, 245
Copes, Dr. Parzival, 451
Cormack, William Epps, 167, 170, 243, 270, 285
Coronation Gulf, 106
Corte Real, 179, 182, 183
crab, 221, 248
Cretaceous, 23, 24, 29
Crosbie Group, 382
Crosbie, John, 332, 382, 437, 463, 477, 501, 534, 537
Crustaceans, 126
Crutchfield, Dr. J.A., 397
Cuba, 391, 428, 429
Cupids, 205, 218, 220
Cyclostomes, 29

D

Dalhousie University, 339, 340, 538
Davies, James, 338
Davis Strait, 210
Davis, Jack, 400, 438
Davis, John, 210, 242
Dawe Commission 1962, 377, 403, 404, 405, 406
de Bané, Pierre, 446
Death on the Ice, 316, 325
Decapods, 129
Demasduit, 243

Denmark, 178, 180, 207, 322, 332, 337, 391, 429
Denys, Jean, 185, 186
Department of Fisheries and Oceans, 340, 447, 453, 459, 460, 461, 462, 463, 467, 469, 470, 524, 525, 537, 540, 541
Department of Natural Resources, 339, 342, 347
Devon, 179
Devonian, 29, 128
Dicuil, 172
Dieppe, 185, 186, 188, 191, 193
Dildo, 323
DNA analyses, 113
dogfish, 120, 135, 328, 329
dolly varden, 123
Dover Fault, 21
Drake, Francis, 210
Drover, Colonel Paul, 412
Duckworth, 243
Dunne report, 463
Dutch, 209, 227, 230, 298

E

Earle, H.R.V., 404
East Germany, 384, 388, 391, 429
echogram, 71, 88, 97, 108, 114
ecosystem, 18, 22, 30, 55, 57, 58, 59, 62, 63, 70, 75, 144, 173, 181, 194, 248, 249, 250, 329, 344, 351, 353, 379, 393, 428, 429, 456, 470, 475, 476, 503, 505, 506, 511, 512, 513, 518, 527, 529, 538, 539, 541, 542
Ecuador, 388
eelgrass, 30, 69, 84
eelpout, 59, 66, 110, 115
eggs, 64, 67, 68, 69, 70, 75, 77, 78, 82, 87, 91, 93, 99, 101, 102, 108, 110, 111, 113, 115, 118, 130, 132, 135, 142, 170, 171, 192, 285, 294, 323, 324, 342, 409, 513, 516, 545
Eminent Panel on Seal Management, 538
English Century, 207
English Shore, 230, 235, 236, 270
Esquiman channel, 73, 75
Estonia, 429
Etchegary, Gus, 426
Euphasiids, 128
European Economic Community, 399, 429, 430, 463
European Union, 328, 429, 470, 507, 524
Evening Telegram, 317
Export price, 289
exports, 284, 288, 294, 308, 324, 347, 447, 529

F

$F_{0.1}$, 455, 456, 463
Fairtry, 384, 385, 422
farming, 222, 272, 274, 401, 402, 542
Faroe Islands, 77, 123, 523, 532
Fermeuse, 218, 220, 221
Ferryland, 209, 213, 217, 220, 221, 222, 223, 230
fires, 239, 275, 382
fish sticks, 383, 423
Fisheries and Marine Institute, 465, 468
Fisheries Broadcast, 537
Fisheries College, 465
Fisheries Conservation Chair, 468
Fisheries Council of Canada, 352, 434
Fisheries Research Board of Canada, 452, 453, 464
Fisheries Resource Conservation Council (FRCC), 524, 525, 538
Fisheries, Food, and Allied Workers (FFAW), 378, 438, 439, 440
Fishery Products, 383, 435, 437, 439, 441, 445, 446, 460, 462, 468, 500, 501, 502, 533, 534, 535, 538, 544
Fishery Products International, 437, 445, 446, 460, 462, 468, 500, 501, 502, 533, 534, 535, 538, 544
Fishery Products International Act, 445, 533, 534, 535
fishing admirals, 235
flatfish, 98, 378, 417
  American plaice, 95, 100, 101, 104, 330, 336, 345, 351, 352, 379, 393, 403, 417, 464, 475, 501, 518, 541
  Atlantic halibut, 98, 99, 100, 103, 282, 301, 302
  Greenland halibut, 36, 59, 99, 100, 103, 110, 308, 330, 336, 352, 393, 403, 429, 505, 518, 528
  winter flounder, 57, 100, 101, 104, 110, 403, 505, 528
  witch flounder, 102
  yellowtail flounder, , 100, 101, 103, 393, 403, 417 501, 513, 518, 533
Flemish Cap, 23, 37, 385, 399, 406, 428, 429, 437, 477, 504, 508, 524
flipper pie, 224
floaters, 292, 312
Flowers, Professor, 212, 279
F_{max}, 455, 456
Fogo Island, 191, 231, 245, 271
food web, 25, 55, 58, 106, 505
foreign fisheries, 277, 337, 416, 419, 442, 443, 469, 507, 510, 532
Fort Amherst, 235
Fort William, 229
Forteau, 187

Fortin, Pierre, 282
Fortune, 283, 304, 305, 336, 432, 508, 534
fossils, 29, 32, 38, 89, 122, 128, 129, 133, 135, 316, 536
French shore, 231, 234, 242, 244, 271, 278, 279, 280, 310, 331, 336
Frost, Nancy, 70, 339, 342, 343
frozen fish, 332, 335, 380, 381, 383, 387, 404, 406, 407, 431, 442
Funk Island, 35, 75, 102, 140, 141, 142, 143, 171, 191, 192, 213, 214, 220, 249, 285, 329, 454

G

gadoids, 32, 45, 64, 65, 68, 81, 84
Gadus atlantica, 470, 471
gaffe cod, 244
Gambier, Governor, 246, 275
gardens, 36, 224, 225, 240, 306
garpike, 29
Gaspé, 232, 242, 283
Gasset, Jose Ortega y, 401
Gastaldi, Giacomodi, 190
Gaultois, 298, 432
Georges Bank, 37, 58, 72, 81, 82, 99, 101, 106, 114, 301, 386, 391, 393, 408
German Hanseatic League, 207
Gilbert, Sir Humphrey, 209, 210, 212, 229
global warming, 511
Globe and Mail, 436
Gloucester, MA, 282, 290, 292, 301, 303, 304, 526
Gordon, Scott, 426, 450
Gosse, Eric, 396, 397, 405
Governors, 182, 212, 215, 218, 219, 225, 237, 238, 239, 243, 245, 246, 249, 271, 274, 275, 278, 324, 346, 348
Grand Banks, 18, 19, 20, 21, 23, 24, 25, 26, 28, 33, 35, 36, 37, 39, 42, 43, 44, 45, 55, 57, 58, 60, 61, 63, 71, 72, 73, 75, 76, 77, 81, 82, 84, 91, 95, 96, 97, 98, 99, 101, 102, 106, 108, 114, 119, 123, 130, 138, 140, 143, 179, 184, 185, 186, 187, 188, 189, 211, 213, 214, 226, 230, 234, 236, 239, 242, 244, 247, 248, 250, 268, 280, 284, 290, 292, 294, 301, 303, 316, 319, 328, 331, 332, 336, 337, 338, 342, 343, 351, 353, 379, 380, 382, 383, 384, 385, 386, 387, 388, 390, 393, 394, 395, 397, 399, 404, 405, 406, 407, 408, 410, 411, 412, 416, 417, 419, 421, 422, 424, 426, 427, 428, 429, 430, 431, 437, 472, 474, 475, 477, 501, 502, 503, 505, 506, 507, 508, 509, 512, 513, 518, 522, 525, 529, 533, 535, 536, 538, 539, 540, 541, 544, 545
Grand Manan, 283, 305

Grates Point, 188
Great International Fisheries Exhibition 1883, 320, 379, 386
Great Lakes, 123
Great Northern Peninsula, 75, 169, 170, 172, 180, 188, 244
Green Bank, 236, 280, 303
green fish, 232
Greenland, 21, 22, 23, 24, 25, 27, 28, 29, 32, 36, 38, 43, 57, 59, 61, 64, 65, 66, 78, 80, 91, 93, 99, 100, 103, 110, 123, 136, 142, 172, 178, 179, 184, 210, 284, 308, 313, 330, 336, 352, 386, 391, 393, 403, 421, 429, 469, 505, 511, 512, 518, 528
Greenland cod, 64, 65, 66, 78, 80, 469
Greenspond, 271
grenadier, 59, 92, 93, 94, 96, 379, 387, 416, 417, 418, 422, 429, 518, 541
Grenfell Mission, 246
Grimes, Roger, 534
Gros Morne, 20
Groswater Bay, 78, 108, 168, 290, 291
growth, 58, 71, 72, 73, 74, 78, 93, 130, 269, 284, 327, 330, 336, 342, 351, 379, 391, 408, 421, 423, 424, 425, 442, 449, 454, 455, 457, 458, 459, 460, 470, 477, 505, 506, 509, 513, 519, 525
Gulf of Mexico, 26, 92
Gulf of St. Lawrence, 38, 40, 43, 44, 58, 60, 61, 73, 75, 81, 88, 91, 98, 99, 108, 114, 116, 117, 118, 133, 143, 172, 180, 185, 186, 189, 208, 214, 231, 242, 244, 249, 279, 282, 283, 285, 289, 301, 303, 314, 322, 342, 343, 413, 421, 426, 432, 504, 506, 520, 523, 525, 526, 541
Gulf Stream, 18, 25, 26, 28, 29, 37, 43, 60, 73, 82, 126, 188
Gulland, Dr. John, 393
gurry fish, 294
Gushue, Dr. Raymond, 350, 352, 382
Guy, John, 218, 219, 220, 222, 235, 243, 249
Gwyn, Richard, 346, 348, 406, 407

H

haddock, 32, 65, 80, 81, 82, 95, 328, 330, 331, 341, 345, 351, 354, 379, 383, 385, 386, 393, 403, 405, 407, 408, 409, 410, 411, 413, 414, 422, 464, 475, 518, 541
hake, 84, 92, 383, 393, 403
  silver, 84
  spotted, 84
  white, 84, 85
Hallock, Charles, 291
Haly. Colonel, 223
Hamilton Bank, 35, 60, 75, 76, 400, 401, 413, 422,

448, 464, 469
Hamilton Inlet, 210, 400, 401, 406, 454
Hampton, W.F., 343, 345
Hann, Allister, 502
Harbour Breton, 271, 328, 432, 499, 502, 503, 527, 534
Harbour Grace, 212, 218, 219, 221, 269, 271, 288, 317
harp seal, 39, 58, 63, 84, 89, 90, 145, 169, 284, 296, 393, 420, 421, 506, 513, 526, 539
Harris report, 462, 463, 537
Harris, Dr. Leslie, 439, 461, 466
Harvey, Reverend Dr. Moses, 266, 322
Hawke Channel, 35, 36, 44, 75, 102, 336, 401, 464, 508, 522
Hawke Harbour, 336
Hayes, Ronald, 453
Hayman, Robert, 212, 219, 221, 224
Hearn, Loyola, 526
Heinrich Hammer, 178
Henley Harbour, 214
Hermitage Bay, 73, 169, 298, 299
Herring Neck, 325
Hibernia, 535
High Liner of Nova Scotia, 534
Hinton, William, 225
History of the New England Fisheries, 308
Hjort, Johan, 321, 326
Hodder, 413
Holdsworth, Arthur, 237
home rule, 267, 275, 276
homing, 18, 30, 32, 60, 64, 75, 115, 123, 125, 126, 138, 171, 179, 193, 208, 211, 218, 222, 224, 232, 242, 246, 275, 276, 280, 282, 288, 292, 295, 301, 319, 325, 331, 346, 385, 386, 388, 421, 429, 476, 515, 536
Honduras, 431
hooded seal, 89, 246, 296, 331, 421, 422, 506, 526
Hopedale Channel, 35, 44
Howley, J.P., 319
Hudson Bay, 43, 106, 335
Humber River, 245
Huntsman, A.G., 339, 340
Huxley, Sir Thomas, 321
hyperaggregation, 471

I

Iapetus Sea, 20, 21, 31
ice, 22, 24, 25, 28, 30, 33, 34, 35, 36, 37, 38, 39, 42, 43, 45, 55, 58, 59, 60, 63, 72, 73, 77, 78, 81, 82, 83, 88, 89, 96, 105, 106, 111, 116, 122, 123, 125, 126, 129, 133, 137, 145, 168, 187, 194, 224, 241, 246, 247, 248, 250, 284, 292, 295, 297, 313, 314, 315, 384, 387, 421, 436, 440, 446, 454, 464, 475, 501, 502, 506, 511, 522, 544, 545
Ice ages, 25, 31, 33, 34, 38, 45, 59, 60, 72, 78, 81, 82, 83, 88, 89, 96, 105, 111, 122, 125, 126, 387
Iceland, 21, 28, 32, 37, 42, 55, 57, 58, 61, 65, 77, 83, 91, 93, 99, 101, 106, 113, 117, 127, 142, 172, 179, 180, 183, 185, 207, 208, 218, 313, 325, 327, 328, 331, 332, 336, 337, 338, 342, 349, 351, 353, 380, 386, 388, 391, 406, 410, 429, 453, 465, 469, 505, 511, 524, 529, 530, 532, 544
Idler, Dr. D.R., 466
immigration, 269
Indian Harbour, 308, 329
Indian Ocean, 29, 178
International Commission of the Northwest Atlantic Fisheries (ICNAF), 353, 377, 385, 386, 391, 392, 393, 394, 397, 401, 409, 410, 411, 413, 415, 416, 417, 419, 421, 427, 428, 429, 431, 453, 455, 458, 459, 462, 464
International Committee on Deep Sea Fisheries, 338
International Council for the Exploration of the Sea, 322
Inuit, 169, 171
Investigator II, 345, 464
Ireland, 23, 172, 179, 184, 185, 188, 269, 270, 425
Irish, 58, 172, 179, 180, 239, 275, 281
Ishkov, 385, 386
Isle au Haut, ME, 303
Isle aux Bois, 187
Italy, 178, 181, 185, 207, 208, 232, 280, 316, 327, 332, 381, 391, 530

J

Jackson, Reverend John, 229
Jamaica, 240, 326, 381
Jamieson, Donald, 427
Japan, 84, 113, 300, 380, 387, 388, 391, 415, 427, 428, 429, 447, 453, 465, 467
jellyfish, 91, 93
jiggers, 286, 290, 292, 308
Job Brothers, 298, 314, 318, 335, 381, 383
John Munn and Company, 288, 299, 317
John Penney and Sons, 446
Johnson, B.G.H., 408
Junger, Sebastian, 469

K

Karlsson, Gunnar, 207
Kean, Abram, 325
Keats, Derek, 447, 459, 460
kelp, 30, 78, 82, 87, 105

Kenya, 123
Keough, Dr. Kevin, 466
keystone species, 57, 58, 60, 505
killick, 312
King William's Act, 237
King, Gordon, 73, 228
Kirby Report, 378, 441, 442, 443, 444, 445, 448
Kirby, Dr. Michael, 441
Kirke family, 209, 216, 217, 221, 223, 230, 235
krill
   see euphasiids, 59, 128, 136, 137
Kurlansky, Mark, 499

L

L'Escarbot, Marcus, 214
La Rochelle, 184, 185, 192, 212, 230
Labrador boats, 290
Labrador cure, 288, 331, 332, 381
Labrador Current, 18, 25, 26, 28, 35, 37, 38, 39, 42, 43, 44, 45, 55, 57, 60, 73, 75, 99, 137, 511
Labrador Sea, 24, 28, 43, 99, 123
Lake Group, 32, 243, 383, 435, 439, 441, 446, 502
Lake, Spencer, 439
landed value, 230, 289, 520, 521
Lankester, Sir Raymond, 321, 411
lanternfish, 23, 59, 96, 97, 123, 144
larvae, 60, 69, 70, 75, 77, 93, 99, 101, 113, 129, 130, 132, 135
Latvia, 429
Law of the Sea, 378, 389, 426, 427, 444
Law on the Scientific Protection of Fishing Grounds on the Shelf of Iceland, 353
Laws, F.A.J., 235, 352
LeBlanc, Romeo, 434, 435, 438, 439, 440
LeMarchant, Governor, 324
Leopold, Aldo, 450, 503, 531, 538
Lindsay, Sheila Taylor, 343
Lithuania, 429
Little Ice Age, 38, 169, 284, 294, 298, 328
liveyers, 205
Lloyd, Thomas, 229
London, 181, 184, 185, 209, 211, 222, 229, 235, 320, 321, 323, 379, 386, 389, 391
Long Harbour, Fortune Bay, 283
longliners, 337, 395, 401, 446
Lower Canada, 233, 281, 282
Lower North Shore of Quebec, 75, 185, 308
lumpfish, 91, 92, 141, 528

M

Mackenzie River, 123
Macpherson, N.L., 332, 343
Magdalene Islands, 116
Magrath, C.A., 346
Maine, 89, 191, 303, 305, 336
Maloney, Aiden, 432, 449
Marblehead, ME, 301
Marconi, 24
Marine Sciences Research Laboratory, 466
Maritime Archaic Indians, 168, 169
Martin, Cabot, 431, 448, 467, 499, 541, 545
Marystown, 407, 534
Mason, John, 212, 219, 220, 221, 223, 248, 282
Matthew, 180, 450, 451, 452
Matthews, David, 450, 451, 452
maximum historic biomass, 72
May, Dr. Arthur, 383, 428, 448, 538
McCallum, Hermitage Bay, 299
McGill University, 282, 340, 341
McGrath, Father Desmond, 438
McIntosh, W.C., 321, 322
McMurrich, J.P., 322
Memorial University Act 1949, 465
Memorial University College, 338, 339, 340, 343, 465
Memorial University Department of Biology, 338, 339, 340, 341, 465, 466, 467
merchants, 179, 181, 184, 189, 207, 208, 209, 211, 215, 216, 217, 221, 223, 224, 239, 240, 241, 242, 245, 249, 269, 270, 271, 272, 274, 275, 276, 278, 281, 288, 295, 299, 314, 315, 317, 318, 319, 326, 327, 337, 348, 350, 431, 532, 533
Merisheen, 187
Mexico, 388, 429
migration, 24, 29, 36, 45, 57, 59, 68, 72, 73, 75, 79, 86, 90, 91, 107, 108, 113, 119, 121, 123, 124, 135, 138, 139, 141, 142, 169, 170, 186, 187, 191, 219, 224, 242, 248, 308, 320, 321, 322, 338, 342, 351, 400, 464, 470, 472, 476, 510, 531, 535, 536
migratory fishery, 211, 215, 217, 221, 240, 269, 274, 276
Millais, John Guille, 319
monkfish, 29
Montagnais, 192
Moody, John, 229
Moores, Frank, 438
moratorium, 421, 428, 430, 457, 463, 464, 472, 501, 502, 503, 507, 508, 509, 510, 513, 527, 537, 539
Moravian Missions, 246, 308
Morgan, Jim, 436
Morocco, 22, 431
Morris, Patrick, 275, 379
Morrow, C.J., 352
Murray, Alexander, 278
mussels, 221

N

Napoleonic war, 274
National Convention 1946, 348
National Sea Products, 431, 436, 437, 441, 445, 446, 460, 461, 502, 535
Natural Sciences and Engineering Research Council (NSERC), 468
Navies, 218, 223, 229, 238, 239, 272, 274, 278
Needler, Dr. Alfred, 391, 394
Nelson, Lord Admiral, 210, 211, 272
New England, 20, 21, 37, 43, 58, 71, 72, 78, 81, 82, 88, 120, 122, 133, 135, 213, 215, 217, 219, 221, 233, 239, 240, 245, 246, 248, 267, 271, 274, 282, 283, 284, 286, 288, 289, 290, 291, 292, 298, 301, 303, 304, 322, 348, 386, 390, 408, 410, 419
New Hampshire, 219, 284
New Perlican, 224
New Zealand, 123, 465, 511, 544
Newburyport, MA, 290
Newfies, 348
New-founde-land, 179, 180, 181, 182
Newfoundland Associated Fish Exporters Limited (NAFEL), 268, 350, 352, 381, 382, 404
Newfoundland Board of Trade, 326
Newfoundland Centralization Program, 424
Newfoundland Company, 218, 219, 220
Newfoundland Department of Fisheries, 299
Newfoundland Department of Marine and Fisheries, 338
Newfoundland Fisheries Commission, 268, 299, 310, 322, 346
Newfoundland Fisheries Research Institute, 332
Newfoundland Fisheries Research Station, 464
Newfoundland Inshore Fishermen's Association (NIFA), 447, 448, 459
Newfoundland Steam Whaling Company, 299
Newman and Company, 271
Nickerson, H.B. & Sons Ltd., 436, 441, 446
Norse, 169, 170, 172, 178, 179, 180, 185
North Atlantic Drift, 28, 29, 43, 77, 126, 184
North Atlantic Oscillation, 42
North Harbour, 319
North Pacific, 25, 30, 32, 45, 57, 64, 83, 86, 89, 93, 98, 99, 101, 106, 110, 111, 113, 114, 115, 116, 126, 128, 129, 132, 141, 142, 145
North Sea, 32, 58, 65, 67, 72, 77, 95, 184, 321, 342, 352, 384
northern cod, 58, 68, 69, 76, 186, 187, 226, 244, 287, 384, 399, 400, 413, 415, 416, 424, 429, 430, 431, 432, 436, 437, 438, 442, 443, 444, 446, 447, 448, 449, 451, 452, 454, 455, 457, 458, 459, 460, 461, 462, 463, 469, 470, 471, 472, 477, 502, 507, 508, 512, 513, 515, 518, 525, 533, 536, 539
Northern cod Adjustment and Recovery Program, 501
Northern Fishermen's Union (NFU), 438
Northwest Atlantic Fisheries Organization (NAFO), 378, 392, 429, 430, 443, 444, 447, 453, 457, 459, 462, 463, 464, 470, 501, 510, 522, 538, 539, 540
Northwest Territories, 123
Norway, 21, 22, 24, 27, 28, 61, 113, 134, 300, 320, 321, 322, 326, 327, 328, 331, 351, 381, 391, 406, 415, 428, 429, 453, 465, 467, 511, 530, 532, 535
Norwegians, 137, 218, 298, 299, 338
Nose and Tail of the Grand Bank, 399, 428, 429, 430, 463, 472, 477, 508
Notre Dame Bay, 78, 88, 231, 243, 244, 245, 299, 325
Nova Scotia, 20, 21, 72, 82, 105, 106, 119, 173, 189, 191, 225, 232, 275, 281, 283, 301, 303, 315, 322, 331, 332, 340, 343, 349, 378, 383, 386, 391, 408, 413, 426, 431, 432, 434, 436, 437, 438, 441, 443, 445, 446, 448, 449, 477, 502, 523, 533, 534, 544

O

Ocean Choice International, 535
Odessa, 384, 385, 386, 422
Old Fort, 185
original species
see aboriginal species, 86
Orphin Basin, 23, 24, 75, 536
Osborne, Henry L., 238, 292, 294
Ougier, Peter, 266, 275
outports, 182, 185, 224, 269, 270, 271, 274, 283, 292, 317, 318, 319, 325, 326, 348, 423, 424, 425, 426, 532, 545
overfishing, 266, 267, 286, 294, 303, 320, 322, 330, 351, 378, 386, 394, 397, 399, 400, 405, 406, 408, 413, 415, 421, 427, 434, 437, 442, 444, 449, 450, 451, 454, 463, 469, 470, 475, 477, 499, 501, 504, 508, 509, 510, 527, 532
Oxford, Thomas, 225

P

Pacific cod, 64, 65, 66, 78
Pacific herring, 86
Pacific ocean perch, 113, 114
Pacific salmon, 122, 125
Palaeo-Eskimos, 168, 169, 172
Palliser, Hugh, 238, 239, 243, 245
Panama, 26, 28, 29, 45, 116, 429, 431
Pandalid shrimp, 57, 64, 93, 129

Parkhurst, Anthony, 208, 209
Parsons, Dr. Scott, 427, 445
pasta, 183
Paton, J.L., 338, 339, 340, 341, 346, 465, 466
Peace of Amiens, 277, 278
Peary, Robert, 298
Peckford, Brian, 440
Peckham, Sir George, 210
Peru, 388
Peters, G.R., 466
Petit Nord, 274, 278
Petite Forte, 187
Pew Charitable Trusts, 536
phytoplankton, 55, 60, 127, 512
Pickersgill, Jack, 433, 465
Pinhorn, Allenby, 442, 454, 455
Pinnaces, 215
Placentia Bay, 43, 45, 73, 76, 77, 95, 187, 218, 225, 228, 230, 231, 238, 305, 311, 312, 314, 319, 323, 331, 336, 504, 508, 524, 536
Plaisance, 187, 230
Planta, Clive, 402, 403
planters, 217, 237, 241, 242
Pointe Riche, 231
Poland, 384, 388, 415, 428, 429, 469
pollock, 32, 64, 65, 82, 83, 95, 119, 403
Polo, Marco, 178
Poole, 209, 271
Port de Grave, 188
Port Saunders, 279
Port-aux-Choix, 73, 168
Portugal, 21, 178, 179, 180, 181, 182, 183, 184, 186, 187, 188, 189, 191, 192, 193, 207, 208, 209, 211, 212, 213, 214, 232, 242, 275, 280, 316, 336, 337, 350, 353, 383, 385, 386, 387, 390, 391, 399, 404, 412, 415, 416, 428, 429, 430, 457, 462, 464, 469, 507, 530
Portugal Cove, 182, 275
Portuguese White Fleet, 337, 387
Pouliot, C.E., 352
prejudice, 379
prey, 57, 58, 63, 65, 71, 74, 82, 92, 95, 98, 99, 111, 118, 130, 137, 140, 142, 298, 428, 476, 505, 506, 512, 513, 519
Prince Edward Island, 232, 301, 313, 432
Prince, E.E., 232, 301, 313, 322, 432, 535, 536
Provincetown, MA, 290
Prowse, Judge D.W., 180, 215, 217, 235, 237, 295, 314
Ptolemy, 177
Puerto Rico, 381
pulse fishing, 393

Q

Québec, 193, 230, 231, 352
Quidi Vidi, 223, 229
quintals, 280, 313, 336
Quirpon, 187, 192, 244
quotas, 391, 392, 393, 395, 397, 427, 428, 429, 430, 436, 445, 446, 448, 449, 455, 457, 460, 467, 470, 526, 537

R

R.G. Reid Company, 324
Raleigh. Sir Walter, 210
Ramea, 246, 432
Random Sound, 328
rationalization, 349, 434, 440, 451, 452
rattails
    see grenadier, 92
Recent Indians
    Beaches Indians, 169
    Beothuks, 169, 170, 171, 192, 243
    Cow Head Indians, 169, 172
    Little Passage Indians, 169
recruitment, 321, 414, 454, 455, 458, 459, 470, 522, 524, 541
redfish, 30, 36, 57, 61, 63, 66, 73, 96, 111, 112, 113, 114, 145, 328, 330, 345, 351, 352, 354, 378, 379, 385, 386, 387, 393, 403, 405, 406, 410, 411, 417, 422, 432, 446, 464, 475, 502, 518, 533, 541
refugo, 183
Renews, 186, 221, 245, 246
Republic of Korea, 429
resettlement, 378, 423, 424, 425
Resources for Tomorrow Conference 1961, 396
Resource-Short plant quotas, 445, 446
Ricker, Dr. William, 393, 397, 450
Rideout, Tom, 437
Riders Harbour, 270
Ridleys, 288
Risley, John, 533, 534
rockfish
    see redfish, 32, 111, 113, 114
Rose Blanche, 327
Rose-au-Rue, 336
Rowe, Derrick, 534
Russels of Bonavista, 383
Russia, 86, 245, 315, 322, 353, 385, 429
Ryan, Daniel, 300
Ryan, Shannon, 295
Ryther, John, 422

S

Sabine, Lorenzo, 183, 282
Sable Island, 116, 209, 285
sack ships, 209, 242
saithe
   see pollock, 32, 83
salinity, 42, 126
Salmon Bay, 245, 282
salt, 21, 26, 29, 183, 185, 186, 188, 189, 207, 208, 210, 217, 224, 242, 269, 271, 272, 275, 277, 279, 280, 287, 288, 289, 290, 291, 313, 325, 327, 331, 332, 334, 335, 337, 346, 349, 350, 351, 353, 380, 381, 382, 383, 395, 404, 406, 407, 442, 446, 447, 465, 526, 530, 531, 532, 545
Salt Codfish Act 1933, 347
salt fish, 183, 185, 188, 207, 208, 217, 224, 269, 271, 272, 275, 277, 279, 280, 287, 288, 289, 313, 324, 327, 331, 332, 334, 337, 346, 349, 350, 351, 353, 380, 381, 382, 383, 395, 404, 406, 407, 442, 446, 447, 465, 530, 531, 532, 545
Salvage, 224, 230
sand lance, 55, 58, 60, 63, 73, 77, 93, 94, 95, 101, 114, 123, 138, 140, 142, 247, 291, 294, 330, 508, 542
Sanfilippo, Angela, 499
Sargasso Sea, 125
Sars, G.O., 321
school fish, 294
schooners, 244, 282, 288, 290, 292, 304, 305, 312, 313, 319, 336, 349
science, 319, 320, 321, 322, 323, 324, 337, 338, 339, 340, 341, 343, 386, 405, 408, 409, 410, 414, 428, 429, 430, 435, 442, 449, 450, 453, 454, 455, 457, 459, 460, 461, 462, 463, 465, 466, 467, 468, 503, 507, 512, 516, 525, 533, 537, 538, 539, 541, 542, 543, 544
Scottish Fisheries Research Board, 321
Scripps Oceanographic Institute, California, 338
sea butterfly, 105
sea ice, 37, 39, 40, 41, 59
sea level, 24, 25, 32, 33, 34, 36, 37, 60, 72, 108
sea urchin, 105, 111, 115, 133, 144, 294, 422, 528
seabirds, 55, 57, 84, 95, 106, 110, 137, 138, 140, 141, 142, 145, 170, 171, 188, 192, 215, 249, 285, 287, 341, 351, 422, 476, 505, 536
   Arctic tern, 140
   Atlantic puffin, 140, 142
   common murre, 140
   dovekie, 184
   great auk, 140, 141, 142, 145, 171, 191, 192, 249, 285
   gull, 99
   murre, 140, 141, 142, 145
   northern gannet, 140, 143, 145, 285
   thick-billed murre, 140, 141
seal hunt, 247, 275, 295, 297, 313, 315, 316, 317, 423, 527
seal oil, 316
Second East Coast Fisheries Conference 1948, 352
seismic surveys, 535
Selachians, 29
Sept Isles, 282
settlement, 211, 212, 214, 215, 216, 217, 218, 220, 221, 223, 224, 225, 226, 229, 238, 239, 240, 243, 246, 269, 270, 271, 283, 285, 295, 346, 530
settlers, 171, 172, 215, 216, 217, 218, 221, 222, 223, 224, 225, 229, 230, 235, 237, 238, 239, 240, 241, 242, 243, 245, 266, 269, 271, 275, 276, 277, 279, 304, 319
Sevastopol, 384, 422
Seven Years War, 231, 232
Shamook, 513
Shanadithit, 167, 180, 243
Shea, Ambrose, 320
Shrank, Dr. William, 435
Siberia, 29, 86
Siddon, Thomas, 461
Sierra Leone, 431
Signal Hill, 24, 231, 235
Sinclair, James, 385, 386
six mile rule, 225
skipper
   see saury, 118, 288, 304, 407, 513
Slade Company, 271
Sleggs, George, 338, 339, 341
slub, 30, 105, 476
Smallwood, Joseph, 272, 326, 348, 377, 382, 390, 402, 403, 404, 406, 407, 423, 424, 441, 533
Smith Sound, 328, 513, 514, 515, 516, 517, 518, 525, 545
Smith, Captain Henry O., 304
Smokey Harbour, 329
snow crab, 30, 36, 57, 66, 126, 129, 132, 133, 284, 294, 476, 500, 503, 512, 518, 519, 520, 521, 522, 523, 537, 544
Sobey family, 436
sole, 98, 168, 271, 324, 447, 524
Southeast Shoal, 23, 37, 60, 82, 108
Southern Shore, 88
Soviet Union, 379, 380, 382, 383, 384, 385, 387, 391, 407, 408, 409, 412, 415, 419, 422, 427, 428, 429, 458, 464, 469
Spain, 21, 22, 178, 179, 180, 181, 182, 183, 186,

189, 191, 193, 207, 208, 209, 210, 211, 232, 240, 247, 272, 280, 285, 316, 327, 332, 334, 336, 337, 350, 351, 353, 383, 385, 386, 387, 390, 391, 401, 407, 408, 410, 412, 415, 416, 428, 429, 430, 431, 457, 462, 464, 469, 507, 530

Spanish Armada, 211

spawning, 33, 60, 61, 65, 67, 68, 69, 70, 73, 75, 76, 77, 81, 84, 87, 88, 91, 99, 101, 106, 108, 109, 110, 118, 122, 125, 126, 130, 142, 186, 188, 244, 245, 250, 278, 287, 290, 294, 320, 322, 323, 324, 342, 351, 386, 400, 401, 407, 411, 413, 414, 419, 422, 427, 431, 447, 448, 452, 454, 458, 459, 463, 469, 470, 472, 474, 504, 505, 506, 507, 508, 511, 512, 515, 524, 525, 535, 536, 539, 541

Species at Risk Act, 540

Spitzbergen, 81, 99

squid, 57, 282, 287, 303, 352, 393, 403, 464, 466

Squires, Richard, 326, 346

St. Barbe, 279

St. John's, 24, 37, 182, 187, 208, 209, 210, 211, 212, 218, 220, 223, 224, 225, 226, 229, 230, 231, 232, 235, 238, 239, 243, 249, 269, 271, 272, 274, 275, 278, 281, 285, 288, 295, 298, 299, 313, 314, 316, 317, 318, 319, 326, 327, 331, 335, 337, 338, 340, 344, 345, 348, 353, 354, 381, 385, 386, 394, 446, 459, 464, 466, 475, 477, 529, 530, 533, 542, 543, 545

St. Malo, 192

St. Mary's Bay, 88

St. Pierre and Miquelon, 187, 231, 233, 234, 235, 244, 274, 277, 279, 280, 426

St. Pierre Bank, 73, 77, 81, 187, 188, 214, 227, 230, 236, 342, 405, 407, 508, 524

stages, 33, 70, 89, 105, 126, 127, 133, 134, 208, 209, 219, 224, 232, 237, 246, 288, 291, 293, 512

stationers, 292, 295, 312

steamers, 313, 314, 315, 316, 317, 318

steelhead trout, 122

Stewart, Sir W., 346, 349, 380, 391, 432, 531

stock area, 72, 508

stock assessment, 339, 413, 447, 453, 455, 456, 457, 458, 459, 460, 462, 463, 467, 470, 524, 525

stockfish, 179, 381

Story, Colin, 396, 405, 469, 500, 533

Strait of Belle Isle, 34, 75, 81, 88, 109, 117, 118, 168, 169, 172, 185, 186, 187, 188, 189, 192, 193, 226, 230, 242, 271, 282, 298, 308

sturgeon, 247

Summers, C.L., 328

surveys, 408, 419, 458, 459, 463, 464, 476, 506, 513, 514, 518, 524, 525, 535, 536, 537

T

tal qual, 331, 334, 350

Tanzania, 467

taverns, 229, 269

taxation, 276

tectonics, 19, 126

Teleost, 29, 514

temperature, 24, 34, 38, 39, 42, 63, 69, 119, 123, 126, 134, 241, 244, 248, 286, 343, 412, 516

temperature-productivity, 241, 286, 412

Templeman, Dr. Wilfred, 328, 339, 340, 341, 343, 345, 393, 394, 395, 396, 405, 407, 408, 409, 410, 464, 466

Terra Nova, 177, 181, 182, 183, 189, 193

territorial limits, 378, 389, 426, 427, 428, 429, 431, 432, 437, 444, 445, 448, 450, 451, 457, 463, 464, 469, 470, 472, 474, 532

Territorial Sea and Fishing Zone Act, 389

The Atlantic Groundfish Strategy, 501, 529

Thompson, Harold, 267, 339, 341, 342, 343, 345, 349, 351, 408

Thoroughfare, 270, 273

Thule, 169, 172

Thule Eskimos, 169

Titanic, 39

toad crab, 133, 294

Tokyo University of Fisheries, 467

Topsail, 275

Torbay, 249, 275

townies, 269

Tragedy of the Commons, 451, 452

trawlers, 280, 316, 332, 336, 337, 349, 351, 352, 353, 380, 383, 384, 385, 386, 387, 388, 393, 395, 404, 406, 407, 408, 410, 411, 412, 413, 416, 419, 422, 428, 430, 431, 432, 437, 445, 446, 448, 449, 454, 457, 463, 464, 469, 511, 522, 523, 533, 539

Traytown, 270

Treaty of Paris, 232, 233, 234, 235, 277

Treaty of Utrecht, 230, 231, 232, 234, 277, 278, 283

Treaty of Versailles, 234, 277

Trepassey, 208, 220, 224, 230, 499, 501, 502, 503

Trois-Rivières, 108

truck, 271, 272, 295, 325, 326, 532

Trudeau, Pierre, 453

Truman, President, 388

turbot
see Greenland halibut, 98, 99, 308

Twillingate, 168, 231, 245, 271, 433

U

U.S.A., 234, 272, 281, 282, 301, 303, 305, 338, 383, 403, 419, 426, 441, 442, 527, 534
underutilized species, 448, 528
Unemployment Insurance Program, 424, 433
Ungava Bay, 45
United Food and Commercial Workers International Union (UFCW), 438
United Nations Conference on the Law of the Sea (UNCLOS I), 389, 426
United Nations Conference on the Law of the Sea (UNCLOS II), 389, 426
United Nations Conference on the Law of the Sea (UNCLOS III), 426
University of Bergen, 467
University of Dar es Salaam, 467
University of Washington, 397, 467
Upper Canada, 225, 281, 282, 283

V

Valcourt, Bernard, 462, 463
Vasco da Gama, 178
Vaughan, William, 220
Venezuela, 429, 431
vento, 183, 208, 242
Verrazzano, Giovanni da, 190, 191
Virgin Rocks, 25, 37, 211, 213, 227, 234, 236, 294
virtual population analysis, 456, 457

W

Waldegrave, Governor, 225
Walsh Commission 1951, 377, 402, 403, 405, 465
Walsh, Sir Albert J., 402
Walter Grieves and Company, 288
War of 1812, 274
Water Street, 317, 345
Wellman, Jim, 537
West Country England, 209, 210, 211, 223, 235, 266, 271, 276
West Germany, 391, 399, 415, 458
West Indies, 332
Western Charter of 1634, 215, 216, 235
whaling, 189, 298, 299, 300, 324
Whitbourne, Richard, 189, 205, 208, 211, 212, 215, 216, 221, 237, 247, 248, 533
White Bay, 88
whitefish, 64, 108
Whiteley, Captain William, 308, 309, 320
Whiteway, William, 283, 299, 311, 325
Whitten, Reuben, 284
William Cox and Company, 271
Williams, Danny, 229, 536
Williams, Lieutenant Griffith, 223, 229, 320
Willoughby, Thomas, 205
Wilson, Anna, 343
winter crews, 209, 211, 226, 423
Winters, George, 459, 460
Wisconsin glaciation, 34, 38, 59, 60, 63, 72, 73, 82, 84, 96, 122, 123, 125, 126, 133, 137, 141, 145, 545
Wolfe, General, 231
wolfish, 110, 112
women, 213, 217, 219, 221, 229, 243, 269, 288, 292, 293, 348
Woods Hole Oceanographic Institute, 96
World War I, 268, 281, 300, 301, 315, 316, 326, 327, 336, 338, 346, 348, 350, 351, 353, 379, 381, 383, 384, 387, 388, 390, 391, 407, 423, 426, 532
World War II, 337, 348, 351, 353, 379, 381, 383, 384, 387, 388, 390, 391, 407, 423, 426, 532
World Wildlife Fund, 540

Y

Young, Neil, 377
Young, Victor, 462, 533

Z

zooplankton, 36, 55, 60, 61, 69, 71, 82, 84, 91, 95, 96, 98, 110, 118, 119, 137, 505, 512, 513